Second Edition

Nanotoxicology
Progress toward Nanomedicine

Second Edition

Nanotoxicology
Progress toward Nanomedicine

Edited by
Nancy A. Monteiro-Riviere and C. Lang Tran

CRC Press
Taylor & Francis Group
Boca Raton London New York

CRC Press is an imprint of the
Taylor & Francis Group, an **informa** business

CRC Press
Taylor & Francis Group
6000 Broken Sound Parkway NW, Suite 300
Boca Raton, FL 33487-2742

First issued in paperback 2016

© 2014 by Taylor & Francis Group, LLC
CRC Press is an imprint of Taylor & Francis Group, an Informa business

No claim to original U.S. Government works

Version Date: 20130923

ISBN 13: 978-1-138-03399-3 (pbk)
ISBN 13: 978-1-4822-0387-5 (hbk)

Library of Congress Cataloging-in-Publication Data

Nanotoxicology (Monteiro-Riviere)
   Nanotoxicology : progress toward nanomedicine / [edited by] Nancy A. Monteiro-Riviere and C. Lang Tran. -- Second edition.
     p. ; cm.
   Includes bibliographical references and index.
   ISBN 978-1-4822-0387-5 (alk. paper)
   I. Monteiro-Riviere, Nancy A., editor of compilation. II. Tran, C. L., editor of compilation. III. Title.
   [DNLM: 1. Nanotechnology--methods. 2. Nanostructures--therapeutic use. 3. Nanostructures--toxicity. QT 36.5]

   R857.N34
   610.28--dc23   2013036688

Visit the Taylor & Francis Web site at
http://www.taylorandfrancis.com

and the CRC Press Web site at
http://www.crcpress.com

Contents

SECTION III   Modeling

SECTION IV   Methodologies and Techniques

SECTION V   Hazards

SECTION VI   Risk Assessment

Preface

Nanotechnology promises new materials for industrial applications by having new or enhanced physicochemical properties that are different in comparison to their micrometer-sized counterparts. The market size for nanotechnology is expected to grow to over $3 trillion by 2015. As in all industrial applications, the potential exposure of humans and the environment to these materials is inevitable. As these new materials go through their life cycle—from development, to manufacture, to consumer usage, to final disposal—different human groups (e.g., workers, bystanders, or consumers), environmental compartments (e.g., air, soil, sediment, or water), and species (e.g., worm, fish, or human) will be exposed to these materials. Emerging data have shown a range of toxic (hazardous) effects from engineered nanoparticles, suggesting that combined with the potential exposure these nanoparticles may result in a risk (product of hazard and exposure) to human health or the environment. Although standard methods exist for hazard and risk analysis of conventional chemicals, these tools need to be modified and verified before being applied to nanomaterials. In a similar way, presently used standard approaches to risk management, control, and reduction need to be rendered relevant for nanomaterials. Thus, the development of nanotechnology-based products must be complemented with appropriate validated methods to assess, monitor, manage, and reduce the potential risks of engineered nanomaterials (ENMs) to human health and the environment. Not only good management tools are important but public awareness is also important for industrial development and acceptance. Public mistrust of any new technology is often high, and demonstrating *safe* products of nanotechnology will enhance the confidence of consumers, workers, and other stakeholders. Hence, efficient communication strategies to the public and stakeholders of significant progress are of high importance.

Many concerns have been expressed as to the safety of ENMs. There has been a concerted effort among government agencies, industrial sponsors, academicians, scientists, and the public to try to identify the potential hazards of nanotechnology. The first edition of this book provided the basic knowledge of nanomaterials safety and discussed the proper use for characterization, nomenclature standards, and physicochemical characteristics of nanoparticles that determined their toxicity in biological and environmental systems. Since the first publication of this book in 2007, the field of nanoscience and nanomedicine continues to grow substantially.

This second edition of *Nanotoxicology: Progress toward Nanomedicine* has been greatly expanded and includes many new authors who offer suggestions on how to conduct nanomaterial toxicology studies and stress the need for proper characterization. The need for developing and ascertaining the tools used for risk assessment and management of ENMs is paramount. In the second publication of this book, our aim is to document the continuing development of the essential tools used for ENM characterization and in vitro and in vivo toxicology testing. Of importance is the contribution of nanosafety research to nanomedicine. The interplay between these two disciplines is going to deliver in the near future the most startling insights into the interaction between ENMs and the biological milieu.

Section I deals with the impacts of nanotechnology on biomedicine. Nanomedicine development has improved over traditional approaches as new ENMs emerge at the nanoscale. There is a great need for appropriate functionalization for tissue-specific targeting or drug delivery to tissues that can be tuned with nanomaterials. Many nanomaterials may degrade once in the body and the toxicity and clearance pathways may be altered. The biointeractions of multifunctional nanoparticle-based therapy are discussed. It was once thought that the primary applications of nanomedicine were for diagnosis and cancer treatment but due to functionalization the treatment of many diseases can now be investigated. The ability to control some of the specific physicochemical properties of nanoparticles such as size, surface area, and shape has allowed for control over the nanoparticle functions so that they can be used for treatment of different diseases.

Section II on exposure demonstrates that there is no consensus on the use of a single metric to characterize exposure to ENMs and that there is no international standard for measuring and characterizing ENMs. This section provides much-needed information on the requirements for proper detection, measurement, and assessment both for workplace exposure and in consumer products. Section III on modeling provides an insight into the development of quantitative endpoints that are useful in predicting the vivo behavior of nanoparticles. Section IV on methodologies and techniques provides the basic tools for assessing nanoparticle toxicity from the early stages of manufacturing to in vivo biodistribution, where size, surface chemistry, and shape play a major role in toxicity in some tissues. Also, an array of detection methods have been implemented for assessing toxicity in in vivo and in vitro systems, as well as at the single cell and tissue levels. Their benefits and limitations will be discussed. Section V on hazards deals with specific organ systems that have been thoroughly investigated by many international experts with different types of ENMs in multiple in vivo and in vitro models. This section assesses not only the toxicity in organ systems but also some of the applications of ENMs for the potential treatment in neuroregeneration and targeted delivery of imaging or therapeutic nanomaterials to tumors. Also, specialized applications have been developed due to the microcidal activity of many nanoparticles for the use in indwelling catheters, orthopedic implants, bandages for diabetic wounds, and tissue engineering. Numerous strategies also exist for additional medical applications ranging from tumor targeting to the treatment of different cancers. Section VI on risk assessment of nanomaterials is complicated due to the limited information obtained on only a few occupational exposures, lack of comprehensive epidemiological studies, and no clear dose metric due to lack of or improper characterization of nanomaterials. Many coordinated research efforts have been implemented to address the potential risks of nanomaterials, and there is even a greater need to address the knowledge gaps especially with long-term studies.

All of these chapters report on exciting science and show great promise for some ENMs to be used in nanomedicine. They reflect a sense of urgency to develop effective targeted nanotheranostics to be used in clinical trials. There is also both a great need to understand the complexities and issues of toxicology of nanomaterials and a need for continued harmonization for risk management.

In this book, we have solicited some of the most crucial research, illustrating the development of nanosafety and nanomedicine, their interplay, and convergence. We hope the readers will find inspiration through the chapters contributed by leading international experts in both fields and appreciate the continuing development in nanosafety and nanomedicine.

Editors

Nancy A. Monteiro-Riviere is a Regents Distinguished Research Scholar and University Distinguished Professor of Toxicology and Director of the new Nanotechnology Innovation Center of Kansas State. She was a professor of investigative dermatology and toxicology at the Center for Chemical Toxicology Research and Pharmacokinetics, North Carolina State University (NCSU) for 28 years. She is also a professor in the Joint Department of Biomedical Engineering at the University of North Carolina (UNC) at Chapel Hill/NCSU and research adjunct professor of dermatology at the UNC-Chapel Hill School of Medicine. She earned a BS (cum laude) in biology from Stonehill College, North Easton, Massachusetts, and an MS and a PhD in anatomy and cell biology from Purdue University, West Lafayette, Indiana, and completed a post-doctoral fellowship in toxicology at Chemical Industry Institute of Toxicology (now Hamner Institutes for Health Sciences) in Research Triangle Park, North Carolina. She was past president of both the Dermal Toxicology and In Vitro Toxicology Specialty sections of the National Society of Toxicology. She is a fellow in the Academy of Toxicological Sciences and in the American College of Toxicology. She was the recipient of the Purdue University inaugural Distinguished Women Scholars Award, Kansas State University Woman of Distinction, and elected to attend the National Academy of Sciences special Keck Futures Initiative Conference on "Designing Nanostructures at the Interface between Biomedical and Physical Systems." She serves as an associate editor for *WIREs Nanomedicine and Nanobiotechnology* and for *Materials Science and Engineering C: Materials for Biological Applications*; and serves on the editorial boards of *Nanomedicine, Nanotoxicology, Journal of Applied Toxicology, Cutaneous and Ocular Toxicology, Research and Reports in Transdermal Drug Delivery*, and *Toxicology In Vitro*. She has served on several national and international expert review panels, including many in nanotoxicology, such as the National Research Council of the National Academies Committee for Review of the Federal Strategy to Address Environmental, Health, and Safety Research Needs for Engineered Nanoscale Materials and the International Council on Nanotechnology. She has served as an invited expert for the National Nanotechnology Initiative on Nanomaterials, and was appointed to the Scientific Committee on Consumer Safety by the European Commission. She has given more than 145 invited presentations and published more than 280 manuscripts in the field of skin toxicology and nanotoxicology and is editor of the books *Nanotoxicology: Characterization, Dosing and Health Effects* and *Toxicology of the Skin—Target Organ Toxicology Series*. Her current research interests involve in vivo and in vitro studies of skin absorption, penetration and toxicity of chemicals, nanoparticles, development of novel scaffolds for tissue engineering, and novel pharmaceutical drug delivery devices.

C. Lang Tran is the head of toxicology at the Institute of Occupational Medicine, Edinburgh, United Kingdom. He earned his PhD in biological sciences from Napier University, Edinburgh, United Kingdom, and has contributed more than 30 peer-reviewed articles on toxicology and nanotoxicology.

Lang Tran has led many European projects on Nanosafety (FP6 PARTICLE_ RISK, FP7 ENPRA) and is currently the coordinator of the large FP7 project, MARINA. He is one of the editors for the journals *Nanotoxicology* and *Particle and Fiber Toxicology*. Since 2012, he was made an Honorary Professor at Heriot-Watt University, Edinburgh, United Kingdom.

Contributors

Matt Basel
Department of Anatomy and Physiology
Kansas State University
Manhattan, Kansas

Dagmar Bilaničová
Department of Environmental Sciences,
   Informatics and Statistics
Ca'Foscari University of Venice
Venice, Italy

and

Qi technologies
Pomezia, Rome, Italy

Sonja Boland
Unit of Functional and Adaptive
   Biology
Laboratory of Molecular and Cellular
   Responses to Xenobiotics
University of Paris Diderot
Paris, France

Diana Boraschi
Laboratory of Innate Immunity and
   Cytokines
Institute of Protein Biochemistry
National Research Council
Naples, Italy

Stefan Bossman
Department of Chemistry
Kansas State University
Manhattan, Kansas

Derk Brouwer
Netherlands Organisation for Applied
   Scientific Research (TNO)
Zeist, the Netherlands

Daniel V. Christophersen
Section of Environmental Health
Department of Public Health
University of Copenhagen
Copenhagen, Denmark

Sara Corradi
Laboratory of Cell Genetics
Department of Biology
Vrije Universiteit Brussel
Brussels, Belgium

Anna L. Costa
Institute of Science and Technology for
   Ceramics
Department of Chemical Science and
   Materials Technology
National Research Council
Faenza, Italy

Pernille H. Danielsen
Section of Environmental Health
Department of Public Health
University of Copenhagen
Copenhagen, Denmark

Heather A. Enright
Biosciences and Biotechnology
   Division
Lawrence Livermore National
   Laboratory
Livermore, California

Bengt Fadeel
Division of Molecular Toxicology
Institute of Environmental
   Medicine
Karolinska Institutet
Stockholm, Sweden

Dorothy Farrell
Office of Cancer Nanotechnology
    Research
Center for Strategic Scientific Initiatives
National Cancer Institute
Bethesda, Maryland

Teresa Fernandes
School of Life Sciences
Heriot-Watt University
Edinburgh, United Kingdom

Denis Fourches
Laboratory for Molecular Modeling
Division of Chemical Biology and
    Medicinal Chemistry
Eshelman School of Pharmacy
University of North Carolina at
    Chapel Hill
Chapel Hill, North Carolina

Laetitia Gonzalez
Laboratory of Cell Genetics
Department of Biology
Vrije Universiteit Brussel
Brussels, Belgium

Piotr Grodzinski
Office of Cancer Nanotechnology
    Research
Center for Strategic Scientific Initiatives
National Cancer Institute
Bethesda, Maryland

Maureen R. Gwinn
Office of Research and Development
National Center for Environmental
    Assessment
U.S. Environmental Protection Agency
Washington, District of Columbia

George Hinkal
Office of Cancer Nanotechnology
    Research
Center for Strategic Scientific Initiatives
National Cancer Institute
Bethesda, Maryland

Danail Hristozov
Department of Environmental Sciences,
    Informatics and Statistics
Ca'Foscari University of Venice
Venice, Italy

Gary R. Hutchison
School of Life, Sport & Social Sciences
Centre for Nano Safety
Edinburgh Napier University
Edinburgh, United Kingdom

Namrata Jain
Department of Clinical Medicine
Institute of Molecular Medicine
Trinity College
Dublin, Ireland

Keld A. Jensen
Danish Nano Safety Centre
National Research Centre for the
    Working Environment
Copenhagen, Denmark

Araceli Sánchez Jiménez
Centre for Human Exposure Science
Institute of Occupational Medicine
Edinburgh, United Kingdom

Micheline Kirsch-Volders
Laboratory of Cell Genetics
Department of Biology
Vrije Universiteit Brussel
Brussels, Belgium

Henrik Klingberg
Section of Environmental Health
Department of Public Health
University of Copenhagen
Copenhagen, Denmark

James F. Leary
Departments of Basic Medical Sciences
    and Biomedical Engineering
Birck Nanotechnology Center
Purdue University
West Lafayette, Indiana

Guoqiang Li
Department of Mechanical & Industrial
  Engineering
Louisiana State University
Baton Rouge, Louisiana

Yiyao Li
Neuroscience Program
School of Science and Engineering
Tulane University
New Orleans, Louisiana

Steffen Loft
Section of Environmental Health
Department of Public Health
University of Copenhagen
Copenhagen, Denmark

Laura MacCalman
Institute of Occupational Medicine
Edinburgh, United Kingdom

Michael A. Malfatti
Biosciences and Biotechnology
  Division
Lawrence Livermore National
  Laboratory
Livermore, California

Antonio Marcomini
Department of Environmental Sciences,
  Informatics and Statistics
Ca'Foscari University of Venice
Venice, Italy

Bashir M. Mohamed
Department of Clinical Medicine
Institute of Molecular Medicine
Trinity College
Dublin, Ireland

Peter Møller
Section of Environmental Health
Department of Public Health
University of Copenhagen
Copenhagen, Denmark

Nancy A. Monteiro-Riviere
Department of Anatomy and
  Physiology
Nanotechnology Innovation Center of
  Kansas State
Kansas State University
Manhattan, Kansas

Jay Nadeau
Department of Biomedical
  Engineering
McGill University
Montreal, Quebec, Canada

Bernd Nowack
Empa: Swiss Federal Laboratories
  for Materials Science and
  Technology
Technology and Society Laboratory
St. Gallen, Switzerland

Adriele Prina-Mello
Department of Clinical Medicine
Institute of Molecular Medicine
Trinity College
Dublin, Ireland

Giulio Pojana
Department of Environmental Sciences,
  Informatics and Statistics
Ca'Foscari University of Venice
Venice, Italy

Tarl W. Prow
Dermatology Research Centre
School of Medicine
University of Queensland
and
Translational Research Institute
Princess Alexandra Hospital
Woolloongabba, Australia

Kenneth L. Reed
DuPont Haskell Global Centers
Newark, Delaware

Jim E. Riviere
Department of Anatomy and Physiology
Institute of Computational Comparative
   Medicine
Kansas State University
Manhattan, Kansas

Bryony L. Ross
Institute of Occupational Medicine
Edinburgh, United Kingdom

Martin Roursgaard
Section of Environmental Health
Department of Public Health
University of Copenhagen
Copenhagen, Denmark

Stefania Sabella
Center for Bio-Molecular
   Nanotechnology
Italian Institute of Technology
Arnesano, Lecce, Italy

Meghan E. Samberg
Regenerative Medicine/Extremity
   Trauma
United States Army Institute of Surgical
   Research
Fort Sam Houston, Texas

Roel Schins
IUF–Leibniz Research Institute for
   Environmental Medicine
Düsseldorf, Germany

Janeck Scott-Fordsmand
Terrestrial Ecology Section
Department of Bioscience
Aarhus University
Copenhagen, Denmark

Tej Shrestha
Department of Anatomy and Physiology
Kansas State University
Manhattan, Kansas

Tobias Stöger
Comprehensive Pneumology
   Center
Institute of Lung Biology and
   Disease
Neuherberg, Germany

Vicki Stone
School of Life Sciences
Heriot-Watt University
Edinburgh, United Kingdom

Fariborz Tavangarian
Department of Mechanical & Industrial
   Engineering
Louisiana State University
Baton Rouge, Louisiana

Treye A. Thomas
U.S. Consumer Product Safety
   Commission
Rockville, Maryland

Alexander Tropsha
Laboratory for Molecular
   Modeling
Division of Chemical Biology and
   Medicinal Chemistry
Eshelman School of Pharmacy
University of North Carolina at
   Chapel Hill
Chapel Hill, North Carolina

Martie Van Tongeren
Centre for Human Exposure
   Science
Institute of Occupational Medicine
Edinburgh, United Kingdom

Navin K. Verma
Lee Kong Chian School of
   Medicine
Nanyang Technological University
Singapore

Lise K. Vesterdal
Section of Environmental Health
Department of Public Health
University of Copenhagen
Copenhagen, Denmark

Yuri Volkov
Department of Clinical Medicine
Institute of Molecular Medicine
Trinity College
Dublin, Ireland

Sandra Vranic
Unit of Functional and Adaptive Biology
Laboratory of Molecular and Cellular
    Responses to Xenobiotics
University of Paris Diderot
Paris, France

Hongwang Wang
Department of Chemistry
Kansas State University
Manhattan, Kansas

David B. Warheit
DuPont Haskell Global Centers
Newark, Delaware

Cao Yi
Section of Environmental Health
Department of Public Health
University of Copenhagen
Copenhagen, Denmark

Section I

Impacts of Nanotechnology on Biomedicine

1 Introduction to Biomedical Nanotechnology

Dorothy Farrell, George Hinkal,
and Piotr Grodzinski

CONTENTS

1.1 PROMISE OF NANOMEDICINE

There is considerable enthusiasm about the potential of nanotechnology to not only improve existing biomedical technologies but also enable entirely new strategies in healthcare. Although this enthusiasm should be balanced with an appreciation for the extended development time and high failure rate for innovative technologies and a recognition that medical nanotechnology or nanomedicine is not yet established in routine clinical practice,[1,2] high-performance nanotechnology-based medical technologies are beginning to translate into clinical use.[3] Recent advances include U.S. Food and Drug Administration (FDA) clearance to market a gold nanoparticle (AuNP)-based nucleic acid test,[4] which is used to identify bacteria linked to bloodstream infections and to detect genes conferring antibiotic resistance,[5] and the release of phase I clinical trial results indicating safe, targeted delivery of siRNA[6,7] and docetaxel[8] to cancer cells through the systemic administration of lipid or polymeric nanoparticles. Although designed only to evaluate safety, these trial results also showed the mechanism of action and therapeutic efficacy of a systematically delivered RNAi nanomedicine and improved therapeutic index for an existing drug (docetaxel) when re-created as a nanoparticle therapeutic. Nanoscale materials function at the same size as biological materials and exhibit size-dependent interactions with cell and tissue components and structures, including efficient intracellular uptake and transcytosis across biological barriers. These characteristics enable systemic delivery of drugs and sensing/imaging agents to target sites throughout the body.[9,10] Particles larger than approximately 10 nm evade renal clearance, so that circulation half-life can be greatly increased over small molecules that are cleared renally within minutes. This results in increased bioavailability and tissue exposure for nanoparticle cargo and concomitantly greater accumulation at target sites. Nanomedicine uptake and drug accumulation can be even greater in diseased or inflamed tissues, such as tumors or arthritic joints, which can exhibit an enhanced permeability and retention (EPR) effect that arises from a combination of leaky vasculature and dysfunctional lymph drainage. Macromolecules and nanoparticles of size 10–200 nm may preferentially deposit in these tissues, as they extravasate from fenestrated vessels but are not cleared from tissue because of the absence of functional lymphatics.[11,12] Exploitation of this phenomenon is known as *passive targeting* of nanomedicines. Appropriate functionalization of nanomedicines with cell- or tissue-specific targeting agents, such as peptides, aptamers, or antibodies, can further improve therapeutic index, a strategy often known as *active targeting*.[13] Whether active targeting increases tissue localization or just cellular internalization remains a somewhat open question, although a body of evidence for tumor-specific delivery seems to support increased cellular internalization as the dominant mechanism for enhanced therapeutic efficacy.[14,15] These studies suggest that the physicochemical characteristics of nanomaterials and the EPR effect dominate biodistribution and tumor accumulation, but active targeting increases internalization and may even be necessary for cytosolic distribution of nanomaterials. This makes active targeting particularly relevant to therapies that must be delivered to the intracellular environment to be effective. Drug delivery to and loading in tissue also depend on vehicle affinity and efficiency of drug release, properties that can be tuned through materials engineering of vehicles such as liposomal and polymeric particles. Liposomal formulations of the anticancer drugs doxorubicin (Doxil®, Janssen Pharmaceuticals Inc.), daunorubicin (DaunoXome®, Galen Ltd.), and cytarabine (DepoCyt®, Sigma-Tau Pharmaceuticals Inc.) and the antifungal medication amphotericin B (AmBisome®, Astellas Pharma Inc.) have been FDA approved for specific indications, as have several lipid–drug and polymer–drug complexes (Oncaspar®, Enzon Pharmaceuticals and Neulasta™, Amgen Inc.) that can also be considered nanomedicines.[1,3] In addition to favorable pharmacokinetics, which improve the solubility, bioavailability, and therapeutic index of their payload, polymeric vehicles can also offer controlled release through erosion of the vehicle, diffusion of the payload through the vehicle, or vehicle degradation in response to biochemical conditions at the delivery site. Nanomedicines generate improved performance over traditional approaches by exploiting distinct material properties that emerge at the nanoscale. Examples of physical phenomena that are unique to the nanoscale and offer significant advantages for biomedical applications include superparamagnetism in magnetic nanoparticles (mNPs),[16] in which the mNPs exhibit strong response to applied magnetic fields but zero hysteresis at room temperature; quantum confinement effects that cause ultraviolet (UV)-excited semiconductor nanoparticles (quantum dots) to exhibit sharp, tunable emission peaks in the visible and near-infrared bands;[17] and surface plasmon resonance (SPR) in noble metal nanoparticles, which results in size-, shape-, and aggregation-dependent absorption and scattering.[18] Superparamagnetic NPs are being investigated as magnetic resonance imaging agents[19] and biomarker probes in multiplex sensor devices,[20,21] and have been approved for clinical use in hyperthermia treatment of cancer in Europe;[22] quantum dots are widely used for in vitro cell studies; AuNPs are being clinically tested as hyperthermia agents for cancer therapy[18] and developed into an enzyme-linked immunosorbent assay for rapid biomarker detection with the naked eye.[23] The optical and electronic properties of nanowires and carbon nanotubes are being explored for use in imaging[24] and sensing applications.[25] Engineering site-specific activity in nanoparticles to decrease toxicity and increase efficacy of therapeutics and imaging reagents is a tantalizing promise in nanomedicine. Nanomedicines can be activated by biochemical recognition or encountering specific physiological conditions, such as enzymes enriched in a tumor or its environment[26] or constricted blood flow.[27] External triggers such as light or applied electromagnetic fields can also be used to activate nanomedicines. Examples include nanoparticle-mediated hyperthermia or laser ablation therapies using gold nanoshells[28] or iron oxide nanoparticles, with nanoparticle temperature increasing because of absorption of light or coupling to an electromagnetic field.[29] A more complex system has also been shown that uses light to heat gold nanorods accumulated within a tumor; the resulting increase in tumor fenestration improves accumulation of a second nanomedicine.[30] A refinement of this concept exploits coagulation signaling resulting from the thermal damage to recruit *clot-targeted* nanomedicines to the tumor.[31] The latter system provides proof of concept of in vivo communication and cooperative behavior between nanomaterials, capabilities that go well beyond traditional drug delivery and that could enable new disease monitoring and treatment strategies. The large surface area to volume ratio of nanoparticles is also advantageous for nanomedicines by enabling surface decoration with multiple targeting ligands or activating agents for in vivo applications and significant signal amplification for high-sensitivity detection of biomarkers in the in vitro setting. Nanomaterials are also capable of carrying large numbers of multiple therapeutics or imaging agents in a single carrier, creating opportunities for multifunctional nanomedicines. Simultaneous delivery of multiple therapeutics could increase potency or prevent or overcome drug resistance through knockdown of compensatory pathways that enable pathogens or diseased cells to survive single-agent therapy. The marriage of therapeutic and diagnostic capabilities in a single material, device, or instrument, popularly termed theranostics, is a widely touted application for nanomedicine because of this unique ability to combine materials with differing properties into a single object. Composite nanomedicines can also be used as reporters in in vivo assays.[32]

Finally, it should be noted that biomedical nanotechnology can take the form of nanoscale features on microscale platforms. Examples include the dissolution of microscale aggregates into bioactive nanoparticles,[27] the use of nanoscale pores in biodegradable mesoscale silicon particles to carry and release nanoparticle imaging agents and therapeutics over time,[33] and the use of mNPs as sensing elements in a micrometer-sized in vivo monitoring device, in which changes in the relaxation properties of mNPs indicate aggregation caused by exposure to cardiac biomarkers.[34]

1.2 CONSIDERATIONS FOR IN VIVO USE OF NANOMATERIALS

Over the past 10 years, a significant body of knowledge has been generated about the behavior of nanomaterials in vivo, and some overarching rules about the effect of size, surface charge, and hydrophobicity on nanomaterial fate in vivo can be articulated, as shown in Figure 1.1, which represents a summary of work done by the Nanotechnology Characterization Laboratory (NCL) and reported by NCL Director Scott McNeil.[35] There is also a growing body of investigations into the clinical use of nanomaterials that supports their safe use when carefully developed

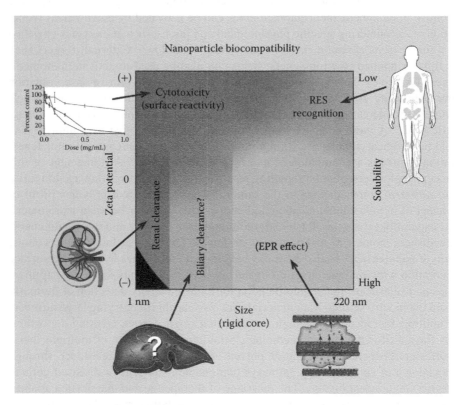

FIGURE 1.1 (**See color insert.**) This phase diagram qualitatively shows trends the NCL has observed in relationships between the independent variables of particle size, particle zeta potential (surface charge), and hydrophobicity with the dependent variable of biocompatibility. Biocompatibility includes route of uptake and clearance (shown in green), cytotoxicity (red), and MPS or reticuloendothelial system (RES) recognition (blue). The physical characteristics of a nanoparticle determine its biocompatibility and ultimately its safety and efficacy as an anticancer drug or diagnostic agent. (From Ling Y et al., *Nat Biotechnol*, 29(3), 273–7, 2011. Reproduced with permission from John Wiley & Sons.)

and tested.[1,6–8] However, nanomaterial fate in vivo is highly variable. Nanomaterial properties, such as size, surface charge, shape, and elastic modulus, along with delivery method (e.g., intravenous, intraperitoneal, or intratumoral) determine nanomedicine transport, biodistribution, and cellular uptake. The efficacy and toxicity of nanomedicines depend on favorable pharmacokinetics/pharmacodynamics (PK/PD), so proper understanding and management of the underlying material parameters to optimize PK/PD is crucial.[8,36–38] A nanoformulation of a drug will have the PK/PD of the carrier, not the free drug, and the two can have very different treatment courses and side effect profiles, behaving in effect as different drugs. For this reason, nanomedicines must undergo thorough preclinical development in appropriate animal models to establish safety, with special attention to uptake of the nanomedicine by compartments inaccessible to the free drug.[39] The conditions underlying an effective treatment regimen must also be established, even when the payload drug is clinically well established. Changes in cellular internalization pathways and the potential for new drug actions based on these should be investigated as well. This latter point is particularly cogent for nanomedicines with active targeting.

1.2.1 IN TISSUES

The interactions between nanomaterials and cells and tissues are complex, and even basic issues surrounding nanomaterial trafficking in vivo and immune responses to nanomaterials remain unresolved. Nanomaterials smaller than approximately 10 nm undergo rapid renal clearance, but a recent study suggests a more complex role for the kidneys in nanomaterial processing than previously realized, with evidence that a component of the renal filtration barrier can disassemble and clear from circulation a cationic polymer-based therapeutic agent with a size greater than 10 nm.[40] It is widely considered that the mononuclear phagocytic system (MPS) is primarily responsible for clearing nanomaterials larger than approximately 10 nm from circulation, accounting for the heavy involvement of MPS-associated organs such as the liver and spleen in nanomaterial clearance, storage, and persistence. To avoid rapid clearance of nanomedicines by the MPS, surfaces are typically sterically stabilized through *stealth* modification such as PEGylation (i.e., coating with polyethylene glycol) that moderates the absorption of plasma proteins and slows uptake by the MPS.[41] However, the full role of the MPS in clearing and delivering nanomedicines is not currently understood. Nanomedicine trafficking and eventual cargo deposition appear to depend on both host (e.g., age, gender) and nanomedicine properties.[42,43] There is some evidence that rather than evading macrophages on their way to disease sites, in some cases nanoparticles are delivered to their destinations by macrophages.[44]

1.2.2 NANOMATERIAL TESTING OF CONSTITUENTS

At present, the determinants of observed nanomaterial toxicity are not well known. For example, the small size and proportionally large surface area have been considered to make nanoparticles, particularly of inorganic materials, inherently reactive and therefore cytotoxic. This could render materials that are generally considered to be biocompatible or *safe*, such as gold, high risk for in vivo use in nanoparticulate form, although the role of surface passivation in ameliorating this risk continues to be investigated. However, detailed investigation can also reveal that cytotoxicity assumed to be due to a nanomaterial instead arises from residual manufacturing components in carrier buffer. For example, cytotoxic effects have been observed from gold nanorod dispersions, but following separation of the gold nanorods from the buffer via filtration, cytotoxicity was observed only in the buffer fraction.[45] The gold nanorods were synthesized using the surfactant cetyltrimethylammonium bromide (CTAB), which is not biocompatible and was responsible for the observed cytotoxicity. The example above highlights how crucial it is to separately test all constituents of a nanomedicine formulation for toxicity during development, along with the intended final product during both in vitro and in vivo testing, and to carefully remove all constituents not necessary for activity, delivery, or stability. Further motivation for this should come from the realization that the FDA will typically require safety information for the nanomedicine as intended for use, all active ingredients and excipients, and any anticipated degradation products as part of an investigational new drug submission package.[46] Scrutiny of degradation products and pathways will be greater for nanomedicines in which degradation or controlled release is part of the therapeutic strategy. Experiments to establish safety and activity should be carefully designed to distinguish which constituent of a nanomedicine is responsible for observed effects, both toxic and therapeutic, by the appropriate use of positive and negative controls. Cytotoxic additives or contaminants could give misleading positive indications of efficacy in in vitro assays, but severely diminish the likelihood of successful use in vivo. To minimize the risk of expensive late-stage failures, as much as possible throughout the development cycle, characterization (physicochemical, in vitro and in vivo toxicity and efficacy) should be done on nanomedicines as they will be used in the clinical setting. Data should be reported for final products including all excipients and active ingredients, such as surface coatings and targeting ligands, and assays should be performed in the intended matrix for use. For example, stability, degradation, and drug release profiles will differ between storage solutions, plasma, and tissue matrices. Data should be collected for each situation the nanomedicine is expected to encounter and at relevant time points. Investigations into drug release profiles should cover the full length of time a nanomedicine is expected to persist in the body and include the possibility of burst release on administration, and drug release occurring at times past expected clearance from the system should not be expected to have therapeutic effect. Care should be taken to appropriately mimic conditions of physiological use, such as pH and nanomedicine concentration.

1.2.3 IMPORTANCE OF NANOMATERIAL CHARACTERIZATION

When reporting on characterization, parameters should be reported in terms most relevant to intended use. For example, whether size is reported as a number or mass average should depend on whether the dispersed or aggregate state causes the dominant medical effect. Surface coating, stability, and aggregation issues must be carefully addressed, with the understanding that environmental conditions including ionic strength, nanomedicine concentration, and the concentration and profile of proteins in contact with or in proximity to the nanomedicine will affect surface charge and coating and therefore stability. Materials should be interrogated to determine if a surface coating, if present, detaches under intended use conditions and, if so, how the properties of nanomedicine change with the loss of the coating. Minimizing material degradation or surface alterations may require careful optimization of storage conditions, including choice of lyophilization agents or storage buffer and concentration. For all measurements, assays should be chosen and carefully monitored to ensure that material properties of the nanomedicine do not interfere with assay results, for example, that light scattering by a nanomaterial does not confound results of an optical absorption assay or that an assay for reactive oxygen species is not confounded by a particle that acts as an oxidant.[47] The NCL, a joint effort between the National Cancer Institute, the FDA, and the National Institute of Standards and Technology, has developed numerous protocols and best practices for characterizing nanomaterials intended for use in cancer care, which can be found on their website.[35,48,49] The knowledge developed by the NCL has broad applicability and is a significant resource for nanomedicine research and development. For biodistribution studies, materials should be carefully labeled so that observations unambiguously reflect whether signal arises from constituents or the intact product. This may require that composite materials be affixed with multiple labels to probe integrity of the product in vivo or the separate pathways taken by degradation products. Clearly, developers must address these issues when nanomedicines intentionally degrade in vivo, as in triggered delivery or controlled release strategies, but they must also investigate degradation of a product over time and the potential for transformation on exposure to in vivo environments, such as rapid dilution on intravenous injection or exposure to plasma proteins. Breakdown products must be investigated for toxicity and circulation half-life on release, and clearance pathways and potential reservoirs should be established. These issues are particularly pertinent for multifunctional nanomedicine strategies, such as targeted nanomedicines in which both an active pharmaceutical ingredient cargo and a targeting ligand may be bioactive, and theranostic materials in which the colocation of imaging and therapeutic agents must be assured for diagnostic information to be reliable. In vivo theranostics face unique challenges arising from the differing requirements for optimal performance for imaging and therapeutics.[50] To minimize background signal, diagnostic agents are best applied at low doses, with high affinity for target tissues; short circulation half-lives further help to limit background. Because diagnostic imaging will cover a patient population that includes healthy people, there is limited tolerance for potential side effects and concerns about repeat exposure and persistence. Accordingly, diagnostic agents are generally inert. Formulation principles for therapeutic agents are quite different. Long circulation half-lives are preferred to increase accumulation at target sites and enable sustained drug release and stable dosing. Optimal dose and affinity vary with therapeutic agent and indication, but in general dosing is higher than for imaging agents. Therapeutic agents will also have specific biological activities that will influence the extent to which persistence can be tolerated. Given these conditions, nanomedicine theranostics may only be viable for select combinations of agents for a limited number of indications. Nanomedicine developers typically use materials that have previously been cleared by the FDA for medical use, when possible. However, because a primary advantage of nanomedicine is altered pharmacokinetics, circulation half-lives can be increased from minutes to hours or days through nanoformulation, radically changing the exposure profile of an agent. This is true of active pharmaceutical agents, excipients, and surface coatings. Active targeting of nanomedicines to specific tissues or cell types can further alter biodistribution and in vivo exposures. This may result in increased cellular uptake of a material, delivery to previously unexposed tissue, or increased deposition of drug in on- or off-target organs. Liver and spleen in particular may see increased exposure, given the established tendency of nanomaterials to accumulate in these organs. Changes in circulation half-life can also alter immune response to a material. Under these conditions, even previously cleared materials must be reevaluated for safety and efficacy.

1.3 ANIMAL MODELS FOR NANOMEDICINE

When conducting in vivo efficacy testing of any agent, using appropriate animal model systems improves the likelihood of successful clinical translation. For nanomedicines this axiom is all the more pressing considering the variety of PK/PD involved, and there is not a one-size-fits-all approach. Rather, there are a series of pros and cons to be considered when planning preclinical animal testing of nanomedicines. For example, to establish efficacy of cancer nanomedicines, subcutaneous xenografts of human cancer cell lines into nude mice are by far the easiest and least expensive mammalian system to study, rapidly establish sizable tumors, and have the potential to approximate pharmacokinetics in humans.[51] However, tumors grown in these systems lack a realistic microenvironment, making drug–tumor interactions of questionable relevance. In addition, xenografted tumors tend to grow aggressively such that they do not reflect the tumor growth kinetics of a human tumor, nor do they have representative vasculature such that the EPR observations are similarly irrelevant for conclusive observations. That being said, it has been repeatedly shown that efficacy in multiple subcutaneous xenograft studies conducted in parallel has a strong likelihood of clinical success.[52,53] Orthotopic xenografts of human cancer cell lines implanted into the tissue of tumor origin provide stromal and inflammatory components that are more similar to human disease, enabling interrogation of biological barrier negotiation by the nanomedicine. They can also enable examination of the efficacy of the nanoparticle's targeting of the tumor or stroma. However, these models do not recapitulate a tumor's inherent heterogeneity and are generated from advanced, disseminated disease, making them inappropriate to locate and treat at the tissue of origin. Patient-derived xenograft models have the advantage of never having been cultured ex vivo such that the tumor best phenocopies and genocopies a patient's heterogeneous tumor.[54] However, these models are technically challenging to use. An additional drawback to any xenograft model is the necessity to implant into immunocompromised mice, which typically misrepresent the clinical setting and lack a functioning MPS, a critical component when considering nanoparticle behavior in vivo. Genetically engineered mouse models have intact immune systems and develop spontaneous tumors that histologically can recapitulate human cancers very well. When selecting animal models it is important to remember that although the FDA does not consider preclinical efficacy testing a predictor of human outcomes, proper animal study design can enable investigators to make go-no-go decisions to enrich the likelihood of downstream success. Such strategies can save valuable time and resources to enable investigators to advance the most viable nanomedicines toward clinical trials. It should be noted that dose, tissue retention, and system clearance factor into the toxicity of any medicine, a fact often not fully appreciated when discussing the use of nanomaterials in medicine. Advantages in delivery efficiency that reduce the amount of toxic drugs administered or reduce off-target effects by increasing the fraction of drug deposited at a target site, the replacement of a highly toxic therapeutic agent altogether with a nanomaterial-based alternative, or efficacy in treating diseases untreatable by standard medicine may sufficiently offset risks arising from the use of a nanomaterial. Proper determination of the relative merits of a nanomedicine formulation or alternative treatment strategy requires meaningful comparison to existing treatments, including free drug for the case of a nanoreformulation approach. An absence of effective treatments for an indication changes the risk–benefit analysis considerably, which is one reason nanomedicines have found their first uses in the most recalcitrant settings, such as metastatic cancers and resistant fungal infections in human immunodeficiency virus (HIV) positive patients.

Currently, there is no consensus on whether or not nanomedicines must fully biodegrade or be cleared from the system to be considered viable options for treatment of acute or lethal disease, but it is generally agreed that this is best practice for nanomedicines intended for chronic indications or as imaging agents. Side effects and persistence that might be acceptable in the context of therapy or therapeutic monitoring cannot be tolerated for screening approaches meant to be used on a healthy population. Similarly, primary diagnosis is considered to be high risk because of the serious consequences of false positives or negatives, and this is true for both imaging and in vitro device strategies. Diagnostics that will guide the course of treatment are also subject to close scrutiny. However, the use of in vitro devices to monitor therapeutic response or progress in the context of clinical studies is not subject to the FDA approval, provided choice of therapy is not influenced by the device's results. These studies can be used both to validate device performance and to gather information about the clinical intervention itself. Finally, manufacturing controls and scale-up are serious concerns for the clinical translation of nanomedicines, as are material purity and sterility. Large batch-to-batch variations that have been observed in nanomedicine production can confound attempts to establish efficacy and are particularly troubling for regulatory agencies. Difficulties with purity, reproducibility, and materials control can be expected to compound when developing and manufacturing complex materials with multiple constituents and functionalities. Sterilization procedures such as heating can alter nanomaterial size and shape, cause aggregation, and deteriorate performance and function by cleaving stealth coatings or targeting ligands from the surface. Filtration procedures can also cause loss of materials, suggesting nanomedicines are best handled in sterile conditions from the outset of production rather than sterilized later in the process. Issues with manufacturing should be considered early in the development cycle and inform choice of materials, synthetic processes, and clinical indications to maximize the feasibility of large-scale production of a sterile nanomedicine.

1.4   RELEVANT EXAMPLES

Sections 1.1 through 1.3 are meant to be a concise introduction to the potential for nanotechnology, and nanomaterials in particular, to improve the practice of medicine and an overview of the complexities that must be addressed when studying the toxicity and efficacy of nanomedicines to gather reliable, actionable preclinical and clinical data. The following examples highlight both the exciting advances being made to realize this potential and how careful study design and implementation can push these advances closer to the patient.

1.4.1   NANOPARTICLE-ENABLED IN VITRO DIAGNOSTIC DEVICES

Nanomaterials offer a number of advantages as elements of sensing devices, including high signal to background ratios, more sensitive readout for lower limit of quantification and greater discrimination between analytes, versatile surface chemistry enabling recognition of multiple targets (including multivalent functionalization) for high-specificity multiplex detection, and the ability to combine sample collection and detection (as in magnetic separation or capillary flow devices). An AuNP-based barcode device that exploits these features for high-sensitivity detection of nucleic acids[55] and proteins[56] is marketed by Nanosphere under the trade name Verigene™ and has been cleared by the FDA for multiple uses.[57] For direct detection of nucleic acids, AuNPs are coated with DNA barcodes and passed over a DNA array that has been previously reacted with genetic material extracted from a sample and washed. The AuNPs will bind to any complementary genetic material from the sample remaining on the array; signal from a geographic address on the array indicates the presence of the complementary nucleic acid in the sample. With the addition of an antibody label to the barcode AuNPs, the system can be used for protein quantification based on measurement of the companion barcodes, which also provide signal amplification. The assays boast extremely high sensitivity and multiplex capability with short measurement times (less than 3 hours) and minimal sample handling. Clinical applications include detection of bacteria that cause blood-borne infections and the presence of genes for antibiotic resistance and a pharmacogenetic test for warfarin metabolism to guide dosing. These indications support the use of nanotechnology to reduce toxicity and improve therapeutic efficacy not only by changing the formulation or delivery of traditional medicines but also by enabling better decision making when following standard clinical practice. Another device in development combines quantitative nuclear magnetic resonance (NMR) packaged with integrated microfluidic sample handling and smartphone interface to rapidly measure proteins in human tumor samples.[20] Cells from tumor samples are labeled with protein-specific mNPs using bioorthogonal chemistry,[58] and the resulting NMR signal is used to quantify protein levels in samples. Results from 60 patients were analyzed and compared to traditional clinical information, including biopsy and imaging, to determine signatures most predictive of malignancy. A four-protein signature was determined to be capable of diagnosing cancer with greater precision and accuracy than conventional immunohistochemistry and pathological analysis. The combination of measurement results for multiple biomarkers through an algorithm to return a single diagnostic value is known as an in vitro diagnostic multivariate index assay (IVDMIA),[59] and nanotechnology-enabled devices that can easily read multiple signals simultaneously are well suited for application in this area. IVDMIAs are expected to become increasingly important in in vitro diagnostic development as clinically relevant single-biomarker assays are fully exploited and represent another area where nanotechnology can improve deployment of traditional treatments.

1.4.2 Targeted Delivery

By preferentially delivering or releasing drugs at a disease site, nanoformulations can reduce systemic toxicity and side effects and increase the effective dose of existing therapeutics while decreasing the total amount of drug used for treatment. Nanoformulation also has the potential to *rescue* drugs that failed because of high toxicity, poor solubility, or unfavorable distribution or release profiles. The altered PK/PD resulting from nanoscale size can be augmented with active targeting strategies tailored for different therapeutics or indications. Preclinical and early-phase clinical results have recently been reported for several targeted nanomedicines.[6–8] In 2012, BIND Biosciences reported early phase I clinical trial results from BIND-014, a polymeric nanoparticle vehicle delivery for docetaxel targeted to prostate-specific membrane antigen, a receptor overexpressed on prostate cancer and the vasculature of most solid tumors.[8] Although this was a study on safe dosing levels and not efficacy, clinical response to treatment with BIND-014 was seen in some patients with advanced or metastatic disease, even for diseases for which free docetaxel is not considered an effective treatment. The activity was seen at doses as low as 20% of the typical effective dose used in the clinic. The success of BIND-014 owes much to a careful particle synthesis and testing strategy. A library of over 100 distinct nanoparticle compositions was prepared by systematically varying synthesis parameters and screened in vivo for pharmacokinetics, biodistribution, toxicity, and efficacy, including optimal targeting ligand density based on maximum tumor uptake, as shown in Figure 1.2. In vitro and in vivo test results were correlated to establish that nanoparticle PK/PD can be controlled through synthetic parameters. Animal model data showed improved efficacy with receptor-mediated uptake of the nanoparticle and the researchers observed agreement between pharmacokinetic profiles in multiple animal models and humans. Optimizing nanomedicine performance across the large number of relevant parameters, variable performance across animal and disease models, and differences between performances in animal models and humans have been major stumbling blocks in nanomedicine development; the consistency and predictability of this platform's performance are key indicators of its potential for future success. These results also illustrate the value of combinatorial screening in nanomedicine development in precision design of nanomedicines, enabling control over delivery and dose specifications for different drugs and indications (e.g., circulation time, target organs, intracellular fate) at the synthesis stage. To reduce systemic toxicity, drug delivery can also be localized by site-specific activation. Although external triggers can be used for this purpose, this requires knowing in advance where it is needed. An alternative approach is to use physiological conditions to activate a nanomedicine only at the disease site;

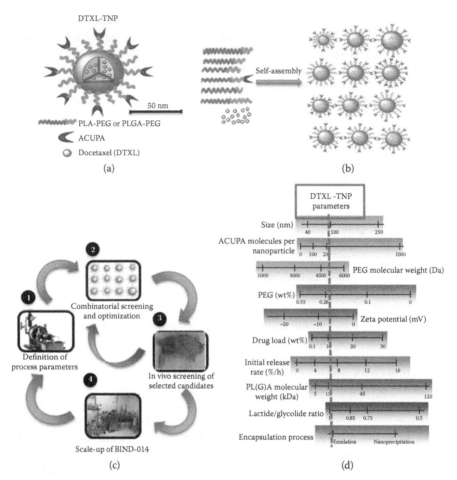

FIGURE 1.2 Combinatorial screening and optimization of docetaxel-loaded targeted polymeric nanoparticles (DTXL-TNPs). (a) Schematic of DTXL-TNP, a PSMA-targeted polymeric nanoparticles composed of a hydrophobic PLA polymeric core encapsulating DTXL and a hydrophilic PEG corona decorated with small-molecule (ACUPA) targeting ligands. (b) Generation of a library of DTXL-TNPs prepared by self-assembly of particles from mixtures of DTXL, PLA, PLGA, PLA- or PLGA-PEG (with varying PLA, PLGA, and PEG block lengths), and PLA-PEG-ACUPA. (c) Development and clinical translation of PSMA-targeted DTXL-TNPs: (1) A nanoemulsion process for efficiently encapsulating DTXL in nanoparticles was developed. (2) Small-scale batches of DTXL-TNPs were prepared and evaluated with respect to drug load and encapsulation efficiency, particle size distribution and reproducibility, and in vitro release kinetics. (3) DTXL-TNPs with promising physicochemical properties were evaluated with respect to pharmacokinetics in rats, and tolerability, tumor accumulation, and efficacy in tumor-bearing mouse models. In vivo results informed additional formulation optimization and led to selection of DTXL-TNP composition and process. (4) The DTXL-TNP manufacturing process was scaled up and used to manufacture sterile clinical supplies under current Good Manufacturing Practices. (d) Range of formulation parameters and nanoparticle physicochemical properties evaluated during development of DTXL-TNPs, with optimized DTXL-TNP parameters and target properties indicated by the dashed line. (Reprinted from Hrkach J et al., *Sci Transl Med*, 4(128), 128ra39, 2012. With permission.)

possible triggers include the low-pH environment associated with tumors[60] and changes in blood flow near obstructed vessels. Recently, clot-specific drug release triggered by the difference in shear flow between healthy and constricted blood vessels was reported for delivery of tissue plasminogen activator (tPA).[27] tPA is used to treat stroke, heart attack, and embolism by dissolving clots and restoring normal blood flow, but is associated with serious side effects caused by bleeding in healthy tissue. Inspired by the activation of platelets in regions of high shear stress, researchers designed shear-responsive microaggregates of biodegradable poly(lactic-co-glycolic acid) (PLGA) nanoparticles coated with tPA. Exposure to shear stress near vessel obstructions caused the aggregates to break apart, releasing the nanoparticles and activating the drug. Tests in three-dimensional microfluidic models of vessel constriction showed the microaggregates disintegrating and releasing drug in constricted areas with clinically relevant levels of shear but not in control areas with shear levels found in healthy vessels. In numerous animal models of vessel blockage, 2–5 μm aggregates preferentially bound to obstructions, released drug, and cleared the obstructions. Neither predissociated tPA-PLGA nanoparticles nor fused aggregates (which did not dissociate on encountering obstructions) dissolved obstructions, and microaggregates were effective even when delivering tPA doses approximately 1/100 of the clinical dose of free tPA. The microaggregates have the potential to significantly improve the therapeutic index for tPA and are a proof of principle for biophysical targeting in nanomedicine. The absence of an additional bioactive material for targeting could simplify design, testing, and production of nanomedicines and reduce the risk of unintended pharmacological activity, for indications in which this strategy is feasible.

1.4.3 NANOMEDICINE DELIVERY

One consideration when designing nanomedicines is how the delivery method and route affect biodistribution, toxicity, and effectiveness. Most nanomedicines currently under development are intended for intravenous administration, possibly because of the dominance of this route of administration for chemotherapeutic applications in the field right now. However, for many other clinical procedures, including imaging, oral or local delivery is more common. Researchers investigating an AuNP contrast agent for use in colonoscopy have performed detailed preclinical studies of biodistribution and toxicity using two routes of administration, rectal and intravenous.[61,62] The contrast agent consists of an AuNP core surrounded by a silica shell that displays enhanced Raman scattering-based detection properties and allows simple surface modification with cancer cell-specific targeting peptides. The nanoparticles are being developed for diagnostic molecular imaging with a dedicated endoscopic probe. When mice were administered nanoparticles rectally, no signs of systemic inflammation, stress, or toxicity were observed, and the nanoparticles were confined to the bowel. This compartmentalization and rapid excretion of the nanoparticles greatly reduces the risk of a systemic response. Even for intravenously administered nanoparticles, although sequestration by the MPS appeared to be rapid and resulted in considerable deposition of the nanoparticles in the spleen and liver, few signs of inflammation or toxicity were observed. These results are encouraging for use in cancer screening, especially with localized administration of the nanoparticles, as in colonoscopy.

REFERENCES

1. Duncan R, Gaspar R. Nanomedicine(s) under the microscope. *Mol Pharmaceutics* 2011;8:2101–41.
2. Venditto VJ, Szoka, Jr. FC. Cancer nanomedicines: so many papers and so few drugs! *Adv Drug Deliv Rev* 2013;65(1):80–8.
3. Etheridge ML, Campbell SA, Erdman AG, Haynes CL, Wolf SM, McCullough J. The big picture on nanomedicine: the state of investigational and approved nanomedicine products. *Nanomedicine* 2013;9:1–14.
4. Taton AT, Mirkin CA, Letsinger RL. Scanometric DNA detection with nanoparticle probes. *Science* 2000;289(5484):1757–60.
5. FDA News Release. *FDA Allows Marketing of First Test to Identify Certain Bacteria Associated with Bloodstream Infections.* Available at: http://www.fda.gov/NewsEvents /Newsroom/PressAnnouncements/ucm309950.htm. (accessed on June 27, 2012)
6. Tabernero J, Shapiro GI, Lorusso PM, Cervantes A, Schwartz GK, Weiss GJ, Paz-Ares L et al. First-in-man trial of an RNA interference therapeutic targeting VEGF and KSP in cancer patients with liver involvement. *Cancer Discov* 2013;3:406.
7. Davis ME, Zuckerman JE, Choi CH, Seligson D, Tolcher A, Alabi CA, Yen Y, Heidel JD, Ribas A. Evidence of RNAi in humans from systematically delivered siRNA via targeted nanoparticles. *Nature* 2010;464(7291):1067–70.
8. Hrkach J, Von Hoff D, Mukkaram Ali M, Andrianova E, Auer J, Campbell T, De Witt D et al. Preclinical development and clinical translation of a PSMA-targeted docetaxel nanoparticle with a differentiated pharmacological profile. *Sci Transl Med* 2012;4(128):128ra39.
9. Davis ME, Chen ZG, Shin DM. Nanoparticle therapeutics: an emerging treatment modality for cancer. *Nat Rev Drug Discov* 2008;7(9):771–82.
10. Farokhzad OC, Langer S. Impact of nanotechnology on drug delivery. *ACS Nano* 2009;3(1):16–20.
11. Maeda H, Wu J, Sawa T, Matsumura Y, Hori K. Tumor vascular permeability and the EPR effect in macromolecular therapeutics: a review. *J Control Release* 2010;65(1–2):271–84.
12. Jain RK, Stylianopolous T. Delivering nanomedicine to solid tumors. *Nat Rev Clin Onol* 2010;7(11):653–64.
13. Goldberg MS, Hook SS, Wang AZ, Bulte JW, Patri AK, Uckun FM, Cryns VL et al. Biotargeted nanomedicines for cancer: six tenets before you begin. *Nanomedicine* 2013;8(2):299–308.
14. Bartlett DW, Su H, Hildebrandt IJ, Weber WA, Davis ME. Impact of tumor-specific targeting on the biodistribution and efficacy of siRNA nanoparticles measured by multimodality in vivo imaging. *Proc Natl Acad Sci USA* 2007;104(39):15549–54.
15. Kirpotin DB, Drummond DC, Shao Y, Shalaby MR, Hong K, Nielsen UB, Marks JD, Benz CC, Park JW. Antibody targeting of long-circulating lipidic nanoparticles does not increase tumor localization but does increase internalization in animal models. *Cancer Res* 2006;66(13):6732–40.
16. Pankhurst QD, Connolly J, Jones SK, Dobson J. Applications of magnetic nanoparticles in biomedicine. *J Phys D: Appl Phys* 2003;36(13):R167.
17. Chan WC, Maxwell DJ, Gao X, Bailey RE, Han M, Nie S. Luminescent quantum dots for multiplexed biological detection and imaging. *Curr Opin Biotechnol* 2002;13(1):40–6.

18. Huang X, Jain PK, El-Sayed IH, El-Sayed MA. Gold nanoparticles: interesting optical properties and recent applications in cancer diagnostics and therapy. *Nanomedicine* 2007;2(5):681–93.

19. Lee JH, Huh YM, Jun YW, Seo JW, Jang JT, Song HT, Kim S et al. Artificially engineered magnetic nanoparticles for ultra-sensitive molecular imaging. *Nat Med* 2007;13(1):95–9.

20. Haun JB, Castro CM, Wang R, Peterson VM, Marinelli BS, Lee H, Weissleder R. Micro-NMR for rapid molecular analysis of human tumor samples. *Sci Transl Med* 2011;3(71):71a16.

21. Osterfeld SJ, Yu H, Gaster RS, Caramuta S, Xu L, Han SJ, Hall DA et al. Multiplex protein assays based on real-time magnetic nanotag sensing. *Proc Natl Acad Sci USA* 2008;105(52):20637–40.

22. Magforce, Fighting Cancer and Improving Quality of Life. Available at: http://www.magforce.de/en/produkte/nanothermr-therapie.html. (accessed on October 18, 2013).

23. De la Rica R, Stevens MM. Plasmonic ELISA for the ultrasensitive detection of disease biomarkers with the naked eye. *Nat Nanotechnol* 2012;7(12):821–4.

24. De la Zerda A, Zavaleta C, Keren S, Vaithilingam S, Bodapati S, Liu Z, Levi J et al. Carbon nanotubes as photoacoustic molecular imaging agents in living mice. *Nat Nanotechnol* 2008;3(9):557–62.

25. Cui Y, Wei Q, Park H, Lieber CM. Nanowire nanosensors for highly sensitive and selective detection of biological and chemical species. *Science* 2001;293(5533):1289–92.

26. Olson ES, Aguilera TA, Jiang T, Ellies LG, Nguyen QT, Wong EH, Gross LA, Tsien RY. In vivo characterization of activatable cell penetrating peptides for targeting protease activity in cancer. *Integr Biol* 2009;1:382–93.

27. Korin N, Kanapathipillai M, Matthews BD, Crescente M, Brill A, Mammoto T, Ghosh K et al. Shear-activated nanotherapeutics for drug targeting to obstructed blood vessels. *Science* 2012;337(6095):738–42.

28. Schwartz JA, Shetty AM, Price RE, Stafford RJ, Wang JC, Uthamanthil RK, Pham K, McNichols RJ, Coleman CL, Payne JD. Feasibility study of particle-assisted laser ablation of brain tumors in orthotopic canine model. *Cancer Res* 2009;69(4):1659–67.

29. Thiesen B, Jordan A. Clinical applications of magnetic nanoparticles for hyperthermia. *Int J Hypertherm* 2008;24(6):467–74.

30. Park JH, von Maltzahn G, Xu MJ, Fogal V, Kotamraju VR, Ruoslahti E, Bhatia SN, Sailor MJ. Cooperative nanomaterial system to sensitize, target and treat tumors. *Proc Natl Acad Sci USA* 2010;107(3):981–6.

31. von Maltzahn G, Park JH, Lin KY, Singh N, Schwöppe C, Mesters R, Berdel WE, Ruoslahti E, Sailor MJ, Bhatia SN. Nanoparticles that communicate in vivo to amplify tumour targeting. *Nature Mater* 2011;10:545–52.

32. Kwong GA, von Maltzahn G, Murugappan G, Abudayyeh O, Mo S, Papayannopoulos IA, Sverdlov DY et al. Mass encoded synthetic biomarkers for urinary monitoring of disease. *Nat Biotechnol* 2013;31(1):63–70.

33. Tasciotti E, Liu X, Bhavane R, Plant K, Leonard AD, Price BK, Cheng MM et al. Mesoporous silicon particles as multistage delivery system for imaging and therapeutic applications. *Nat Nanotechnol* 2008;3(3):151–7.

34. Ling Y, Pong T, Vassiliou CC, Huang PL, Cima MJ. Implantable magnetic relaxation sensors measure cumulative exposure to cardiac biomarkers. *Nat Biotechnol* 2011;29(3):273–7.

35. McNeil SE. Nanoparticle therapeutics: a personal perspective. Wiley *Interdiscip Rev Nanomed Nanobiotechnol* 2009;1:264–271.

36. Liu Y, Tan J, Thomas A, Ou-Yang D, Muzykantov VR. The shape of things to come: the importance of design in nanotechnology for drug delivery. *Ther Deliv* 2012;3(2):181–94.

37. Godin B, Driessen WH, Proneth B, Lee SY, Srinivasan S, Rumbaut R, Arap W, Pasqualini R, Ferrari M, Decuzzi P. An integrated approach for the rational design of nanovectors for biomedical imaging and therapy. *Adv Genet* 2010;69:31–64.

38. Wang J, Byrne JD, Napier ME, DeSimone JM. More effective nanomedicines through particle design. *Small* 2011;7(14):1919–31.

39. Sharma A, Madhunapantula SV, Robertson GP. Toxicological considerations when creating nanoparticle-based drugs and drug delivery systems. *Exper Opin Drug Metab Toxicol* 2012;8(1):47–69.

40. Zuckerman JE, Choi CH, Han H, Davis ME. Polycation-siRNA nanoparticles can disassemble at the kidney glomerular basement membrane. *Proc Natl Acad Sci USA* 2012;109(8):3137–42.

41. Gref R, Lück M, Quellec P, Marchand M, Dellacherie E, Harnisch S, Blunk T, Müller RH. 'Stealth' corona-core nanoparticles surface modified by polyethylene glycol (PEG): influences of the corona (PEG chain length and surface density) and of the core composition on phagocytic uptake and plasma protein absorption. *Colloids Surf B Biointerfaces* 2000;18(3–4):301–13.

42. Song G, Wu H, Yoshino K, Zamboni WC. Factors affecting the pharmacokinetics and pharmacodynamics of liposomal drugs. *J Liposome Res* 2012;22(3)177–92.

43. Caron WP, Song G, Kumar P, Rawal S, Zamboni WC. Interpatient pharmacokinetic and pharmacodynamic variability of carrier-mediated anticancer agents. *Clin Pharmacol Ther* 2012;91(5):802–12.

44. Toraya-Brown S, Sheen MR, Baird JR, Barry S, Demidenko E, Turk MJ, Hoopes PJ, Conejo-Garcia JR, Fiering S. Phagocytes mediate targeting of iron oxide nanoparticles to tumors for cancer therapy. *Integr Biol* 2013;5:159–71.

45. Crist RM, Grossman JH, Patri AK, Stern ST, Dobrovolskaia MA, Adiseshaiah PP, Clogston JD, McNeil SE. Common pitfalls in nanotechnology: lessons learned from NCI's Nanotechnology Characterization Laboratory. *Integr Biol* 2013;5(1):66–73.

46. Tyner K, Sadrieh N. Considerations when submitting nanotherapeutics to FDA/CDER for regulatory review. *Methods Mol Biol* 2011;697:17–311.

47. Lyon DY, Brunet L, Hinkal GW, Wiesner MR, Alvarez PJ. Antibacterial activity of fullerene water suspensions (nC60) is not due to ROS-mediated damage. *Nano Lett* 2008;8(5):1539–43.

48. Nanotechnology Characterization Laboratory, http://ncl.cancer.gov/. (accessed on October 18, 2013).

49. Zamboni WC, Torchilin V, Patri AK, Hrkach J, Stern S, Lee R, Nel A, Panaro NJ, Grodzinski P. Best practices in cancer nanotechnology: perspective from NCI Nanotechnology Alliance. *Clin Cancer Res* 2012;18:3229–41.

50. Cheng Z, Al Zaki A, Hui JZ, Muzykantov VR, Tsourkas A. Multifunctional nanoparticles: cost versus benefit of adding targeting and imaging capabilities. *Science* 2012;338(6109):903–10.

51. Kerbel RS. Human tumor xenografts as predictive preclinical models for anticancer drug activity in humans: better than commonly perceived—but they can be improved. *Cancer Biol Ther* 2003;2(4 Suppl 1):S134–9.

52. Johnson JI, Decker S, Zaharevitz D, Rubinstein LV, Venditti JM, Schepartz S, Kalyandrug S et al. Relationships between drug activity in NCI preclinical in vitro and in vivo model and early clinical trials. *Br J Cancer* 2001;84(10):1424–31.

53. Peterson JK, Houghton PJ. Integrating pharmacology and in vivo cancer models in preclinical and clinical drug development. *Eur J Cancer* 2004;10(6):837–44.

54. Julien S, Merino-Trigo A, Lacroix L, Pocard M, Goéré D, Mariani P, Landron S et al. Characterization of a large panel of patient-derived tumor xenografts representing the clinical heterogeneity of human colorectal cancer. *Clin Cancer Res* 2012;18(19):5314–28.

55. Alhasan AH, Kim DY, Daniel WL, Watson E, Meeks JJ, Thaxton CS, Mirkin CA. Scanometric microRNA array profiling of prostate cancer markers using spherical nucleic acid-gold nanoparticle conjugates. *Anal Chem* 2012;84(9):4153–60.

56. Thaxton CS, Elghanian R, Thomas AD, Stoeva SI, Lee JS, Smith ND, Schaeffer AJ et al. Nanoparticle-based bio-barcode assay redefines "undetectable" PSA and biochemical recurrence after radical prostatectomy. *Proc Natl Acad Sci USA* 2009;106(44):18437–42.

57. Nanosphere, http://www.nanosphere.us/products. (accessed on October 18, 2013).

58. Haun JB, Devaraj NK, Hilderbrand SA, Lee H, Weissleder R. Bio-orthogonal chemistry amplifies nanoparticle binding and enhances the sensitivity of cell detection. *Nat Nanotechnol* 2010;5:660–5.

59. Zhang Z, Chan DW. The road from discovery to clinical diagnostics: lessons learned from the first FDA-cleared in vitro diagnostic multivariate index assay of proteomic biomarkers. *Cancer Epidemiol Biomarkers Prev* 2010;19(12):2995–9.

60. Lee E, Bae YH. Recent progress in tumor pH targeting using nanotechnology. *J Control Release* 2008;132(3):164–70.

61. Thakor AS, Luong R, Paulmurugan R, Lin FI, Kempen P, Zavaleta C, Chu P, Massoud TF, Sinclair R, Gambhir SS. The fate and toxicity of Raman-active silica-gold nanoparticles in mice. *Sci Transl Med* 2011;3(79):70ra33.

62. Zavaleta CL, Hartman KB, Miao Z, James ML, Kempen P, Thakor AS, Nielsen CH, Sinclair R, Cheng Z, Gambhir SS. Preclinical evaluation of Raman nanoparticle biodistribution for their potential use in clinical endoscopy imaging. *Small* 2011;7(15):2232–40.

2 Impact of Bionanointeractions of Engineered Nanoparticles for Nanomedicine

Stefania Sabella

CONTENTS

2.1 INTRODUCTION

This chapter will review the use of engineered nanoparticles (NPs) for nanomedicine providing an overview of design, properties, and functions of a variety of NPs along with their medical applications (especially focusing on innovative drug delivery strategies). Examples provided will span from NPs, which have been already translated in clinics (NPs therapeutics) to complex multifunctional and biological responsive nanostructures (biomimetic and bioactive NPs) whose medical translation is still in its infancy. The enormous impact of multifunctional NPs-based therapy will be highlighted taking in mind the regulatory requirements for nanomedicine.

In the last decades, scientific progresses in nanotechnology have led to the creation of a variety of nanomaterials that can be classified as engineered NPs, nanocomposites, and nano- and microdevices containing NPs. Reasons for this tremendous growth can be searched in the special properties (optoelectronic, magnetic, mechanic) that nanomaterials have as compared to the conventional macroscale materials. Many branches of science, including material science, physics, biotechnology, biology, and medicine, enthusiastically study the promising applications of nanomaterials and, some years ago, a new discipline—nanomedicine—was defined. Nanomedicine is a multidisciplinary science defined as "the applications of nanotechnology for treatments, diagnosis, monitoring, and control of biological systems."[1] The primary application of nanomedicine was initially a diagnosis and treatment of cancer;[2] however, an increasing number of functional NPs are currently being tested for the treatment of many other diseases such as cardiovascular diseases, Alzheimer's disease, and neurological diseases, which when considered together with cancer are the primary causes of deaths worldwide.[3,4] Other important applications of nanomedicine are the development of hybrid nanomaterials for in vitro and in vivo diagnostics, and tissue engineering. Theranostics—a combination of therapy and diagnosis—is ultimately the final challenging frontier of nanomedicine.[5] Engineered NPs are considered the beating heart of nanomedicine. Commonly designed as single or multiple components of a therapy or as a diagnostic assay, NPs can also act as smart ingredients forming novel composites.[6–8] Whatever the application for which the particles are produced, they in general ensure advanced or even novel performances compared to the conventional therapies or diagnostic assays or materials.[9] Ranging from 1 to 100 nm in size, NPs can be broadly grouped by chemical composition of their cores. They are made of organic materials, inorganic materials, or a hybrid combination of more of them (e.g., polymers, proteins, metals, metal oxides, semiconductors, and rare earth metals), presenting a size and, in some cases, a tridimensional structure that make them very similar (in the scale size) and highly reactive toward biological components such as DNA, proteins, and cellular organelles. Moreover, as a good chef who properly combines ingredients, the same way, by properly mixing ingredients during the fabrication process, the nanotechnologist has enormous possibilities to mold NPs in a variety of sizes and almost in any shapes (e.g., spheres,[10] cubes,[11] long sticks,[12] nanostar-like NPs,[13] particles shaped like viruses,[14] and particles shaped like red blood cells[15] [RBCs]) (Figure 2.1). Notably, the ability to make NPs to exact specifications and to control their size and shape properties at the nanoscale level enables nanotechnologists to control their functions. Generally obtained as colloidal solutions or by top-down nanofabrication processes,[16,17] NPs present a higher surface-to-volume ratio than that exposed by microparticles. Size and surface area are important material characteristics to explain specific NP properties. Indeed there is an inverse relationship between the particle size and its surface area so that a decrease in the size of a particle results in an increase in the number of molecules/atoms displayed on the external surface, which, if reactive, are the potential reactive groups on the particle surface too (Figure 2.2a).[18] Exposed chemical groups determine the NP surface charge that is the property that relates to the degree of hydrophilicity or hydrophobicity, amphipilicity, and catalytic activity of an NP. A practical example of how a high surface area may impact on a specific biological response is the concept of NP multivalency in cancer therapy.[19] By carrying multiple ligands on their surface (due to chemical engineering of their surface), NPs may enhance the affinity of the ligand against multiple receptor molecules due to the highest ligand density. This type of engineering is useful in the case of low affinity ligands.[19] Moreover, it has been demonstrated that multivalency may favor recruiting the highest number of expressed receptors, thus leading to a greater amount of drug for tumor penetration.

Physical–chemical properties (PCs) such as size, surface charge, chemical composition, shape, and roughness define the NP interaction modes with living systems and regulate the biological responses. In particular, when NPs encounter

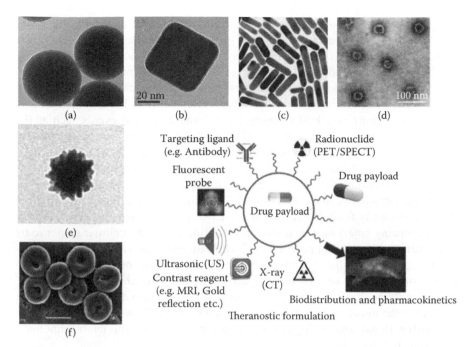

FIGURE 2.1 Multipanel figure shows representative images of synthetic NPs with different shapes: (a) spheres,[10] (b) cubes,[11] (c) rods,[12] (d) virus-like NPs,[14] (e) nanostar-like NPs,[13] and (f) RBC-like NPs.[15] (Panel b is reprinted with permission from Li et al.; panel c is reproduced with permission from Petrova et al.; panel d is reproduced from Schmidt et al.; panel e is reproduced from Maiorano et al.; panel f is reprinted with permission from Doshi et al.) The cartoon of NPs used as theranostic formulation represents a variety of properties, necessary to gain multifunctionality. (From Ding H and Wu F, *Theranostics* 2, 11, 1040–53, 2012.)

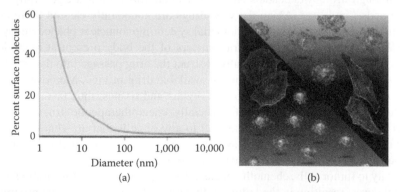

FIGURE 2.2 (a) Inverse relationship between the particle size and the number of molecules/atoms displayed on the external surface of an NP (from Nel A, Xia T, Madler L, Li N. *Science* 2006;311(5761):622–7); (b) cartoon of protein–NP complexes formed in biofluids and confocal images of AuNPs internalized by HeLa cells. (From Sabella S. et al., *Nanotechnology 2011: BioSensors, Instruments, Medical, Environment and Energy*, NSTI, Boston, MA, 2011. With permission.)

biological entities (proteins, membranes, cells, DNAs, and organelles), a series of NP/biological interactions, defying specific interfaces, may occur. Their nature depends on (1) colloidal characteristics (the intrinsic properties of pristine NPs), (2) thermodynamic exchanges between nanomaterial surface and biological elements, and (3) intrinsic properties of biological elements.[20] These complex and dynamic interactions have been referred to as bionanointeractions. Bionanointeractions may lead to many wanted and unwanted events such as the formation of the protein corona[21–23] (Figure 2.2b), particle wrapping, protein unfolding,[24] active intracellular internalization, and biocatalysis, which could have specific adverse or beneficial outcomes.[20] Importantly, understanding bionanointeractions could allow for the development of a quantitative and, possible predictive, relationship between the PCs of an NP and the triggered biological response. This knowledge is fundamental for at least two principal reasons. First, it may allow producing smart bionanomaterials, enabling a specific cellular response by simply tuning the physical properties of NPs/nanomaterials (i.e., size, shape, texture, roughness).[25] In this framework, novel designs are currently investigated with the aim of producing smart nanocomposites, showing enhanced properties (i.e., antibacterial, mechanical, magnetic properties).[6,7,26–30] Second, an in-deep understanding of the interfacial properties between NPs and cells will enable the design of multifunctional and biological responsive nanosystems for nanotechnology-based targeted therapy.[31]

2.2 NANOPARTICLES FOR CANCER THERAPY

The therapeutic effect of a drug depends on many pharmacokinetic factors, which may affect its biodistribution, influencing the achievement of the therapeutic dose at the target site, where the beneficial effect is desired. Once administered, drugs are widely distributed in body organs and tissues through different distribution paths that, although strongly dependent on the route of administration, are generally blood, lymph nodes, and liver. Along the way, drugs are abundantly cleared by first-pass renal filtering and by the reticuloendothelial system/mononuclear phagocytes system (RES/MPS).[32] Yet, numerous natural barriers of the body present at many levels (both tissue and cellular) mechanically obstruct the drug passage into the action site, further reducing the absolute amount of available drug in vivo. Moreover, conventional drugs are poorly selective toward organs and tissues, thus favoring the risk of systemic toxicity. To this respect, especially, chemotherapeutic drugs, which are potent cytotoxic drugs,[33,34] result harmful to any of cells that are characterized by enhanced proliferation rates, such as bone marrow, gastrointestinal tract, and hair follicles.[32,33] As a consequence of such high toxicity against normal tissues and low selectivity to tumor cells, chemotherapeutic drugs are normally administered at suboptimal doses, resulting in the failure of therapy and, very often, in the progression of cancer in metastatic forms. Finally, it has been clearly demonstrated that reduction of doses is generally related to the occurrence of multi-drug resistance (MDR).[35]

To enhance target selectivity, many studies have been conducted with new therapies such as ligand-targeted therapeutics (LTTs)[36] and genetically encoded, therapeutic PEGylated proteins.[37,38] In both cases, selective tissue targeting and/or increased circulation half-life of antibody- or protein-based drugs have been successful. However, despite these advances, the scenario is complex and the "magic bullet" to perform the perfect delivery of a drug has not been found yet. With this regard, enormous efforts are continuously made toward the direction of creating the perfect drug that may enable multiple actions such as (1) selectivity for cancerous tissue while sparing the normal cells (even in the absence of enhanced permeability and retention [EPR] effect), (2) higher tumor/target tissues penetration, and (3) efficient intracellular delivery of a drug with consequent targeting of cellular organelles (when required).

While the organ/cellular target selectivity can be considered a common feature between the NPs and the above-mentioned cancer therapeutics, the NPs present innovative and distinguishing properties that make them particularly suitable for responding to the characteristics of an ideal drug (Figure 2.1). Payload transport (multiple substances may be loaded), nanoscale size (and possibility to precisely control the size with nanoscale sensitivity), potential to overcome drug resistance (due to abundant and active cellular uptake), potential to intracellularly deliver drugs, to target cellular organelles, ligand multivalency, and temporal control of drug release (by controlling chemical nature and biodegradability of NP constituents) are some of these unique features.

For the development of more effective clinical cancer therapeutics, the capacity to transport a payload is certainly the first characteristic that was deeply exploited by NP chemical and physical design. Polymer NPs in the form of nanospheres, nanocapsules, and nanoscale self-assembled polymers can bring a large payload of drug and protect them from biodegradation during their blood circulation. Examples of nanocarriers transporting conventional drugs, interfering RNA molecules (siRNAs), peptides, proteins, combination of more molecules in layer-by-layer architectures, and even ions have been described. Importantly, the surface PC traits of these nanosystems are, in general, not affected by the presence of the payload (even when different molecules are entrapped inside). This has been seen as a great advantage because biodistribution and pharmacokinetic parameters of NPs may not be affected as well.[39] Moreover, transport of internal payload may allow vehicle molecules that, although having a good therapeutic index, are, in general, very difficult to administer due to their PC instability and/or hydrophobicity. Liposomes, micelles, and polymeric NPs are, in general, widely used for drug transporting. Many examples are already transferred in clinics and used as effective therapeutics for many cancer diseases, showing passive and active selective tissue targeting.[40–42] Furthermore, by using biodegradable polymers and by finely tuning the chemical nature of various layers composing the polymeric NPs, a controlled release of drugs may be gained.

The EPR effect is a phenomenon that defines a combination of factors as high permeability of tumor vasculature (due to leaky and incomplete vasculature), lack of tumor drainage (due to a poorly defined lymphatic system), high presence of vascular endothelial growth factor and bradykinin, and reduced presence of nitric oxide. Put all together, they may explain the passive accumulation of biomacromolecules (MW > 50 kDa) at the tumor site.[43] As for biomacromolecules with certain traits in size, the NPs can take advantage of the EPR effect for passively targeting tumors. Moreover, considering the size range of abnormal vasculature fenestrations (50–100 nm) and combining this physical aberration to the rapid renal and RES clearance (NPs of dimensions superior of 200 nm are cleared much faster than those of inferior sizes[44]), the NP size range mostly ensuring longer circulation time and increased accumulation in the tumor interstitium is 10–150 nm. This range is of particular interest when considering the very good ability of nanotechnology to tune the size of the produced NPs around 10–50 nm.[16] Notably, there is the opportunity to control the degree of particle dispersion, the shape, the surface charge, and even the degree of surface roughness for the colloids in solutions. These latter points may be of particular importance in view of producing therapeutics with properties of biomimicry (see Section 2.3).

Size, shape,[45] and chemical coating of NPs are key parameters for controlling cellular uptake. It is well known that the NPs decorated or not by targeting ligands are internalized into the cells by processes that are mainly energy mediated, and such cellular internalization is more abundant as compared to that activated by conventional drugs or small molecules.[46] Of course, internalization efficiency and specific processes involved (clathrin and/or caveolae-mediated mechanisms, or others)[47] may vary depending on the specific NP considered. This phenomenon has been compared to a Trojan horse mechanism,[48] since once taken up by cells, the NPs may exert their biological effect (e.g., release of the carrying drug and even intrinsic toxicity) (Sabella et al. unpublished data). The NP-enhanced internalization combined to the passive or active (due to ligand selectivity) accumulation at the tumor site are factors of paramount importance as they allow for greatest tumor penetration, increasing of the effective intracellular therapeutic dose without producing side effects, overcoming of the efflux-mediated MDR[35] and, finally, cellular organelle drug targeting. This latter point is therefore challenging and strategies for NP endosomal escaping must be encouraged by suitable chemical engineering of NPs by design. Davis et al. recently described an attractive example of nanocarrier enabling the intracellular delivery of siRNA in humans by using a multifunctional polymeric NP of 70 nm. The NP is composed of a linear cyclodextrin-based polymer electrostatically self-assembled to siRNA molecules. A transferrin-linked hydrophilic polymer (polyethylene glycol) ensures active targeting and NP colloidal stability in biofluids. After internalization by endocytosis, due to the acidic pH conditions in the lysosomes, the NPs dissociate, thus releasing siRNAs in the solid tumor at doses that correlate with dose levels of the NPs administered.[49] Another elegant example of endosomal escaping is represented by an artificial recombinant elastin-like polypeptides supported by acid labile linkers.[50] Due to the peculiar behavior of self-assembling in the presence of very hydrophobic drug molecules, these polypeptides self-assemble in micelles of size less than 50 nm, carrying the chemotherapeutic drug. Moreover, in this case, the release of the drug is mediated by the acidic pH (that breaks acidic linkers) and occurs very close to the nuclear target. Noteworthy, a temporal controlled release of the drug at a precise subcellular site is another important condition that should be achieved. To this regard, advancements in polymer technology and colloidal chemistry of inorganic NPs are opening up interesting and novel routes toward this direction. Examples of NPs releasing drug in a pH- and ion-responsive manner have been described. For example, when the drug target is nuclear DNAs, the release in proximity of the cell nucleus is important and, at the same time, it is necessary to protect the drug from cytoplasmic resistance effects or degradation. By a rationale design of NP, Murakami developed a nanosystem, whose drug release was activated only by a combination of an acidic environment and high concentrations of Cl⁻, conditions known to be exclusively present in the late endosomes and lysosomes, which are organelles in close proximity to the nucleus.[51] Although with more difficulties (mostly due to the lower payload capacity of inorganic NPs), intriguing examples enabling a kinetic control of drug release have been recently reported for porous silica NPs. Park produced fluorescent porous silicon NPs of 150 nm in size that can be loaded with therapeutic doses of doxorubicin.[52] By controlling the biodegradation of the silica core in derivatives of orthosilicic acid ($Si(OH)_4$), the release of doxorubicin has been demonstrated both in vitro and in vivo. Importantly, silicic acid derivatives have been demonstrated to be biocompatible components (they have been found as trace elements in humans and as components of numerous tissues in which they are naturally adsorbed or excreted) (Figure 2.3a). Ashley has proposed another

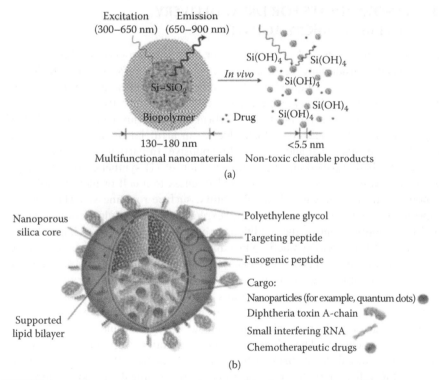

FIGURE 2.3 **(See color insert.)** (a) Biocompatible fluorescent porous silicon NPs of size 150 nm loaded with doxorubicin. The drug release may be controlled by biodegradation of the particle core in derivatives of orthosilicic acid. (Reprinted with permission from Park JH et al., *Nat. Mater.*, 8(4), 331–336, 2009.) (b) Schematic of a hybrid nanosystems composed of liposomes and nanoporous silica NPs referred to as a protocell. This system synergistically combines the properties of both silica NPs and liposomes and shows nearly all the ideal properties of an NP-based therapeutics. (Reprinted by permission from Macmillan Publishers Ltd. *Nat Mater*, Ashley CE et al., 10(5), 389–397, copyright 2011.)

elegant example of hybrid nanosystems composed of liposomes and nanoporous silica NPs referred to as protocell (Figure 2.3b). This system synergistically combines properties of both silica NPs and liposomes and shows nearly all the ideal properties that a targeted nanocarrier should present, such as capability of carrying a high level of multiple and diverse drugs, long circulating half-life, selective tumor receptor targeting, endosomal escaping, and selective nuclear intracellular drug delivery.[53] Although the clinical translation of inorganic NPs as therapeutics is still in its infancy, these examples may pave the way to new chemical design strategies for drug delivery by using inorganic NPs. Moreover, porous and amorphous silicon NPs may be very attractive nanosystems for future developments of multifunctional nanocarriers. There are several reasons for this: (1) simple fabrication methods (with scale-up possibilities for various synthetic methods), (2) high control of nanoscale size range, (3) high control of colloidal dispersion, (4) easy and quantitative conjugation with biomacromolecules, (5) biodegradability, (6) biocompatibility, and (7) payload capacity.[54]

2.3 NANOPARTICLES FOR DRUG DELIVERY WITH BIOMIMICRY ACTIVITY

The nascent field of drug delivery based on NPs with properties of biomimicry has advanced considerably in recent years. Taking inspiration from biological functionalities associated with morphology of cells (bacteria, viruses, proteins), several examples of drug delivery by bioinspired and biomimetic NPs have been recently reported.[31] The focus is on fabricating hybrid nanosystems (NPs and biological components) for drug delivery, which, due to an artificial or natural biomimetic camouflage, are functional and biological responsive. In particular, two main strategies can be distinguished aiming at the creation of (1) synthetic NPs showing a morphology-based or mode of action-based likeness to a cell or natural biological components. In the first case, shape, flexibility, surface roughness, surface protein composition, and density are the properties exploited, while cellular communication, signaling, and movement are the conditions to be considered in mimicking molecular mechanisms;[55] (2) hybrid synthetic NPs camouflaged as biological components or cells (e.g., natural protein NPs and leukocyte) by the presence of bioactive, natural, and stable layers of biomacromolecules/cell membranes on the NP surface.

NPs with biomimicry activity generally present a higher degree of molecular architecture, with consequent molecule orientation, recognition capability, and, most important, they, in principle, can replicate biological functions and interactions with living entities. They show advanced recognition properties mostly driven by chemotaxis-based mechanisms leading to site-specific NP accumulation (even without EPR effect[56]) or to the activation of biochemical cascade pathways.[55] Moreover, they may be multifunctional (they can contemporaneously diagnose and treat a disease). Medical applications of these nanosystems are in cancer therapy, atherosclerosis, cardiovascular diseases, and neurological diseases.[4,31]

Cellular flexibility is a fundamental property of many cells that nanotechnologists reproduced synthetically. Mammalian RBCs are natural example of flexible cells. They tune the degree of their flexibility to control many functions such as their half-lives in the blood, their passage through the restrictions in the vasculatures (usually smaller than the RBCs' diameters) and, becoming less flexible, they accelerate their removal from the circulation via splenic filtration. Inspired by this natural example, Doshi synthesized RBC-like particles that mimic the key structural and functional features of these cells by means of polymer technology coupled to protein self-assembly (Figure 2.1f).[15] They demonstrated the feasibility of the synthetic procedure to produce stable RBC-like particles that possess the ability to carry oxygen and flow through capillaries showing diameters smaller than their own diameter. Further, by means of this nanosystem, they can also encapsulate drugs and agents for imaging.[15] By using PRINT(Particle Replication in Non-Wetting Templates) technology,[17] DeSimone and coworkers have recently developed long-circulating RBC-like microparticles showing tunable levels of deformation (Figure 2.4a). Importantly, it has been demonstrated that the different degrees of flexibility (by varying the elastic modules of the particles) is related to the half-lives of the particles in the blood stream.[57] By applying a similar approach, several examples of viral-like NPs have been reported.[14,31] Interestingly, in this framework, Cumbo has recently proposed an innovative chemical imprinting strategy leading to the production of an artificial organic-silica NPs on which the surface features of a nonenveloped icosahedral virus has been printed (virus-imprinted nanoparticles [VIPs]).[58] It has been demonstrated that VIPs possess geometrical and chemical recognition capability toward low copies of the template virus in water (picomolar sensitivity). Although the primary application of VIPs is currently for viruses detection assay, the chemical imprinting strategy represents a milestone in the production of remove purely synthetic materials with virus biomimicry. This will be also helpful for future research in drug delivery.

RBC-like or leukocyte-like NPs have been obtained by the second approach (hybrid synthetic NPs) (Figure 2.4b). Hu and coworkers developed a method to recover stable RBC membranes. Afterward, they wrapped them like a camouflaging cloak around a sub-100 nm biodegradable polymer NP, which was previously loaded by a cocktail of small molecule drugs. After intravenous injection in mice, these NPs showed a blood half-life that was superior compared to the control. This suggests that the main clearance systems are eluded thanks to the biomimetism of the NPs.[59] Porous silica NPs have been masked with leukocyte membranes. Parodi et al. have produced a nanocarrier system called leukolike vectors that were composed of nanoporous silicon NP coated with a leukocyte cellular membranes.[56] This system was demonstrated to efficiently recognize inflamed endothelium, elude opsonization, and delay the mononuclear phagocyte system, transport and release doxorubicin across the endothelium, while eluding the lysosomal pathway. Importantly, stable leukocyte-modified NPs have shown the potential to naturally recognize and bind the tumor endothelium in an active, nondestructive manner, thus improving the tumoritropic activity of a drug carrier even in the absence of irregular endothelium (absence of EPR effect).

Examples of biomacromolecules (e.g., enzymes, proteins, signaling or transduction molecules) inspired NPs have been reported recently. Owing to their potential medical use for the diagnosis and therapy of cardiovascular diseases and atherosclerosis,[4,60] lipoprotein-like NPs have recently raised great interests. Lipoproteins are natural NPs with a size range between 10 and 20 nm and mainly composed of a core of apolar triglycerides and cholesteryl esters. They are surrounded by phospholipid monolayers containing forms and amounts of apolipoproteins and cholesterol molecules that have

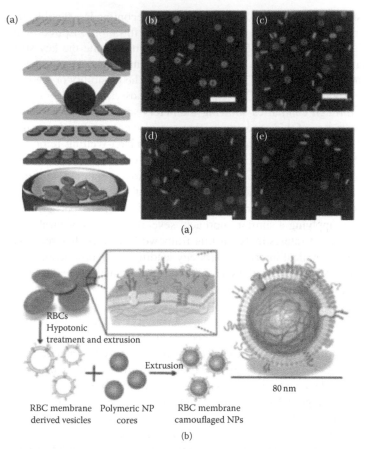

FIGURE 2.4 **(See color insert.)** Examples of RBC-like NPs obtained by different fabrication methods leading to (a) synthetic NPs showing likeness to RBC cells. (Reprinted with permission from Merkel TJ et al., *Proc. Natl. Acad. Sci. USA*, 108(2), 586–591, 2011.) (b) Hybrid synthetic NPs covered by natural RBC membranes. (Reprinted with permission from Hu CM et al., *Proc. Natl. Acad. Sci. USA*, 108(27), 10980–10985, 2011.)

not yet esterified. Lipid NPs were fabricated to mimic the behavior and functions of lipoproteins. With this regard, a variety of inorganic NPs (gold, iron oxide, and quantum dots) mimicking the chemical composition, size, and functions of the high-density lipoproteins (HDLs) were fabricated in the previous years,[61,62] demonstrating that they may act as very efficient multimodal imaging agents of atherosclerotic plaques. Moreover, iron oxide high-density lipoproteins (FeO-HDL) NPs have been tested in vivo, finding that they allow for a selective targeting of HDL receptors leading to a regulation of cellular cholesterol efflux. Luthi and coworkers have used gold NPs (AuNPs) as a template to generate biomimetic nanostructures that are similar to HDLs. Importantly, the physical properties (sizes and surface chemistries) of these NPs can be tailored to have a precise control of cholesterol binding and efflux from macrophages, as for natural HDL.[63] Finally, thanks to the NP cell-specific targeting afforded by HDL biomimicry, these NPs were applied for gene therapy.[64]

2.4 CONCLUSION

As mentioned earlier, polymer-based NPs have been successfully translated into clinics. They represent the first generation of NPs that usually are referred to as "therapeutic NPs."[39] However, although intriguing and promising, the multifunctional and bioactive NPs have not been translated into the clinics because several regulatory issues must be considered.[65,66] The lack of well-defined toxicity assessment of both the single components of a multifunctional nanosystem and the entire nanosystem itself is undoubtedly a key issue. All communities (research institutions, regulatory bodies, and industries) have raised concerns and urgently argued the need for nanoregulatory and risk assessment of NPs and nanomaterials. These requirements must be considered with priority especially for those classes of NPs which have been recently reported to be toxic both by in vitro and in vivo studies.[66-73] Moreover, other serious limitations may be represented by difficulties in industrial scale-up processes (that usually require the application of rigid good manufacturing practice rules), lack of well-defined chemistries, and lack of intellectual-property regulation. Finally, important considerations about the possible environmental impact of large-scale productions of nanotechnology-enabled medical devices must be taken into account in the future.[74]

REFERENCES

1. Moghimi SM, Hunter AC, Murray JC. Nanomedicine: Current Status and Future Prospects. *FASEB J* 2005;19(3):311–30.
2. Blanco E, Hsiao A, Mann AP, Landry MG, Meric-Bernstam F, Ferrari M. Nanomedicine in Cancer Therapy: Innovative Trends and Prospects. *Cancer Sci* 2011;102(7):1247–52.
3. Brambilla D, Le Droumaguet B, Nicolas J, Hashemi SH, Wu LP, Moghimi SM, Couvreur P, Andrieux K. Nanotechnologies for Alzheimer's Disease: Diagnosis, Therapy, and Safety Issues. *Nanomedicine* 2011;7(5):521–40.
4. Lobatto ME, Fuster V, Fayad ZA, Mulder WJ. Perspectives and Opportunities for Nanomedicine in the Management of Atherosclerosis. *Nat Rev Drug Discov* 2011;10(11):835–52.
5. Ding H, Wu F. Image Guided Biodistribution and Pharmacokinetic Studies of Theranostics. *Theranostics* 2012;2(11):1040–53.
6. Cingolani R, Athanassiou A, Pompa PP. Modulating Antibacterial Properties Using Nanotechnology. *Nanomedicine (Lond)* 2011;6(9):1483–5.
7. Rizzello L, Galeone A, Vecchio G, Brunetti V, Sabella S, Pompa PP. Molecular Response of Escherichia coli Adhering onto Nanoscale Topography. *Nanoscale Res Lett* 2012;7(1):575.
8. Kao J, Thorkelsson K, Bai P, Rancatore BJ, Xu T. Toward Functional Nanocomposites: Taking the Best of Nanoparticles, Polymers, and Small Molecules. *Chem Soc Rev* 2012;42(7):2654–78.
9. Talapin DV, Lee JS, Kovalenko MV, Shevchenko EV. Prospects of Colloidal Nanocrystals for Electronic and Optoelectronic Applications. *Chem Rev* 2010;110(1):389–458.
10. Malvindi MA, Brunetti V, Vecchio G, Galeone A, Cingolani R, Pompa PP. SiO_2 Nanoparticles Biocompatibility and Their Potential for Gene Delivery and Silencing. *Nanoscale* 2012;4(2):486–95.
11. Li JF, Tian XD, Li SB, Anema JR, Yang ZL, Ding Y, Wu YF et al. Surface Analysis Using Shell-Isolated Nanoparticle-Enhanced Raman Spectroscopy. *Nat Protoc* 2013;8(1):52–65.

12. Petrova H, Perez-Juste J, Zhang Z, Zhang J, Kosel T, Hartland GV. Crystal Structure Dependence of the Elastic Constants of Gold Nanorods. *J Mater Chem* 2006;16(40):3957–63.
13. Maiorano G, Rizzello L, Malvindi MA, Shankar SS, Martiradonna L, Falqui A, Cingolani R, Pompa PP. Monodispersed and Size-Controlled Multibranched Gold Nanoparticles with Nanoscale Tuning of Surface Morphology. *Nanoscale* 2011;3(5):2227–32.
14. Schmidt U, Gunther C, Rudolph R, Bohm G. Protein and Peptide Delivery Via Engineered Polyomavirus-Like Particles. *FASEB J* 2001;15(9):1646–815.
15. Doshi N, Zahr AS, Bhaskar S, Lahann J, Mitragotri S. Red Blood Cell-Mimicking Synthetic Biomaterial Particles. *Proc Natl Acad Sci USA* 2009;106(51):21495–9.
16. Parak WJ. Materials Science. Complex Colloidal Assembly. *Science* 2011;334 (6061):1359–60.
17. Euliss LE, DuPont JA, Gratton S, DeSimone J. Imparting Size, Shape, and Composition Control of Materials for Nanomedicine. *Chem Soc Rev* 2006;35(11):1095–104.
18. Nel A, Xia T, Madler L, Li N. Toxic Potential of Materials at the Nanolevel. *Science* 2006;311(5761):622–7.
19. Montet X, Funovics M, Montet-Abou K, Weissleder R, Josephson L. Multivalent Effects of Rgd Peptides Obtained by Nanoparticle Display. *J Med Chem* 2006;49(20): 6087–93.
20. Nel AE, Madler L, Velegol D, Xia T, Hoek EM, Somasundaran P, Klaessig F, Castranova V, Thompson M. Understanding Biophysicochemical Interactions at the Nano-Bio Interface. *Nat Mater* 2009;8(7):543–57.
21. Lynch I, Salvati A, Dawson KA. Protein-Nanoparticle Interactions: What Does the Cell See? *Nat Nanotechnol* 2009;4(9):546–7.
22. Maiorano G, Sabella S, Sorce B, Brunetti V, Malvindi MA, Cingolani R, Pompa PP. Effects of Cell Culture Media on the Dynamic Formation of Protein-Nanoparticle Complexes and Influence on the Cellular Response. *ACS Nano* 2010;4(12):7481–91.
23. Sabella S, Maiorano G, Rizzello L, Kote S, Cingolani R, Pompa PP. Framing the Nano-Biointeractions by Proteomics. In *Proc. SPIE 8232 2012;Colloidal Nanocrystals for Biomedical Applications VII, 82320S.* San Francisco, CA, 2012.
24. Deng ZJ, Liang M, Monteiro M, Toth I, Minchin RF. Nanoparticle-Induced Unfolding of Fibrinogen Promotes Mac-1 Receptor Activation and Inflammation. *Nat Nanotechnol* 2011;6(1):39–44.
25. Mitragotri S, Lahann J. Physical Approaches to Biomaterial Design. *Nat Mater* 2009;8(1):15–23.
26. Rizzello L, Sorce B, Sabella S, Vecchio G, Galeone A, Brunetti V, Cingolani R, Pompa PP. Impact of Nanoscale Topography on Genomics and Proteomics of Adherent Bacteria. *ACS Nano* 2011;5(3):1865–76.
27. Brunetti V, Maiorano G, Rizzello L, Sorce B, Sabella S, Cingolani R, Pompa PP. Neurons Sense Nanoscale Roughness with Nanometer Sensitivity. *Proc Natl Acad Sci USA* 2010;107(14):6264–9.
28. Bayer IS, Fragouli D, Attanasio A, Sorce B, Bertoni G, Brescia R, Di Corato R et al. Water-Repellent Cellulose Fiber Networks with Multifunctional Properties. *ACS Appl Mater Interfaces* 2011;3(10):4024–31.
29. Rizzello L, Shankar SS, Fragouli D, Athanassiou A, Cingolani R, Pompa PP. Microscale Patterning of Hydrophobic/Hydrophilic Surfaces by Spatially Controlled Galvanic Displacement Reactions. *Langmuir* 2009;25(11):6019–23.
30. Rizzello L, Cingolani R, Pompa PP. Nanotechnology Tools for Antibacterial Materials. *Nanomedicine (Lond)* 2013;8(5):807–21.
31. Yoo JW, Irvine DJ, Discher DE, Mitragotri S. Bio-Inspired, Bioengineered and Biomimetic Drug Delivery Carriers. *Nat Rev Drug Discov* 2011;10(7):521–35.

32. Steichen SD, Caldorera-Moore M, Peppas NA. A Review of Current Nanoparticle and Targeting Moieties for the Delivery of Cancer Therapeutics. *Eur J Pharm Sci* 2012;48(3):416–27.

33. Fu D, Calvo JA, Samson LD. Balancing Repair and Tolerance of DNA Damage Caused by Alkylating Agents. *Nat Rev Cancer* 2012;12(2):104–20.

34. Ziegler CJ, Silverman AP, Lippard SJ. High-Throughput Synthesis and Screening of Platinum Drug Candidates. *J Biol Inorg Chem* 2000;5(6):774–83.

35. Gottesman MM, Fojo T, Bates SE. Multidrug Resistance in Cancer: Role of ATP-Dependent Transporters. *Nat Rev Cancer* 2002;2(1):48–58.

36. Allen TM. Ligand-Targeted Therapeutics in Anticancer Therapy. *Nat Rev Cancer* 2002;2(10):750–63.

37. Gao W, Liu W, Christensen T, Zalutsky MR, Chilkoti A. In Situ Growth of a Peg-Like Polymer from the C Terminus of an Intein Fusion Protein Improves Pharmacokinetics and Tumor Accumulation. *Proc Natl Acad Sci USA* 2010;107(38):16432–7.

38. Hubbell JA, Chilkoti A. Chemistry. Nanomaterials for Drug Delivery. *Science* 2012;337(6092):303–5.

39. Davis ME, Chen ZG, Shin DM. Nanoparticle Therapeutics: An Emerging Treatment Modality for Cancer. *Nat Rev Drug Discov* 2008;7(9):771–82.

40. Duncan R. Polymer Conjugates as Anticancer Nanomedicines. *Nat Rev Cancer* 2006;6(9):688–701.

41. Peer D, Karp JM, Hong S, Farokhzad OC, Margalit R, Langer R. Nanocarriers as an Emerging Platform for Cancer Therapy. *Nat Nanotechnol* 2007;2(12):751–60.

42. Webster DM, Sundaram P, Byrne ME. Injectable Nanomaterials for Drug Delivery: Carriers, Targeting Moieties, and Therapeutics. *Eur J Pharm Biopharm* 2013;84(1):1–20.

43. Matsumura Y, Maeda H. A New Concept for Macromolecular Therapeutics in Cancer Chemotherapy: Mechanism of Tumoritropic Accumulation of Proteins and the Antitumor Agent Smancs. *Cancer Res* 1986;46(12 Pt 1):6387–92.

44. Alexis F, Pridgen E, Molnar LK, Farokhzad OC. Factors Affecting the Clearance and Biodistribution of Polymeric Nanoparticles. *Mol Pharm* 2008;5(4):505–15.

45. Champion JA, Katare YK, Mitragotri S. Particle Shape: A New Design Parameter for Micro- and Nanoscale Drug Delivery Carriers. *J Control Release* 2007;121(1–2):3–9.

46. Sahay G, Alakhova DY, Kabanov AV. Endocytosis of Nanomedicines. *J Control Release* 2010;145(3):182–95.

47. Conner SD, Schmid SL. Regulated Portals of Entry into the Cell. *Nature* 2003;422(6927):37–44.

48. Service RF. Nanotechnology. Nanoparticle Trojan Horses Gallop from the Lab into the Clinic. *Science* 2010;330(6002):314–5.

49. Davis ME, Zuckerman JE, Choi CH, Seligson D, Tolcher A, Alabi CA, Yen Y, Heidel JD, Ribas A. Evidence of RNAi in Humans from Systemically Administered siRNA Via Targeted Nanoparticles. *Nature* 2010;464(7291):1067–70.

50. MacKay JA, Chen M, McDaniel JR, Liu W, Simnick AJ, Chilkoti A. Self-Assembling Chimeric Polypeptide-Doxorubicin Conjugate Nanoparticles That Abolish Tumours after a Single Injection. *Nat Mater* 2009;8(12):993–9.

51. Murakami M, Cabral H, Matsumoto Y, Wu S, Kano MR, Yamori T, Nishiyama N, Kataoka K. Improving Drug Potency and Efficacy by Nanocarrier-Mediated Subcellular Targeting. *Sci Transl Med* 2011;3(64):64ra2.

52. Park JH, Gu L, von Maltzahn G, Ruoslahti E, Bhatia SN, Sailor MJ. Biodegradable Luminescent Porous Silicon Nanoparticles for in vivo Applications. *Nat Mater* 2009;8(4):331–6.

53. Ashley CE, Carnes EC, Phillips GK, Padilla D, Durfee PN, Brown PA, Hanna TN et al. The Targeted Delivery of Multicomponent Cargos to Cancer Cells by Nanoporous Particle-Supported Lipid Bilayers. *Nat Mater* 2011;10(5):389–97.

54. Tang F, Li L, Chen D. Mesoporous Silica Nanoparticles: Synthesis, Biocompatibility and Drug Delivery. *Adv Mater* 2012;24(12):1504–34.
55. von Maltzahn G, Park JH, Lin KY, Singh N, Schwoppe C, Mesters R, Berdel WE, Ruoslahti E, Sailor MJ, Bhatia SN. Nanoparticles That Communicate in vivo to Amplify Tumour Targeting. *Nat Mater* 2011;10(7):545–52.
56. Parodi A, Quattrocchi N, van de Ven AL, Chiappini C, Evangelopoulos M, Martinez JO, Brown BS et al. Synthetic Nanoparticles Functionalized with Biomimetic Leukocyte Membranes Possess Cell-Like Functions. *Nat Nanotechnol* 2013;8(1):61–8.
57. Merkel TJ, Jones SW, Herlihy KP, Kersey FR, Shields AR, Napier M, Luft JC et al. Using Mechanobiological Mimicry of Red Blood Cells to Extend Circulation Times of Hydrogel Microparticles. *Proc Natl Acad Sci USA* 2011;108(2):586–91.
58. Cumbo A, Lorber B, Corvini PF, Meier W, Shahgaldian P. A Synthetic Nanomaterial for Virus Recognition Produced by Surface Imprinting. *Nat Commun* 2013;4:1503.
59. Hu CM, Zhang L, Aryal S, Cheung C, Fang RH. Erythrocyte Membrane-Camouflaged Polymeric Nanoparticles as a Biomimetic Delivery Platform. *Proc Natl Acad Sci USA* 2011;108(27):10980–5.
60. Ng KK, Lovell JF, Zheng G. Lipoprotein-Inspired Nanoparticles for Cancer Theranostics. *Acc Chem Res* 2011;44(10):1105–13.
61. Cormode DP, Skajaa T, van Schooneveld MM, Koole R, Jarzyna P, Lobatto ME, Calcagno C et al. Nanocrystal Core High-Density Lipoproteins: A Multimodality Contrast Agent Platform. *Nano Lett* 2008;8(11):3715–23.
62. Skajaa T, Cormode DP, Jarzyna PA, Delshad A, Blachford C, Barazza A, Fisher EA, Gordon RE, Fayad ZA, Mulder WJ. The Biological Properties of Iron Oxide Core High-Density Lipoprotein in Experimental Atherosclerosis. *Biomaterials* 2011;32(1):206–13.
63. Luthi AJ, Zhang H, Kim D, Giljohann DA, Mirkin CA, Thaxton CS. Tailoring of Biomimetic High-Density Lipoprotein Nanostructures Changes Cholesterol Binding and Efflux. *ACS Nano* 2012;6(1):276–85.
64. McMahon KM, Mutharasan RK, Tripathy S, Veliceasa D, Bobeica M, Shumaker DK, Luthi AJ et al. Biomimetic High Density Lipoprotein Nanoparticles for Nucleic Acid Delivery. *Nano Lett* 2011;11(3):1208–14.
65. Wagner V, Dullaart A, Bock AK, Zweck A. The Emerging Nanomedicine Landscape. *Nat Biotechnol* 2006;24(10):1211–7.
66. Anon. Regulating Nanomedicine. *Nat Mater* 2007;6(4):249.
67. Vecchio G, Galeone A, Brunetti V, Maiorano G, Rizzello L, Sabella S, Cingolani R, Pompa PP. Mutagenic Effects of Gold Nanoparticles Induce Aberrant Phenotypes in Drosophila Melanogaster. *Nanomedicine* 2012;8(1):1–7.
68. Vecchio G, Galeone A, Brunetti V, Maiorano G, Sabella S, Cingolani R, Pompa PP. Concentration-Dependent, Size-Independent Toxicity of Citrate Capped AuNPs in Drosophila Melanogaster. *PLoS One* 2012;7(1):e29980.
69. Pompa PP, Vecchio G, Galeone A, Brunetti V, Maiorano G, Sabella S, Cingolani R. Physical Assessment of Toxicology at Nanoscale: Nano Dose-Metrics and Toxicity Factor. *Nanoscale* 2011;3(7):2889–97.
70. Sabella S, Brunetti V, Vecchio G, Galeone A, Maiorano G, Cingolani R, Pompa PP. Toxicity of Citrate-Capped AuNPs: An In Vitro and In Vivo Assessment. *J Nanopart Res* 2011;13(12):6821–35.
71. Brunetti V, Chibli H, Fiammengo R, Galeone A, Malvindi MA, Vecchio G, Cingolani R, Nadeau JL, Pompa PP. Inp/Zns as a Safer Alternative to Cdse/Zns Core/Shell Quantum Dots: In Vitro and In Vivo Toxicity Assessment. *Nanoscale* 2013;5(1):307–17.

72. Galeone A, Vecchio G, Malvindi MA, Brunetti V, Cingolani R, Pompa PP. In vivo Assessment of Cdse-Zns Quantum Dots: Coating Dependent Bioaccumulation and Genotoxicity. *Nanoscale* 2012;4(20):6401–7.
73. Pompa PP, Vecchio G, Galeone A, Brunetti V, Sabella S, Maiorano G, Falqui A, Bertoni G, Cingolani R. In vivo Toxicity Assessment of Gold Nanoparticles in Drosophila Melanogaster. *Nano Research* 2011;4(4):405–13.
74. Mahapatra I, Clark J, Dobson PJ, Owen R, Lead JR. Potential Environmental Implications of Nano-Enabled Medical Applications: Critical Review. *Environ Sci: Proc & Impacts* 2013;15(1):123–44.

3 Rational Approach for the Safe Design of Nanomaterials

Anna L. Costa

CONTENTS

3.1 INTRODUCTION

The promising technological challenges offered by *nanostructured materials* are a consequence of their three important features: (1) at least some of phase heterogeneity occurs in the size range of nanostructures (~1 to 100 nm); (2) such heterogeneity is crucial in determining properties; and (3) at least in part nanostructures are synthesized, distributed (or organized) by design (atom by atom, molecule by molecule, particle by particle).[1]

The huge amount of nano-research, carried out in the past three decades, answers fundamental questions such as "what nanostructures are interesting," "how can they be synthesized," and "how can they be introduced into materials." It is now necessary to address the more challenging goal of understanding "how can nanostructures morphology, composition, interface features affect the properties of the material that incorporate them." The achievement of such an objective, in fact, would allow for the control of material performances by design, offering to material scientists a powerful tool for the development of high-impact nanotechnological applications. Such a challenge becomes even more of a concern if the properties controlled by design have potential risks when generated and introduced into the environment.

The understanding of the mechanism that govern both the adverse effects of nanomaterials (NMs) on biological system and the emission/exposure potential in terms of fate from nano-aereosolization to bio-uptake is fundamental when implementing a rational approach for the safe design of NMs. This chapter discusses the role of *safety by design* (SbyD) as preventive and effective risk control measurement for the risk management of NMs; the conceptual framework to identify the key features that drive the design of safe NMs; and the approach developed, accordingly, within the European Union collaborative project SANOWORK.

3.2   RISK MANAGEMENT OF NANOMATERIALS

Risk management programs for nanoparticles (NPs) should be seen as an integral part of the overall occupational safety and health program for any workplace manufacturing NMs.

The hierarchy of exposure controls shown in Figure 3.1 is an integral part of a nanomaterial risk management program. Traditional hierarchy of controls emphasizes reducing the hazard as close to the source as possible, by using (1) *elimination or substitution*, (2) engineering controls, (3) administrative or work practice controls, and (4) personal protective equipment (PPE), in order of decreasing priority.[2]

Control technologies in use today, as well as studies reporting the efficiency of these controls when applied to NMs risk management program, address the use of engineering controls (fume hoods, local exhaust ventilation, enclosed glove boxes, filtration of the exhaust air), administrative controls (high efficiency particulate air [HEPA]-filtered vacuum cleaner or wet wiping methods), and PPE (respirators), for controlling workers' exposure.

Actual reported environment, health and safety (EHS) practices, including selection of engineering controls, PPE, cleanup methods, and waste management, do not significantly depart from conventional safety practices for handling chemicals. In fact, practices were occasionally described based on the properties of the bulk form or the solvent carrier and not specifically on the properties of the NM. Reported practices in the handling of NMs, with some exceptions, are based on criteria unrelated to any perceived risks stemming specifically from working with nanoscale materials. The *by default* use of conventional practices for handling NMs appears to

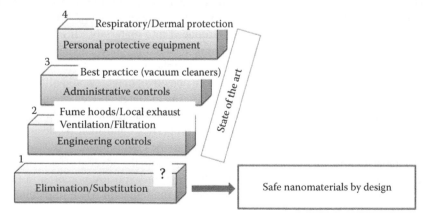

FIGURE 3.1   Components of a nanomaterial risk management program: control measures that are currently in use.

stem from a lack of information on the toxicological properties of NMs and nascent regulatory guidance on EHS practices. So, the nearly half of the organizations that reported implementing a nano-specific EHS program described it as a precaution against unknown hazards.

Elimination, that is, avoid using the hazardous substance or the process that causes exposure, is unlikely to be an option if the NM has been selected for its specific properties; however, consideration should be given as to whether the improved properties of the NM justify any enhanced risks associated with its use. Although it might not be possible to substitute that material, it might be possible to reduce the likelihood of exposure by, for example, binding powder NMs in liquid or solid media or increasing the agglomeration state. Dispersions, pastes, or pelletized forms should be used instead of powder substances wherever this is technically feasible. Also, surface modification must be addressed to decrease the hazard properties by preserving the nanoscale reactivity, as elimination or substitution as a primary prevention strategy.

The development and integration within real processes of "elimination" control measurements opens the doors to *prevention through design* (PtD) as a proactive tool to prevent possible hazard and exposure potential. PtD is an approach (and in the United States, a national initiative) to design out the hazards rather than address them later during exposures. Such an approach is particularly applicable to NMs at the molecular and process scales. At the molecular scale, there is potential for modification of molecules to retain commercial and scientific functionality while reducing toxicity. At the process scale, companies can look to the pharmaceutical industry for engineering controls that could be adopted for potentially hazardous NMs.

3.3 KEY FEATURES TO DRIVE THE DESIGN OF SAFE NANOMATERIALS

The development and implementation of SbyD control strategies with its *primary* prevention value of risk management represents one of the biggest challenges of nanotechnology that should guarantee its sustainable development. Nevertheless, two main factors affect its diffusion: the knowledge gaps, still existing on exposure and hazard features in relation to the process, and the NM physicochemical characteristics and the control of costs that may increase without an advantage in competitiveness.

The *design* approach to safe nanotechnology is well described by Morose,[3] which recalls the *design for X* concepts that is well known to design engineers, industrial designers, material scientists that are involved in the initial design of a product. The safe *nano design* principles will supplement *design for environment* strategies by addressing the unique properties of NPs. The integration of such principles during the design stage for products that contain NPs has a chance to keep out risks rather than address them after they occur. To be successfully addressed, a key edge challenge needs to be supported by robust systems for predicting and evaluating the environmental and human health impacts of NPs over their entire life cycle, both during the development of design principles and during their necessary evaluation, when integrated within nanoproducts.

Hazard to environmental and human health and its exposure potential are the risk determining factors that should be mitigated by design control measures. The hazard of a substance is assessed by understanding the relationship between the dose of the substance and the acute and chronic responses to the substance. The exposure potential of an NP is a function of its bioavailability to humans through inhalation, oral, and dermal pathways as well as its ability to accumulate, persist, and translocate within the environment and within the human body.[4] Maynard et al.[5] identified five challenges that should be addressed to promote safe and sustainable nanotechnology. Most of them are knowledge tools that allow for the mechanisms that drive the nano–bio interaction. While, another challenge, within the next 5 years, is the focus on the development of predictive models that support the introduction of safe-by-design NMs.

As well explained in the book by Puzyn and Leszczynski,[6] a synergy between computational (chemo-informatic) methods and experimental techniques is necessary for exploring experimental data and discovering nano–bio interaction mechanisms.

An exhaustive conceptual framework to guide the exploration of bio-physicochemical interactions occurring at the nano–bio interface has been reviewed.[7] There are three dynamically interacting components that drive nano–bio interactions: (1) nano surface characterized by its physicochemical properties, generally on the basis of its nanoscale reactivity; (2) biological media, characterized by the intercellular components; and (3) biological substrates, characterized by cellular membrane and intracellular components. These dynamic domains undergo continuous transformations because of the interactions between species that affect the impact on the biological receptor structure and function. Traditional interfacial and surface characterization tools exploited in colloidal and catalyst science (colloidal stability, zeta potential, particle size distribution, wetting, interfacial tension, specific surface area, crystal structure, band-gap energy, scanning electron microscopy/transmission electron microscopy [SEM/TEM]) need to be integrated by nano–bio interface specific characterizations (TEM cryomicroscopy, fluorescently labeled NPs and corresponding imaging techniques, nanobiosensors for the detection of reactive oxygen species [ROS], surface-enhanced Raman scattering) toward the creation of hazard-specific characterization tool, where nanosurface physicochemical properties could provide useful information to predict toxicity and possible design features to mitigate toxicity.

A very impressive simplification of concepts expressed earlier is reached by Donaldson et al.[8] where the biologically effective dose (BED) issues are discussed. The BED is used to identify and quantify the actual components of the total dose that cause the adverse effects (particle size distribution, specific surface area, aspect ratio, zeta potential, oxidative potential, soluble toxins released from NPs). NPs can take the form of one or more physicochemical characteristics associated with them as soluble or ionic species released from the NP, ROS produced, or can be integral to the particle surface or the particle shape. The applications and importance of the BED are suggested in terms of different uses: as the most relevant exposure metric, as key component driving the structure–reactivity relationships, as well as NPs categorization for hazard evaluation. Furthermore, a BED-based classification of NPs can have a primary role in the development of SbyD strategies. Design solutions affecting BED dose in the direction to decrease adverse effects, in fact, have the chance to represent an effective primary prevention strategy to NM risk management.

On the basis of the simplification in Nanoinformatics for Safe-by-Design Engineered Nanomaterials,[6] the components of nano design framework toward the design of a new generation of *safe* engineered NMs are described in Figure 3.2.

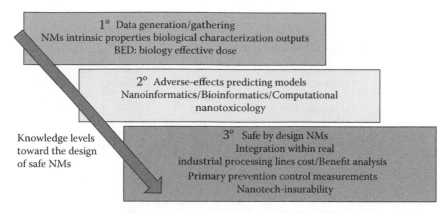

FIGURE 3.2   Components of nano design framework.

3.4 APPROACH DEVELOPED WITHIN EUROPEAN UNION COLLABORATIVE PROJECT SANOWORK

The SANOWORK (Safe Nano Worker Exposure Scenarios, NMP4-SL-2011-280716, www.sanowork.eu) project is built around the promotion, development, and implementation of elimination or substitution control strategies and proposes to fill the gaps that already delay their diffusion. To address the needs of nano-manufacturing companies, the SANOWORK project proposes sustainable risk remediation strategies with a balanced approach between design for manufacturing and design for safety.

The evaluation proposed by risk remediation strategies passes through a globally harmonized analysis and reporting of process/NMs performances, hazard data, emission/exposure collection, in relation to operational conditions and NMs physicochemical properties. The process of NM performances and risk-specific evaluations are performed offline (NMs as delivered, provided by companies or surface engineered during the project, exposure scenarios at lab-scale level) and on-site (exposure scenarios at pilot-scale level). The process and NMs performances, as well as exposure and hazard profiles, are assessed before and after the introduction of risk control extra-steps.

NM design strategies are developed and integrated within target processing lines as extra-steps for improving the efficiency of the process while preserving and/or increasing NM performances. Health risk, depending on the intrinsic hazard and exposure frequency/concentration level, is evaluated according to physicochemical properties of NMs.

Five strategies are proposed, based on NMs surface engineering, so that exposure can be reduced, but if exposure has accidentally occurred the health hazard would still be decreased (Figure 3.3).

On the basis of the knowledge of NMs dispersion behavior, a combination of self-assembled monolayer coating and tailored aggregation processes are developed to decrease the hazard and/or emission potential of target NMs (strategies I, II, III, V). The described strategies are accompanied by a process of immobilization of NMs with an expected decrease in exposure potential strategy (IV).

	Strategies	
I	Organic/Inorganic coating	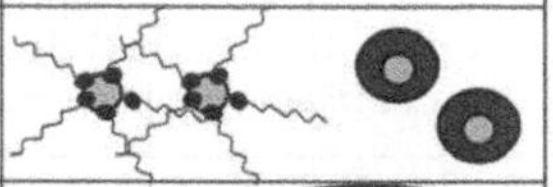
II	Controlling colloidal forces and disintegration of nanoparticles	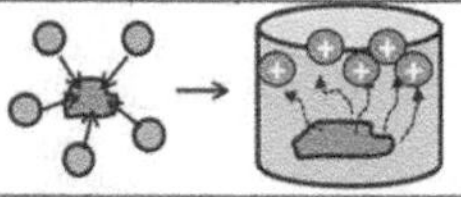
III	Integration of spray-drying process	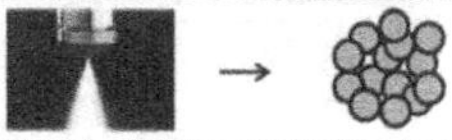
IV	Immobilization by organic film coating deposition	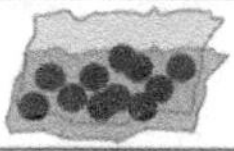
V	Recovery by wet dispersion and milling	

FIGURE 3.3 SANOWORK-proposed five strategies based on nanomaterial surface engineering.

The SANOWORK project proposed strategies that are industry-driven and will comply with the following criteria:

- Satisfying the production requirements
- Cost-effectiveness and suitability for large-scale production
- Easy processing-line implementation for manufacturing nano-structured components
- Decreasing exposure potential and/or health impacts, while preserving nanoscale properties

Manufacturing processes, relevant for different industrial sectors, were identified and the proposed primary prevention of risk has been integrated within six processing lines and implemented at a pilot-scale level by companies involved in SANOWORK.

The expected result is the development of a cost-efficient benefit *design options*-based risk remediation strategies, providing practical tools for

- Developing potentially useful safe design features
- Preventing NM-related worker injuries
- Reducing the need of expensive risk management measures for implementing safe manufacturing processes

3.5 CONCLUSION AND FUTURE RESEARCH

Nano design concepts open new challenges toward the creation of nanotechnology that satisfies performance and safety requirements. In the context of EHS vision to 2020, SbyD takes center stage as an emerging strategy to face NMs occupational risk management.

The conceptual framework that identifies key features that drive the design of safe NMs has been discussed. The SANOWORK project approach toward the promotion, development, and implementation, within real processing lines, of design option-based risk remediation strategies, has been presented. Cost-efficient benefits are expected in terms of developing useful safe design features that prevent NM-related worker injuries and reduce the need of expensive risk management for implementing safe manufacturing processes.

REFERENCES

1. Whitesides GM, Kriebel JK, Mayers BT. "Self-assembly and Nanostructured Materials 2005." http://gmwgroup.harvard.edu/pubs/pdf/936.pdf.
2. Schulte PA, Geraci C, Zumwalde R, Hoover M, Kuempel E. "Occupational risk management of engineered nanoparticles." *Journal of Occupational and Environmental Hygiene* 2008, 5(4), 239–249.
3. Morose G. "The five principles of 'design for safer nanotechnology.'" *Journal of Cleaner Production* 2010, 18, 285–289.
4. Oberdörster G, Maynard A, Donaldson K, Castranova V, Fitzpatrick J, Ausman K, Carter J et al. "Principles for characterizing the potential human health effects from exposure to nanomaterials: Elements of a screening strategy." *Particle and Fibre Toxicology* 2005, 2(8), 1–35.
5. Maynard AD, Aitken RJ, Butz T, Colvin V, Donaldson K, Oberdorster G, Philbert MA et al. "Safe handling of Nanotechnology." *Nature* 2006, 444(16), 267–269.
6. Puzyn T, Leszczynski J eds. *Towards Efficient Designing of Safe Nanomaterials—Innovative Merge of Computational Approaches and Experimental Techniques.* RCS Nanoscience and Nanotechnology No. 25, RSC, Cambridge, UK, 2012.
7. Nel AE, Madler L, Velegol D, Xia T, Hoek EM, Somasundaran P, Klaessig F et al. "Understanding biophysicochemical interactions at the nano–bio interface." *Nature Materials* 2009, 8, 543–557.
8. Donaldson K, Schinwald A, Murphy F, Cho WS, Duffin R, Tran L, Poland C. "The biologically effective dose in inhalation nanotoxicology." *Accounts of Chemical Research* 2013, 46, 723–732.

4 Characterization of Manufactured Nanomaterials, Dispersion, and Exposure for Toxicological Testing

Keld A. Jensen, Giulio Pojana,
and Dagmar Bilaničová

CONTENTS

4.1 INTRODUCTION

After the onset of the nanotoxicology research area, it was almost immediately identified that understanding the physicochemical characteristics of manufactured nanomaterials (MNs) and their interactions with biological systems could be essential for understanding their toxicological mechanisms.[1,2] High demands first developed for validation of MNs and further understanding their characteristics and the exposure. Today, detailed physicochemical characterization of MNs, as well as their in vitro and in vivo exposure, is requested for most scientific publications.[3–8] Currently, an increasing number of papers perform detailed characterization of MN exposure and provide their dose to organs,[9–11] their chemical reactivity (e.g., formation of reactive oxygen species and dissolution), and interaction with biomolecules in both the preparation and test media and the tissue.[5,12–14] Recently, dispersion methods, which can be suitable for both in vitro and in vivo toxicity testing, have received increased attention due to experimental and regulatory needs for reliable testing and repeatability of test results as described in the Organization for Economic Co-operation and Development (OECD) report ENV/JM/MONO(2012)40. Consequently, today the physicochemical characterization that is requested by nanotoxicology can be a rather wide and demanding activity and requires the application of several different analytical techniques.

In the previous edition of this book, Chapter 3 offered an elaborate overview of the main characterization methods with primary focus on the physicochemical characterization of MN in dry and liquid suspensions.[15] This chapter just offers a summary table of the methods typically applied for the physicochemical characterization of MNs and a brief discussion on selected new techniques, which may be of importance in the future. The main purpose of this chapter is to introduce the reader to some of the challenges, considerations, and typically applied procedures for airborne and liquid dispersions in mammalian nanotoxicology and methods for exposure characterization, including measurements of dispersions and their stabilities.

4.2 PHYSICOCHEMICAL CHARACTERIZATION OF NANOMATERIALS

The typical definition of an MN is that it is an intentionally manufactured condensed chemical substance where each object (nano-object; ISO/TS 80004-1:2010) has a physical dimension approximately between 1 and 100 nm in at least one direction. MNs already include a wide range of both organic and inorganic materials and mixtures (hybrids) thereof, and the list is still expanding as technology develops. Typical industrial MNs include synthetic amorphous silica (SAS) (SiO_x); titanium dioxide (TiO_2); calcium carbonate ($CaCO_3$); aluminum silicate (Al_2O_3); iron oxide (Fe_xO_y); cadmium selenide (CdS); lead sulfide (PbS); selenium (Se); silver (Ag); lead (Pd); platinum (Pt); gold (Au); and different allotropes of carbon (C) including carbon black, fullerene, graphite, and various types of carbon nanotubes (CNTs). Examples of six commercial MNs are shown in Figure 4.1, which illustrates their morphological variability. MNs may also have organic and/or inorganic chemical surface modifications, which are not always reported in the producers' technical information sheets. Consequently, MNs may vary from being simple to chemically and structurally complex nanoscale materials and full physicochemical characterization clearly requires a range of different techniques. Several books now offer elaborate introductions to the types, surface modifications, and applications of MN.[16–19]

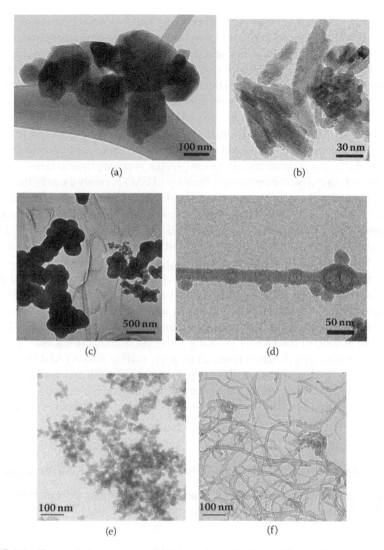

FIGURE 4.1   Transmission electron microscopy images of a manufactured nanomaterial with different nano-object size distributions and NOAA structures: (a) assumed aggregate of a mainly pigment-size rutile (NRCWE-004; RDI-S) showing the presence of nanometer-size crystallites, (b) agglomerated aggregates of UV-Titan L181 consisting of rutile coated with silicon, aluminum, zirconium, and polyalcohol. The crystallites vary from being equidimensional to elongated (c) partially agglomerated chain aggregates of Flammrüss 101 carbon black. The material has a very wide size distribution, ranging from a few nanometers to several hundred nanometers. (e) Dense to open-structured aggregate of synthetic amorphous silica (NM201 from the sample suite established under the Organization for Economic Co-operation and Development (OECD) Working Party on Manufactured Nanomaterials [WPMN]). (f) Entangled multiwalled carbon nanotubes (MWCNTs) (NM402 from the sample suite established under the OECD WPMN) with large variation in aspect ratio. The lengths of the MWCNT generally exceed the width of the image. (Images [a–d] are contributions from KA Jensen [NRCWE]. Images [e] and [f] are courtesy of Pieter-Jan De Temmermann and Jan Mast [CODA-CERVA].)

Currently, there are no specific requirements for or general consensus on using any specific measurement technologies to obtain specific physicochemical characteristics or properties. It is clear that several different techniques can provide similar information on an MN.[20–22] This is also emphasized in the proposed guidance for specifying the type of nano-object (International Organization for Standardization [ISO] TS 12805). For example, more than 10 techniques can be used to analyze particle size, each applying different physical approaches and algorithms, leading to different values for the same MN. But the use of different techniques can lead to different method-dependent results and, therefore, cause some confusion in the physicochemical characterization data. For example, the average physical size of a particle analyzed by transmission electron microscopy (TEM) is rarely exactly the same as the average hydrodynamic size as determined by photon correlation spectroscopy or disk centrifugation (DC). The limits and constraints (degrees of comparability) of different data from different techniques can vary significantly and should not be underestimated. For each type of material and matrix, the most adequate methods need to be identified and validated with quantification of technical and methodological uncertainty to be adopted into existing or even new recommendations and guidelines. It would be beneficial to establish a generic characterization paradigm with documented recommendations on most suitable methods and procedures or performance requirements. Such activities are in progress under CEN/TC 352, under ISO TC229, in the OECD Working Party on Manufactured Nanomaterials, and in connection with different research projects such as nanoSTAIR (http://www.nanostair.eu-vri.eu/) and NANoREG (http://nanoreg.eu/). This is not least sparked by the regulatory issues and definitions established by some countries,[23,24] which will require high reliability (comparability, accuracy, and precision) in specific measurements.

4.2.1 CHARACTERIZATION REQUIREMENTS AND TOOLS IN NANOTOXICOLOGY

In nanotoxicological research studies, the typical minimum information requirements for a test material include the following:

- Identification of chemical phase (e.g., TiO_2)
- Amorphous or crystalline polymorph (e.g., rutile)
- Average TEM size and/or size distribution of primary nano-objects (e.g., 25 nm or 12–60 nm, average: 25 nm)
- Average size and/or size distribution of agglomerates at best dispersed state (e.g., 110 nm)
- Morphology (shape) (e.g., equidimensional or spherical)
- Specific surface area (e.g., 60 m^2/g)
- Impurity phases (purity) (e.g., none detected)
- Identification and quantification of inorganic coatings (e.g., 2 wt% silicon and 4 wt% aluminum)
- Identification and quantification of molecular functionalization or coatings (e.g., 2 wt% alkyl-silane; total mass by thermogravimetric analysis [TGA])

In addition to these characteristics, the chemical nature of the surface, catalytic and photocatalytic properties, the surface charge (zeta potential [ζ-Pot]) in specific media, and pH values can sometimes be required to support an investigation. Some of the listed characteristics are also mentioned in the guidance on recommended characteristics for toxicological assessments given by ISO/TR13014 and in a broader context by OECD (ENV/JM/MONO(2012)40).

Table 4.1 lists the typical techniques used for characterization of MNs and measurement of exposures. For a more in-depth introduction, see the previous edition of this book and scientific literature. It should be noted that even though different measurement techniques may give the same type of information, they may be only suitable for analyzing the material in specific states, such as in a powder or dispersed in a liquid or a gas/air.

As expected, high-resolution imaging techniques are the only methods that can be used for the determination of primary particle size distribution and morphology in the approximately 1–100 nm size range. However, most of these analyses are a result of one-dimensional or two-dimensional (2D) measurements. For example, TEM and scanning electron microscopy (SEM) have high 2D resolution in the "nanosize range." Three-dimensional (3D) nanoscale analyses are still not generally available, but cross-sectional analysis is possible to obtain highly reliable 3D information by TEM and SEM. For some nanomaterials, advanced electron tomographic analysis is also a possibility to obtain 3D information.[25,26] Atomic force microscopy (AFM) may have an atomic resolution in vertical analysis but relatively poor resolution with significant edge effects of measurements in the horizontal plane. Therefore, AFM is most suitable for size measurement of equidimensional materials or well-oriented objects. However, recent developments in probe tips and additions of physicochemical sensors enable much greater resolutions and a range of possibilities for the in situ analysis of specific properties.[27–29] Many other sizing techniques do not have morphological resolution and/or are not suitable for the entire size range. For example, only a few aerosol instruments are capable of measuring the size of airborne particles below 5–10 nm and by default report, similar to, for example, dynamic light scattering (DLS), nanoparticle tracking analysis (NTA), and DC, on the spherical equivalent particle size.

Developments in the ability to perform detailed 3D analysis on the nanoscale would be highly beneficial. This is especially because nano-objects in reality often occur in agglomerates and aggregates. Aggregates and agglomerates are likely to increase the size range, number of size modes, and types of morphologies to be measured and characterized. Consequently, the measurement variability and uncertainty is expected to increase with increasing complexity of these agglomerates. As an example, it may not be trivial to establish reliable size distribution data on aggregates of, for example, TiO_2 (Figure 4.1a and b), carbon black (Figure 4.1c), and SAS, which can vary from being highly dispersed (Figure 4.1d) to complex 3D aggregate structures (Figure 4.1e), and CNT materials (Figure 4.1f) that often consist of heavily entangled or aggregated tubes with extreme aspect ratios and the presence of both organic impurities and inorganic catalysts. Subsequent and increasing challenges arise "downstream" the experimental route through, for example, dispersion, exposure, deposition on a cell or in a primary organ, uptake, translocation, and fate.

TABLE 4.1

Summary Table of Typical Instruments Applied for the Characterization of Nanomaterials in Dry and Liquid States

	Average Size/Size Distribution	Shape	Surface Area	Porosity	Atomic Structure	Chemical Composition	Purity	Surface Chemistry	Surface Charge	Agglomeration/ Aggregation State
SEM, FE-EMPA	♣◇	♣				♣	♣			♣
TEM	♣(◇*)	♣(◇*)	♣		♣	♣	(♣)	*		♣(◇*)
AFM, STM, STIM	♣	♣*	*					*		♣
GPC, DC, APC, PAC	◇									◇
DLS, SLS, NTA	◇	*								◇
FFF-MALS	◇									◇
SP-ICP-MS	◇					◇	(◇)			◇
SMPS, FMPS, ELPI	℘	℘*							℘*	℘*
UV-Vis	*						*			
XRD	(♣◇)*	(♣◇)*			♣◇					
SAXS	♣◇*		◇* ♣							♣◇*
LIBS	♣◇℘					♣◇℘				♣
BET			♣	♣						◇

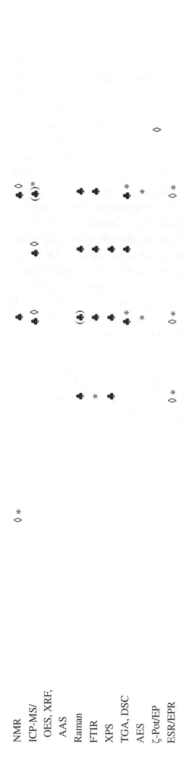

NMR

ICP-MS/
OES, XRF,
AAS

Raman

FTIR

XPS

TGA, DSC

AES

ζ-Pot/EP

ESR/EPR

Note: For acronym description, please refer to the list of abbreviations. ♣, Applicable to powders; *, applicable in some cases or in combination with other analyses; ◊, applicable in liquid dispersion; ℘, applicable in aerosols; ∙, applicable to biological matrices; (#), possible to a certain degree. AAS, atomic absorption spectroscopy; AES, atomic emission spectrometry; APC, antigen-presenting cell; DSC, differential scanning calorimetry; EPR, electron paramagnetic resonance; ESR, electron spin resonance; FE-EMPA, field emission electron microprobe analyzer; FTIR, Fourier transform infrared; GPC, gel permeation chromatography; OES, optical emission spectrometry; STIM, scanning transmission ion microscopy; STM, scanning tunneling spectroscopy; XPS, x-ray photoelectron spectroscopy; XRD, x-ray diffraction; XRF, x-ray fluorescence.

4.2.2 EMERGING APPLICATIONS FOR GENERAL MANUFACTURED NANOMATERIAL CHARACTERIZATION

For sizing new complementary procedures are developing, such as single-particle (SP) inductively coupled plasma mass spectrometry (ICP-MS), whereas other established methods for general estimation, such as small-angle x-ray scattering (SAXS) and nuclear magnetic resonance (NMR) analysis, have been shown to determine average particle size and specific surface area measurements in both dry and suspended states.[30–32] SAXS has, for example, been applied in analysis for the monomodal certified reference material ERM-FD100.[33] Data from the certification of this standard reveal greater variability in the determination of average size by TEM/SEM analysis than DLS, centrifugal liquid sedimentation, and SAXS (Figure 4.2). More advanced determinations of nanoscale bimodal size distributions have been demonstrated for SAXS combined with asymmetrical flow field-flow fractionation (FFF).[34]

Further developments in light scattering equipment for size distribution analyses of liquid dispersions, such as NTA, have been established as an alternative to DLS and static light scattering (SLS). The physical approach (i.e., the Stokes–Einstein diffusion equation) is the same, while the sensor is a high-speed video camera instead of a photodetector. NTA has already been tested in various biological systems and compared with DLS.[35,36] Compared to DLS, one of the advantages of NTA is the visualization of the Brownian motion of dispersed agglomerates in real time. The limitations of NTA are a relatively restricted applicable size range (~10–1000 nm in current models, depending also on the type of MN) and the need to perform analysis on relatively diluted samples (a few hundreds of tracked nanoparticles per analysis) compared with DLS.

SP-ICP-MS is based on a new statistical approach that enables the handling of signal spikes generated by ICP-MS when analyzing insoluble MNs. This method

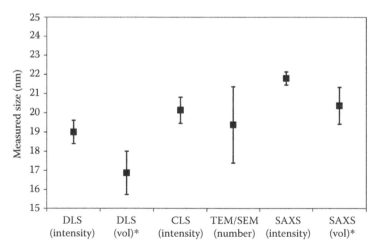

FIGURE 4.2 Size measurement data from the establishment of the certified reference material ERM-FD100 (spherical colloidal silica) using different measurement techniques and units applicable for the specific techniques: standard deviations include measurement uncertainty and interlaboratory variability from round robin test. The "*" denotes that the value in the specific unit is not certified.

generates data for evaluating the number and mass size distribution of agglomerates in an injected sample within certain limitations.[37] The SP-ICP-MS technique requires a relatively complex statistical data treatment and is strongly instrument dependent in terms of sensitivity and selectivity. It has been applied mainly to environmental and unknown samples, and only a few biological applications have been reported.[38–41] Although this is still at the research and development level, it can be considered as an analytical approach with great potential in the field of nanotoxicology.

Laser-induced breakdown spectroscopy (LIBS) is another potentially valuable technique for size determination and quantification of MNs in liquids and aerosols. LIBS exploits the plasma ion formation of solid particles generated by a pulsed laser beam focused in the dispersion. After decades of experimentation and validation with prototypes applied mainly to environmental samples,[42–45] the first commercial units have recently become available.

The integration of resonant mass measurement and microelectromechanical systems (MEMS) technology has been applied in a commercial device called suspended nanochannel resonator for the analysis of agglomerates in liquid dispersions. This instrument can count particles down to 50 nm (in the current model) while measuring their mass, in various liquids, potentially including biological fluids.[46] From mass analysis, it may be possible to distinguish these particle agglomerates from other materials of similar size (such as protein agglomerates) concurrently present in the sample. Its prospective applications, although limited by the lowest size and sensitivity, are expected to advance the knowledge of MN behavior in biological media, where agglomerates in the >100 nm size range are usually present. Recently, string-based MEMS technology has proven to be applicable for aerosol measurements with count efficiencies at 65% for 28 nm silica spheres, which are currently in the preprototype stage.[47,48]

Finally, recent improvements in magnet technology have allowed the reduction of both size and cost of NMR instrumentation to enable commercial benchtop units dedicated to the measurement of specific surface area of MNs in hydrous dispersions.[30] The technique is based on the hydrogen proton relaxation times of water molecules surrounding a dispersed MN. Since NMR is based on a different physical approach than the Brunauer–Emmett–Teller (BET) nitrogen adsorption method, the values obtained through these two techniques are not fully comparable. The same accounts for the more mathematically challenging determination of specific surface area by SAXS, although a good agreement was found between the results obtained by SAXS and those by conventional BET on several TiO_2, SAS, and CNT powders.[32] The advantage of SAXS is that analyses can be completed on both relatively small and large powder samples as well as liquid dispersions. For both the NMR- and SAXS-based methods, the capability of surface area analysis of MNs dispersed in liquids is of high interest for toxicological and medical studies.

4.3 DISPERSION AND EXPOSURE MEASUREMENTS

One of the key aspects in nanotoxicological testing is proper dispersion and exposure to an MN. For selecting the most suitable procedures, one may consider the exposure situation or hypothesis to be investigated. The test materials may be dispersed into agglomerates as small as possible such as primary particles or agglomerates, as realistic as possible in a submicrometer size with wide size distributions to simulate exposure during powder handling, or just small enough to allow dosing or inhalation in the test species. The choice of exposure method may have an important impact on the toxicological results.

For pulmonary toxicology, the substance may be delivered either airborne for inhalation or dispersed in a biologically acceptable liquid medium for intratracheal instillation or aspiration. For intravenous injection and oral gavage, the substance is normally predispersed in a biologically compatible liquid. As an alternative procedure and for mechanistic studies, in vitro studies are frequently used and require either dispersion of powders in a liquid medium or use of aerosolization and exposure using an air-to-liquid deposition system.[49,50] Exposure to MNs dispersed in liquids is very popular as it in principle facilitates well-controlled dosing (in vivo or into the test system) and high-throughput screening using less manpower. However, exposure using liquid dispersions also has some drawbacks. For example, different dispersion procedures can give different size distributions and, therefore, potentially different possibilities for the type of biological interaction. In addition, constituents in the different media as well as potentially added ingredients and dispersants may interact with the test materials, whereby they may facilitate or inhibit dispersion stability and/or specific biological reactions. Finally, the MN may partially dissolve during preparation and storage, whereby the test material may transform or alter, resulting in different physicochemical characteristics and properties of the test material compared to the pristine material. These potentially important reactions should be considered during selection of methods and assessment of experimental data.

4.3.1 AIRBORNE EXPOSURE

One of the most critical issues in aerosol exposure is control of particle concentrations to avoid extensive agglomeration between the particle generator and the inhalation chamber. In aerosol physics, this mechanism is called coagulation and its rate varies with both particle size and concentrations. For monomodal nanoparticles, coagulation starts to become important at concentrations above 1×10^6 particles/cm^3 and coagulation occurs almost immediately at 1×10^9 particles/cm^3. For 100 nm size particles and initial concentrations at 1×10^{10} particles/cm^3, the coagulation will be half the concentration in 0.5 second and double the size in 1.4 seconds.[51] This change in particle number concentration can be described as a function of time, $N(t)$, by Equation 4.1, in which N_0 is the initial particle concentration, K is the size-dependent coagulation coefficient, β is a correction factor for the size-specific thermal diffusion, and t is the time (after Ref. 51):

$$N(t) = \frac{N_0}{1 + N_0 \times K_0 \times \beta \times t} \tag{4.1}$$

The size-dependent values of the corrected coagulation factor, $K_0 \times \beta$, is 9.5×10^{-16}, 7.2×10^{-16}, 3.4×10^{-16}, and 3.0×10^{-16} m^3/s for 10 nm, 100 nm, 1 μm, and 10 μm size particles, respectively.

The time-dependent change in average size, $d(t)$, of an aerosol can be described using the initial monomodal size, d_0, and the same other parameters as for $N(t)$ and is given by Equation 4.2 (after Ref. 51):

$$d(t) = d_0 \left(1 + N_0 \times K_0 \times \beta \times t\right)^{1/3} \tag{4.2}$$

From Equations 4.1 and 4.2, it is clear that the reduction in particle number concentration is faster than the growth in particle size. This is easily understood if one imagines a spherical agglomerate of size 100 nm growing by the continuous addition of 5 nm singlet nanoparticles.

Continuing with this example, the coagulation rate can actually be strongly increased if the airborne particles have different sizes. This phenomenon is called heterogeneous coagulation or scavenging, and the increased coagulation rates are due to the increased possibility of collision between particles with small and large cross-sectional areas than between two particles of the same size. Based on data calculated by Hinds,[51] Figure 4.3 shows the variation in coagulation coefficients at different combinations of primary and secondary particle sizes. It is clear that a significant increase in coagulation rate is observed in mixtures compared to the homogeneous coagulation rate (encircled data points). For example, the coagulation rate for mixtures between 10 and 100 nm size particles is 12.7 times faster than for 10 nm particles alone. For mixtures between 10 nm and 10 μm size particles, the rate is 1770.8 times faster than for the 10 nm particles alone.

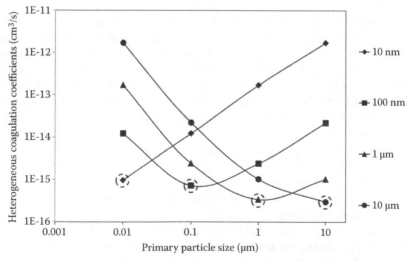

FIGURE 4.3 Coagulation rates of aerosols with different initial monomodal to bimodal particle size distributions: the encircled data points are the coagulation rates for monomodal aerosols. When mixtures are bimodal, the coagulation is heterogeneous and increased due to the scavenging of small particles by larger particles. (Data from Hinds, W.C., *Aerosol Technology: Properties, Behavior, and Measurement of Airborne Particles*, John Wiley & Sons, Inc., New York, 1999.)

Setting the risk of coagulation into the perspective of inhalation exposure studies, it was noted that homogeneous coagulation starts to be important at 10^6 particles/cm^3 and is almost immediate at 10^9 particles/cm^3. A concentration of 10^9 particles/cm^3 corresponds to a mass concentration of approximately 2.2 mg/m^3 100 nm size rutile (specific density = 4.2 g/cm^3), which is at the lower end of the mass concentrations used in most inhalation exposure studies (e.g., 0.8–28.5 mg gas-phase-synthesized TiO$_2$/m^3,[52] 42 mg UV-Titan/m^3,[9] 88 mg P25 TiO$_2$/m^3,[53] and 0.5–10 mg P25 TiO$_2$/m^3).[54] Clearly, different levels of agglomeration by coagulation must be expected as a function of exposure concentration in these studies. Moreover, dust generated from resuspended powder normally contains a significant mass fraction of coarser micrometer-size particles, despite containing high numbers of nanosize or very fine sub-micrometer-size particles. This is clearly demonstrated by Jacobsen et al.,[55] who performed a study on resuspended Printex 90 (carbon black). As seen from Figure 4.3, such dust characteristics are likely to dramatically speed up the coagulation rate.

As a consequence of these challenges and considering the typical exposure conditions, inhalation toxicologists at an OECD export workshop recommended that the mass median diameter in airborne exposures should not exceed 3 μm and a maximum reasonable effort should be put to aerosolize and deagglomerate a powder into its smaller aggregates (ENV/JM/MONO(2012)14). This is of course a pragmatic solution.

For high-quality dose–response inhalation exposure studies, the main experimental challenges are clearly to test relevant size distributions and to avoid excessive coagulation while still maintaining exposure concentrations suitable for hazard assessment. This problem can be somewhat reduced by lowering the exposure concentration and increasing the exposure duration, which, on the other hand, increases the experimental cost significantly and may raise additional ethical issues.

In-line airborne exposure may be performed directly during particle synthesis or indirectly by complete resuspension or dust generation from powders or dispersions. In-line exposure to MNs may be achieved in many different ways, including evaporation–condensation or hot wire,[56] arc or spark discharge,[57] and flame synthesis[58] and is probably the procedure resulting in the most well-controlled aerosol exposures. Lately, advanced techniques have been developed where aggregated primary particles were synthesized by either hot-wire or spark discharge and then focused to monomodal aggregate size using a differential mobility analyzer (DMA) and subsequently exposed as they are or reshaped (recrystallized) into monomodal "spherical" SP exposures using thermal treatment.[59] Following this procedure, one may test the effect of high surface area aggregates versus low surface area spherical particles while still maintaining the mass.

Airborne exposure to ready-made MNs may be done by resuspending dust from powders or aerosolizing the test material from a liquid suspension. However, it is known that relatively high energies are required to deagglomerate MN powders and that simulation of handling processes such as rotating drum dustiness and continuous drop testing of powders generate agglomerated dust particles with typical maximum number peak size located between 100 and several hundred nanometers and another in the μm-size.[60–63] Methods typically used to aerosolize dust particles from MN powders for inhalation exposure include microfeeders combined with venturi aspirators,[64,65] acoustic membrane systems combined with venturi aspirators,[65,66] the vortex shaker,[67] brush generators combined with fluidized bed aerosolization,[68] or fluidized bed aerosol generators alone combined with a cyclone for size selection.[69] Typically used procedures for aerosolization of nanomaterials from liquid dispersions mainly include aerosol spray,[70,71] nebulizers,[72] and electrospray.[73] Methods such as vibrating orifice aerosol generators[74] and atomizers[75] are also possible but are less used at this point in time. Aerosolization from dispersions, however, also requires suitable liquids where the MN is subjected to limited degrees of agglomeration. As for direct exposure to dispersions in liquid media discussed in Section 4.3.3, one must consider potential changes of the MN and the exposure due to, for example, dissolution, dispersion media, and presence of stabilizers or dispersants.

The current OECD guidance for safety testing of MNs considers a risk-oriented (agglomerates > 100 nm) approach and a hazard-oriented (primary nano-objects and small agglomerates < 100 nm) approach (ENV/JM/MONO(2012)40). The highlighted procedure for the risk-oriented approach is aerosolization of powders using a venturi device and further deagglomeration of the aerosol using a jet mill. But a rotating brush generator; a Wright Dust Feeder; and, specifically for CNTs, an acoustic membrane system were highlighted as examples for the aerosolization of powders. From liquid dispersions a jet nebulizer was mentioned, and the dispersions could be in a phosphate buffer. For liquid dispersions with CNTs, a nebulizer was highlighted and dispersions could be made by sonic treatment in water with, for example, dipalmitoylphosphatidylcholine, bovine serum albumin (BSA), or glucose as dispersants. For hazard identification, it was suggested to use a spark generator to produce primary particles and aggregates. However, not all MNs can be produced using this technique.

Clearly, harmonization of criteria for the exposure characteristics and optimization of resuspension technologies would enable better comparability in inhalation toxicology. Optimization of the technologies may be achieved by empirical documentation of methods and further development by enabling better control of the energy added to deagglomerate the powders in the systems. At the same time, it is clear that the required deagglomeration energies will be highly material dependent because agglomeration in powders may be caused by a number of different mechanisms (Figure 4.4).[76] Therefore, if similar levels of deagglomeration are required, if at all possible, it is necessary to enable the addition of variable energies for deagglomeration. The alternative is the harmonization of the added energy or generation principles to mimic specific exposure scenarios.

4.3.2 MEASUREMENT OF AIRBORNE EXPOSURE

Ideally, airborne exposures should give a complete range of online and offline exposure characteristics. ISO 10808:2010(E) requests the reporting of agglomerate mass concentrations (possible as a size distribution) as well as online number size distributions and/or the number concentrations to assess the stability of exposure conditions. The instrumentation proposed for measuring the size distribution of the sub-100 nm range of agglomerates is a differential mobility particle analyzing system, which includes a fast mobility particle sizer (FMPS) as well as stepping, sequential and scanning mobility particle sizer (SMPS). Agglomerates larger than 100–200 nm may be measured using

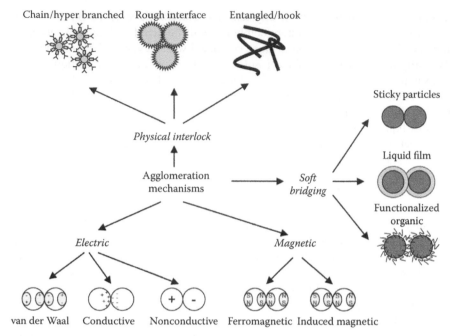

FIGURE 4.4 Different mechanisms of agglomeration in powders. (Modified from Schneider, T., and K. Jensen, *J. Nanopart. Res.*, 2009; 11/7:1637–1650.)

aerodynamic, optical, or electrical properties. However, the exact selection of online monitors should be carefully considered. Instruments such as the electrostatic low-pressure impactor (ELPI), which use a combination of aerodynamic size fractionation and number counts by electrometers, are also good candidates for online measurements of size distributions, as suggested in the OECD report ENV/JM/MONO(2012). The advantage of ELPIs is a wide measurement range in both size distribution and number concentrations. The advantages of SMPS-type systems are a high size resolution and a more accessible measurement principle for particle number concentrations and size, especially for nanoobjects and their aggregates and agglomerates (NOAA) smaller than 100 nm. Results from both systems are relatively comparable (typically within 10% uncertainty). However, the results can be affected by the presence of aggregates and agglomerates, which at least to some extent can be corrected using additional data treatment to correct for the charge distributions in agglomerates.[77–79]

Online instruments such as nanoparticle surface area monitors and aerosol particle mass (APM) analyzers are also of interest for deriving information on the inhaled or deposited surface area and the effective density of the dust particles, respectively. Surface area monitors are all based on the same unipolar diffusion charging principles, but studies on different commercial instruments have shown that they may give different results. Compared with both FMPS and SMPS, differences up to 30% and several hundred percentages were observed depending on whether agglomerate size distributions were within or outside the specified limits of the instruments, respectively.[80,81] For surface area monitors, the presence of large, but still respirable, agglomerates would not contribute to the measurements causing an underestimation of the alveolar deposited fraction given by some instruments.

The APM analyzer is a relatively new instrument and has the capacity to measure the mass of particles online at a mass resolution of 0.01–100 fg (~30–580 nm and unit density).[82] The method is based on a combination of centrifugal forces and electrostatic forces. In combination with a DMA, the system can be used to determine the effective density of agglomerates and consequently the number of primary particles in individual agglomerate.[59]

For specific materials such as carbon-based MNs, alternative methods should be considered in future, such as online monitoring devices for the CNT coming out as a product from the EU FP7 project NANODEVICE (http://www.nano-device.eu/). The potential applicability of aerosol carbon monitors has also been demonstrated where the CNT can be quantified online using sequential heating programs.[83] Currently, there is limited information on the accuracy and reliability of these methods.

To date, offline measurements of size-selective filter-collected mass concentrations probably still give the most reliable measurement of an exposure. Despite the uncertainties, additional information from online monitors gives important and specific information on the nature of the exposure. In any case, the measurement technology should always be considered in the assessment of the data, and electron microscopy analysis of airborne agglomerates will always give strong support to the analysis.

4.3.3   EXPOSURE USING LIQUID DISPERSIONS

One may consider three principal ways of dispersing and stabilizing particles in liquid media: electrostatic, steric, and polymeric stabilizations (Figure 4.5). In reality, stabilization may often occur due to a mixture of different mechanisms called

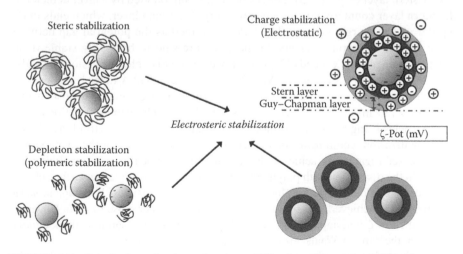

FIGURE 4.5 Principal mechanisms for the stabilization of particles in liquid suspensions: the right side of the image illustrates the electrostatically stabilized manufactured nanomaterial.

"electrosteric stabilization." Independently of the mechanisms of stabilization, it has to prevent agglomeration due to the van der Waals forces between the suspended objects. The attraction energy between two small objects depends on the radii of the two objects and the distance between them. Assuming the same radius, R, for two nano-objects and d as the center-to-center distance, the van der Waals attraction force ($\varphi_{vw}(d)$) between the two particles is given by Equation 4.3, where A is the Hamacker constant ($\sim 10^{-20}$ J) (after Ref. 84):

$$\varphi_{vw}(d) = \frac{A}{6}\left[\frac{2R^2}{d^2 - 4R^2} + \frac{2R^2}{d^2} + \ln\left(\frac{d^2 - 4R^2}{d^2}\right)\right] \tag{4.3}$$

From Equation 4.3, it is clear that $\varphi_{vw}(r)$, where r is the interparticle distance, increases as the distance between the particles diminishes and that small nano-objects are subject to less van der Waals attraction forces than coarser nanoparticles.

For the establishment of electrostatic stabilization, relatively high repulsive surface charges (coulombic energy barriers) are required for the individual objects in the medium. This energy can be described by Equation 4.4 as a function of the interparticle distance (r as earlier); the ζ-Pot of the particles; the Boltzmann constant, k; and a material constant, c (after Ref. 85):

$$\varphi_c(d) = \frac{\zeta^2}{c} \times \ln\left[1 - \exp(-k \times d)\right] \tag{4.4}$$

Electrostatic stabilization can be achieved by adjusting the pH, ionic strength, and/or composition of the medium suitable for a given nanoparticle. In this way, a material-dependent positive or negative charge can be established at the particle surface with countercharge-balancing ions concentrated at the surface, the so-called Stern layer (Figure 4.5). The Stern layer is surrounded by a layer depleted in the Stern layer counterbalancing ions, the Guy–Chapman layer, which ends in the regular mixed charge medium. The ζ-Pot is defined as the potential gap between Stern and Guy–Chapman layers. Dispersions are said to be highly stable if the ζ-Pot of their objects exceeds |30| mV. However, metastable dispersions can also be obtained at lower, say, $\geq$|20| mV, ζ-Pot values. Yet, NOAA in such suspensions may have a tendency to rapidly form small agglomerates compared to suspensions at ζ-Pots > |30| mV. High surface charges can also be achieved by functionalizing nanoparticles using highly polar molecules, such as polyvinylpyrrolidone. Such a functionalization would result in an electrosteric stabilization of the dispersion.

Steric stabilization is achieved by coating or chemically functionalizing the objects in the dispersion with polymers or proteins that neutralize the surface charge and prevent agglomeration by physically blocking the particles. If complete steric stabilization is achieved, the surface charge will be zero or close to zero. However, despite the low ζ-Pot, agglomeration is still prevented if the coating is thick enough to prevent the van der Waals attraction forces from interacting.

In polymeric (or depletion) stabilization, the dispersants, for example polymers and proteins, do not attach to the surface of the objects. Instead, they prevent the objects from agglomerating by creating repulsion or physically blocking the interaction between the van der Waals forces. Polymeric stabilization can occur when the NOAA have neutral or similar charge as the objects in the suspension.

As the mechanisms by which dispersions can be achieved open several possibilities, the type of medium plays an equally significant role. Until recent years, it was the impression that the media used for toxicological testing are almost a default for each specific laboratory and the resulting dispersibility is of secondary importance. Pure water, saline, phosphate-buffered water, phosphate-buffered saline, and specific cell media are among the types that have previously been widely used to make batch dispersions in particle toxicology and also for MNs. However, hydrophobic nanomaterials are poorly dispersible in these media and require specific treatment to disperse in water-based systems.

Similar to the preferred media, many laboratories previously followed their own preferred dispersion procedures. Typical procedures have involved the use of vigorous manual shaking, magnetic stirring, vortex shaking, ultrasound bath treatment, and probe-horn or cup-horn sonication. Normally, a higher degree of dispersion is achieved by increasing the directly added energy and/or the duration of the treatment.

In recent years, a general shift in paradigm has occurred toward the use of highly dispersed test materials. Therefore, several different MN dispersion protocols have been used and/or investigated. Most of these methods have been targeted at specific MNs or groups of MNs or types of assay and are based on the use of proteins or the like such as human serum albumin, BSA, mouse serum albumin, or fetal bovine serum (FBS) as biocompatible dispersants.[86–88] The use of nonbiological dispersants such as citrate and the addition of hydroxyl groups have also been pursued.[89–91] Generic probe-sonicator dispersion protocols have been developed over the last few years. For example, two methods were developed as part of the EU FP7 project ENPRA and the Joint Actions project NANOGENOTOX. The ENPRA protocol utilizes the dispersion of nanomaterials in 2% w/v FBS water by probe sonication and prewetting of hydrophobic samples with 0.5% w/v ethanol and has been applied in recent work.[92,93] The NANOGENOTOX protocol uses dispersion in sterile-filtered 0.05% w/v BSA instead of FBS and prewetting of all samples with 0.5% w/v ethanol.[94] The downside of generic dispersion protocols is that they represent compromises and will not result in complete dispersion for all MNs.

The pH-induced electrostatic stabilization has not been considered much in toxicological studies, but it could be applicable for maximizing dispersibility and stability in the initial steps for some nanomaterials. This was demonstrated for TiO_2 and SAS by Guiot and Spalla,[95] who initially predispersed nanomaterials in pH-buffered water, followed by the addition of suitable amounts of BSA to "just" cover the specific surface area, and then finally adjusted the pH to the required physiological condition (pH = 7.4). However, this procedure is not applicable to all MNs and does involve different test conditions. Therefore, different procedures are necessary to enable reasonable dispersion of hydrophobic nanomaterials.

4.3.4 CONSIDERATIONS AND POTENTIAL ARTIFACTS IN EXPERIMENTS USING LIQUID DISPERSIONS

When using a liquid dispersion, one has to take into consideration the fact that any material will interact physically and hydrochemically with the liquid and all its constituents. This results in a closely interlinked list of consequences between the interaction of the test material with the medium and the exposure characteristics (or dose). Some of the immediate reactions to consider are listed in Figure 4.6.

As described in Section 4.3.3, MNs are intrinsically unstable toward agglomeration when dispersed in liquid media. When agglomeration exceeds a certain size, sedimentation can also occur. Kinetics of both phenomena depend on MN characteristics such as size; density; shape; surface charge; and properties of the medium, such as viscosity, type and concentration levels of salts and additives, ionic strength, and temperature. It follows that physicochemical features of dispersions prepared for toxicological investigations, both in vitro and in vivo, can change, even dramatically, over the experimental time frame, and their dose–response relationships can be strongly affected by this behavior.[96,97] A computational model for the evaluation of nanoparticle sedimentation and diffusion in in vitro systems has been proposed,[98] but this model is valid only for fully dispersed agglomerates and does not take into account possible surface modifications due to absorption of other chemicals.

It has been reported that exposure to agglomerated MNs in dispersions could lead to different conclusions than exposure to well-dispersed MNs.[97,99–103] Agglomerates 50–200 times larger than their primary sizes have been reported when dispersing MNs in physiological media, depending on their concentration levels and the ionic strength of the media.[87,100,104,105]

Changes in media pH may be caused by hydration or hydroxylation at the surfaces of the MN as well as a result of dissolution. This issue is rarely addressed in toxicological studies using liquid dispersions. The pH levels differing from the normal equilibrium conditions in a cell medium or localized in an organ may cause a deleterious toxicological response. The redox potential, E_h, in liquid dispersions is also rarely reported, but it is an important parameter that can provide a better understanding of the electron transfer reactions or oxygen concentrations in the liquids. At the same time, redox activity may be used to determine whether a liquid dispersion is in a reasonable pH–E_h regime compared to a specific biological compartment as well as to describe the type and extent of toxicologically relevant reactivity a specific MN can facilitate. Research data in this area are slowly emerging, at least as a result of recently completed European research projects.

**Assessment of hydrochemical
reactivity of MN vs. exposure and dosimetry**

pH
Redox activity
Alteration/dissolution (rate)
Secondary phase formation
Change in surface chemistry
Interaction with biomolecules
Agglomeration/aggregation

How much and of what?
Dispersed or agglomerated
Mass/number/surface/reaction product
Dispersed or direct contact (in vitro)
Sedimentation rate (in vitro)

FIGURE 4.6 Overview of some potential mechanisms and reactivities of a manufactured nanomaterial in liquid dispersions and considerations on consequences for exposure characterization and dosimetry.

Interaction of MNs with biological fluids is a complex topic. Due to their high specific surface area and/or surface reactivity, MNs in such environments are known to adsorb molecules such as salts, sugars, amino acids, or macromolecules such as proteins and oligonucleotides through electrostatic adsorption, hydrophobic interactions, and even specific covalent bonding.[14,106–111] Specifically, the interactions of MNs with living systems may be controlled mainly by protein–agglomerate interactions.[112,113] As expected, each modification of the nanomaterial surface could significantly affect their cellular uptake and toxicity, thus leading to invalid results.[104,114–121]

Protein binding to agglomerates has been shown to affect both their interaction with cells and their toxicity.[122–126] High-power probe sonication was reported to be the most efficient technique for breaking down nanoparticle agglomerates in aqueous dispersions in comparison with milder techniques such as vortexing and bath sonication.[88,104,127] The application of high-intensity ultrasounds was found to cause conformational changes in and denaturation of proteins.[128–131]

4.3.5 MEASUREMENT OF DISPERSIBILITY AND SEDIMENTATION RATES IN EXPOSURE MEDIA

Characterization of dispersion stabilities and sedimentation behavior of agglomerates in batch dispersions and exposure media for in vitro toxicological studies has become a typical request in nanotoxicology. Usually, this information is obtained by DLS or SLS measurements. In some cases, similar techniques, such as disk centrifugation (DC) or photo analytical centrifugation (PAC), SAXS, wide-angle x-ray diffraction, and small-angle neutron scattering, may be used (Table 4.1). NTA and more advanced techniques such as FFF, multiangle light scattering (MALS), and ICP-MS may also be applied to determine the size distributions in dispersions, but these principles are not suitable for assessing sedimentation rates.

As for any measurement technology, the methods mentioned here rely on assumptions and conventions. For example, reported sizes are the equivalent "hydrodynamic" spherical sizes (e.g., the reported size is an equivalent spherical unit of isotropic density moving in a fluid). This is analogous to the reported equivalent spherical aerodynamic diameter given by some aerosol monitors. Consequently, the sizes derived for aggregates, agglomerates, and high-aspect-ratio MNs are not the true physical dimensions observed by, for example, TEM. These techniques may, however, still be applied to assess MN dispersibility in liquid dispersions. Moreover, for SLS and DLS high-quality measurements require knowledge on the viscosity of the dispersions, the optical indices of both the media and the MN, and the dielectrical constant of the liquid. Such a set of information is not always available and must be approximated. Therefore, the obtained data may have significant uncertainty and should be evaluated and understood according to the measurement principles and the available information.

An important aspect for analysis of agglomeration and sedimentation by, for example, DLS and SLS, is that instrumental sensitivity varies with agglomerate size ($I \propto r^6$). Therefore, if the size distribution is broad, or multimodal distributions are present due to dynamic agglomeration, the autocorrelation function, applied to derive the intensity fluctuation of scattered light, may be gradually dominated by signals from larger sizes of the agglomerate. If sedimentation, on the other hand, leaves small agglomerates in suspension, the intensity loss may suggest a higher deposited fraction than that de facto occurs in terms of volume (mass). Another issue is the size range that may develop as a function of time in stability and sedimentation studies. All DLS algorithms can correctly detect particles only up to a few micrometers in size; so if very large agglomerates are present, they cannot be quantified or possibly even not detected at all. In these studies, interpretation of the obtained data requires caution.

Only very few techniques have been developed specifically to investigate sedimentation processes with time and include DC,[132] the manometric centrifuge,[133] and PAC.[134,135] Among them, the latter seems the most promising for toxicological applications. Although conceived for investigating kinetic processes (such as settling, deliquoring, phase separation, and cake elasticity) that finally provide temporal stability of industrial and commercial (including food) slurries, the technique has been successfully applied in the development of MN dispersion protocols for toxicological investigations. Figure 4.7 shows an example of the inferred sedimentation rate of zinc oxide NOAA in three cell media as determined by a PAC. As observed, the

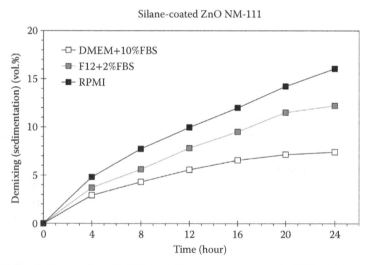

FIGURE 4.7 Sedimentation of NOAA of silane-coated zinc oxide (NM-111 from the sample suite established under the Organization for Economic Co-operation and Development Working Party on Manufactured Nanomaterials) in different cell media as a function of time as determined by analytical photocentrifugation: the technique allows the measurement of space- and time-resolved extinction profiles according to a patented algorithm (STEP™ Technology). The main advantages of this technique are that temperature control is possible and that the vials, when filled under sterile conditions, maintain their sterility for a relatively long period. The experiments were conducted with a MN concentration of 0.256 mg/mL in all cell media (DMEM: Dulbecco's modified Eagle's medium; F12: Dulbecco's modified Eagle medium nutrient mixture F-12; RPMI: Roswell Park memorial institute medium) some of them containing also FBS. The temperature was set to 37°C and an accelerated stability using 1000 rpm speed, simulating 24 hours lifetime at $g = 1$. (The data were generated by D. Bilaničová and G. Pojana [University of Venice]. Assistance by R. Cossi, Qi-tech s.r.l. [Pomezia, Rome, Italy], is gratefully acknowledged.)

apparent deposited fraction varies considerably depending on the type of cell media. Such information is crucial for the assessment of toxicological test results if the hazard is associated with direct physical contact between the agglomerates and the cells.

As all of the online techniques rely on indirect methods, microscopic techniques such as SEM, environmental scanning electron microscopy, and TEM should be concurrently applied for visual confirmation and interpretation of agglomeration. Electron microscopy is a very valuable tool for the assessment of sizes and shapes of agglomerates. Still, further development in sample preparation techniques for electron microscopy is needed with the aim to ensure no artefacts due to agglomeration and structural modifications of the dispersed MN occurring on the sample substrate. For such analysis, cryotechniques might be of interest.

4.4   OUTLOOK

The 6-year period between the first edition and the second edition of this book has resulted in significant advancement in the characterization of MNs and test materials and exposure scenarios in nanotoxicology. New applications of existing analytical techniques as well as new instruments have emerged, such as surface area and average size analysis in dispersions using NMR and SAXS, size distribution analysis and the use of SP-ICP-MS (liquid dispersions), MEMS technology (liquid dispersions and aerosols), and APM (aerosols). The in situ analysis of surface reactivity or chemistry of MNs may in future be facilitated by the development of advanced AFM probes or tip sensors in MEMS technology. In general, combinations of different analytical methods are likely to open up new frontiers in nanoscale characterization, as seen in experiments with correlated AFM and Raman for imaging, as well as with SP-ICP-MS and asymmetrical FFF-MALS-ICP-MS for size, mass and elemental analysis.

This chapter does not exhaustively investigate the progress of all state-of-the art analytical methods. In general, there is growing confidence in the characteristics obtained on dry MN, but there are still concerns on analytical uncertainty and complexity when the MN are agglomerated, aggregated, or have high aspect ratios. Further developments are required in these areas and especially for analyses of liquid and airborne dispersions where the uncertainty and effects of specific materials and morphologies appear to play a major role. Reliable procedures for detection and quantification of inorganic and organic coatings and functionalizations on MNs are needed. It is likely that most of the required methods exist, but procedures need to be established and validated.

A proper and relevant dispersion of an MN is of key importance in nanotoxicological testing. Any experiment should include characterization during an exposure or, in the case of in vitro assays, usually parallel to the exposure. The variations in size distributions and surface charge are of high interest and can be assessed with good reliability. Greater challenges arise for determining the suspension stability and the actual concentration levels in the dispersions and at the cellular interface in an experiment. Similarly, there is a rapidly growing interest in understanding the physicochemical interaction between MNs and the media, biomolecules, cells, and organs. It is expected that this will be an area where the knowledge level will increase significantly over the next 5–10 years. New techniques and methods are needed and are expected to develop along with the need to understand these phenomena.

Finally, only a very few, if any, MN dispersion protocols have been thoroughly tested for application in nanotoxicology. Usually, the tests focus on the ability to disperse and stabilize an MN and, in some cases, the role of specific variations on in vitro toxicological effects. Dispersion media and/or conditions have been demonstrated to have an influence on the toxicological response after exposure. The lack of standardized exposure procedures may be one of the main reasons for the variability observed in nanotoxicology in addition to the limitations of in vitro assay systems that have shown that nanomaterials can interfere with dyes and adsorb the dye and dye products of an assay.[136–137] To build confidence, there is an urgent need to systematically investigate the role of different dispersion procedures (both dry and wet), exposure media, use of dispersants or not, and the resulting exposure characteristics on toxicological test results. To achieve these new steps, strong collaboration is needed among experts in, for example, nanomaterials, biophysicochemical characterization, biochemistry, and nanotoxicology.

REFERENCES

1. Nel A, Xia T, Madler L, and Li N. Toxic Potential of Materials at the Nanolevel. *Science* 2006; 311:622–627.
2. Powers KW, Brown SC, Krishna VB, Wasdo SC, Moudgil BM, and Roberts SM. Research Strategies for Safety Evaluation of Nanomaterials. Part VI. Characterization of Nanoscale Particles for Toxicological Evaluation. *Tox Sci* 2006; 90:296–303.
3. Poland CA, Duffin R, Kinloch I, Maynard A, Wallace WAH, Seaton A, Stone V, Brown S, MacNee W, and Donaldson K. Carbon Nanotubes Introduced into the Abdominal Cavity of Mice Show Asbestos-Like Pathogenicity in a Pilot Study. *Nat Nanotech* 2008; 3:423–428.
4. Porter DW, Hubbs AF, Mercer RR, Wu N, Wolfarth MG, Sriram K, Leonard S et al. Mouse Pulmonary Dose- and Time Course-Responses Induced by Exposure to Multi-Walled Carbon Nanotubes. *Toxicol* 2010; 269:136–147.
5. Loeschner K, Hadrup N, Qvortrup K, Larsen A, Gao XY, Vogel U, Mortensen A, Lam HR, and Larsen EH. 2011. Distribution of Silver in Rats Following 28 Days of Repeated Oral Exposure to Silver Nanoparticles or Silver Acetate. *Part and Fibre Toxicol* 2008; 8:14–18.
6. Palomaki J, Valimaki E, Sund J, Vippola M, Clausen PA, Jensen KA, Savolainen K, Matikainen S, and Alenius H. Long, Needle-Like Carbon Nanotubes and Asbestos Activate the NLRP3 Inflammasome through a Similar Mechanism. *ACS Nano* 2011; 5:6861–6870.
7. Saber AT, Jensen KA, Jacobsen NR, Birkedal R, Mikkelsen L, Møller P, Loft, S, Wallin H, and Vogel U. Inflammatory and Genotoxic Effects of Nanoparticles Designed for Inclusion in Paints and Lacquers. *Nanotoxicol* 2012; 6:453–471.
8. Li R, Wang X, Ji Z, Sun B, Zhang H, Chang CH, Lin S et al. Surface Charge and Cellular Processing of Covalently Functionalized Multiwall Carbon Nanotubes Determine Pulmonary Toxicity. *ACS Nano* 2013; 7:2352–2368.
9. Hougaard KS, Jackson P, Jensen KA, Sloth JJ, Loeschner K, Larsen EH, Birkedal RK et al. Effects of Prenatal Exposure to Surface-Coated Nanosized Titanium Dioxide (UV-Titan). A Study in Mice. *Part and Fibre Toxicol* 2010; 7:16.
10. Schinwald A, Chernova T, and Donaldson K. Use of Silver Nanowires to Determine Thresholds for Fibre Length-Dependent Pulmonary Inflammation and Inhibition of Macrophage Migration in Vitro. *Part and Fibre Toxicol* 2012; 9:47.

11. Petitot F, Lestaevel P, Tourlonias E, Mazzucco C, Jacquinot S, Dhieux B, Delissen O et al. Inhalation of Uranium Nanoparticles: Respiratory Tract Deposition and Translocation to Secondary Target Organs in Rats. *Toxicol Let* 2013; 217:217–225.
12. Cedervall T, Lynch I, Foy M, Berggad T, Donnelly SC, Cagney G, Linse S, and Dawson KA. Detailed Identification of Plasma Proteins Adsorbed on Copolymer Nanoparticles. *Angew Chem Int Ed* 2007; 46:5754–5756.
13. Jacobsen NR, Møller P, Cohn CA, Loft S, Vogel U, and Wallin H. Diesel Exhaust Particles Are Mutagenic in FE1-Muta(TM)Mouse Lung Epithelial Cells. *Mutat Res* 2008; 641:54–57.
14. Monopoli MP, Walczyk D, Campbell A, Elia G, Lynch I, Bombelli FB, and Dawson KA. Physical-Chemical Aspects of Protein Corona: Relevance to in vitro and in vivo Biological Impacts of Nanoparticles. *J Am Chem Soc* 2011; 133:2525–2534.
15. Zuin S, Pojana G, and Marcomini A. Effect-oriented physicochemical characterization of nanomaterials. In Monteiro-Riviere NA and Tran CL, eds. *Nanotoxicology—Characterization, Dosing and Health Effects*. New York: Informa Healthcare, 2007 19–57.
16. Gunter Schmid, Ed. *Nanoparticles—From Theory to Application*. Weinheim, Germany: WILEY-VCH Verlag GmbH & Co. KGaA 2010.
17. Guldi DM and Martín N, Eds. *Carbon Nanotubes and Related Structures—Synthesis, Characterization, Functionalization, and Applications*. Weinheim, Germany: WILEY-VCH Verlag GmbH & Co. KGaA, 2010.
18. Cademartiri L and Ozin GA, Eds. *Concepts of Nanochemistry*. Weinheim, Germany: WILEY-VCH Verlag GmbH & Co. KGaA, 2009.
19. Monthioux M, Ed. *Carbon Meta-Nanotubes—Synthesis, Properties and Applications*. Chichester, UK: John Wiley & Sons Ltd, 2012.
20. Hassellov M and Kaegi R. Analysis and characterization of manufactured nanoparticles in aquatic environments. In Lead, JR and Smith, E, eds. *Nanoscience and Nanotechnology: Environmental and Human Health Implications*. 211–256. Chichester, UK: John Wiley & Sons, Ltd. Library of Congress Cataloging-in-Publication Data, 2009.
21. Bleeker EAJ, de Jong WH, Geertsma RE, Groenewold M, Heugens EHW, Koers-Jacquemijns M, van de Meent D et al. Considerations on the EU Definition of a Nanomaterial: Science to Support Policy Making. *Regul Toxicol and Pharmacol* 2013; 65:119–125.
22. Hall JB, Dobrovolskaia MA, Patri AK, and Mcneil SE. Characterization of Nanoparticles for Therapeutics. *Nanomedicine* 2007; 2:789–803.
23. Health Canada. *Policy Statement on Health Canada's Working Definition for Nanomaterial*. 6-10-2011. Ottawa, Ontario, Health Canada.
24. Potocnik J. Commission Recommendation of 18 October 2011 on the Definition of Nanomaterial. *Off J Eur Union L* 2011; 275:38–40.
25. Frederik PM and Sommerdijk N. Spatial and Temporal Resolution in Cryo-Electron Microscopy—A Scope for Nano-Chemistry. *Curt Opinion in Colloid & Interface Sci* 2005; 10:245–249.
26. Van Doren EAF, De Temmerman PJRH, Francisco MAD, and Mast J. Determination of the Volume-Specific Surface Area by Using Transmission Electron Tomography for Characterization and Definition of Nanomaterials. *J Nanobiotech* 2011; 9:17.
27. Rodriguez RD, Lacaze E, and Jupille J. Probing the Probe: AFM Tip-Profiling via Nanotemplates to Determine Hamaker Constants from Phase-Distance Curves. *Ultramicroscopy* 2012; 121:25–30.
28. Hovelmann J, Putnis CV, Ruiz-Agudo E, and Austrheim H. Direct Nanoscale Observations of CO_2 Sequestration During Brucite [$Mg(OH)_2$] Dissolution. *Envirl Sci & Technol* 2012; 46:5253–5260.
29. Hong SS, Cha JJ, and Cui Y. One Nanometer Resolution Electrical Probe via Atomic Metal Filament Formation. *Nano Let* 2011; 11:231–235.

30. anonymous. Acorn Area—Particle Surface Area Analyzer. Bethlehem, Pennsylvania, XiGO nanotools 2013.
31. Marega C, Causin V, Saini R, and Marigo A. A Direct SAXS Approach for the Determination of Specific Surface Area of Clay in Polymer-Layered Silicate Nanocomposites. *J Phys Chem B* 2012; 116:7596–7602.
32. Nanogenotox partners. *Nanogenotox—Facilitating the Safety Evaluation of Manufactured Nanomaterials by Characterising Their Potential Genotoxic Hazard.* Bialec 2013. ISBN: 978-2-11-138272-5.
33. Braun A, Kestens V, Franks K, Roebben G, Lamberty A, and Linsinger TPJ. A New Certified Reference Material for Size Analysis of Nanoparticles. *J Nanoparticle Res* 2012; 14:1021.
34. Thunemann AF, Rolf S, Knappe P, and Weidner S. In Situ Analysis of a Bimodal Size Distribution of Superparamagnetic Nanoparticles. *Anal Chem* 2009; 81:296–301.
35. Boyd RD, Pichaimuthu SK, and Cuenat A. New Approach to Inter-Technique Comparisons for Nanoparticle Size Measurements; Using Atomic Force Microscopy, Nanoparticle Tracking Analysis and Dynamic Light Scattering. *Colloid Surface A* 2011; 387:35–42.
36. Cho EJ, Holback H, Liu KC, Abouelmagd SA, Park J, and Yeo Y. Nanoparticle Characterization: State of the Art, Challenges, and Emerging Technologies. *Mol Pharma* 2013; 10:2093–2110.
37. Degueldre C and Favarger PY. Colloid Analysis by Single Particle Inductively Coupled Plasma-Mass Spectroscopy: A Feasibility Study. *Colloid Surface A* 2003; 217:137–142.
38. Pace HE, Rogers NJ, Jarolimek C, Coleman VA, Higgins CP, and Ranville JF. Determining Transport Efficiency for the Purpose of Counting and Sizing Nanoparticles via Single Particle Inductively Coupled Plasma Mass Spectrometry. *Anal Chem* 2011; 83:9361–9369.
39. Scheffer A, Engelhard C, Sperling M, and Buscher W. ICP-MS As a New Tool for the Determination of Gold Nanoparticles in Bioanalytical Applications. *Anal Bioanal Chem* 2008; 390:249–252.
40. Mitrano DM, Barber A, Bednar A, Westerhoff P, Higgins CP, and Ranville JF. Silver Nanoparticle Characterization Using Single Particle ICP-MS (SP-ICP-MS) and Asymmetrical Flow Field Flow Fractionation ICP-MS (AF4-ICP-MS). *J Anal Atom Spectrom* 2012; 27:1131–1142.
41. Laborda F, Jimenez-Lamana J, Bolea E, and Castillo JR. Selective Identification, Characterization and Determination of Dissolved Silver(I) and Silver Nanoparticles Based on Single Particle Detection by Inductively Coupled Plasma Mass Spectrometry. *J Anal Atom Spectrom* 2011; 26:1362–1371.
42. Scherbaum FJ, Knopp R, and Kim JI. Counting of Particles in Aqueous Solutions by Laser-Induced Photoacoustic Breakdown Detection. *Appl Phys B-Lasers and Optics* 2011; 63:299–306.
43. Jung EC, Yun JI, Kim JI, Park YJ, Park KK, Fanghanel T, and Kim WH. Size Measurement of Nanoparticles Using the Emission Intensity Distribution of Laser-Induced Plasma. *Appl Phys B-Lasers and Optics* 2006; 85:625–629.
44. Walther C, Buchner S, Filella M, and Chanudet V. Probing Particle Size Distributions in Natural Surface Waters from 15 Nm to 2 Mu m by a Combination of LIBD and Single-Particle Counting. *J Colloid and Interface Sci* 2006; 301:532–537.
45. Kaegi R, Wagner T, Hetzer B, Sinnet B, Tzuetkov G, and Boller M. Size, Number and Chemical Composition of Nanosized Particles in Drinking Water Determined by Analytical Microscopy and LIBD. *Water Res* 2008; 42:2778–2786.
46. Lee J, Shen WJ, Payer K, Burg TP, and Manalis SR. Toward Attogram Mass Measurements in Solution with Suspended Nanochannel Resonators. *Nano Let* 2010; 10:2537–2542.

47. Schmid S, Dohn S, and Boisen A. Real-Time Particle Mass Spectrometry Based on Resonant Micro Strings. *Sensors* 2010; 10:8092–8100.
48. Schmid S, Kurek M, Adolphsen J, and Boisen A. Real-Time Single Airborne Nanoparticle Detection with Nanomechanical Resonant Filter-Fiber. *Scientific Reports* 2013; 3:1–5.
49. Lenz AG, Karg E, Lentner B, Dittrich V, Brandenberger C, Rothen-Rutishauser B, Schulz H, Ferron GA, and Schmid O. A Dose-Controlled System for Air-Liquid Interface Cell Exposure and Application to Zinc Oxide Nanoparticles. *Part and Fibre Toxicol* 2009; 6:32.
50. Elihn K, Cronholm P, Karlsson HL, Midande K, Wallinder IO, and Moller L. Cellular Dose of Partly Soluble Cu Particle Aerosols at the Air-Liquid Interface Using an In Vitro Lung Cell Exposure System. *J Aerosol Med and Pulmon Drug Del* 2013; 26:84–93.
51. Hinds WC. *Aerosol Technology: Properties, Behavior, and Measurement of Airborne Particles*. New York, USA: John Wiley & Sons, Inc., 1999.
52. Koivisto AJ, Makinen M, Rossi EM, Lindberg HK, Miettinen M, Falck GCM, Norppa H et al. Aerosol Characterization and Lung Deposition of Synthesized TiO_2 Nanoparticles for Murine Inhalation Studies. *J Nanopart Res* 2011; 13:2949–2961.
53. van Ravenzwaay B, Landsiedel R, Fabian E, Burkhardt S, Strauss V, and Ma-Hock L. Comparing Fate and Effects of Three Particles of Different Surface Properties: Nano-TiO_2, Pigmentary TiO_2 and Quartz. *Toxicol Lett* 2009; 186:152–159.
54. Bermudez E, Mangum JB, Wong BA, Asgharian B, Hext PM, Warheit DB, and Everitt JI. Pulmonary Responses of Mice, Rats, and Hamsters to Subchronic Inhalation of Ultrafine Titanium Dioxide Particles. *Tox Sci* 2004; 77:347–357.
55. Jacobsen NR, Moller P, Jensen KA, Vogel U, Ladefoged O, Loft S, and Wallin H. Lung Inflammation and Genotoxicity Following Pulmonary Exposure to Nanoparticles in ApoE(-/-) Mice. *Part and Fibre Toxicol* 2009; 6:2.
56. Boies AM, Lei PY, Calder S, Shin WG, and Girshick SL. Hot-Wire Synthesis of Gold Nanoparticles. *Aerosol Sci and Technol* 2011; 45:654–663.
57. Kreyling WG, Biswas P, Messing ME, Gibson N, Geiser M, Wenk A, Sahu M et al. Generation and Characterization of Stable, Highly Concentrated Titanium Dioxide Nanoparticle Aerosols for Rodent Inhalation Studies. *J Nanopart Res* 2011; 13:511–524.
58. Demokritou P, Buchel R, Molina RM, Deloid GM, Brain JD, and Pratsinis SE. Development and Characterization of a Versatile Engineered Nanomaterial Generation System (VENGES) Suitable for Toxicological Studies. *Inhal Toxicol* 2010; 22:107–116.
59. Messing ME, Svensson CR, Pagels J, Meuller BO, Deppert K, and Rissler J. Gas-Borne Particles with Tunable and Highly Controlled Characteristics for Nanotoxicology Studies. *Nanotoxicol* 2012; 7:1052–1063.
60. Schneider T and Jensen KA. Combined Single-Drop and Rotating Drum Dustiness Test of Fine to Nanosize Powders Using a Small Drum. *Ann of Occupatl Hyg* 2008; 52:23–34.
61. Jensen KA, Koponen IK, Clausen PA, and Schneider T. Dustiness Behaviour of Loose and Compacted Bentonite and Organoclay Powders: What Is the Difference in Exposure Risk? *J Nanopart Res* 2009; 11:133–146.
62. Tsai CJ, Wu CH, Leu ML, Chen SC, Huang CY, Tsai PJ, and Ko FH. Dustiness Test of Nanopowders Using a Standard Rotating Drum with a Modified Sampling Train. *J Nanopart Res* 2009; 11:121–131.
63. Dahmann D and Monz C. Determination of Dustiness of Nanostructured Materials. *Gefahrstoffe—Reinhaltung der Luft* 2011; 71:481–487.
64. Saber AT, Halappanavar S, Folkmann JK, Bornholdt J, Boisen AMZ, Moller P, Williams A et al. Lack of Acute Phase Response in the Livers of Mice Exposed to Diesel Exhaust Particles or Carbon Black by Inhalation. *Parte and Fibre Toxicol* 2009; 6:12.
65. Schmoll LH, Elzey S, Grassian VH, and O'Shaughnessy PT. Nanoparticle Aerosol Generation Methods from Bulk Powders for Inhalation Exposure Studies. *Nanotoxicol* 2009; 3:265–275.

66. Adamcakova-Dodd A, Stebounova LV, O'Shaughnessy PT, Kim JS, Grassian VH, and Thorne PS. Murine Pulmonary Responses After Sub-Chronic Exposure to Aluminum Oxide-Based Nanowhiskers. *Part and Fibre Toxicol* 2009; 9:22.
67. Tsai CJ, Lin GY, Liu CN, He CE, and Chen CW. Characteristic of Nanoparticles Generated from Different Nano-Powders by Using Different Dispersion Methods. *J Nanopart Res* 2012; 14:777.
68. Myojo T, Oyabu T, Nishi K, Kadoya C, Tanaka I, Ono-Ogasawara M, Sakae H, and Shirai T. Aerosol Generation and Measurement of Multi-Wall Carbon Nanotubes. *J Nanopart Res* 2009; 11:91–99.
69. Ma-Hock L, Treumann S, Strauss V, Brill S, Luizi F, Mertler M, Wiench K, Gamer AO, van Ravenzwaay B, and Landsiedel R. Inhalation Toxicity of Multiwall Carbon Nanotubes in Rats Exposed for 3 Months. *Toxi Sci* 2009; 112:468–481.
70. Chen BT, Afshari A, Stone S, Jackson M, Schwegler-Berry D, Frazer DG, Castranova V, and Thomas TA. Nanoparticles-Containing Spray Can Aerosol: Characterization, Exposure Assessment, and Generator Design. *Inhal Toxicol* 2010; 22:1072–1082.
71. McKinney W, Jackson M, Sager TM, Reynolds JS, Chen BT, Afshari A, Krajnak K et al. Pulmonary and Cardiovascular Responses of Rats to Inhalation of a Commercial Antimicrobial Spray Containing Titanium Dioxide Nanoparticles. *Inhal Toxicol* 2012; 24:447–457.
72. Ma-Hock L, Gamer AO, Landsiedel R, Leibold E, Frechen T, Sens B, Linsenbuehler M, and van Ravenzwaay B. Generation and Characterization of Test Atmospheres with Nanomaterials. *Inhal Toxicol* 2007; 19:833–848.
73. Kim SC, Chen DR, Qi CL, Gelein RM, Finkelstein JN, Elder A, Bentley K, Oberdorster G, and Pui DYH. A Nanoparticle Dispersion Method for In Vitro and In Vivo Nanotoxicity Study. *Nanotoxicol* 2010; 4:42–51.
74. Biskos G, Vons V, Yurteri CU, and Schmidt-Ott A. Generation and Sizing of Particles for Aerosol-Based Nanotechnology. *KONA Powder and Part Jl* 2008; 26:13–35.
75. Zoccal JVM, Fabio de Arouca D, Coury JR, and Goncalves JAS. Size-Distribution of TiO$_2$ Nanoparticles Generated by a Commercial Aerosol Generator for Different Solution Concentrations and Air Flow Rates. *Mat Sci Forum* 2012; 727–728:861–866.
76. Schneider T and Jensen K. Relevance of Aerosol Dynamics and Dustiness for Personal Exposure to Manufactured Nanoparticles. *J Nanopart Res* 2009; 11:1637–1650.
77. Lall AA and Friedlander SK. On-Line Measurement of Ultrafine Aggregate Surface Area and Volume Distributions by Electrical Mobility Analysis: I. Theoretical Analysis. *J Aerosol Sci* 2006; 37:260–271.
78. Lall AA, Seipenbusch M, Rong WZ, and Friedlander SK. On-Line Measurement of Ultrafine Aggregate Surface Area and Volume Distributions by Electrical Mobility Analysis: II. Comparison of Measurements and Theory. *J Aerosol Sci* 2006; 37:272–282.
79. Shin WG, Wang J, Mertler M, Sachweh B, Fissan H, and Pui DYH. The Effect of Particle Morphology on Unipolar Diffusion Charging of Nanoparticle Agglomerates in the Transition Regime. *J Aerosol Sci* 2010; 41:975–986.
80. Asbach C, Kaminski H, Fissan H, Monz C, Dahmann, D, Mulhopt S, Paur H et al. Comparison of Four Mobility Particle Sizers with Different Time Resolution for Stationary Exposure Measurements. *J Nanopart Res* 2009; 11:1593–1609.
81. Kaminski H, Kuhlbusch TAJ, Rath S, Gotz U, Sprenge, M, Wels D, Polloczek J et al. Comparability of Mobility Particle Sizers and Diffusion Chargers. *J Aerosol Sci* 2013; 57:156–178.
82. Kanomax. Aerosol Particle Mass Analyzer Model APM 36012013.
83. Ono-Ogasawara M and Myojo T. Characteristics of Multi-Walled Carbon Nanotubes and Background Aerosols by Carbon Analysis; Particle Size and Oxidation Temperature. *Adv Powder Technol* 2013; 24:263–269.

84. Hamaker HC. The London–Van Der Waals Attraction Between Spherical Particles. *Physica* 1937; 4:1058–1072.
85. Wiersema PH, Loeb AL, and Overbeek JT. Calculation of Electrophoretic Mobility of A Spherical Colloid Particle. *J of Colloid and Interface Sci* 1966; 22:78.
86. Bihari P, Vippola M, Schultes S, Praetner M, Khandoga AG, Reichel CA, Coester C, Tuomi T, Rehberg M, and Krombach F. Optimized Dispersion of Nanoparticles for Biological In Vitro and In Vivo Studies. *Part and Fibre Toxicol* 2008; 5:14.
87. Murdock RC, Braydich-Stolle L, Schrand AM, Schlager JJ, and Hussain SM. Characterization of Nanomaterial Dispersion in Solution Prior to In Vitro Exposure Using Dynamic Light Scattering Technique. *Toxicol Sci* 2008; 101:239–253.
88. Allouni ZE, Cimpan MR, Hol PJ, Skodvin T, and Gjerdet NR. Agglomeration and Sedimentation of TiO$_2$ Nanoparticles in Cell Culture Medium. *Colloids Surface B* 2009; 68:83–87.
89. Roebben G, Ramirez-Garcia S, Hackley V, Roesslein M, Klaessig F, Kestens V, Lynch I et al. Interlaboratory Comparison of Size and Surface Charge Measurements on Nanoparticles Prior to Biological Impact Assessment. *J Nanopart Res* 2011; 13:2675–2687.
90. Ramirez-Garcia S, Chen L, Morris MA, and Dawson KA. A New Methodology for Studying Nanoparticle Interactions in Biological Systems: Dispersing Titania in Biocompatible Media Using Chemical Stabilisers. *Nanoscale* 2011; 3:4617–4624.
91. Sayes CM, Fortner JD, Guo W, Lyon D, Boyd AM, Ausman KD, Tao YJ et al. The Differential Cytotoxicity of Water-Soluble Fullerenes. *Nano Let* 2004; 4:1881–1887.
92. Kermanizadeh A, Pojana G, Gaiser BK, Birkedal R, Bilaničová D, Wallin H, Jensen KA et al. In Vitro Assessment of Engineered Nanomaterials Using a Hepatocyte Cell Line: Cytotoxicity, Pro-Inflammatory Cytokines and Functional Markers. *Nanotoxicol* 2013; 7:301–313.
93. Corradi S, Gonzalez L, Thomassen LC, Bilaničová D, Birkedal R K, Pojana G, Marcomini A, Jensen K A, Leyns L, and Kirsch-Volders M. Influence of Serum on In Situ Proliferation and Genotoxicity in A549 Human Lung Cells Exposed to Nanomaterials. *Mutat Res* 2012; 745:21–27.
94. Jensen KA, Kembouche Y, and Christiansen E. *Final protocol for producing suitable manufactured nanomaterial exposure media—The generic NANOGENOTOX dispersion protocol.* Maisons-Alfort, Cedex, France: French Agency for Food, Environmental and Occupational Health & Safety(ANSES) Final protocol for producing suitable manufactured nanomaterial exposure media. http://www.nanogenotox.eu/index.php?option = com_content&view = article&id = 136&Itemid = 158 2011.
95. Guiot C and Spalla O. Stabilization of TiO$_2$ Nanoparticles in Complex Medium through a pH Adjustment Protocol. *Environ Sci & Technol* 2013; 47:1057–1064.
96. Teeguarden JG, Hinderliter PM, Orr G, Thrall BD, and Pounds JG. Particokinetics In Vitro: Dosimetry Considerations for In Vitro Nanoparticle Toxicity Assessments. *Tox Sci* 2007; 95:300–312.
97. Drobne D, Jemec A, and Tkalec ZP. In Vivo Screening to Determine Hazards of Nanoparticles: Nanosized TiO$_2$. *Environ Pollution* 209; 157:1157–1164.
98. Hinderliter PM, Minard KR, Orr G, Chrisler WB, Thrall BD, Pounds JG, and Teeguarden JG. ISDD: A Computational Model of Particle Sedimentation, Diffusion and Target Cell Dosimetry for In Vitro Toxicity Studies. *Part and Fibre Toxicol* 2010; 7:36.
99. Li JA, Li QN, Xu JY, Li J, Cai XQ, Liu RL, Li YJ, Ma JF, and Li WX. Comparative Study on the Acute Pulmonary Toxicity Induced by 3 and 20 Nm TiO$_2$ Primary Particles in Mice. *Environ Toxicol and Pharmacol* 2007; 24:239–244.
100. Warheit DB, Webb TR, Reed KL, Frerichs S, and Sayes CM. Pulmonary Toxicity Study in Rats with Three Forms of Ultrafine-TiO$_2$ Particles: Differential Responses Related to Surface Properties. *Toxicol* 2007; 230:90–104.

101. Kobayashi N, Naya M, Endoh S, Maru J, Yamamoto K, and Nakanishi J. Comparative Pulmonary Toxicity Study of Nano-TiO$_2$ Particles of Different Sizes and Agglomerations in Rats: Different Short- and Long-Term Post-Instillation Results. *Toxicol* 2009; 264:110–118.

102. Adams LK, Lyon DY, and Alvarez PJJ. Comparative Eco-Toxicity of Nanoscale TiO$_2$, SiO$_2$, and ZnO Water Suspensions. *Water Res* 2006; 40:3527–3532.

103. Plumley C, Gorman EM, El-Gendy N, Bybee CR, Munson EJ, and Berkland C. Nifedipine Nanoparticle Agglomeration As a Dry Powder Aerosol Formulation Strategy. *Int J Pharm* 2009; 369:136–143.

104. Jiang JK, Oberdorster G, and Biswas P. Characterization of Size, Surface Charge, and Agglomeration State of Nanoparticle Dispersions for Toxicological Studies. *J Nanoparticle Res* 2009; 11:77–89.

105. Warheit DB, Hoke RA, Finlay C, Donner EM, Reed KL, and Sayes CM. Development of a Base Set of Toxicity Tests Using Ultrafine TiO$_2$ Particles As a Component of Nanoparticle Risk Management. *Toxicol Let* 2007; 171:99–110.

106. Ellingsen JE. A Study on the Mechanism of Protein Adsorption to TiO$_2$. *Biomat* 1991; 12:593–596.

107. Yang YZ, Cavin R, and Ong JL. Protein Adsorption on Titanium Surfaces and Their Effect on Osteoblast Attachment. *J Biomed Mater Res Part A* 2003; 67A:344–349.

108. Sousa SR, Moradas-Ferreira P, and Barbosa MA. TiO$_2$ Type Influences Fibronectin Adsorption. *J Mater Sci Mater Med* 2005; 16:1173–1178.

109. Patil S, Sandberg A, Heckert E, Self W, and Seal S. Protein Adsorption and Cellular Uptake of Cerium Oxide Nanoparticles As a Function of Zeta Potential. *Biomat* 2007; 28:4600–4607.

110. Aubin-Tam ME and Hamad-Schifferli K. Structure and Function of Nanoparticle-Protein Conjugates. *Biomed Mat* 2008; 3:034001.

111. Lynch I and Dawson KA. Protein-Nanoparticle Interactions. *Nano Today* 2008; 3:40–47.

112. Cedervall T, Lynch I, Lindman S, Berggard T, Thulin E, Nilsson H, Dawson KA, and Linse S. Understanding the Nanoparticle-Protein Corona Using Methods to Quantify Exchange Rates and Affinities of Proteins for Nanoparticles. *Proceedings of the Nat Acad of Sci of the U S Am* 2007; 104:2050–2055.

113. Klein J. Probing the Interactions of Proteins and Nanoparticles. *Proceedings of the Nat Acad of Sci of the U S Am* 2007; 104:2029–2030.

114. Goodman CM, McCusker CD, Yilmaz T, and Rotello VM. Toxicity of Gold Nanoparticles Functionalized with Cationic and Anionic Side Chains. *Bioconjug Chem* 2004; 15:897–900.

115. Lemarchand C, Gref R, Passirani C, Garcion E, Petri B, Muller R, Costantini D, and Couvreur P. Influence of Polysaccharide Coating on the Interactions of Nanoparticles with Biological Systems. *Biomat* 2006; 27:108–118.

116. Yan AH, Lau BW, Weissman BS, Kulaots I, Yang NYC, Kane AB, and Hurt RH. Biocompatible, Hydrophilic, Supramolecular Carbon Nanoparticles for Cell Delivery. *Adv Mat* 2006; 18:2373.

117. Dutta D, Sundaram SK, Teeguarden JG, Riley BJ, Fifield LS, Jacobs JM, Addleman SR, Kaysen GA, Moudgil BM, and Weber TJ. Adsorbed Proteins Influence the Biological Activity and Molecular Targeting of Nanomaterials. *Tox Sci* 2007; 100:303–315.

118. Thevenot P, Cho J, Wavhal D, Timmons RB, and Tang LP. Surface Chemistry Influences Cancer Killing Effect of TiO$_2$ Nanoparticles. *Nanomedicine* 2008; 4:226–236.

119. Alberola AP and Radler JO. The Defined Presentation of Nanoparticles to Cells and Their Surface Controlled Uptake. *Biomat* 2009; 30:3766–3770.

120. Bajaj A, Samanta B, Yan HH, Jerry DJ, and Rotello VM. Stability, Toxicity and Differential Cellular Uptake of Protein Passivated-Fe$_3$O$_4$ Nanoparticles. *J Mats Chem* 2009; 19:6328–6331.

121. Tahara K, Sakai T, Yamamoto H, Takeuchi H, Hirashima N, and Kawashima Y. Improved Cellular Uptake of Chitosan-Modi Fled PLGA Nanospheres by A549 Cells. *Int J Pharma* 2009; 382:198–204.
122. Magdolenova Z, Bilaničová D, Pojana G, Fjellsbo LM, Hudecova A, Hasplova K, Marcomini A, and Dusinska M. Impact of Agglomeration and Different Dispersions of Titanium Dioxide Nanoparticles on the Human Related In Vitro Cytotoxicity and Genotoxicity. *J Environ Monit* 2012; 14:455–464.
123. Ruh H, Kuhl B, Brenner-Weiss G, Hopf C, Diabate S, and Weiss C. Identification of Serum Proteins Bound to Industrial Nanomaterials. *Toxicol Let* 2012; 208:41–50.
124. Sabuncu AC, Grubbs J, Qian SZ, Abdel-Fattah TM, Stacey MW, and Beskok A. Probing Nanoparticle Interactions in Cell Culture Media. *Colloid Surface B* 2012; 95:96–102.
125. Gualtieri M, Skuland T, Iversen TG, Lag M, Schwarze P, Bilaničová D, Pojana G, and Refsnes M. Importance of Agglomeration State and Exposure Conditions for Uptake and Pro-Inflammatory Responses to Amorphous Silica Nanoparticles in Bronchial Epithelial Cells. *Nanotoxicol* 2012; 6:700–712.
126. Ehrenberg MS, Friedman AE, Finkelstein JN, Oberdorster G, and McGrath JL. The Influence of Protein Adsorption on Nanoparticle Association with Cultured Endothelial Cells. *Biomat* 2009; 30:603–610.
127. Mandzy N, Grulke E, and Druffel T. Breakage of TiO_2 Agglomerates in Electrostatically Stabilized Aqueous Dispersions. *Powder Technol* 2005; 160:121–126.
128. Stathopulos PB, Scholz GA, Hwang YM, Rumfeldt JAO, Lepock JR, and Meiering EM. Sonication of Proteins Causes Formation of Aggregates That Resemble Amyloid. *Protein Sci* 2004; 13:3017–3027.
129. Povey MJ, Moore JD, Braybrook J, Simons H, Belchamber R, Raganathan M, and Pinfield V. Investigation of Bovine Serum Albumin Denaturation Using Ultrasonic Spectroscopy. *Food Hydrocolloids* 2011; 25:1233–1241.
130. El Kadi N, Taulier N, Le Huerou J, Gindre M, Urbach W, Nwigwe I, Kahn P, and Waks M. Unfolding and Refolding of Bovine Serum Albumin at Acid pH: Ultrasound and Structural Studies. *Biophyl Jl* 2006; 91:3397–3404.
131. Barnes C, Evans JA, and Lewis TJ. Low-Frequency Ultrasound Absorption in Aqueous-Solutions of Hemoglobin, Myoglobin, and Bovine Serum-Albumin—the Role of Structure and Ph. *J Acoust Soc Am* 1988; 83:2393–2404.
132. Schauer T. Actual Trends in the Development of Methods and Apparatus for Measuring the Size of Particles in Powder Materials. *Polimery* 1996; 41:9–14.
133. Bickert G and Stahl W. Settling Behavior Characterization of Submicron Particles in Dilute and Concentrated Suspensions. *Part & Part Systems Charact* 1997; 14:142–147.
134. Detloff T, Sobisch T, and Lerche D. Particle Size Distribution by Space or Time Dependent Extinction Profiles Obtained by Analytical Centrifugation (Concentrated Systems). *Powder Technol* 2007; 174:50–55.
135. Sobisch T and Lerche D. Application of a New Separation Analyzer for the Characterization of Dispersions Stabilized with Clay Derivatives. *Colloid and Polymer Sci* 2000; 278:369–374.
136. Monteiro-Riviere NA and Inman AO. Challenges for Assessing Carbon Nanomaterial Toxicity to the Skin. *Carbon* 2006; 44: 1070–1078.
137. Monteiro-Riviere NA, Inman AO, and Zhang LW. Limitations and Relative Utility of Screening Assays to Assess Engineered Nanoparticles Toxicity in a Human Cell Line. *Toxicol Appl Pharmacol* 2009; 234:222–235.

Section II

Exposure

5 Workplace Inhalation Exposure to Engineered Nanomaterials

Detection, Measurement, and Assessment

Araceli Sánchez Jiménez, Derk Brouwer, and Martie Van Tongeren

CONTENTS

5.1 INTRODUCTION

The global market for engineered nanomaterials (ENMs) is forecast to grow 3 trillion euros and employ 6 million workers by 2020.[1] It is still dominated by materials that have been in use for decades, such as carbon black (mainly used in tires) or synthetic amorphous silica (used not only as a polymer filler but also in toothpaste or as an anticoagulant in food powders).

In the past years, a growing number of consumer products have been introduced in the market such as electronic (e.g., computer hard drives), magnetic, medical imaging, drug delivery, cosmetic and sunscreen, catalytic, stain-resistant fabric, dental bonding, and corrosion-resistance and coating applications. Major future applications are expected to be in motor vehicles, electronics, personal care products, cosmetics, along with household and home improvement products.[2]

Human exposure can occur at any stage during the life cycle of ENMs (synthesis, product manufacture, use, and disposal). Once they are released into the environment they can be transported through air, soil, water, and sediments, affecting organisms and leading to further human exposure through contaminated water and food. The extent of exposure is subject to large uncertainties partly because there are no adequate methods to measure the concentrations of these materials in the air or other environmental media. Occupational exposure is potentially greater than consumer or environmental exposure because of the potentially large amounts of ENMs handled in their pure (usually as a powder) form.

Engineered nanoparticles (ENPs) refer to intentionally engineered and produced particles with specific properties. It is important to remember that particles in the nanorange are not only found in the nanotechnology sector. Incidental release can occur in any sector where there are hot processes, either as combustion products or from the saturated vapors arising from sources such as the melting or ionization of metals. Conventional powders are also likely to contain a fraction of particles in the nano-range. In addition, nanoparticles can also occur naturally in the general environment such as volcanic ashes and photochemical smog as well as in cigarette smoke and diesel engine exhaust. This has important implications for the measurement of ENPs. There are many available measurement instruments but they are generally not specific for the type of nanoparticles under investigation, but will also include incidental nanoparticles from other sources.

This chapter describes the key aspects that should be considered in the exposure assessment of ENMs in the workplace.

5.2   MEASUREMENT OF ENGINEERED NANOMATERIALS

5.2.1   REASONS FOR MEASURING

There are various reasons for carrying out inhalation exposure measurements in the workplace. Measurements may be required for risk assessment and risk management, to contribute to epidemiological studies or health surveillance. For example, exposure measurements may be required to test compliance with occupational exposure limits (OELs) or other benchmark values such as derived no effect levels under the European Union (EU) REACH (Registration, Evaluation, Authorization and Restriction of Chemicals) regulations. Risk management requires routine assessment of exposure concentrations and the evaluation of the effectiveness of exposure controls for which measurement data may be required. The design of a measurement strategy will depend on the aims of measurements.

There are only a few OELs for ENMs. The National Institute for Occupational Safety and Health (NIOSH)[3] in the United States has reviewed animal and other toxicological data relevant to assessing the potential nonmalignant adverse respiratory effects of carbon nanotubes (CNTs) and carbon nanofibers (CNFs) and have established recommended exposure limits (RELs) for these two ENMs of 1 $\mu g \ m^{-3}$ elemental carbon in the respirable fraction as an 8-hour time-weighted average (TWA). For ultrafine, including engineered nanoscale, titanium dioxide (TiO_2) a REL of 0.3 mg m^{-3} has been proposed by NIOSH.[4]

The British Standard Institute (BSI) has developed benchmark values for four types of ENMs.[5] For fibrous materials, a benchmark value of 0.01 fibers ml^{-1} was proposed, which is the clearance limit in the United Kingdom in asbestos removal activities. For other ENMs, the proposed benchmark values are based on the OEL (expressed as mass concentration) of the corresponding micro-sized bulk material: $0.066 \times$ OEL for insoluble, $0.5 \times$ OEL for soluble, and $0.1 \times$ OEL for ENMs that have been shown to be a carcinogenic, mutagenic, asthmagenic, or reproductive toxin in bulk form. These are intended to be guidance levels only and should not be assumed to be safe workplace exposure limits. They are based in each case on the assumption that the hazard potential of the nanoparticle form is greater than the large particle. Van Broekhuizen[6] adopted the concept of benchmark values and proposed the so called non-substance-specific nanoreference values. In this scheme, if the exposure exceeds an *action level* more detailed measurements or control measures are needed. Some countries (e.g., the Netherlands) have proposed to implement programs to collect and store in exposure registers high-quality exposure data that could serve any future epidemiological studies.[7] This would clearly require a planned programme of measurement at the relevant industries.

Very few studies have been published on the effectiveness of engineering controls for nanoparticles. Therefore, assessment of the effectiveness at the workplace is key to ensure that workers are protected from inhaling ENMs. These investigations would be best facilitated by appropriate measurements.

5.2.2 Metrics

There is currently no consensus on the use of a single metric to characterize inhalation exposure to ENMs and to date there is no international standard for measuring and characterizing ENMs. International Organization for Standards (ISO) has issued a guidance document[8] entitled *Workplace Atmospheres—Ultrafine, Nanoparticle and Nano-structured Aerosols. Inhalation Exposure Characterization and Assessment.*

For ENMs, the metrics such as airborne particle number concentration or surface area are considered to be more relevant for determining the potential for human health risk than mass-based metrics. Therefore, exposure surveys for ENMs often measure size-resolved particle number, mass, and surface area concentration.

As a result of various aerosol dynamic processes, such as coagulation, diffusion, and deposition, the particle size distribution and total particle number concentration will change over time. For example, coagulation between nanoparticles and *scavenging* by larger (background) particles can lead to agglomerates in the micron-size range. Changes in the surface charge and as a consequence of the particle reactivity are also likely to occur because of interaction with other particles. Therefore, the distribution and physicochemical properties of ENPs at the source are likely to be different from those at the breathing zone of a worker.

ENMs are not usually found as single particles in the workplace but as agglomerates or aggregates of pure ENPs and in combination with background particles. The relevance of these agglomerated forms to health, including potential for dissolution, or disaggregation, needs to be considered also from the toxicological perspective in the risk characterization. Therefore, measurements of exposure to ENMs should be extended beyond measuring isolated nanoparticles, and should also include measurements of coarser, respirable sized particles.

Morphological and chemical characterization is often required to confirm the presence of ENMs, as distinguished from background particles, and to obtain information on the particle shape, which is another potentially important aspect driving toxicity.

Most instruments used for measuring size-resolved airborne particle concentrations are calibrated with spherical, compact, nonporous particles of a specific density. However, nano-sized particles and their agglomerated/aggregated forms tend to have a fractal-like structure. Consequently, the instruments can only provide an equivalent diameter: electrical equivalent mobility diameter when sizing is by an electric field, diffusive (or thermodynamic) equivalent diameter, thermophoretic equivalent diameter, and aerodynamic equivalent diameter for separation by impaction. Therefore, comparison of measurements collected with instruments using different principles has to be done carefully as nonspherical particles may have very different equivalent diameters.

The size distribution of particles can also change during inhalation. Hygroscopic particles will grow in size because of the saturated air in the airways, which will affect the deposition pattern of the particles in the lung. Once deposited, particles can also change their size depending on their physicochemical characteristics.

A group of exposure experts have agreed to a harmonized approach to measure and report exposure to ENM. With regard to risk assessment a multimetric approach has been proposed.[9]

5.2.3 MEASUREMENT METHODS

There are various techniques available to detect, measure, and characterize airborne nanoparticles. Diffusion chargers (DCs) or electrical mobility analyzers are often used for size-resolved nanoparticle measurements below 500 nm. Size-resolved measurements of particles above 500 nm are usually based on aerodynamic properties. Each technique provides a different type of information.

The most common techniques are described in Sections 5.2.3.1 and 5.2.3.2. They have been divided into real-time techniques (online) and integrated sampling (offline).

5.2.3.1 Online Techniques

Online monitors allow continuous data logging of real-time concentrations. Some instruments have time resolutions of 1 second making them very suitable to evaluate temporal variations during processes or tasks.

5.2.3.1.1 Optical Detectors and Condensation Particle Counters

Optical detectors use the light scattered by airborne particles and transform this information into a total particle number concentration, which may be converted into mass concentration based on a number of assumptions. Particles must be at least half the size of the wavelength of the light used (~500 nm). Therefore, the technique fails to detect particles below 200–300 nm. Optical detectors are also susceptible to sizing errors resulting from variations in particle shape and refractive index.

This size limitation can be overcome by using a condensation technique. Alcohol or water are used as condensation fluids to enlarge particles to a size that can be detected by an optical detector. Condensation particle counters (CPCs) are available as portable and handheld devices. The lower detectable size can be 2.5 nm and the upper size limit usually goes up to 10 μm depending on the instrument model. The upper concentration limit of the single particle count mode (counting of individual pulses for each particle) is commonly between 10^4 and 10^6 particles cm^{-3}. At higher concentrations, some CPC models can switch to a photometric mode where the total light scattered by all particles is measured and compared with calibration data. Although the accuracy of the single particle count mode is usually very good (± 10), the photometric mode is known to be less accurate (± 20). If water is used as the condensation fluid, the accuracy of the photometric mode will depend on the growth rate of the particles that is significantly higher for hygroscopic than for hydrophobic particles. When used in combination with a differential mobility analyzer (DMA), it can be used to measure size-resolved particle number concentrations.

The EU Framework 7th program, the NANODEVICE project, has developed a real-time CNT monitor, able to detect CNTs at very low concentrations using certain physical/chemical characteristics. Individual CNT objects are counted using an optical detector. The instrument is close to commercialization by Naneum (http://www.nano-device.eu).

5.2.3.1.2 Electrical Mobility Analyzers

The electrical mobility of particles is the most common property used to measure the size-resolved particle number concentration. This technique is not susceptible to variations in the refractive index, although it is sensitive to variations in particle shape.

In this technique particles are charged and introduced into a DMA. By applying different voltages either smoothly or in steps, particles can be scanned and classified according to their electrical mobility and subsequently counted. The particle electrical mobility is proportional to the ratio of particle charge and size, that is, for particles with the same charge smaller particles will have a higher mobility.

The two main DMAs are the scanning mobility particle sizer (SMPS) and the fast mobility particle sizer (FMPS). The SMPS uses a bipolar charger or neutralizer to establish a known equilibrium charge before particles enter the DMA, where particles are counted directly using a CPC according to their electrical mobility. The bipolar charge is obtained by exposing the aerosol to a radioactive source which poses some significant problems from a regulatory perspective when used in workplaces. The time resolution to perform a complete scan of the particle size distribution is between 2 and 6 minutes. A new version, which features a bipolar diffusion charging using soft X-ray as an alternative to traditional radioactive sources (TSI model 3087), has been commercialized recently. Another alternative without a radioactive source is the Naneum Nano-ID NPS500, which uses a novel unipolar corona charger system.

In the FMPS the aerosol flow passes first through a negative charger to prevent overcharging, and then through a positive charger where a fixed charge is applied using a corona unipolar diffusion charger. The aerosol flow containing positively charged particles is subsequently introduced into a high-voltage electrode that repels the particles toward the electrometers. The particles transfer their charges to the electrometers generating currents that are inverted to produce a particle size distribution in a size range from 5.6 to 560 nm with a time resolution of 1 second, a much faster response than the SMPS. However, measurements from the FMPS are generally considered to be less accurate than those from the SMPS.

A recent development, part of the NANODEVICE project, is the NanoGuard. This new instrument allows online measurement of particle number size distribution and lung-deposited surface area concentration for short-term (1–30 minute). It also allows offline full shift (8 hour) sampling to assess particle morphology and for chemical analysis of particles between 20 and 450 nm. However, the instrument is not commercially available.

5.2.3.1.3 Diffusion Chargers

DCs are used to measure the surface area concentration of particles less than 1000 nm. The response of the DC can be calibrated against a monodisperse aerosol in conjunction with a DMA to measure the electrical mobility of the particles and a CPC to measure their number.

Diffusion charging occurs when particles come in contact with unipolar ions and the charge is transferred to the particle. The particles are collected on an electrometer to measure the current produced by the charged particles. The amount of charge a particle can hold is dependent on the particle's active surface area, defined as the portion of the particle that interacts with the surrounding gas, as opposed to the physical surface area of the particle.

In the DC, an ion trap collects the ions that did not come in contact with a particle. Some DCs have the ion trap voltage set such that the electrometer only measures the current that is proportional to the surface area concentration that deposits in the alveolar (gas-exchange) or the thoracic region of the human lungs.

Depending on the model, measurements can be logged every second. The upper limit concentration is usually 10^6 particles cm^{-3}. The particle size fraction measured is between 10 and 1000 nm, depending on the model. Some studies have shown large errors for particles >400 nm because of the squared dependency of the particle surface area.[10]

A limitation of this technique for measuring ENMs is that fractal structures acquire a slightly higher charge than compact particles and therefore, the size based on the charge for fractal particles is less reliable.

DCs have the advantage over CPCs of not using fluids, so they require less maintenance and can be used for weeks without special attention. They are therefore suitable as permanent monitors. There are several commercially available devices that use this technique (Table 5.1).

5.2.3.1.4 Aerodynamic Diameter Classification

The electrostatic low pressure impactor (ELPI) provides a real-time size-selective (6 nm–10 μm, aerodynamic diameter) active-surface-area concentration and from

TABLE 5.1

Summary of the Characteristics of Instrument Devices Used in the Exposure Assessment of ENMs

Technique/ Instrument	Metric Measured	Size Range	Concentration	Accuracy
		Optical Monitors		
OPC	PSD (Number)	0.3–20 µm	4×10^6 (cm^{-3})	50% for 0.3 µm 100% > 0.45%
		Condensation Particle Counters		
CPC	Number	2.5–3,000 nm	10^6 (cm^{-3})	±20%–30%
		Electrical Mobility Analyzers		
SMPS	PSD (Number)	3–1,000 nm	Aerosol can be a concentrated sample of 10^6–2.4×10^6 (cm^{-3})	±2%–3% Sizing Number < 5%–20%,
FMPS	PSD (Number)			NA (lower than the SMPS)
		Electrostatic/Centrifugal Forces		
APM	PSD (Mass)	30–580 nm (1 g cm^{-3} density)	0.01~100 fg (femtogram)	±10% Sizing ±5%–10% Mass
		Diffusion Charges		
miniDISC	SA	10–300 nm	10^3–10^6 (cm^{-3})	±30% for the con.
nanoTracer		20–120 nm	0–10^6 (cm^{-3})	± 1,500 particles cm^{-3} ± 10 nm
NanoCheck		25–300 nm	5×10^3–5×10^6 (cm^{-3})	±5
NSAM AeroTrak		10–1,000 nm	0–10,000 (µm^2 cm^{-3})	±20%
		Oscillating Microbalance		
TEOM	Mass	Depends on cyclone used	0–10^6 (µg m^{-3})	±0.75%
		Electrostatic Low Pressure Impactor		
ELPI	SA	7 nm–10 µm	Depends of the stage. For D50% = 6 nm the minimum mass detected is 0.0004 µg m^{-3}	NA
		Electron Microscopy		
TEM	PSD, SA, shape, agglomeration, structure			
SEM	PSD, shape, agglomeration, structure			

(*Continued*)

TABLE 5.1 (*Continued*)
Summary of the Characteristics of Instrument Devices Used
in the Exposure Assessment of ENMs

Technique/ Instrument	Metric Measured	Size Range	Concentration	Accuracy
		Electron Microscopy		
X-ray Diffraction				
XRD	Structure Crystalline samples (>1 mg required)	1 nm		

PSD: particle size distribution; SA: surface area concentration.

this the mass concentration can be determined if the particle charge and density are known. It uses a combination of three techniques: particle charging in a unipolar corona charger, size classification in a cascade impactor, and electrical detection with sensitive electrometers.

The particles are first charged to a known level and then enter a cascade low pressure impactor with electrically insulated collection stages. The particles are collected in the different impactor stages according to their aerodynamic diameter, and the electric charge carried by particles into each impactor stage is measured in real time by sensitive electrometers. This measured current signal is directly proportional to particle number concentration and size. The use of impaction substrates allows chemical characterization of the size-classified samples.

5.2.3.1.5 Particle Mass Measurement Techniques

In contrast to real-time instruments that measure particle number concentration there are fewer instruments available to directly measure the mass ENPs in real time.

The Aerosol Particle Mass Analyzer (APM) classifies nanoparticles according to their mass-to-charge ratio based on the balance between centrifugal force and electrostatic force. The output is independent of particle size and shape. The APM uses two cylindrical electrodes rotating about a common axis at the same angular speed. Charged particles enter the annular gap rotating at the same speed as the electrodes. A voltage is applied to the inner electrode creating opposing centrifugal and electrostatic forces. Particles of a specific mass-to-charge ratio will pass through the APM.

The tapered element oscillating microbalance (TEOM) with a suitable size-selective inlet can be used to measure ENM mass concentration. The instrument comprises a tapered element that incorporates a filter on to which the sampled aerosol is deposited. The tapered element is fixed at one end while at the other end it is free to oscillate at its natural frequency. As mass is added to the filter, the frequency of oscillation of the tapered element will change, which is monitored electronically.

5.2.3.2 Offline

5.2.3.2.1 Filter Sampling

Airborne ENMs can be collected on filter samples for subsequent morphological characterization and/or chemical analysis. Samples can be collected using conventional samplers that follow the CEN/ISO/ACGIH convention[11] for collection of the inhalable (e.g., open-face cassettes, IOM head) and respirable (e.g., cyclone) fractions. Open-faced cassettes fitted with an electrically conducting cowl to minimize electrostatic effects are suitable for collection of fiber-like ENMs (e.g., CNTs and CNFs). The filter media will depend on the analytical methods. Polycarbonate filters or silicon wafers are commonly used in scanning electron microscopy (SEM) or transmission electron microscopy (TEM) techniques.

Recently, a few instruments have been developed that allow collection of size-resolved mass concentration. The principles used to separate particles from the air stream are diffusion, electrostatic forces, thermophoretic, impaction, and interception.

The Nano-ID Select provides information on the size-resolved mass concentration for particles ranging in diameter from 2 nm to 20 μm. Particles are collected simultaneously in 12 size stages using both inertial deposition (for aerosols between 250 nm and 20 μm) and diffusion (for aerosols between 2 and 250 nm). The collection substrates are prepared for use with TEM grids.

The ELPI (see Section 5.2.3.1.4) also allows collection of sized fraction filter samples for further characterization.

New devices for mass sampling (but not yet commercially available) include the NanoBadge, which allows collection of personal samples daily or weekly (developed by CEA), the personal Nano-sampler (Naneum), which collects particles from 2 nm to 5 μm in eight size fractions, and the personal thermal precipitator (developed by IUTA and BAuA) for collection of low concentration of particles up to 300 nm. Both were developed as part of the NANODEVICE project.

5.2.3.2.2 Electron Microscopy

Electron microscopy can be used to characterize nanoparticles according to their structure, size, and morphology. This technique can provide information on the particle surface area with respect to size and also on the agglomeration state of the aerosol. The technique is also useful to confirm the presence of ENMs because real-time instruments measure all aerosols present in the workplace atmosphere and do not distinguish between ENMs and background aerosols.

TEM provides direct information on the projected area of collected particles, which may be related to geometric area for some particle shapes. It also provides information on the particle size distribution, structure, and agglomeration state. It uses transmitted electrons that can penetrate the sample enabling determination at the atomic level. The sample must be able to withstand the electron beam and also the high vacuum chamber that the sample is put into. The sample preparation can be difficult as a thin sample on a support grid must be prepared. TEM analysis is often used together with energy dispersive X-ray spectroscopy (EDXS) to obtain information on the elemental composition of individual particles.

SEM uses a high-energy electron beam but the beam is scanned over the surface and the back scattering of the electrons. The sample must be under a vacuum and be electrically conductive at least on the surface. Alternatively, the sample can be coated with an electrically conductive material. SEM can provide information on the particle size distribution, shape, agglomeration state, and structure. SEM is also used in combination with EDXS to obtain both morphological and elemental analysis information.

5.2.3.2.3　Energy Dispersive X-Ray Spectroscopy

EDXS is an analytical technique used for the elemental analysis or chemical characterization of a sample. It relies on the investigation of a sample through interactions between electromagnetic radiation or particles and matter. X-rays emitted by the matter as a result of the interaction are analyzed. The underlying fundamental principle is that each element has a unique atomic structure allowing emission of X-rays that are characteristic of the element's atomic structure. To stimulate the emission of characteristic X-rays from a specimen, a high-energy beam of charged particles such as electrons or protons, or a beam of X-rays, is focused into the sample being studied.

5.2.4　Instrument Devices

Table 5.1 summarizes the characteristics of the most common instruments and techniques used in exposure assessment of ENMs. Most of these instruments are large and heavy and are not suitable as personal monitors.

5.2.5　Measurement Strategies

To date there is no international standard for measuring and characterizing ultrafine particles or nanoparticles. ISO has issued a guidance document[8,12] although it should be noted that this document is not a standard, but a technical report for guidance only. Any monitoring strategy has to be designed so that it responds to the sampling objectives. There are a number of reasons to undertake a sampling survey including the following:

For risk assessment purposes: to assess compliance with an OEL or to estimate exposure dose for health risk assessment

For risk management: to identify emission sources, to evaluate the effectiveness of exposure controls measures, to indicate the need of health surveillance

To contribute to epidemiological studies

For routine monitoring

When sampling for ENMs some considerations have to be taken into account regarding the spatial and temporal variability, the background contribution of aerosols, and the selection of equipment.

In workplaces there is usually a high spatial and temporal variability of aerosols with higher concentrations near the source.[13] Concentrations can also vary rapidly over time because of deposition, diffusion, and the effects of ventilation. This is especially important for ENPs because of their high diffusion rates and the effects of agglomeration and scavenging by background particles, which results in changes in the aerosol concentration and size distribution. Therefore, measurements near the source or at fix points do not represent workers exposure. However, given the lack of personal monitors the assessment of personal exposure becomes quite challenging.

Most studies limit the assessment to the characterization of aerosol release at the source point[14] or try to locate the instrument inlet at breathing height. Although these approaches inform on potential personal exposures they are subject to large errors if an exposure is to be estimated for health risk assessment.

Background aerosols concentrations in workplaces can also show large temporal variations because of emissions from nearby processes or movement of workers around the process area. There are different approaches to estimate background concentrations, which are as follows:

- Identical activity and materials but without using the ENM of concern: background measurements collected in this way allow determination of the contribution of the ENM to the total aerosol concentration and therefore this is the preferred method. This is especially important during processes where other ENMs are generated. However, this approach is not often feasible.
- Far-field background: background measurements collected simultaneously during the activity at a site where no contribution of ENM is expected. This background concentration does not allow differentiating process-generated ENMs.
- Before and after the activity: background measurements collected in this way cannot differentiate between process-generated nanoparticles and any other aerosol source present during the activity but absent before or after the activity (e.g., aerosols from a passing vehicle).

Given the multiple sources of NMs in the workplace, it is crucial to collect contextual information on incidental sources that could give rise to aerosols in the nanorange (e.g., substances used, process characteristics, presence of compressors, and forklifts). Currently, there is still no consensus in what is the best metric to characterize exposure to ENM. Current guidance advises a multimetric approach including samples for morphological characterization to confirm the presence of ENMs.[15]

Data analysis may include the analysis of time series data and reporting of summary statistics (maximum, minimum, geometric mean, and arithmetic mean) during either the measuring period or averaged over the exposure period of interest (task, shift). This is undertaken for the activity and for the background measurements. There is no agreement on the best method to account for background aerosol concentrations. Most studies show comparisons of the particle size distribution and time series data collected during the activity and the background. Activity and background concentrations are presented separated or subtracted.[16] Estimation of the ratio and *t*-test for the differences between background and activity has also been suggested.[17] However, statistical analysis including the background can be difficult in situations with high temporal variations in the background aerosol. The information generated from real-time instruments is supported by the results from the chemical and morphological characterization because none of the real-time instruments is able to differentiate ENPs from incidental nanoparticles.

Finally, harmonization of sampling strategies and data storage has been high-lighted as an urgent need.[9,15] The Partnership for European Research in Occupational Safety and Health is currently working on the construction of a database to store exposure information.

5.3   EXPOSURE ASSESSMENT

On the basis of data collected during a measurement survey (as well as data collected elsewhere), usually in conjunction with some kind of model (either a statistical or a deterministic model), it may be possible to estimate an individual's exposure. Exposure can be defined as the airborne concentration in the breathing zone that is ultimately inhaled over a specified period (e.g., averaged over 8 hours) as opposed to the estimation of airborne concentration in the workplace. Exposure may also be estimated without the use of measurement data using simple or sophisticated exposure models. For ENMs, some simple models are available, such as Stoffenmanager-nano[18] and NanoSafer,[19] although these have not yet been validated thoroughly.[20]

5.3.1   Exposure Scenarios

Exposure for workers is usually greater than that for consumers, because they are likely to be exposed to the free ENM in larger amounts, especially in the agrifood and chemical/materials sectors, although in other sectors (e.g., health sector) exposure may be greater for consumers.[21]

Exposure assessment plays a crucial role for the risk assessment and for the development of effective risk management strategies. The presence of a hazardous substance does not lead to a risk if there is no exposure. Furthermore, the harmful effects are in most cases caused by the dose. Therefore, the estimation of exposure measurements that are toxicologically relevant is a key element in any risk assessment.

Exposure scenarios are useful tools for framing the exposure assessment and they can form the basis of the quantitative exposure estimation. Exposure scenarios within the REACH regulations are defined as sets of information describing the conditions under which the risk associated with the identified use(s) of a substance can be controlled, including operational conditions and risk management measures.

Developing a life-cycle map to identify all the ENM uses, applications, and transport can be useful to guide the development of exposure scenarios and identify potential exposure groups.[22]

The information provided in the exposure scenario must comply with the objectives of the exposure assessment. Minimum information requirements for any exposure assessment are the main parameters determining the probability of release of the ENM (i.e., substance and activity emission potential) and the severity of the exposure (i.e., amount of ENM handled, or contained in the formulation/product, frequency and duration of the exposure, and exposure pathway). Risk management measures aimed to limit/prevent release are also required (e.g., type of ventilation, personal protection equipment, and level of enclosure). The information should be collected in a comprehensive and standard way to allow for proper interpretation and facilitate comparison of exposure estimates across different exposure scenarios.

5.3.2 Exposure Assessment Strategies

A number of organizations have developed exposure assessment strategies for ENMs. The nanoparticle emission assessment technique (NEAT),[23] OECD (Organisation for Economic Co-operation and Development),[16] BSI,[5] the French Approach,[24] and those developed as part of the project NANOSH[17] and NanoGEM based on a number of approaches by German institutions.[10] NIOSH[3,4] has also developed specific guidance to assess exposure to TiO_2 and CNT.

The OECD *Nanomaterial Emission Assessment Guidance* aims to determine whether any airborne releases of ENMs occurs. It is focused on collection of static particle count measurements and therefore it is acknowledged that results from the assessment should not be taken as representative of the worker exposure.

The BSI guidance, which is not regarded as a British Standard,[5] broadly follows the recommendations given for exposure assessment by NIOSH[3,4] and those developed as part of the NANOSH[17] project. The document highlights the importance of obtaining information on the morphology and elemental composition as it is likely, especially with CNTs, that ENMs are present in the air as agglomerates.

The EU-sponsored NANOSH project aimed to harmonize the approach for measurement strategy, data analysis, and reporting of exposure data for ENMs. A review of the methodologies followed for different studies was undertaken as well as collection of new data. The exposure assessment strategy is very comprehensive including guidance on how to undertake the monitoring survey, methodology for data analysis, and data interpretation.[17]

The NIOSH approach (NEAT)[23] recommends that if the ENMs can be chemically identified (e.g., metal oxides), then mass-filter measurements for chemical analysis are recommended.

In general all these strategies follow a typical occupational hygiene approach for evaluating exposure to chemical hazards (Figure 5.1). They are based on a tiered approach or decision logic flow diagram where in each subsequent tier exposure information is collected in more detail to reduce the uncertainty in the exposure assessment.

In the first tier, information to establish the need for exposure monitoring is gathered. In this tier, the ENM, uses and applications are identified along with the potential release sources for each process. If release of ENM cannot be excluded, then the next tier should be implemented.

The aim of tier two is to confirm the presence of ENM in the workplace air. During this assessment it is important to take account of any background particles. Measurements during activities involving ENMs and background particles are carried out using basic monitoring equipment (e.g., optical particle counter [OPC], CPC). Some approaches (NEAT[23] and French approach[24]) include this assessment

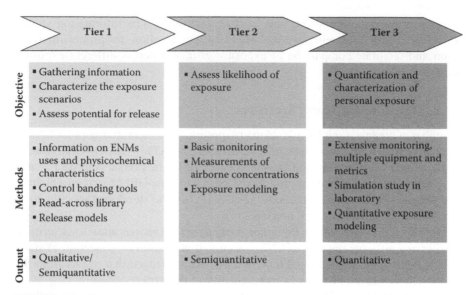

FIGURE 5.1   Flow diagram of a general tiered approach for exposure to ENMs.

in the first tier. Filter samples for SEM or TEM analysis are also recommended. If concentrations of ENMs are above background levels then the next tier is triggered, where a detailed measurement survey is carried out with the aim to collect data that will allow inhalation exposure in the breathing zone to be estimated. This involves the use of more sophisticated equipment (e.g., FMPS and SMPS) for collection of particle size distribution data and lung-deposited surface area.

At the end of each tier a decision needs to be made whether to move to the next tier. Apart from the NIOSH-recommended limits for CNTs and nano-TiO$_2$ there are no ENM-specific exposure limits. Therefore, the criteria for deciding to move to a higher tier are not based on assessment of risk, but on controlling exposure. For example, the NanoGEM approach[10] recommends going to tier three when nanomaterial concentrations are larger than three times the standard deviation of the background levels. However, this means that in workplaces with high and/or highly variable background levels (and consequently high standard deviations), it is unlikely that the contribution of the ENM will be classified as significant, even though the absolute levels of ENM may be high.

The French approach acknowledges that in some situations a measurement campaign might not be possible and suggests carrying out laboratory measurements.[24] However, it is unclear how results from the laboratory experiments could be extrapolated to estimate personal exposures.

The NIOSH approach for assessment of TiO$_2$ and CNT is different as it is based on the RELs established for these two ENMs. Figure 5.2 shows a flow diagram for the exposure assessment of TiO$_2$, which combines the assessment of both nano- and total TiO$_2$. This mass-based approach does not specify a tiered measurement approach, but has a tiered analytical approach for respirable dust samples, involving gravimetric, chemical, and electron microscopic analyses.

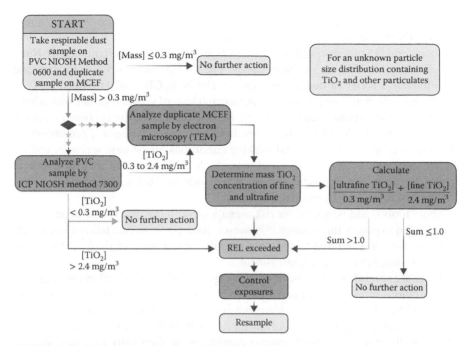

FIGURE 5.2 NIOSH strategy for TiO_2 exposure assessment. (From NIOSH, *Current Intelligent Bulletin 63: Occupational Exposure to Titanium Dioxide*, U.S. Department of Health and Human Services, Centers for Disease Control and Prevention, National Institute for Occupational Safety and Health, 2011. With permission.)

In general, all approaches are designed for exposure management purposes. The criteria proposed to progress to a higher tier are based only on the likelihood of exposure. No consideration is given to the hazardous nature of the ENM and therefore their use for human health risk assessment is limited.

5.3.3 EXPOSURE MODELING

Exposure assessment for ENMs using monitoring instruments is quite challenging because of the limitations of the current equipment to track changes in the physicochemical characteristics of the particles and the few personal monitors commercially available. In addition, monitoring surveys are expensive and data analysis is complex. Therefore, modeling of particle behavior can be very useful. Further, modeling allows us to estimate exposure for varying conditions within a scenario, providing necessary data are available.

In recent years, simple models in the form of control banding or risk prioritization tools have been developed specifically to manage the potential risk from occupational exposure to ENMs. They generally use limited physicochemical and exposure scenario information to put the substance of interest into a hazard and exposure band, which subsequently is used to classify them into risk categories with recommended control measures.

NanoSafer was developed by National Research Center for the Working Environment (NRCWE) in Denmark and provides a risk output in the near- and far-field for short-term (15 minutes) and long-term (8-hour) exposure.[19] Stoffenmanager-Nano was developed by TNO in the Netherlands. The tool provides a risk evaluation and recommends a series of control measures.[18] The NanoCB was developed by Paik et al.[25] and Zalk et al.[26] and provides exposure and hazard scores together with advice on whether the exposure control should be upgraded. The precautionary matrix (PM) developed by TEMAS[27] in Switzerland advises on whether a precautionary approach is required under normal working conditions, worst-case scenario, and for the environment. Brouwer[20] reviewed these tools and concluded that the different approaches to assign exposure and hazards bands and the different applicability domain can result in different risk outcomes.

These tools could be useful for risk management in the absence of toxicological and detailed exposure information.[28] However, their performance has not been comprehensively evaluated and they do not provide quantitative exposure information that can be used in health risk assessment.

There are several mechanistic inhalation exposure models that allow estimation of concentrations of micro-sized particles in occupational settings. For example, those developed by Tielemans et al.,[29] Marquart et al.,[30] and Fransman et al.,[31] which are used in Stoffenmanager and the Advanced REACH Tool (ART). Extension of these models to the nanorange needs careful consideration. Generally they describe the movement of particles from an emission source to a receptor (typically a worker) and take account of the ventilation in the room and size of the room. These models tend to assume a specific particle size distribution that does not change over time and only include the effects of ventilation, dilution, and gravitational settling. However, for nanoparticles the effects of electrostatic forces and coagulation and thermophoretic losses are also important and, depending on the specific scenario, cannot be ignored.

Coagulation is especially important in situations with high concentrations[32] and presence of background microparticles.[33] Coagulation results in a shift of the size distribution over time toward larger particle sizes. In addition, other adjustments are required as coagulation of solid nanoparticles tends to result in fractal-like agglomerates, which is a behavior associated to their structure, in contrast to liquid particles, which follow coalescence coagulation and result in spherical particles.

Schneider et al.[33] developed a conceptual model that includes coagulation of particles across all sizes using the Fuchs interpolation formula that allows modifications to take into account the fractal dimension and interparticle forces. The temporal evolution can be calculated for duration of up to 1 hour, based on an algorithm developed by Miikka Dal Maso from the University of Helsinki, following the UHMA (University of Helsinki Multicomponent Aerosol) atmospheric dynamic model.[34]

Maynard and Zimmer[35] described and validated in a laboratory experiment a numerical model for estimating the size-distribution time-evolution of compact and fractal-like aerosols within workplaces resulting from coagulation, diffusional deposition, and gravitational settlings. The model appeared well-suited to estimating the relationship between the size distribution of emitted well-mixed ultrafine aerosols, and the aerosol that is ultimately inhaled where diffusion losses are small.[35] A modified version of this model has been used as part of the Risk Assessment of

Engineered Nanoparticles (ENPRA) project to estimate the temporal evolution of the particle number concentration of carbon nanofibers in a manufacturing site.

Inclusion of more complex aerosol dynamic processes (e.g., thermophoretic effects, condensation/evaporation) leads to complicated numerical models. The general dynamic equation includes mathematical descriptions for these mechanisms. By solving the equation for initial and boundary conditions, the size distribution of the aerosol can be estimated.[36] However, there is a lack of published information on the application of these models to estimate the temporal evolution of particle size distribution of ENMs in the workplace.

Computational fluid dynamic methods (CFDs), traditionally used in chemical engineering applications, can also be applied to study the temporal evolution of nanoaerosols. CFD examines the flow fields and contaminant dispersions. In the case of a room, the internal volume is divided into a grid and the equations governing the flow of fluids are solved.[37] The Nanocare Project applied the commercial CFD software package FLUENT to estimate the flow patterns and dispersion of nanoaerosols leaked from a reactor into a simulated workplace.[38] Results showed that temperature changes may have a significant impact on the particle dispersion and deposition because of thermophoresis and buoyancy effects.

5.4 CONCLUSIONS

Exposure assessment for ENMs is currently still very challenging because of the limitations in current instrument monitors and the uncertainties surrounding the metrics that best represent the toxic effects. Current instruments do not differentiate background from ENPs and are often not appropriate for long-term personal monitoring. A number of personal monitors have recently come on the market, but because of the need to do multimetric measurements a combined approach of personal and stationary measurements is still required.

Further developments in instrumentation will allow for future collection of long-term personal measurement, for specific nanomaterials (without the need for background correction). However, the development of personal monitors will depend on a better understanding the metrics that are most closely related to any health outcome. Before such measurement and sampling methods are available, a multi-instrument, tiered approach will often be required for measurement surveys.

Quantitative exposure modeling needs to be developed for nanomaterials that can be used for regulatory risk assessment, such as under the REACH regulations. Development of such models requires a better understanding of critical exposure determinants for exposure to airborne ENMs. Such understanding can be developed only by collecting detailed exposure and contextual information from real exposure scenarios, using standard measurement protocols and storing the data in a harmonized way, thus allowing for combined analyses of the data. In addition, specific laboratory experiments are required to determine the release of nanomaterials during the use of products containing ENMs and end-of-life exposure scenarios.

It is likely that in the coming years, we will see huge advances in occupational exposure science for ENMs that will lead to better risk assessment approaches and improved health protection of workers.

REFERENCES

1. Rocco MC, Mirkin CA, Hersam MC. (2010). WTEC Panel Report on *Nanotechnology Research Directions for Societal Needs in 2020*. Available at http://www.wtec.org /nano2 (last accessed May 17, 2013).
2. Yokel RA, MacPhail RC. (2011). Engineered nanomaterials: Exposures, hazards, and risk prevention. *Journal of Occupational Medicine and Toxicology*, 6:7.
3. NIOSH. (2013). *Current Intelligent Bulletin 65: Occupational Exposure to Carbon Nanotubes and Nanofibers*. US Department of Health and Human Services, Centers for Disease Control and Prevention, National Institute for Occupational Safety and Health.
4. NIOSH. (2011). *Current Intelligent Bulletin 63: Occupational Exposure to Titanium Dioxide*. US Department of Health and Human Services, Centers for Disease Control and Prevention, National Institute for Occupational Safety and Health.
5. BSI. (2010). *Nanotechnologies. Guide to Assessing Airborne Exposure in Occupational Settings Relevant to Nanomaterials*. PD 6699-3:2010. British Standard Institute, London, UK.
6. van Broekhuizen P, van Broekhuizen F, Cornelissen R, Reijnders L. (2012). Workplace exposure to nanoparticles and the application of provisional nanoreference values in times of uncertain risks. *Journal of Nanoparticle Research*, 14:770. http://link .springer.com/article/10.1007%2Fs11051-012-0770-3.
7. Health Council of the Netherlands. (2012). *Working with Nanoparticles: Exposure Registry and Health Monitoring*. The Hague: Health Council of the Netherlands, publication no. 2012/31E.
8. ISO/TR 27628. (2007). *Workplace Atmospheres—Ultrafine, Nanoparticle and Nano-Structured Aerosols. Inhalation Exposure Characterization and Assessment*. International Organization for Standardization (ISO), Geneva.
9. Brouwer D, Berges M, Virji MA, Fransman W, Bello D, Hodson L, Gabriel S, Tielemans E. (2012). Harmonization of measurement strategies for exposure to manufactured nano-objects; report of a workshop. *The Annals of Occupational Hygiene*, 56:1–9.
10. Asbach C, Kuhlbusch TAJ, Kaminski H, Stahlmecke B, Plitzko S, Götz U, Voetz M, Kiesling HJ, Dahmann D. (2012). *Standard Operation Procedures for Assessing Exposure to Nanomaterials, Following a Tiered Approach*. NanoGEM.
11. ACGIH. (1999). *Threshold Limit Values for Chemical Substances and Physical Agents and Biological Exposure Indices*. American Conference of Governmental Industrial Hygienists, Cincinnati, Ohio.
12. ISO Standard 7708. (1995). *Air Quality—Particle Size Fraction Definition for Health Related Sampling*. International Organization for Standardization (ISO), Geneva.
13. Cherrie JW, MacCalman L, Frasman W, Tielemans E, Tischer M, van Tongeren M. (2011). Revisiting the effect of room size and general ventilation on the relationship between near- and far-field concentrations. *The Annals of Occupational Hygiene*, 55(9):1006–1015.
14. Kuhlbusch TAJ, Asbach C, Fissan H, Göhler D, Stintz M. (2011). Nanoparticle exposure at nanotechnology workplaces: A review. *Particle and Fibre Toxicology*, 8:22.
15. Brouwer D, Berges M, Virji MA, Fransman W, Bello D, Hodson L, Gabriel S, Tielemans E. (2011). Harmonization of measurement strategies for exposure to manufactured nano-objects; report of a workshop. *Commentary: The Annals of Occupational Hygiene* 56(1):1–9.
16. OECD ENV/JM/MONO. (2009). 11 Emission Assessment for Identification of Sources and Release of Airborne Manufactured Nanomaterials in the Workplace: Compilation of Existing Guidance. Organization for Economic Co-operation and Development, Paris, France.

17. Brouwer D, Van Duuren-Stuurman B, Berges M, Jankowska E, Bard D, Mark D. (2009). From workplace air measurement results toward estimates of exposure? Development of a strategy to assess exposure to manufactured nano-objects. *Journal of Nanoparticle Research*, 11:1867–1881.

18. Van Duuren-Stuurman B, Vink SR, Verbist KJ, Heussen HG, Brouwer DH, Kroese DE, Van Niftrik MF, Tielemans E, Fransman W. (2012). Stoffenmanager Nano version 1.0: A web-based tool for risk prioritization of airborne manufactured nano objects. *The Annals of Occupational Hygiene*, 56(5):525–541.

19. Jensen KA, Saber AT, Kristensen HV, Koponen IK, Ligouri B, Wallin H. (2013). NanoSafer vs 1.1: Nanomaterial risk assessment using first order modelling. Inhaled Particles XI, 23–25 September 2013, Nottingham, UK.

20. Brouwer DH. (2012). Control banding approaches for nanomaterials. *The Annals of Occupational Hygiene*, 56(5):506–514.

21. Nowack B, Brouwer C, Geertsma RE, Heugens EH, Ross BL, Toufektsian MC, Wijnhoven SW, Aitken RJ. (2012). Analysis of the occupational, consumer and environmental exposure to engineered nanomaterials used in 10 technology sectors. *Nanotoxicology*, 7:1152–1156.

22. Owen R, Handy R. (2007). Viewpoint: Formulating the problems for environmental risk assessment of nanomaterials. *Environmental Science and Technology*, 41(16):5582–5588.

23. Methner M, Hodson L, Geraci C. (2010). Nanoparticle emission assessment technique (NEAT) for the identification and measurement of potential inhalation exposure to engineered nanomaterials–Part A. *Journal of Occupational and Environmental Hygiene*, 7(3):127–132.

24. Witschger O, Le Bihan O, Reynier M, Durand C, Marchetto A, Zimmerman E, Charpentier D. (2012). Recommendations for characterizing potential of emissions and potential occupational exposure to aerosols during operations involving released from nanomaterials in workplace operations. *Hygiene et Securite du Travail*, 226:41–55.

25. Paik SY, Zalk DM, Swuste P. (2008). Application of a pilot control banding tool for risk level assessment and control of nanoparticle exposures. *The Annals of Occupational Hygiene*, 52(6):419–428.

26. Zalk DM, Paik SY, Swuste P. (2009). Evaluating the control banding nanotool: A qualitative risk assessment method for controlling nanoparticle exposures. *Journal of Nanoparticle Research*, 11:1685–1704.

27. Höck J, Epprecht T, Hofmann H, Höhner K, Krug H, Lorenz C, Limbach L et al. (2010). *Guidelines on the Precautionary Matrix for Synthetic Nanomaterials*. Federal Office of Public Health and Federal Office for the Environment, Berne, Version 2.

28. Zalk DM, Nelson DI. (2008). History and evolution of control banding: A review. *Journal of Occupational and Environmental Hygiene*, 5(5):330–346.

29. Tielemans E, Schneider T, Goede H, Tischer M, Warren N, Kromhout H, Van Tongeren M, Van Hemmen J, Cherrie JW. (2008). Conceptual model for assessment of inhalation exposure: Defining modifying factors. *The Annals of Occupational Hygiene*, 52(7): 577–586.

30. Marquart H, Heussen H, Le Feber M, Noy D, Tielemans E, Schinkel J, West J, Van Der Schaaf D. (2008). 'Stoffenmanager', a web-based control banding tool using an exposure process model. *The Annals of Occupational Hygiene*, 52(6):429–441.

31. Fransman W, Cherrie JW, Van Tongeren M, Schneider T, Tischer M, Schinkel J, Marquart H et al. (2009). Development of a mechanistic model for the Advanced REACH Tool (ART), Beta release. TNO report V 8667, Zeist, the Netherlands, pp 34–45.

32. Koivisto AJ, Yu M, Hämeri K, Seipenbusch M. (2012). Size resolved particle emission rates from an evolving indoor aerosol system. *Journal of Aerosol Science*, 47:58–69.

33. Schneider T, Brouwer DH, Koponen IK, Jensen KA, Fransman W, Van Duuren-Stuurman B, Van Tongeren M, Tielemans E. (2011). Conceptual model for assessment of inhalation exposure to manufactured nanoparticles. *Journal of Exposure Science and Environmental Epidemiology*, 21(5):450–463.
34. Korhonen H, Lehtinen KEJ, Kulmala M. (2004). Multicomponent aerosol dynamics model UHMA: Model development and validation. *Atmospheric Chemistry and Physics*, 4:757–771.
35. Maynard AD, Zimmer AT. (2003). Development and validation of a simple numerical model for estimating workplace aerosol size distribution evolution through coagulation, settling and diffusion. *Aerosol Science and Technology*, 37(10):804–817.
36. Kommu S, Khomami B, Biswas P. (2004). Simulation of aerosol dynamics and transport in chemically reacting particulate matter laden flows. Part I: Algorithm development and validation. *Chemical Engineering Science*, 59(2):345–358.
37. Nielsen PV. (2004). Computational fluid dynamics and room air movement. *Indoor Air*, 14(s7):134–143.
38. NanoCare Project Partners. (2009). *NanoCare, Health related Aspects of Nanomaterials Final Scientific Report*. Kuhlbusch TAJ, Krug HF, Nau K. (eds). DECHEMA e.V., Frankfurt, Germany.

6 Nanotechnology in Consumer Products

Addressing Potential Health and Safety Implications for Consumers

Treye A. Thomas

CONTENTS

6.1 INTRODUCTION

The emergence and growing use of nanotechnology is expected to have a significant impact on consumer products in the United States and across the globe. Proponents of this technology point to benefits and improvements in product function and performance and the commercialization of nanotechnology has become a strategic goal for the U.S. government through the National Nanotechnology Initiative.[1] Nanotechnology is defined by the ability to manipulate matter at the nanoscale, where materials of approximately 1–100 nm in at least one dimension may possess unique physicochemical properties that are not expected based on size alone. The unique physicochemical properties, such as increased strength, reactivity, or conductivity, can be exploited by incorporating these materials into a wide range of matrices and finished products. Thus, a critical feature of nanotechnology is the development of new materials (nanomaterials) that may have fundamentally different properties from their non-nanoscale counterparts. These materials can be used in a wide range of products to improve their performance and provide benefits to the consumer. The introduction of any new technological innovation into the public domain will inevitably raise questions regarding the potential health implications for those that use the product, the general public, and the environment. Understanding the role of potential stressors, like new chemicals in the environment, is a key factor that impacts the health and well-being of the individual. The need may be more pronounced for new classes of materials such as those produced through nanotechnology. As nanotechnology has developed as a science and nanomaterials are introduced into commerce, stakeholders, including consumer and environmental organizations, have emphasized the importance of understanding the potential health and safety impacts of nanomaterials to avoid any unintended health consequences for the public, especially sensitive receptors such as young children. Decision-makers including regulators and industry health scientists have also acknowledged the need for responsible development of nanotechnology.[2] Because consumer products are an important potential source of exposure to nanomaterials for the public, it is important critical to understand the types of materials the public is most likely to encounter, the factors that are likely to impact consumer exposures, and approaches to understanding and quantifying the potential health impacts of these exposures.

6.2 COMMERCIALIZATION OF NANOTECHNOLOGY

The use of nanomaterials in consumer products is a global phenomenon. A primary challenge in characterizing the potential health and safety concerns of nanomaterials in consumer products is understanding the wide scope of consumer products that may contain nanomaterials and the specific materials in question. The production volume of nanomaterials is growing and will expand considerably in the near future; estimating the number of products in the marketplace that will contain the materials is challenging. Products containing nanomaterials may be produced in a number of countries across the globe. For consumers in the United States, a large portion of the consumer products in the market is imported. A significant challenge for manufacturers and importers across the globe is to be able to identify and characterize potential hazards in all materials that are contained within a product. Estimates of the revenues for products containing nanomaterials exceeded \$1.5 billion in 2009.[3] These markets are expected to increase significantly over the next 5–10 years with estimated values of over \$5.3 billion.[3] The United States is expected to have the highest demand for raw nanomaterials of any region of the world followed by Western Europe and Japan (Table 6.1; Figure 6.1). Estimates of the growth of the nanomaterials market suggest that by 2015, electronics containing nanomaterials will be the

TABLE 6.1
World Nanomaterial Demand by Region (Million Dollars)

	2001	2006	2011	2016	2021
Regional demand	340	890	2,030	5,500	15,500
The Americas	124	360	805	2,350	6,050
United States	118	337	755	2,220	5,620
Other Americas	6	23	50	130	430
Western Europe	75	203	460	1,110	2,700
Asia/Pacific	132	305	720	1,900	6,150
China	10	29	130	445	1,880
Japan	72	150	285	670	1,900
Other Asia/Pacific	50	126	305	785	2,370
Eastern Europe	5	15	27	85	400
Africa/Mideast	4	7	18	55	200

Source: Data from Fredonia Group, Inc., Cleveland, Ohio.

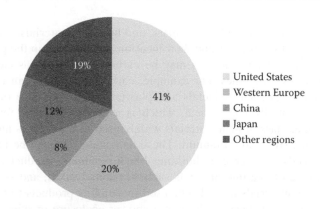

FIGURE 6.1 World nanomaterial demand by region, 2016 ($5.5 billion). (Courtesy of The Freedonia Group, 2012.)

product category with the largest revenue (over $2 billion), followed by household and personal care (Figure 6.2). The Woodrow Wilson Center Project on Emerging Nanotechnologies maintains a database of consumer products that contain nano-materials, and as of March 2011, they estimate over 1300 nanomaterial-enhanced products existed in the market.[4]

6.3 USE OF NANOMATERIALS IN PRODUCTS

The primary purpose of incorporating nanomaterials into products is to improve their performance and function. Consumer products include a vast array of items that are typically used in and around the home, including toys, clothing, furniture,

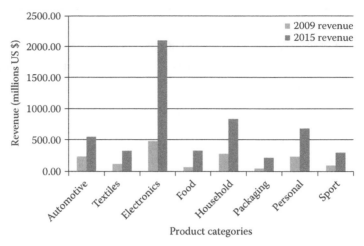

FIGURE 6.2 World market for nanotechnology and nanomaterials in customer productions, 2009–2015. (Courtesy of Future Markets Inc.)

appliances, electronics, household cleaners, and building materials.[5] The types of nanomaterials used in products and their location and function in the product may vary considerably.[6,7] Nanomaterials may be carbon based such as carbon nanotubes, graphene, or fullerenes. For example, carbon nanotubes can improve the strength of materials such as metals and plastics, reduce the size of integrated circuits, and potentially reduce heat, which is a major challenge for electronic devices.[3] Another class of nanomaterials widely used in products is metal oxides, particularly zinc, silica, and titanium. These nanomaterials may be incorporated into raw materials that serve as building blocks in many product applications including a range of coating materials such as paints, waxes, and floor finishes. Metal oxide nanomaterials may also be used in cleaning products to facilitate the cleaning of material surfaces such as in a shower enclosure or window, making these surfaces *self-cleaning*.[8] An example of this property is the photocatalytic reaction of nanoscale titanium dioxide in the presence of ultraviolet radiation, which may break down organic material and kill microorganisms.[9] Nanosilver is a widely used antimicrobial material that can be incorporated into plastics, textiles, and other materials resulting in products as diverse as toys, clothing, and food containers. Nanomaterials may have different product applications with different use patterns, durations, storage, and disposal that will shape the safety profile of each material, including the release potential of nanomaterials from the product. Other factors may alter the physicochemical properties and behavior of these materials in vivo or in the environment, thus affecting the health and safety profile of nano-enabled products. For example, the presence of coatings on the product, such as in metal-based nanomaterials; the presence of added chemical functional groups (e.g., hydroxyl group), such as in carbon nanotubes and fullerenes; and presence of any contaminants in the raw materials (e.g., iron) used to produce nanomaterials or

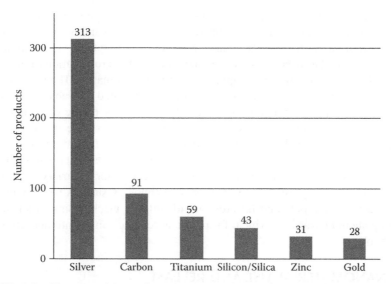

FIGURE 6.3 Nanomaterial use in customer products. (Courtesy of Woodrow Wilson Center, The Project on Emerging Nanotechnologies.)

adsorbed to materials as they are produced will alter their behavior.[10] Despite the range of nanomaterials in use, there are some materials that are of greater interest, because of the volume of material in commerce and the number of product applications. Some researchers suggest that carbon nanotubes, nanosilver, and nanotitanium dioxide are among those most likely to be encountered by consumers (Figure 6.3). These materials may be of greatest interest for consumer exposures, although their functional uses and physicochemical properties vary considerably, and illustrate the differences and challenges when making generalizations about *nanomaterials* in products. Given this array of product types and nanomaterials, a framework approach must be taken to assessing where nanomaterial releases and exposures may occur, the quantity and characteristics of the release, the receptors, and the likelihood that any adverse health impact may arise from the nanomaterial interaction with that receptor.

6.4 CONSUMER PERCEPTION OF NANOMATERIALS

Globally the perception of nanomaterials has varied considerably with many consumers anxious to use products that will improve their lives to those that are concerned about the potential exposure to and subsequent effects that may arise from using products containing these materials.[11] In the United States, the use and impact of nanomaterials in products remain somewhat novel concepts to a considerable portion of the general public. Some studies of the public attitudes toward nanotechnology suggest that initial reactions to the use of nanomaterials are generally positive.[12] Concerns regarding nanomaterial use may not be as great as other risks they may encounter.[13,14] In a study comparing the perceived risks of nanomaterials to other hazards found in the environment, the public does not appear to perceive nano-materials as presenting a greater health risk than other potential stressors such as pesticides in food, nuclear power plants, or alcohol consumption.[13] Communicating risks to the public has become more complex because the proliferation of electronic and social media facilitates more rapid exchanges of information. If a negative inci-dent occurs or report released, this information can be posted on websites, e-mailed, and texted, subsequently reaching significant numbers of the public very rapidly and potentially changing the opinions of a substantial portion of the population. The public reaction will include a search for reliable sources of information to determine the validity of such claims. Robust and comprehensive assessments that adequately address concerns, and are easily accessible, are needed for a variety of risk fac-tors. This is especially true for an *exotic* new technology such as nanotechnology. Thus, understanding potential hazards and identifying mechanisms to adequately address potential hazards is critical for public acceptance and continued growth of the technology.

6.5   NANOMATERIAL USE AND RELEASE ACROSS THE PRODUCT LIFE CYCLE

The emergence of nanotechnology and a new class of materials, nanomaterials, provides the scientific community including toxicologists, exposure scientists, epi-demiologists, and risk assessors with the opportunity and the challenge to use exist-ing approaches to addressing chemical hazards and utilizing new tools to identify potential health risks, which is critical if the public is to accept the use of this new technology. Addressing the potential health implications of nanomaterials can occur through an analysis of the life cycle of the materials from the time they are produced, during incorporation into a product, transportation, consumer use, and disposal (Figure 6.4). One of the primary questions is the magnitude of the exposure poten-tial for the general public, and specifically for consumers who may use products that contain nanomaterials.

The life cycle of a nanomaterial, as with any chemical, begins with the produc-tion of the raw material, which is subsequently assimilated into a material matrix of some type such as a plastic or textile. These intermediate materials are incorporated into a finished product to improve some feature of the product (Figure 6.3). This finished product is purchased and used by a consumer, and ultimately the product through its use or disposal is returned to the environment where it may be disposed of in a variety of ways, including sequestration within a landfill. As the material moves through the life cycle, questions regarding the potential effects and target populations change. During manufacturing of raw materials and finished goods, the primary population of concern is the worker. Because these populations are engaged in production, their potential exposure may be expected to be higher than other population cohorts. However, individuals in these settings are usually trained to engage in these activities to maximize quality and minimize potential injury. There are also institutional efforts such as engineering controls and the prescribed use of personal protective equipment that may minimize exposure. The worker is usually an adult that is in relatively good health and less susceptible to the effects

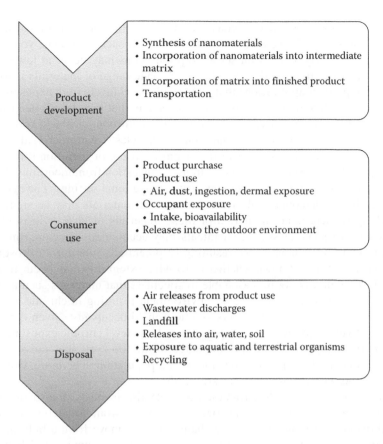

FIGURE 6.4 Nanomaterial releases across the life cycle.

of chemical exposures. There has been a considerable amount of work assessing and identifying approaches to controlling workplace exposures compared to consumer exposures.[15] There are many studies on the fate and transport of nanomaterials through the environment; their movement and concentrations through air, water, and soil; and their potential impact on aquatic and terrestrial organisms.[16] Products that contain nanomaterials such as food, drugs, and pesticides are subject to premarket approval, where their hazards are typically quantified and subject to review.[17] For products that do not fall within these product classes, which encompass thousands of types of products that may be found in the home environment, there is a dearth of information on the potential exposure to nanomaterials during use and subsequent health risks.[18,19]

6.6 HEALTH IMPLICATIONS OF NANOMATERIALS

Determining health risks that may arise from consumer product exposures is challenging, but critical given the large size of the potentially exposed population where consumer products may serve as the primary source for nanomaterial exposures.

This exposed subgroup may vary considerably from those in occupational settings due to the presence of potentially sensitive receptors such as infants, young children, and the elderly. Effects on sensitive populations may occur at lower levels compared to a healthier worker population. A wide range of materials is used in a number of product applications that may result in relatively high exposures. For example, nano-titanium dioxide is used in products that may be applied directly to the body (sunscreens) or sprayed into the air (bathroom cleaners) resulting in potential dermal penetration or inhalation of particles.[20] Antimicrobial applications of silver to clothing will result in dermal exposure or ingestion through the mouthing of the material by young children.[21,22] Ingestion from mouthing of products is a route that is not expected to occur in occupational settings. The aggregate exposures from a nanomaterial or similar class of materials may be significant, particularly if there is accumulation of these materials in vivo or in the home. It is known that some classes of chemicals may accumulate in compartments in the home such as the house dust, resulting in potentially long-term exposures to inhabitants.[23,24] It is unknown whether or to what extent this may occur in nanomaterials, though it is not unreasonable to speculate that materials such as metal oxides may accumulate in the indoor environment resulting in chronic low-level exposures to consumer product users and household occupants. Given the potential for relatively high-level acute exposures and low-level chronic exposures, consumer products represent an important stage of the life cycle that can apply to a large and potentially vulnerable portion of the population. Therefore, characterizations of these potential exposures are needed.

Examining nanomaterials across the life cycle also involves an investigation of the changes in the material that may occur as it moves from production to product use and disposal. Nanomaterials may change as they move from a bulk material to a material matrix. Further changes may occur as this becomes a finished product. Depending on where it is placed in a product and the type of product, various chemical transformations can occur. For example, if the material is located in a surface coating it may be exposed to ultraviolet light, moisture, and abrasion that may result in pyrolysis, oxidation, or the attachment of chemical functional groups. Any surface coatings may be removed or chemically altered. Once released during product use, the material may undergo additional reactions with the plethora of chemical constituents typically present in the indoor environment or agglomerate into larger particles no longer in the scale of 1–100 nm. If nanomaterials agglomerate into larger particles, research suggests that they may retain physicochemical characteristics that differ from similar sized particles that were not initially engineered nanomaterials, particularly on the surfaces of these agglomerated materials.[25] Toxicological studies suggest that agglomeration of nanomaterials may have a significant impact on their effects and deposition in the lungs and ultimately, the body.[26] Characterizing these changes is challenging due to a dearth of information on the release of nanomaterials. It is also critical that toxicological information be available that is appropriate for assessing the potential by-products of these reactions.

6.7 APPROACHES FOR ASSESSING NANOMATERIAL RELEASE AND EXPOSURE

As nanomaterials enter the marketplace, in a wide range of products, utilizing reliable methods to address potential safety and health risks becomes critical. A paradigm in the United States and other countries for identifying health risks involves conducting a formal risk assessment of the product in question involving hazard identification, dose/response, exposure assessment, and risk characterization.[27–29] This approach has been widely used for several years and involves identifying a particular chemical constituent in a product that may pose hazards, determining the health effects that may occur through use of the product and release of the chemical, and determining the range of doses where these effects are likely to occur. The paradigm also includes identifying a threshold for exposure and uptake into the body below which there is no unreasonable risk of health effects, and determining the actual exposure and uptake of a material from the foreseeable use of a product by the consumer and those in the household. The challenge is to determine whether these approaches can be used for nanomaterials, and what special considerations must be made to adequately quantify nanomaterial risks. To develop reliable thresholds for exposure and uptake, robust studies of nanomaterial bioavailability and toxicity are needed.

6.8 DEVELOPING RELIABLE EXPOSURE LIMITS

Nanomaterials are typically classified based on size and physicochemical behavior, thus characterization of these materials during toxicological studies is critical. The question for those using data derived from such studies are (1) whether the material has been well characterized, (2) whether the investigator understands the form of the nanomaterial used in dosing laboratory animals or cell cultures, and (3) what is the relevance to materials to which consumers may be exposed.[30] The urgent need for nanomaterial health and safety data has led to an increased reliance on in vitro methods of toxicity testing; however, there are challenges in using in vitro study results. The size, properties (e.g., high adsorption capacity, surface charge, and catalytic activity), and behavior of nanomaterials in a culture medium may impact the results of these studies. These studies may also lack the sensitivity and reliability needed to sufficiently predict nanomaterial behavior in the body.[31,32] Efforts to improve the reliability of these in vitro assays are underway, including studying these compounds under exposures that may be relevant to those encountered in the environment.[33] Validation of these methods will be an important factor if the results of these studies are to be used in robust assessments of nanomaterial health implications.

Toxicity studies must also consider issues of bioavailability and distribution in the body to determine the extent to which an exposure results in a dose to a target endpoint. For example, do these materials cross the dermal barrier? And if so, to what extent? If not, do they cause irritation or other effects to skin? If they enter the body how does their absorption, distribution, metabolism, and effects differ from other chemicals? Studies must consider target organ endpoints such as the liver and kidney and consider reproductive and developmental effects. Understanding the relationship of dose to these endpoints and their relationship to the unique characteristics of nanomaterials is critical in developing exposure limits that are protective of the public health. If relatively reliable estimates of an exposure threshold are developed, the next issue is to determine if there are exposures and then characterize them in terms of quantity, duration, uptake into the body, bioavailability, disposition in vivo, quantity of the material at the site of action, and the ultimate effect. The accumulation of information regarding the toxicity of nanomaterials may also lead to the development of route-specific limits. For example, the primary focus of toxicity data for carbon nanotubes has been their potential pulmonary effects. Given the amount of data on this specific endpoint, there may be an exposure limit in weight per volume in air (e.g., $\mu g/m^3$). Other limits may include the amount that may be applied to the skin or ingested. Given the limitations in data, the aggregate amount of nanomaterial that enters the body from all routes of exposure can be summed to determine the average daily exposure (ADE) from expected product used and compared to the calculated exposure limit, the acceptable daily intake (ADI). The ADI is the level of daily exposure where no adverse effects are expected. The ADI can be calculated by applying appropriate uncertainty factors to results of dose–response studies such as the no-observed-adverse-effect level, or by benchmark dose estimates.[28]

6.9　QUANTIFYING NANOMATERIAL RELEASE, EXPOSURE, AND UPTAKE FROM PRODUCT USE

Once exposure limits are estimated, it is important to determine the release of nanomaterials from a product and the subsequent exposure to the consumer. The wide range of nanomaterials and products has been discussed, and this significant variability in nanomaterial characteristics, product applications, and release makes exposure assessment for nanomaterials challenging. New approaches for characterizing nanomaterial exposure and risks may be required.[34] Given that nanomaterials are incorporated into a product, fundamental questions are the exact location in the product, the durability of the matrix, and the conditions of use. The product application provides opportunities for a wide array of use/activity patterns, releases, and exposure routes. An assessment of the *human factors* involved with the use of nanomaterials is a critical step in developing use scenarios that may lead to human exposure. Human factors assessment of use scenarios will identify the frequency of use (e.g., days per week), the duration of use, specific use behaviors (e.g., mouthing), and potential routes of exposure.

A primary challenge for estimating exposure is the potential for matrix interference.[35] The environment into which a nanomaterial is released may impact the behavior of the material and result in changes. When nanomaterials enter the media such as air and water and routes of human exposure such as saliva (ingestion) or other body fluids, there may be changes to the material. Metrologists have begun to tackle the questions of how to accurately and thoroughly characterize nanomaterials in a variety of media. For consumer products, understanding the form of the material that comes in contact with the body is a particularly important analytical need. Thus, reliable measurement tools and methods to characterize nanomaterials released from products and that come in contact with various compartments of the human body must be developed and validated.

6.10 ESTIMATING ROUTE-SPECIFIC EXPOSURE POTENTIAL

Accurate and thorough assessments of nanomaterial consumer exposure through various routes of exposure and uptake are challenging when the wide array of exposure scenarios and routes is considered.[6] To characterize these exposures, scientists may require examination of exposures on actual human subjects that can be challenging because of the ethical and legal questions. Another approach is the development of laboratory studies that mimic the expected use patterns of the product and collect chemical residues in materials that simulate components of the human body such as the hand, mouth, or skin.[36] For example, wipe studies using a fabric to mimic the human hand were used to estimate exposure to lead and other metals released from products.[37] Questions have arisen regarding the applicability of these methods for nanomaterials, and whether these can be used or modified to account for the unusual properties of nanomaterials. The potential for ingestion in some nanomaterial applications such as in food may be relatively straightforward to determine, whereas other applications may be more challenging in quantifying and characterizing exposure, for instance, nanomaterials used in textile applications (e.g., nanosilver). If these nanomaterials are applied to a product such as clothing, there is the possibility for ingestion of nanomaterials through direct mouthing of the product by children. An examination of the human factors involved in this interaction would require an estimate of the duration of the mouthing, the area contacted, and the amount of saliva released.[38] Another pathway may include direct hand-to-mouth contact that involves the release of nanomaterials onto the hand of the child and the direct placing of the hands into the mouth. For adults, nanomaterials may adsorb to the surface of the hand through contact and subsequently ingested with food, such as a sandwich, that is touched during a meal. A laboratory-based exposure study may involve a simulation of this contact through the use of an extraction study that involves some form of agitation that mimics mastication by children and release into a surrogate saliva. Introducing the nanomaterials into the saliva surrogate raises questions regarding its effects on the materials. For example, the chemicals used to create the saliva surrogate may change the level of agglomeration, which may interfere with certain analytical techniques. Because the agglomeration state is an important potential factor in behavior, any such changes would not be experienced in actual use scenarios. Other factors include the removal or alteration of any coating materials and the addition of any functional groups to the material.

6.11 AIRBORNE CONCENTRATIONS OF NANOMATERIALS

Use patterns for consumer products often result in the release of airborne particles. A variety of techniques may be used to agitate product samples that mimic consumer behavior. The agitation often occurs in some type of exposure chamber where release materials may be collected through a wide range of techniques such as liquid impingers or solid filter matrices. Given their small size and physicochemical properties, nanomaterials may be more difficult to collect in these media, and the collection, storage, and subsequent processing of collected air samples may impact nanomaterials. Once these materials are extracted from a filter media, they must be analyzed, and the questions regarding matrix interference for analytical techniques are valid for the extraction media. Mathematical models are used to predict the behavior of compounds in the air, and new types of models may be needed to assess the shape, surface area, and characteristics of the nanomaterials.[39] Absorption in the lung may be predicted through the use of physiologically based models. The validity of the models and their ability to provide adequate estimates of nanomaterial in the air and subsequent exposures to consumers is not clear. Once measurements have been made, the concentrations must be extrapolated to estimate concentrations in a typical home. Factors such as room sizes, air exchange rates, and deposition potential will impact nanomaterial concentrations once released from a product. The concentrations should be in a mass per volume metric (e.g., $\mu g/m^3$), although some scientists have suggested that particle surface area may be a more useful metric.

6.12   DERMAL EXPOSURES

Nanomaterials that are released from treated products may deposit on the skin. The skin is a relatively impermeable barrier to a wide range of compounds and some studies suggest this is the case for nanomaterials,[40] whereas others suggest that some nanomaterials may be able to penetrate through the stratum corneum and eventually enter the bloodstream.[41] Deposition of nanomaterials may occur from agitation of the product and subsequent contact with dry skin, or it may be mediated by some liquid, such as sweat. Factors to consider from dermal applications include the amount of surface area contacted by the material and the concentration per unit area of skin. If the contact is liquid mediated, the media may impact the behavior of the nanomaterial and result in agglomeration, deagglomeration, a change in surface chemistry, or other phenomena that may impact the bioavailability of the material. Analytical methods must be used to identify the amount of the nanoparticles that can penetrate the dermal barrier and enter the circulatory system. As in the case of inhalation, mathematical models or other relevant tools may be used to estimate the bioavailability of nanomaterials that deposit on the skin. Quantifying the amount of a nanomaterial that may enter the body from a product is challenging, and many assumptions may be used to derive a reasonable estimate of the amount of material entering the body through various exposure routes (Figure 6.5). The route-specific absorption of a nanomaterial from a product during foreseeable daily use is aggregated into a value that can be referred to as the ADE. This value is compared to the amount that may enter the body with minimal risk of health effects referred to previously as the ADI. The hazard quotient (HI) compares the ADE to the ADI. If the ADE exceeds the ADI, resulting in an HI greater than 1, there may be potential health effects from the nanomaterial in question that is released during the use of the product.[28]

$$HI = \frac{ADE}{ADI} \quad \text{if } HI > 1 \text{ there is a potential for adverse effects}$$

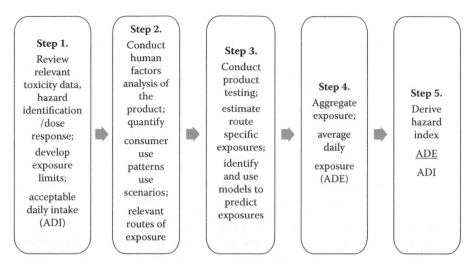

FIGURE 6.5 Stepwise approaches for assessing health risks.

The derivation of the hazard index can be used to provide a reasonable estimation of the potential risks associated with a specific nanomaterial released from a product. Permutations of this approach may be used to estimate risks from multiple nanomaterials released from the same product. The reliability of this estimate is based on the accuracy of the inputs, and as improvements in methods for toxicity assessment, metrology, and exposure science are made, the accuracy of the HI will increase.

6.13 CONCLUSIONS

The growth in the use of nanomaterials provides the consumer with the possibility of significantly improved product performance and new features and functions that can improve daily life. It is critical that the introduction of these new materials into commerce is conducted in a responsible manner and that any potential safety risks are understood and adequately addressed.[42] Existing tools in toxicology, exposure science, epidemiology, and other fields can be used to understand the potential health impacts and consumer exposure to nanomaterials. Nanotechnology also provides the opportunity for stakeholders from around the globe to develop new tools and approaches to understand the behavior of these materials in the environment.[43,44] Applying adequate resources to address these questions will allow the public, particularly in an information age, to better understand the products they use and be assured of their ability to use them safely.

DISCLAIMER

This document was not prepared in the authors' official capacity. The views expressed in this document are those of the author and have not been reviewed or approved by, and may not necessarily reflect the views of, the Commission.

REFERENCES

1. National Science and Technology Council, Committee on Technology, Subcommittee on Nanoscale Science, Engineering and Technology. *National Nanotechnology Initiative 2011 Strategic Plan*. National Science and Technology Council, Committee on Technology, Subcommittee on Nanoscale Science, Engineering and Technology, Washington, DC. February, 2011.

2. Morris J, Willis J, De Martinis D, Hansen B, Laursen H, Sintes JR, Kearns P, Gonzalez M. Science policy considerations for responsible nanotechnology decisions. *Nat Nanotechnol*. 2011;6(2):73–77.

3. Future Markets Inc. The World Market for Nanotechnology and Nanomaterials in Consumer Products, 2010–2015. 2010.

4. Woodrow Wilson Center (WWC) The Project on Emerging Nanotechnologies. Available at: http://www.nanotechproject.org/inventories/consumer/analysis_draft/. 2013.

5. U.S. Consumer Product Safety Commission (CPSC). Nanomaterial Statement. Available at: www.cpsc.gov/PageFiles/84703/Nanotechnology.pdf. 2005.

6. Hansen S, Michelson E, Kamper A, Borling P, Stuer-Lauridsen F, Baun A. Categorization framework to aid exposure assessment of nanomaterials in consumer products. *Ecotoxicology*. 2008;17:438–447.

7. Thomas T, Thomas K, Sadrieh N, Savage N, Adair P, Bronaugh R. Research strategies for safety evaluation of nanomaterials, part VII: Evaluating consumer exposure to nanoscale materials. *Toxicol Sci*. 2006;91:14–19.

8. Parkin I, Palgrave R. Self-cleaning coatings. *J Mater Chem*. 2005;15:1689–1695.

9. Mellott NP, Durucan C, Pantano CG, Guglielmi M. Commercial and laboratory prepared titanium dioxide thin films for self-cleaning glasses: Photocatalytic performance and chemical durability. *Thin Solid Films*. 2006;502:112–120.

10. Chen W, Duan L, Zhu D. Adsorption of polar and nonpolar organic chemicals to carbon nanotubes. *Environ Sci Technol*. 2007;41:8295–8300.

11. Handy R, Shaw B. Toxic effects of nanoparticles and nanomaterials: Implications for public health, risk assessment and the public perception of nanotechnology. *Health Risk Soc*. 2007;9:125–144.

12. Cobb M, Macoubrie J. Public perceptions about nanotechnology: Risks, benefits and trust. *J Nanopart Res*. 2004;6:395–405.

13. Berube DM, Cummings CL, Frith JH, Binder A, Oldendick R. Comparing nanoparticle risk perceptions to other known EHS risks. *J Nanopart Res*. 2011;13:3089–3099.

14. Siegrist M, Keller C, Kastenholz H, Frey S, Wiek A. Laypeople's and experts' perception of nanotechnology hazards. *Risk Anal*. 2007;27:59–69.

15. National Institute for Occupational Safety and Health (NIOSH). Filling the Knowledge Gaps for Safe Nanotechnology in the Workplace: A Progress Report from the NIOSH Nanotechnology Research Center, 2004–2011. 2012

16. Savage N, Thomas T, Duncan J. Nanotechnology applications and implications research supported by the US Environmental Protection Agency STAR grants program. *J Environ Monit*. 2007;9:1046–1054.

17. U.S. Food and Drug Administration. Nanotechnology Task Force Report 2007. Rockville, MD.

18. Lioy P, Nazarenko Y, Han T, Lioy MJ, Mainelis G. Nanotechnology and exposure science: What is needed to fill the research and data gaps for consumer products. *Int J Occup Environ Health*. 2010;16:378–387.

19. Shatkin JA, Abbott L, Bradley A, Canady RA, Guidotti T, Kulinowski KM, Löfstedt RE et al. Nano risk analysis: Advancing the science for nanomaterials risk management. *Risk Anal*. 2010;30:1680–1687.

20. Chen B, Afshari A, Stone S, Jackson M, Schwegler-Berry D, Frazer DG, Castranova V, Thomas TA. Nanoparticles-containing spray can aerosol: Characterization, exposure assessment, and generator design. *Inhal Toxicol*. 2010;22:1072–1082.
21. Benn T, Westerhoff P. Nanoparticle silver released into water from commercially available sock fabrics. *Environ Sci Technol*. 2008;42:4133–4139.
22. Benn T, Cavanaugh B, Hristovski K, Posner JD, Westerhoff P. The release of nanosilver from consumer products used in the home. *J Environ Qual*. 2010;39:1875–1882.
23. Kato K, Calafat AM, Needham LL. Polyfluoroalkyl chemicals in house dust. *Environ Res*. 2009;109:518–523.
24. Zota A, Rudel R, Morello-Frosch R, Brody JG. Elevated house dust and serum concentrations of PBDEs in California: Unintended consequences of furniture flammability standards? *Environ Sci Technol*. 2008;42:8158–8164.
25. Karakoti AS, Hench L, Seal S. The potential toxicity of nanomaterials—The role of surfaces. *JOM*. 2006;58:77–82.
26. Card JW, Zeldin, DC, Bonner JC, Nestmann ER. Pulmonary applications and toxicity of engineered nanoparticles. *Am J Physiol Lung Cell Mol Physiol*. 2008;295:L400–L411.
27. National Research Council. *Risk Assessment in the Federal Government: Managing the Process*. National Academy Press, Washington, DC. 1983.
28. U.S. Consumer Product Safety Commission (CPSC). Chronic Hazard Guidelines, 16 CFR Part 1500. 1992.
29. European Commission, Scientific Committee on Emerging and Newly-Identified Health Risks (SCENIHR). *Opinion on the Appropriateness of the Risk Assessment Methodology in Accordance with the Technical Guidance Documents for New and Existing Substances for Assessing the Risks of Nanomaterials*. European Commission. Health & Consumer Protection DG, Brussels, Belgium. 2007.
30. MINChar Initiative. Characterization Matters. Available at: http://characterizationmatters .org/. Accesses November, 2008.
31. Jones C, Grainger D. In vitro assessments of nanomaterial toxicity. *Adv Drug Deliv Rev*. 2009;61:438–456.
32. Kroll A, Pillukat M, Hahn D, Schnekenburger J. Current in vitro methods in nanoparticle risk assessment: Limitation and challenges. *Eur J Pharm Biopharm*. 2009;72:370–378.
33. Kathawala M, Xiong S, Richards M, Ng KW, George S, Loo SC. Emerging in vitro models for safety screening of high-volume production nanomaterials under environmentally relevant exposure conditions. *Small*. 2013;9:1504–1520. doi:10.1002/smll .201201452.
34. Abbott L, Maynard A. Exposure assessment approaches for engineered nanomaterials. *Risk Anal*. 2010;30:1634–1644.
35. Stamm H, Gibson N, Anklam E. Detection of nanomaterials in food and consumer products: Bridging the gap from legislation to enforcement. *Food Addit Contam Part A Chem Anal Control Expo Risk Assess*. 2012;29:1175–1182.
36. Washburn ST, Bingman TS, Braithwaite SK, Buck RC, Buxton LW, Clewell HJ, Haroun LA, Kester J, Rickard RW, Shipp AM. Exposure assessment and risk characterization for perfluorooctanoate in selected consumer articles. *Environ Sci Technol*. 2005;39:3904–3910.
37. Fenske RA. Dermal exposure assessment techniques. *Ann Occup Hyg*. 1993;37:687–706.
38. Reed K, Jimenez M, Freeman N, Lioy P. Quantification of children's hand and mouthing activities through a videotaping methodology. *J Expo Anal Environ Epidemiol*. 1999;9:513–520.
39. Kumar P, Fennell P, Robins A. Comparison of the behaviour of manufactured and other airborne nanoparticles and the consequences for prioritising research and regulation activities. *J Nanopart Res*. 2010;12:1523–1530.

40. Monteiro-Riviere NA, Riviere JE. Interaction of nanomaterials with skin: Aspects of absorption and biodistribution. *Nanotoxicology*. 2009;3:188–193.
41. Wu J, Liu W, Xue C, Zhou S, Lan F, Bi L, Xu H, Yang X, Zeng FD. Toxicity and penetration of TiO_2 nanoparticles in hairless mice and porcine skin after subchronic dermal exposure. *Toxicol Lett*. 2009;191:1–8.
42. Morris J, Pena C, Bronaugh R, Corley E, Frankel M, Geraci C, Hansen M et al. Risk Management Methods & Ethical, Legal, and Societal Implications of Nanotechnology. 2009–2010 NNI Series of EHS Workshops and Reports, Washington, DC. 2010.
43. Organization for Economic Co-operation and Development Working Party on Manufactured Nanomaterial (OECD WPMN). Important Issues on Risk Assessment of Manufactured Nanomaterials. 2012. Available at: //www.oecd.org/env/ehs/nanosafety.
44. National Science and Technology Council, Committee on Technology, Subcommittee on Nanoscale Science, Engineering and Technology. *National Nanotechnology Initiative 2011 Environmental, Health, and Safety Research Strategy*. National Science and Technology Council, Committee on Technology, Subcommittee on Nanoscale Science, Engineering and Technology, Washington, DC. 2011.

Section III

Modeling

7 Quantitative Nanostructure–Activity Relationships

From Unstructured Data to Predictive Models for Designing Nanomaterials with Controlled Properties

Denis Fourches and Alexander Tropsha

CONTENTS

7.1 INTRODUCTION

Designing nanomaterials with controlled properties represents a new frontier at the interface of experimental and theoretical research. Nanotechnology is evolving rapidly as a standard, multipurpose engineering platform in different industrial areas.[1–3] It is not surprising that both the number of marketed products[2,4] based on manufactured nanoparticles (MNPs) and their variety (from semiconductors to constituents of sunscreen lotions or drug delivery systems) are steadily increasing. MNPs offer unique properties due to their size, surface/size ratio, and infinite possibilities for modifying their surface chemistry.

A plethora of useful applications notwithstanding, the same unique properties of MNPs may be also the cause of potential toxic effects. There is indeed an increasing amount of evidence suggesting detrimental effects of MNPs on living organisms[5–10] directly caused by nanomaterials. Meanwhile, experimental in vivo toxicological studies are costly and time-consuming, making it impractical to test all marketed consumer products incorporating MNPs.[11] As a consequence, both manufacturers and regulators aim to develop methods to design MNPs with controlled and safe bioprofiles.[4] In vitro short-term cell-based assays that are massively used in chemical toxicology (e.g., Tox21[12] and ToxCast projects[13]) represent a fast and inexpensive way to rapidly assess chemicals' bioprofiles that may be useful for predicting in vivo toxicity. However, the overall reliability and prediction abilities of such assays are still under the criticism by the research community[14,15] due to the general lack of in vitro–in vivo concordance and significant experimental variability.

Computer simulation approaches have been widely explored, especially for organic compounds[16–18] to quickly assess their potential safety concerns (e.g., carcinogenicity,[19] aquatic toxicity,[20] liver toxicity[21,22]) with reasonable accuracy. Similarly, computer-based approaches could be used to (1) predict the biological (including toxicological) effects of MNPs solely from their physical, geometrical, and chemical properties (these properties being either measured experimentally or computed); (2) screen in silico libraries of virtual nanomaterials and prioritize those with the most promising predicted properties; and thus (3) guide experimental investigations by focusing costly toxicological studies on a small number of selected and/or rationally designed MNPs. We and others have already voiced in the scientific literature[23–25] the need to develop and further explore such computational methods and tools.

As many more MNPs are expected to enter the market in the next few years, the research community needs to develop novel and efficient approaches, both experimental and computational, for evaluating the portfolio of potential health risks associated with different levels of exposure to MNPs. In this chapter, we discuss (1) the difficulty of curating and integrating nanomaterial-related data from different sources, (2) the challenges related to computational modeling of NPs, (3) the use of quantitative nanostructure–activity relationships (QNARs) modeling to predict the biological effects caused by diverse types of MNPs, and (4) the perspectives of computational nanotoxicology that naturally depend on the intense collaborations between experimental and computational scientists.

The structure of this chapter follows what we consider to be the general workflow for developing publicly accessible QNAR models to enable rational design of nanomaterials. As illustrated in Figure 7.1, a study starts with the compilation, curation, and integration of nanomaterials-related data to create a centralized knowledge base of published characteristics and bioactivities for various types of MNPs (e.g., quantum dots, carbon nanotubes, iron oxides). As detailed in Section 7.2.1, the lack of centralized database of MNPs characteristics and properties is one of the reasons why predictive computational nanotechnology techniques have not yet fully emerged. Analyzing the data accumulated in the literature requires an integrated and synergistic knowledge mining workflow. Such analysis should become a

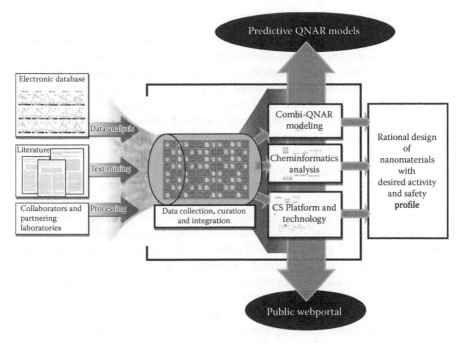

FIGURE 7.1 Overall strategy for developing publicly accessible quantitative nanostructure–activity relationship (QNAR) models to enable rational design of nanomaterials.

community-wide effort for developing and delivering a unique database as a searchable online repository of data, knowledge, and computational models (including QNAR models and other computer simulation approaches). If successful, this overall strategy will boost the field of computational nanotoxicology by enabling large-scale QNAR studies relevant for designing MNPs with safe and controlled bioprofiles. In this chapter, we are describing the challenges, the main steps, and some preliminary studies that all fit into this overall project strategy.

7.2 DATA ON NANOMATERIALS: COMPILATION, CURATION, AND INTEGRATION ISSUES

The lack of a centralized and searchable data repository severely limits our capability to explore published information about the biological effects caused by nanomaterials. Specifically, in the absence of such a repository, the research community cannot fully comprehend the whole amount and diversity of MNPs' properties measured experimentally that are spread within the published literature. Also it is harder to develop predictive tools for prognosticating nanotoxicity in advance of manufacturing, which ultimately hampers the development of nanomaterials that are environmentally benign and safe for human exposure. In this section, we describe the rationale and the different types of difficulties for building such large-scale database of MNP-related characteristics and properties.

7.2.1 CURRENT SITUATION AND CHALLENGES

Currently, all published MNP-related data are diverse, nonsearchable, and spread among numerous sources of information. To illustrate this point, let us consider the following example of human lung adenocarcinoma A549 cells (see Figure 7.2). In the past few years, various MNPs have been tested on A549 cells to study their cellular sensitivity depending on MNPs' types. However, there is no simple tool that can be used to instantly access a list of all tested MNPs and their measured properties let alone get a sense of the overall "big picture." Indeed, Pulskamp et al.[26] reported that several carbon MNPs (multiwalled, single-walled, carbon black, quartz) increased reactive oxygen species (ROS) and decreased mitochondrial membrane potential in a dose- and time-dependent manner in rat macrophages and human A549 lung cells. Moreover, polyvinylpyrrolidone (PVP)-coated silver NPs were reported to induce ROS and damage DNA in A549 cells depending on their doses, as well as increase gap junctional intercellular communication.[27,28] Nanodiamonds were also tested by Liu et al.[29] and found to significantly decrease A549 cell viability and stop the tumor growth in mice. We can also cite the work of Tahara et al.[30] showing the potential role of poly(lactic-*co*-glycolic acid) (PLGA) nanospheres as drug carriers because these MNPs did not cause any noticeable cytotoxicity when tested in presence of A549 cells.

Adenocarcinoma A549 cells represent a very interesting case because unlike most studies testing the chemosensitivity toward MNPs of normal cell lines from different tissues and organisms, A549 are malignant cancer cells. Cytotoxicity in cancer cells

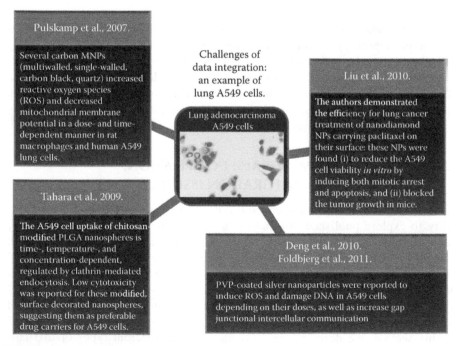

FIGURE 7.2 Data integration concerning different types of manufactured nanoparticles (MNPs) tested against human A549 lung cells.

is therefore a positive result as long as it is selective toward A549 cells (cell survival for normal tissue must be high). In the context of data integration, this case study also shows how complex the data extracted from literature can be: completely different MNPs with heterogeneous compositions, sizes, surface chemistry are applied to the same cell line following different protocols (different solutions, time period, doses and controls). In cases like this, the integration of reported experimental results is very difficult. Even the comparison of results when the same MNPs are tested toward A549 cells in two studies conducted by two research teams is not obvious.

The creation of a curated database of physical, chemical, and biological properties of nanomaterials is thus in high demand. Obviously, the first goal of this effort is to provide the research community with easy access to a repository of filtered experimental observations reported in diverse peer-reviewed journals. This repository could facilitate the research collaboration between different teams as well as provide a platform for better and safer data sharing. Furthermore, it would also enable the computational modeling of nanomaterials by providing compiled sets of integrated and curated data. Overall, the database would benefit both experimental scientists and modelers by enabling the access to curated data collection and in-depth analysis/modeling of experimental data on biological effects of MNPs.

Efforts in this direction have already started but as of now there are very few MNP database projects that we are aware of that should be cited. We can certainly point to the Nanomaterial Registry database (https://www.nanomaterialregistry.org/) developed at RTI (http://www.rti.org/) in the Research Triangle Park, North Carolina. This publicly accessible dataset contains hundreds of records, the most popular entry being the silver-based MNPs (~200 records). Different types of information are stored in the database: particles' size, size distribution, zeta potential, aggregation properties, purity, and so on. Importantly, a great value is given to (1) quality control of data before their actual integration into the repository, (2) consistency of MNPs' naming and description ontology, (3) storing protocols for enabling proper data sharing according to Nano-Tab[31,32] recommendations.

7.2.2 CASE STUDY: CERIUM OXIDE

To illustrate how integrating literature data can potentially help with the overall understanding of complex phenomena linking MNPs and biological systems, we have conducted a rapid analysis of the different reported results obtained for CeO_2. These MNPs are mainly used as diesel fuel additive, as a constituent of catalytic converters or in solution for polishing various materials, or as a UV-blocking agent,[33] thus making the systematic evaluation of their environmental effects highly important. There is a large compendium of literature data describing various types of experiments in which CeO_2 NPs have been tested in different cell-based assays and animal models using different media, particle sizes, pH, ionic strength (IS), with or without natural organic matter (NOM), and so on. In this section, we summarize some of the findings of these studies (Figure 7.3) to underline how difficult it is to find consensus and overall trends in the literature data for a given MNP.

Van Hoecke et al.[33] showed the significant influence of the abiotic factors pH, NOM concentration, and IS on CeO_2 NP aggregation and toxicity toward the

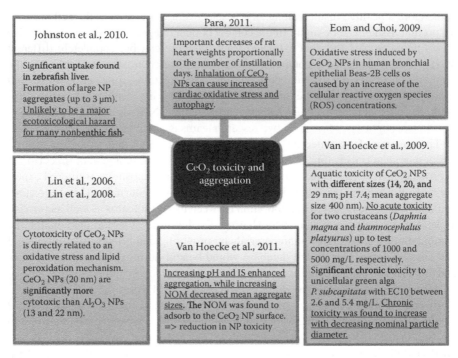

FIGURE 7.3   Literature assertions retrieved for CeO$_2$ nanoparticles (NPs).

unicellular alga *Pseudokirchneriella subcapitata*. Compared to the Organisation for Economic Co-operation and Development (OECD) measure of 3.5 mg CeO$_2$/L 48h-EC20 obtained in standard test medium at pH 7.4, Van Hoecke et al. obtained dramatically higher EC20 values (between 4.7 and 395.8 mg CeO$_2$/L) varying pH (6.0–9.0), NOM (0–10 mg C/L), and IS (1.7–40 mM) on 14 nm CeO$_2$ NPs. They also calculated the mean of experimental NP aggregate sizes ranging between 200 and 10,000 nm. They concluded that increasing pH and IS enhanced aggregation, whereas increasing NOM decreased mean aggregate sizes. The NOM was found to adsorb to the CeO$_2$ NP surface, which could explain the reduction in NP toxicity through a decrease in bioavailability of the particles.

Li et al.[34,35] studied the aggregation kinetics of CeO$_2$ NPs in KCl and CaCl$_2$ solutions. As the initial hydrodynamic radius of CeO$_2$ NPs was approximately 95 nm, the authors measured a critical coagulation concentration of CeO$_2$ NPs at pH 5.6 being approximately 34 mM for KCl and 9.5 mM for CaCl$_2$.

Johnston et al.[36] conducted a very detailed study to evaluate in vivo bioavailability of TiO$_2$, CeO$_2$, and ZnO to fish at different aqueous exposures (50, 500, or 5000 µg/L). Significant NP uptake was found only for CeO$_2$ in the zebrafish liver exposed via water, and ionic titanium in the gut of trout exposed via the diet. There was no measurable uptake of metal oxides in fish tissues following short-term exposures in the water column (up to a nominal exposure concentration of 5000 µg/L) or following a 21-day feeding exposure (up to 300 mg/g NPs) in the food. For the aqueous exposures undertaken, formation of large NP aggregates (up to 3 µm) occurred resulting in limited NP bioavailability. Johnston's results showed that metal oxide MNPs, in the absence of NOM, are likely to have low bioavailability in high-cation environments and thus are unlikely to be a major ecotoxicological hazard for many nonbenthic fish.

Lin et al.[37] showed that cytotoxicity of CeO_2 NPs was directly related to an oxidative stress mechanism and lipid peroxidation, whereas in a recent study[38] they found that CeO_2 NPs (20 nm) were significantly more cytotoxic than Al_2O_3 NPs (13 and 22 nm). Thill et al.[39] also investigated the cytotoxicity of CeO_2 NPs for *Escherichia coli*. They notably found that a large amount of CeO_2 NPs can be adsorbed on the outer membrane of *E. coli*, modifying NP speciation by chemical reduction leading to significant cytotoxicity.

Roh et al.[40] studied CeO_2 biological effects on the soil nematode *Caenorhabditis elegans*. The worms were exposed to 15 and 45 nm CeO_2 NPs, leading to a noticeable increase in the expression of the *cyp35a2* gene and decrease in fertility and survival parameters.

Para[41] evaluated the toxicological effects of intratracheal instilled CeO_2 NPs (7 mg/kg body weight) on the heart of male Sprague–Dawley rats. Animals were killed 1, 3, 14, and 28 days after instillation of NPs. Para noticed important decreases of rat heart weights proportionally to the number of instillation days. The inhalation of CeO_2 NPs also led to elevations in the amount of Beclin-1 and LC3, suggesting that CeO_2 NPs exposure can induce autophagy in the rat heart. Overall, Para's results suggest that the inhalation of CeO_2 NPs can cause increased cardiac oxidative stress and autophagy.

Van Hoecke et al.[42] studied the aquatic toxicity of CeO_2 NPs with different sizes (14, 20, and 29 nm; pH 7.4). The mean aggregate size in the considered aqueous medium was approximately 400 nm. The unicellular green alga *P. subcapitata*, two crustaceans (*Daphnia magna* and *Thamnocephalus platyurus*), and embryos of *Danio rerio* were exposed to NPs. No acute toxicity was observed for the two crustaceans and *D. rerio* embryos, up to test concentrations of 1000, 5000, and 200 mg/L. In contrast, significant chronic toxicity to *P. subcapitata* with 10% effect concentrations (EC10) between 2.6 and 5.4 mg/L was observed. Chronic toxicity was found to increase with decreasing nominal particle diameter.

Eom and Choi[43] showed that the oxidative stress of CeO_2 NPs in human bronchial epithelial cell Beas-2B was caused by an increase of the cellular ROS concentrations, subsequently leading to the strong induction of heme oxygenase-1 via the p38–Nrf-2 signaling pathway.

Recently, Zhang et al.[44] studied the biological effects of four types of NPs (ZnO, TiO_2, SiO_2, and Al_2O_3) with similar sizes (~20 nm) on human fetal lung fibroblasts (HFL1). Cellular mitochondrial dysfunction, morphological modifications, and apoptosis at the concentration range of 0.25–1.50 mg/mL NPs were observed.

Finally, we can mention the study conducted by Oliveira et al.[45] The authors used molecular dynamics simulations to study the aggregation of CeO_2 NPs in water. Different water boxes involving four NPs were simulated using the GROMACS program for 5 ns. Simulations showed NPs starting to agglomerate including counter-ions (Na^+ for positively charged NPs and Cl^- for negatively charged NPs in different boxes).

In summary, there are several recently published experimental studies describing biological effects potentially caused by CeO_2 MNPs in different well-determined and controlled conditions. Cellular oxidative stress induced by CeO_2 seemed to be the major biological event found in vitro. This oxidative stress was shown to lead to moderate-to-high cytotoxicity depending on CeO_2 concentration. Other experiments showed that the oxidative stress directly depends on the bioavailability of CeO_2, which is influenced by the type of media, pH, IS, and importantly, the presence of solubilized organic matters. All these latter parameters tend to influence how CeO_2 particles aggregate, leading to a broad distribution of size and bioavailability of MNPs. Para's study results (using rats) suggest that the inhalation of CeO_2 NPs can cause increased cardiac oxidative stress and autophagy. The potential toxicity of CeO_2 NPs is thus very much dependent on how cells are exposed to the particles (time, dose, media) and what actual size these particles have.

We recognize that the reliability of some of the aforementioned literature-extracted results may vary: indeed, there is a significant interlaboratory experimental variability which is, in the case of nanomaterials, amplified by the fact that composition/purity/initial size of MNP lots can be quite different from one study to another. Nevertheless, we found a remarkable agreement concerning the average particle size for CeO_2 in water at pH 7: in the study by Johnston et al.[36] cited in Figure 7.3, the authors noticed the formation of large NP aggregates (up to 3 μm in water), making CeO_2 NPs unlikely to be a major ecotoxicological hazard for many nonbenthic fish.

An increasingly popular trend for retrieving literature data is the use of automatic text-mining algorithms. Building on our expertise in text mining resulting in the development of the ChemoText methodology[46] and the use of data resulting from text mining for building predictive quantitative structure–activity relationships (QSARs) models,[21] we have used the ChemoText platform developed at University of North Carolina (beta version is available at http://chemotext.wordpress.com/) to search for biomedical assertions in literature involving MNPs.

ChemoText has been built with the overall objective to create a repository of chemicals associated with terms extracted from the literature and representing chemical's bioprofile. To this end, ChemoText was built on top of the Medline annotation database containing over 19 million references to journal articles in life sciences. Compared to other databases, a particular advantage of MEDLINE is that the records are indexed using NLM MedicalSubjectHeadings (MeSH) exploited by ChemoText to rapidly query, mine, and extract relevant chemocentric information. Three categories of annotations were considered: MeSH *effects* annotations, MeSH *disease* annotations, and the proteins listed in MeSH section of the Medline record. MeSH or medical subject headings are well defined, with precise annotations assigned by indexers at the National Library of Medicine.

After the series of manual searches for CeO_2-related data, we used ChemoText to extract CeO_2-related assertions from the scientific literature. We were particularly interested in retrieving the assertions concerning the following properties: (1) constitutional, structural, physical, and chemical characteristics (e.g., notes concerning aggregation, size distribution, solubility); (2) known protein targets or other biological receptors (e.g., cyclooxygenase 2, superoxide dismutase); (3) reported biological effects at the cellular level (e.g., apoptosis, oxidative stress, necrosis); (4) reported biological effects at the whole tissue/organ/organism level (e.g., pulmonary fibrosis, skin sensitization).

Our search for CeO_2-related literature assertions using our ChemoText text-mining resulted in an ensemble of 502 assertions that directly linked CeO_2 to a biomedical term. These 502 assertions represent 217 unique biomedical terms related to CeO_2. With more than 20 assertions each, *particle size* and *oxidative stress* are the two most common CeO_2-related MeSH terms. This result confirms our previous observations, that is, CeO_2 size distribution and oxidative stress as the main biological effect CeO_2 are directly associated. The other terms retrieved by ChemoText reflect the consequences of the oxidative stress on tested cells with *apoptosis* (i.e., cell death) and *cell survival*. Importantly, among the *cell lines* investigated in the retrieved papers, we found a group of studies related to bronchial *epithelial cells* of human and rat *lung*: it indicates that many researchers focused on studying the biological effects (mainly oxidative stress) of CeO_2 NPs in lung cells to investigate whether the inhalation of CeO_2 NPs is likely to induce significant effects.

Overall, manual and automatic text mining approaches led to similar observations concerning CeO_2-induced biological effects reported in the literature. It shows the usefulness of automatic text-mining approaches (such as ChemoText) as a rapid and efficient method to retrieve valuable nanomaterial-related information spread among hundreds of articles. Moreover, any new experimental study on CeO_2 should take the available literature information into account to select the appropriate biological assays/experimental protocols to avoid retesting the same particles in the same assays over and over again or simply to have a general sense of what assays, protocols, or endpoints have not been considered yet for these MNPs.

Lastly, we have examined the distribution and the evolution of the retrieved assertions over the past eight years. The results are illustrated in Figure 7.4. One can see several time-dependent trends in the evolution of a given assertion. CeO_2-induced

CeO_2 assertions	2005	2006	2007	2008	2009	2010	2011	2012
Particle size	1		1	1	2	5	7	6
Oxidative stress		2	2	2	1	1	9	3
Cell survival		2	1	1	1	4	5	3
Macrophages			2	2	2		4	2
Cell line			1	3	2	3		3
Apoptosis	1			2		2	3	2
Lung		1		1		2	4	3
Epithelial cells				2	1	3		1
Time factors				1	1	1	1	1
Cell line, tumor		2				2	2	
Cell proliferation						4	2	
Inflammation					2		3	1
Liver							2	3
Oxidation–reduction		1		1	1	1	2	
Phagocytosis							2	4
Superoxide dismutase				1	1	2	2	1
Surface properties				1		3	2	
Cytokines					1	1	2	1

FIGURE 7.4 Chronological distribution of ChemoText-identified biomedical terms associated to CeO_2 manufactured nanoparticles (MNPs).

oxidative stress has been present since 2006 (as potential explanation of cellular toxicity) but it seems that only recently (2011, 9 assertions; 2012, 3 assertions) have there been several published studies reporting the experimental confirmations of this assertion. Results given in Figure 7.4 can also help to identify the new research trends: for instance, the term *liver* only appeared in the past two years. One can note a similar trend for *phagocytosis*, this biological process being done by *macrophages* to internalize CeO_2 NPs into their cytoplasm.

7.3 CHALLENGES IN MODELING NANOPARTICLES

MNPs are characterized by high structural complexity and diversity.[24,47,48] They are complex, multilayer assemblies of inorganic and/or organic elements, sometimes mixed and coated with diverse organic compounds. Within the same sample of MNPs, the exact stoichiometry of the different constitutive elements may vary from one particle to another as well as the actual composition and complexity of coating molecules. Thus, this is a particularly difficult system for computer simulations that need well-defined molecular compositions and complete knowledge of the chemical entities under investigation. From compact, spherical iron oxide particles to heterogeneous, multifunctional carbon nanotubes with tunable surface chemistry, there is a vast variety of MNP categories leading to numerous potential applications and ranges of desired and undesired physical, chemical, and biological activities. In addition to the fact that the exact composition of a given MNP is not precisely known, three-dimensional nanostructures are highly complex too. As a consequence, the development of quantitative parameters capable of characterizing the structural and chemical properties of MNPs is lacking.

As far as we know, there is no standard list of characteristics (also called *descriptors*) that can describe every single MNP. Recently, Puzyn et al.[24] emphasized the following experimentally derived physicochemical properties: size, size distribution, shape, surface area and structure, surface chemistry (especially for MNPs decorated with organic ligands), chemical composition, overall charge, solubility, agglomeration state. Other types of experimental descriptors have been proposed by Glotzer and Solomon[49] using scanning electron microscopy images. Thomas et al.[31] published a unique MNP ontology for a various scope of nanomaterials used in cancer research. This detailed ontology applied to a relatively large database of MNPs could be extremely useful to develop novel descriptors for characterizing MNPs, and clear hierarchical classifications according to MNPs' constitutional and structural properties.

Next, there are no systematic physicochemical, geometrical, structural, and biological studies of MNPs available in the literature.[50] For instance, aforementioned studies reported experimental investigations on one or just a few MNPs. Importantly, we should underline the great variety of in vitro measurement techniques and biological endpoints. As discussed in Section 7.2, every laboratory seems to have its own criteria and protocols in terms of cell lines, assays, MNP concentrations, and so on, to determine whether a given NP may induce some biological (especially, toxicological) effects. Also, it is not surprising to obtain low interlaboratory assay reproducibility due to the high expertise required to perform such analysis involving NPs that can easily aggregate, possess impurities with their own associated hazards, and have different characteristics from one sample to another. Intralaboratory reproducibility should be evaluated as well. In a recent study, Shaw et al.[51] estimated that the correlation coefficient between biological activity vectors of MNPs (64-feature vector: 4 cell lines × 4 assays × 4 doses) associated with the independent replicates for the same NP was as high as 0.93. However, this result is data specific and cannot be generalized for all available MNPs. Such knowledge about data variability is essential for any modeling study because training a model with a large fraction of unreliable data is pointless.

For all these reasons, it makes both development and validation of statistically significant computational models difficult as these procedures require relatively large amounts of reliable data. It is extremely important to emphasize the need of consistent and massive experimental data where series of MNPs would be characterized by a set of physicochemical properties and tested in well-defined assays. This lack of available data explains why there is almost no literature reporting the use of computational modeling techniques applied to MNPs, especially in the area of nanotoxicology.[52] It is important to underline that popular ab initio quantum chemistry methods are inadequate for such large MNP systems comprising hundreds or thousands of atoms. Recently, molecular dynamics simulations were used by Liu and Hopfinger[53] (1) to reveal the overall changes in the structure of cellular membranes caused by the insertion of carbon nanotubes and (2) to estimate the affinity of drug-like molecules for carbon nanotubes in an aqueous environment.[54] These studies show the utility of molecular dynamics to assess perturbations created by MNPs in biological environment such as cellular membranes. As mentioned earlier, Shaw et al.[51] tested 51 MNPs in vitro against four cell lines in different assays to study their induced biological effects. Using different techniques such as clustering, they identified structure–property relationships linking the biological activity profiles of MNPs and their structural characteristics. Following the principle that similar compounds should possess similar biological functions, Puzyn et al.[24] introduced the term *nano-QSAR* (which is essentially equivalent to the term QNAR used in this review) referring to the use of QSAR modeling for the analysis of NPs. In a comprehensive review, the authors suggested that nano-QSAR models could be built to link characteristics of MNPs with the biological properties they cause in vitro and in vivo. Owing to structural complexity, they concluded that there is a strong need to develop *local* models for each class of particles. Unfortunately, the authors reported some QSAR models built using very small datasets (usually less than 20 MNPs) of carbon nanotubes and fullerenes to assess their solubility and lipophilicity; these models were insufficiently validated according to common QSAR modeling practices (such as OECD principles[55]).

7.4 QUANTITATIVE NANOSTRUCTURE–ACTIVITY RELATIONSHIPS

Molecular modeling has a long history of successful applications for the elucidation and analysis of structure–activity relationships of organic molecules.[56] In a recent paper,[57] we introduced the terminology of QNAR modeling that uses machine-learning methods for establishing quantitative links between descriptors characterizing MNPs and their measured activity (Figure 7.5). Such computational models are built with significantly large sets of MNPs and a well-defined protocol for model selection and validation. Beyond the quality of each single experimental measurement (reflected by both precision and reproducibility of each data point), there should be sufficiently large amount of data to make QNAR studies possible.

To enable MNP modeling, it is suggested that every particle should be characterized by numerical parameters called descriptors. As previously mentioned, properties such as size, shape, zeta potential, morphology, surface area, chemical reactivity, chemical composition, and aspect ratio are often measured experimentally as these characteristics may be critical to determine the behavior of MNPs. However, systematic characterization of nanostructures using experimentally measured properties is almost absent in the public data. Thus, there is an obvious need to develop and use computational descriptors of MNP structure. Classical molecular descriptors (e.g., constitutional, topological, electrostatic, fragmental)[58] were developed originally for small organic molecules. Unfortunately, none of these commercially available software can handle MNPs because (1) the exact composition and three-dimensional structures of MNPs are not available for most datasets, and (2) novel descriptors still need to be developed to appropriately describe MNPs (in particular, to take into account the mixture of organic and inorganic patterns in the same molecule, repetition of structural motifs, etc.). For all these reasons, there is a significant advantage to use experimentally measured descriptors that describe a given MNP and express this information through numerical values precious for further modeling.

Once MNPs are characterized by their (computational or experimentally derived) descriptors, classical QSAR modeling workflow and techniques are directly

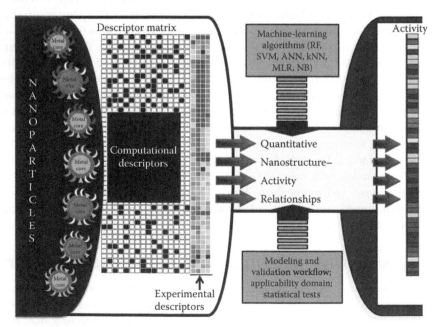

FIGURE 7.5 Quantitative nanostructure–activity relationships.

applicable to model MNPs. QSAR modeling is based on the empirical hypothesis that similar compounds have similar chemical and biological properties. In brief, QSAR models establish quantitative relationships between chemical structures characterized by molecular descriptors and a target property, for example, the biological activity of chemicals tested using a well-defined experimental assay. Models are built using complex machine-learning algorithms such as multilinear regression (MLR), artificial neural networks (ANNs), support vector machines (SVMs), random forest (RF), or k-nearest neighbors (kNNs). These techniques take the descriptor matrix of compounds as inputs and they output a predicted value for the modeled property. Externally predictive models can be applied to screen virtual chemical libraries to identify compounds with desired properties and bias the design of new molecules.[59]

The QSAR modeling workflow[60] comprises three major steps: (1) data preparation/analysis (selection of compounds and descriptors), (2) model building, and (3) model validation/selection, including the evaluation of the model's applicability domain (AD). The entire set of compounds with known experimental activity is randomly split into several training and test sets. Importantly, models are built using training set compounds only and then applied to test set compounds to assess their properties. Rigorous tests, such as leave-one-out, n-fold cross-validation, and Y-randomization (models obtained for the modeling set with randomized activities as negative controls of the overall modeling significance), are applied to evaluate both predictivity and robustness of models. On the basis of these statistical tests, certain models are selected only if they can reasonably predict both training sets as assessed by cross-validation procedures and the test sets. In the end, the selected models are applied to the external validation set(s) of compounds. The AD is defined to limit the ability of the model to extrapolate beyond the chemistry space occupied by the training set molecules. We typically define the AD as a threshold distance between a query compound and its nearest neighbors in the training set but other definitions have been considered as well.[20,61] If the distance of the test compound from any of its kNNs in the training set exceeds the threshold, the prediction is considered unreliable.

As a proof-of-concept QNAR study, we[23,57] modeled 109 cross-linked iron oxide MNPs decorated with small organic ligands to predict their uptakes by PaCa2 pancreatic cancer cells. Unlike other cell types, PaCa2 cell line was selected for in-depth QNAR study because of the significant variance of cellular uptakes among all tested MNPs. To enable QNAR modeling, we represented each individual MNP by the structure of a single organic molecule modifying the MNP's surface. We assumed that because the core MNP structure is the same for all particles, the difference in activity between different MNPs is mainly due to the differences in the structure of the surface-modifying molecules. This approximation allowed us to use classical, computational descriptors for organic surface modifiers. Thus, an ensemble of 150 Molecular Operating Environment (MOE)[62] descriptors was calculated for all 109 organic compounds. Again, each MNP was represented by a unique set of descriptor values determined for the conjugated small molecule.

Then, we developed statistically robust kNN QSAR models relying on chemical descriptors and NP cellular uptakes. The kNN approach predicts the activity of a given MNP by averaging the known activities of its k most similar NPs. Models' external prediction power was shown to be significant ($R^2 = 0.72$, correlation coefficient between experimental and predicted cell uptakes) using fivefold external validation procedure. According to this rigorous statistical procedure, our QNAR models were capable of blindly predicting the PaCa2 cell uptake of the entire set of 109 MNPs with a correlation coefficient equal to 0.72 and a mean absolute error (MAE) of 0.18. Taking into account the models' AD, these prediction performances led to an R^2 as high as 0.77 (MAE = 0.17) at the cost of a moderate loss of external dataset coverage: only 20% of NPs were indeed considered to be outside the models' AD. It is interesting to note that recently, the PaCa2 dataset of 109 MNPs has been modeled again by two research teams[63,64] who obtained very similar results using MLR and CORAL approach, respectively.

Other research teams reported on the development of QNAR models for predicting the biological effects of different types of MNPs. For instance, Winkler et al.[65] recently wrote an interesting review on the current status of QSAR modeling for MNPs and its potential roles for accurately predicting the bioprofiles of MNPs. The authors notably insist on the importance of choosing the right assay/endpoint for assessing the potential toxicity of MNPs, thus increasing the appeal of QNAR models built using these experimental data. A recent study[66] concerned the development of QNAR models for assessing the cellular toxicity of 24 metal oxide MNPs using different linear and nonlinear machine-learning techniques. Prediction performances of obtained models were as high as 94% classification accuracy, but of course, the dataset was very small.

The overall current situation makes us consider two categories of MNP datasets:

1. Those comprising MNPs with diverse metal cores and different surface-modifying organic molecules: in those cases, we recommend to use experimentally measured properties (e.g., size distribution, aggregation states, zeta potential) as MNP descriptors for building QNAR models. We[57] showed the validity of this approach in an earlier proof-of-concept study with 51 diverse MNPs.
2. Those involving MNPs possessing the same core but different surface-modifying organic molecules: as a first approximation, the differences in MNPs bioprofiles can be directly correlated with those in the structure of the decorating surface modifiers. In this case, chemical descriptors can be calculated for a single representative of the decorative molecule and used in QNAR modeling. The aforementioned studies on 109 MNPs tested in PaCa2 cells represent a good illustration of those cases. One could also mention our recent QNAR modeling study[23] involving 83 carbon nanotubes with different surface modifiers.

7.5 PERSPECTIVES

In this chapter, we described some of the latest advances related to computational approaches applied to nanotoxicology, especially QNAR modeling. We believe in the important role that QNAR models could play in the future for evaluating MNPs' bioprofiles prospectively and for influencing the design of safe nanomaterials. Meanwhile, we also underlined the main challenges that computational modelers are facing in developing and applying those approaches to MNPs. Notably, we emphasized the absence of (1) systematic experimental data that characterize constitutional, physical, and chemical properties of large series of MNPs; (2) standard set of biological assays (and associated protocols) clearly defined as the most relevant to strictly and objectively evaluate the overall in vitro (and in vivo) effects of tested MNPs; and (3) absolute MNPs' ontology that would dramatically facilitate the comparison of MNPs data from different sources and their potential integration into a centralized repository.

Pursuing the development and the actual use of computational methods in nanotoxicology represents a concerted effort between researchers in different fields (chemistry, cheminformatics, toxicology, and biology). It requires establishing a set of good practices to characterize MNPs and test them in appropriate biological assays under fixed conditions and protocols. Such multidisciplinary synergized guidelines and action plan would dramatically enhance the long-term development of nanotoxicology and would, in turn, make cleaned and standardized data available to the community, which will further boost applications of molecular modeling approaches in nanotechnology.

As previously mentioned, computational techniques are likely to play a central role in the future of MNPs' design but only if they can predict valuable biological endpoints that are truly indicative of any key activity, property, or toxicological effects caused by these MNPs. In this context, QNAR models are among the most interesting techniques due to their great prediction performances, modest CPU requirements, and the fact that they can be obtained for a very diverse panel of activities. The widespread implementation of this modeling workflow requires intensive collaboration between experimentalists, toxicologists, modelers, and so on, from both academia and industry. A good example of such collaborative effort is the SRC/SEMATECH Engineering Research Program (http://www.erc.arizona.edu/) that offers a dynamic framework where numerous research teams from different U.S. Universities and industrial companies can share not only their knowledge about NP characterization or toxicological assays for MNPs but also samples of MNPs to be tested in different laboratories, datasets of MNPs and corresponding QNAR models, and so on. Such kind of joint effort from different organizations is definitely a good way to go for leveraging the costs of such research as well as the overall number and scale of the different systems tested or modeled by the different research teams.

In summary, it is clear that critical progress of both experimental and computational nanotoxicology requires the understanding and, *in fine*, predictive knowledge of the relationships between the structure and biological activities of MNPs. Obtaining these relationships is crucial if one wants to establish computational models that will help us in designing environmentally benign nanomaterials and prioritizing existing and future MNPs for in vivo toxicological testing. Ongoing improvements for the characterization of MNPs will ultimately enable further development of predictive QNAR models for increasingly larger datasets. Moreover, systematically acquired in vitro data could enable the development of predictive QNAR models to correlate descriptors of MNPs (in silico and in vitro) with clinically important in vivo endpoints. As nanomaterials continue to proliferate within many different fields, there is no doubt that computational methodologies, such as QNAR modeling, are expected to provide critical support to experimental studies aimed to identify safer NPs with the desired properties.

REFERENCES

1. Jones R. Nanotechnology, energy and markets. *Nat Nanotechnol* 2009;4:75.
2. Demirdjian ZS. Nanotechnology: The new frontier for business and industry. *J Am Acad Bus, Cambridge* 2006;8:1–2.
3. Luther W. *Industrial Application of Nanomaterials—Chances and Risks.* Dusseldorf (Germany): Future Technologies Division of VDI Technologiezentrum; 2004.
4. van Zijverden M, Sips AJAM. *Nanotechnology in Perspective. Risks to Man and the Environment* RIVM Report 601785003. National Institute of Public Health; the Netherlands: Bilthoven; 2009; 138 pp.
5. Qu G, Bai Y, Zhang Y, Jia Q, Zhang W, Yan B. The effect of multiwalled carbon nanotube agglomeration on their accumulation in and damage to organs in mice. *Carbon* 2009;48:2060–9.
6. Bai Y, Zhang Y, Zhang J, Mu Q, Zhang W, Butch ER. Repeated administrations of carbon nanotubes in male mice cause reversible testis damage without affecting fertility. *Nat Nanotechnol* 2010;5:683–9.
7. Maynard AD, Aitken RJ, Butz T, Colvin V, Donaldson K, Oberdörster G, Philbert MA et al. Safe handling of nanotechnology. *Nature* 2006;444:267–9.
8. Service RF. Nanotoxicology. Nanotechnology grows up. *Science* 2004;304:1732–4.
9. Nel A, Xia T, Mädler L, Li N. Toxic potential of materials at the nanolevel. *Science* 2006;311:622–7.
10. Zhang LW, Bäumer W, Monteiro-Riviere NA. Cellular uptake mechanisms and toxicity of quantum dots in dendritic cells. *Nanomedicine* 2011;6:777–91.
11. Savolainen K, Pylkkänen L, Norppa H, Falck G, Lindberg H, Tuomi T, Vippola M et al. Nanotechnologies, engineered nanomaterials and occupational health and safety—A review. *Safety Sci* 2010;48:957–63.
12. Kavlock RJ, Austin CP, Tice RR. Toxicity testing in the 21st century: Implications for human health risk assessment. *Risk Anal* 2009;29:485–7.
13. Dix DJ, Houck KA, Martin MT, Richard AM, Setzer RW, Kavlock RJ. The ToxCast program for prioritizing toxicity testing of environmental chemicals. *Toxicol Sci* 2007;95:5–12.
14. Benigni R, Bossa C, Giuliani A, Tcheremenskaia O. Exploring in vitro/in vivo correlation: Lessons learned from analyzing phase I results of the US EPA's ToxCast Project. *J Environ Sci Health C: Environ Carcinog Ecotoxicol Rev* 2010;28:272–86.
15. Thomas RS, Black MB, Li L, Healy E, Chu T-M, Bao W, Andersen ME, Wolfinger RD. A comprehensive statistical analysis of predicting in vivo hazard using high-throughput in vitro screening. *Toxicol Sci* 2012;128:398–417.
16. Arts JHE, Muijser H, Jonker D, van de Sandt JJM, Bos PMJ, Feron VJ. Inhalation toxicity studies: OECD guidelines in relation to REACH and scientific developments. *Exp Toxicol Pathol* 2008;60:125–33.
17. European Parliament EC. REACH regulation. *Official Journal of the European Union* 2007;L136:3–280.
18. Hartung T. Evidence-based toxicology—The toolbox of validation for the 21st century? *ALTEX* 2010;27:253–63.
19. Zhu H, Rusyn I, Richard A, Tropsha A. Use of cell viability assay data improves the prediction accuracy of conventional quantitative structure–activity relationship models of animal carcinogenicity. *Environ Health Perspect* 2008;116:506–13.
20. Zhu H, Tropsha A, Fourches D, Varnek A, Papa E, Gramatica P, Oberg T, Dao P, Cherkasov A, Tetko IV. Combinatorial QSAR modeling of chemical toxicants tested against *Tetrahymena pyriformis. J Chem Inf Model* 2008;48:766–84.
21. Fourches D, Barnes JC, Day NC, Bradley P, Reed JZ, Tropsha A. Cheminformatics analysis of assertions mined from literature that describe drug-induced liver injury in different species. *Chem Res Toxicol* 2010;23:171–83.

22. Rodgers AD, Zhu H, Fourches D, Rusyn I, Tropsha A. Modeling liver-related adverse effects of drugs using k-nearest neighbor quantitative structure–activity relationship method. *Chem Res Toxicol* 2010;23:724–32.
23. Fourches D, Pu D, Tropsha A. Exploring quantitative nanostructure–activity relationships (QNAR) modeling as a tool for predicting biological effects of manufactured nanoparticles. *Comb Chem High Throughput Screen* 2011;14:217–25.
24. Puzyn T, Leszczynska D, Leszczynski J. Toward the development of "nano-QSARs": Advances and challenges. *Small* 2009;5:2494–509.
25. Burello E, Worth A. Computational nanotoxicology: Predicting toxicity of nanoparticles. *Nat Nanotechnol* 2011;6:138–9.
26. Pulskamp K, Diabaté S, Krug HF. Carbon nanotubes show no sign of acute toxicity but induce intracellular reactive oxygen species in dependence on contaminants. *Toxicol Lett* 2007;168:58–74.
27. Foldbjerg R, Dang DA, Autrup H. Cytotoxicity and genotoxicity of silver nanoparticles in the human lung cancer cell line, A549. *Arch Toxicol* 2011;85:743–50.
28. Deng F, Olesen P, Foldbjerg R, Dang DA, Guo X, Autrup H. Silver nanoparticles up-regulate Connexin43 expression and increase gap junctional intercellular communication in human lung adenocarcinoma cell line A549. *Nanotoxicology* 2010;4:186–95.
29. Liu K-K, Zheng W-W, Wang C-C, Chiu Y-C, Cheng C-L, Lo Y-S, Chen C, Chao JI. Covalent linkage of nanodiamond-paclitaxel for drug delivery and cancer therapy. *Nanotechnology* 2010;21:315106.
30. Tahara K, Sakai T, Yamamoto H, Takeuchi H, Hirashima N, Kawashima Y. Improved cellular uptake of chitosan-modified PLGA nanospheres by A549 cells. *Int J Pharm* 2009;382:198–204.
31. Thomas DG, Pappu RV, Baker NA. NanoParticle Ontology for cancer nanotechnology research. *J Biomed Inform* 2011;44:59–74.
32. Thomas DG, Klaessig F, Harper SL, Fritts M, Hoover MD, Gaheen S, Stokes TH et al. Informatics and standards for nanomedicine technology. *Wiley Interdiscip Rev Nanomed Nanobiotechnol* 2011; 3:511–32.
33. Van Hoecke K, De Schamphelaere KAC, Van der Meeren P, Smagghe G, Janssen CR. Aggregation and ecotoxicity of CeO_2 nanoparticles in synthetic and natural waters with variable pH, organic matter concentration and ionic strength. *Environ Pollut* 2011;159:970–6.
34. Li K, Chen Y. Effect of natural organic matter on the aggregation kinetics of CeO(2) nanoparticles in KCl and CaCl(2) solutions: Measurements and modeling. *J Hazard Mater* 2012;209–210:264–70.
35. Li K, Zhang W, Huang Y, Chen Y. Aggregation kinetics of CeO_2 nanoparticles in KCl and $CaCl_2$ solutions: Measurements and modeling. *J Nanopart Res* 2011;13:6483–91.
36. Johnston BD, Scown TM, Moger J, Cumberland SA, Baalousha M, Linge K, van Aerle R, Jarvis K, Lead JR, Tyler CR. Bioavailability of nanoscale metal oxides TiO(2), CeO(2), and ZnO to fish. *Environ Sci Technol* 2010;44:1144–51.
37. Lin W, Huang Y-W, Zhou X-D, Ma Y. Toxicity of cerium oxide nanoparticles in human lung cancer cells. *Int J Toxicol* 2006;25:451–7.
38. Lin W, Stayton I, Huang Y, Zhou X-D, Ma Y. Cytotoxicity and cell membrane depolarization induced by aluminum oxide nanoparticles in human lung epithelial cells A549. *Toxicol Environ Chem* 2008;90:983–96.
39. Thill A, Zeyons O, Spalla O, Chauvat F, Rose J, Auffan M, Flank AM. Cytotoxicity of CeO_2 nanoparticles for *Escherichia coli*. Physico-chemical insight of the cytotoxicity mechanism. *Environ Sci Technol* 2006;40:6151–6.
40. Roh J-Y, Park Y-K, Park K, Choi J. Ecotoxicological investigation of CeO(2) and TiO(2) nanoparticles on the soil nematode *Caenorhabditis elegans* using gene expression, growth, fertility, and survival as endpoints. *Environ Sci Pharmacol* 2010;29:167–72.

41. Para R. Evaluation of Toxicological Effects of Intra Tracheal Instilled CeO_2 Nanoparticles on the Heart of Male Sprague–Dawley Rats. PhD Thesis. Available at: http://mds.marshall.edu/etd/48; 2011; in press.

42. Van Hoecke K, Quik JTK, Mankiewicz-Boczek J, De Schamphelaere KAC, Elsaesser A, Van der Meeren P, Barnes C et al. Fate and effects of CeO_2 nanoparticles in aquatic ecotoxicity tests. *Environ Sci Technol* 2009;43:4537–46.

43. Eom H-J, Choi J. Oxidative stress of CeO_2 nanoparticles via p38-Nrf-2 signaling pathway in human bronchial epithelial cell, Beas-2B. *Toxicol Lett* 2009;187:77–83.

44. Zhang XQ, Yin LH, Tang M, Pu YP. ZnO, TiO(2), SiO(2,) and Al(2)O(3) nanoparticles-induced toxic effects on human fetal lung fibroblasts. *Biomed Environ Sci* 2011;24:661–9.

45. Oliveira OV. Atomistic Molecular Dynamics Simulation of the CeO_2 Nanoparticle Aggregation. 2009 International Conference on Advanced Materials, Rio de Janeiro, Brazil. Available at: http://www.sbpmat.org.br.

46. Baker NC, Hemminger BM. Mining connections between chemicals, proteins, and diseases extracted from Medline annotations. *J Biomed Inform* 2010;43:510–19.

47. Oberdörster G. Safety assessment for nanotechnology and nanomedicine: Concepts of nanotoxicology. *J Intern Med* 2010;267:89–105.

48. Oberdörster G, Oberdörster E, Oberdörster J. Nanotoxicology: An emerging discipline evolving from studies of ultrafine particles. *Environ Health Perspect* 2005;113:823–39.

49. Glotzer SC, Solomon MJ. Anisotropy of building blocks and their assembly into complex structures. *Nat Mater* 2007;6:557–62.

50. Stone V, Nowack B, Baun A, van den Brink N, Kammer Fv, Dusinska M, Handy R et al. Nanomaterials for environmental studies: Classification, reference material issues, and strategies for physico-chemical characterisation. *Sci Total Environ* 2010;408:1745–54.

51. Shaw SY, Westly EC, Pittet MJ, Subramanian A, Schreiber SL, Weissleder R. Perturbational profiling of nanomaterial biologic activity. *Proc Natl Acad Sci USA* 2008;105:7387–92.

52. Meng H, Xia T, George S, Nel AE. A predictive toxicological paradigm for the safety assessment of nanomaterials. *ACS Nano* 2009;3:1620–7.

53. Liu J, Hopfinger AJ. Identification of possible sources of nanotoxicity from carbon nanotubes inserted into membrane bilayers using membrane interaction quantitative structure–activity relationship analysis. *Chem Res Toxicol* 2008;21:459–66.

54. Liu J, Yang L, Hopfinger AJ. Affinity of drugs and small biologically active molecules to carbon nanotubes: A pharmacodynamics and nanotoxicity factor? *Mol Pharm* 2009;6:873–82.

55. Group QE. The report from the expert group on (Quantitative) Structure–Activity Relationships [(Q)SARs] on the principles for the validation of (Q)SARs. Organisation for Economic Co-operation and Development 2004;49:206.

56. Kortagere S, Lill M, Kerrigan J. Role of computational methods in pharmaceutical sciences. *Methods Mol Biol* 2012;929:21–48.

57. Fourches D, Pu D, Tassa C, Weissleder R, Shaw SY, Mumper RJ, Tropsha A. Quantitative nanostructure–activity relationship modeling. *ACS Nano* 2010;4:5703–12.

58. Todeschini R, Consonni V. *Molecular Descriptors for Chemoinformatics*. Weinheim: Wiley-VCH; 2009.

59. Zhang L, Fourches D, Sedykh A, Zhu H, Golbraikh A, Ekins S, Clark J et al. Discovery of novel antimalarial compounds enabled by QSAR-based virtual screening. *J Chem Inf Model* 2013;53:475–92.

60. Tropsha A. Best Practices for QSAR model development, validation, and exploitation. *Mol Inf* 2010;29:476–88.

61. Sushko I, Novotarskyi S, Körner R, Pandey AK, Cherkasov A, Li J, Gramatica P et al. Applicability domains for classification problems: Benchmarking of distance to models for Ames mutagenicity set. *J Chem Inf Model* 2010;50:2094–3111.

62. MOE. Chemical Computing Group. http://www.chemcomp.com/index.htm. Accesses September 2013; 2010.

63. Epa VC, Burden FR, Tassa C, Weissleder R, Shaw S, Winkler DA. Modeling biological activities of nanoparticles. *Nano Lett* 2012;12:5808–12.
64. Toropov AA, Toropova AP, Puzyn T, Benfenati E, Gini G, Leszczynska D, Leszczynski J. QSAR as a random event: Modeling of nanoparticles uptake in PaCa2 cancer cells. *Chemosphere* 2013;92:31–7.
65. Winkler DA, Mombelli E, Pietroiusti A, Tran L, Worth A, Fadeel B, McCall MJ. Applying quantitative structure–activity relationship approaches to nanotoxicology: Current status and future potential. *Toxicology* 2012; in press.
66. Liu R, Zhang HY, Ji ZX, Rallo R, Xia T, Chang CH, Nel A, Cohen Y. Development of structure–activity relationship for metal oxide nanoparticles. *Nanoscale* 2013;5:5644–53.

8 Pharmacokinetics and Biodistribution of Nanomaterials

Jim E. Riviere

CONTENTS

8.1 INTRODUCTION

Since this chapter was published in the first edition of this text,[1] a significant amount of research has been conducted on nanoparticle (NP) disposition that has begun to frame the nature of mathematical models needed to describe NP absorption, distribution, metabolism, elimination (ADME) processes in the body. The author believes that the most fundamental finding in this field over this period is the intricate association between NPs and biomolecules in vivo, which result in the formation of NP biomolecular coronas that define the biological identity of NPs in vivo.[2–4] The property of NPs that makes them *unique*, relative to predicting their activity or toxicity relative to small molecules, is that to a large extent their restricted pattern of biodistribution inherent to their particulate structure, define the nature of their pharmacokinetic properties, and thus therapeutic and adverse effects. Small molecules traverse biological membranes via diffusion or small-molecule transport systems. Particles have different bioenergetics and are transported across cellular membranes often by vesicular systems. Numerous reviews[5–9] have discussed the biodistribution and pharmacodisposition of nanomaterials that make this statement true. This includes, for some specific NPs, detailed analyses of their biodistribution and unique protein interactions. Despite these developments, few actual pharmacokinetic models that quantitate these processes or that take specific interaction mechanisms into consideration have been published. A recent review of NP drug delivery strategies confirms this notion that classic pharmacokinetic parameters have not been determined for most nanomaterials[10] except for the more conventional liposomes and polymeric NPs.[7]

The focus of this chapter will be to take into consideration what is qualitatively known about NP biodistribution and elimination, and use this information to start defining the model structure and parameters needed to develop true physiologically based nanoparticle pharmacokinetic (PBNPK) models that are relevant to predict NP ADME behavior in vivo.

8.2   WHAT MAKES NANOMATERIAL ABSORPTION, DISTRIBUTION, METABOLISM, AND EXCRETION DIFFERENT?

8.2.1   ABSORPTION

The first step in a proper ADME pharmacokinetic study is to define the mechanism of NP entry into the body; that is, when exposure occurs by nonparenteral routes where an absorption phase is evident, usually after oral, transdermal, or inhalational administration. These topics have been addressed in detail elsewhere in this book. When oral exposure is considered, early studies in rats have demonstrated that even large micrometer-scale particles can be absorbed across the intestines via Peyer's patches or by intestinal enterocytes emptying into the lymphatic system.[11] Such work has concentrated on vaccine delivery and targeting of local lymphatic organs and will not be extensively discussed in this chapter because it is well known that a particulate pathway exists for oral absorption.

Similarly, many studies have been conducted on topical NP exposure to the skin although the ability to detect transdermal flux sufficient for a systemic effect has not been reported, even in sunburned, damaged skin.[12,13] However, a small amount of particles do manage to get lodged in the outer stratum corneum where they may be available to be taken up by resident dendritic cells and get mobilized to the lymphatic system. This route has potential applications for vaccine delivery. Finally, a considerable number of toxicology studies have focused on inhalational exposure, a primary route of occupational concern. In fact, a number of original nanotoxicology studies directly grew out of methods used to study inhaled particle toxicology, where translocation from the lungs to the systemic circulation was recognized.[14] One point that is important in this discussion is that if a particle associates with a biomolecule after oral or inhalational administration (e.g., pulmonary surfactant), a biocorona may form that could dictate the subsequent path of system biodistribution.

A great deal of work in nanomedicine drug development has to do with using nanotechnology as a component of the pharmaceutical formulation itself. This may involve taking advantage of large surface area to mass ratio of an NP to control drug release, or of unique surface properties to promote drug solubility and subsequent controlled release. Alternatively, nanosized delivery structures (e.g., needles, liposomal topical formulations, and complex emulsions) may be used to facilitate topical delivery of small molecules or to target delivery to hair follicle shafts.[15] Although such applications are all major drivers of nanomedicine development, from the perspective of this chapter, they do not directly impact on the fundamental nature of ADME properties of the nanoscaled materials themselves that determines their pharmacokinetics once they enter the systemic circulation.

8.2.2 Distribution

Once absorbed into the systemic circulation, the biodistribution properties now become important and the concept of a biocorona discussed in Section 8.2.1 becomes critical. The unique and possibly even defining characteristic of nanomaterials is their strong propensity to adsorb and interact with a wide variety of biomolecules, the most important being proteins. This interaction has been termed biocorona formation and often determines both the half-life of the NP's survival in the systemic circulation and the tissues to which it would distribute. We will discuss the kinetics of this process in Section 8.3, as it becomes evident that understanding this phenomenon is central to designing relevant pharmacokinetic models.

The process by which any foreign particulate material or pathogen is bound to an antibody or complement system protein in the systemic circulation is termed opsonization. It is a prerequisite for phagocytosis and ultimate removal from the circulation through cells of the reticuloendothelial system (RES) or other mononuclear phagocytes. Many NP–protein interactions involve this classic opsonization pathway and often dictate the pattern of biodistribution. Initial attempts to modify NP surfaces with polyethylene glycol (PEG) polymers, so-called *PEGylation*, were specifically related to prevent recognition and binding to opsonization factors.

However, the concept of biocorona formation is much broader, represents a dynamic event, and involves protein interactions (e.g., apolipoproteins like high- or low-density lipoprotein, albumin, acute phase proteins, ferritin, and coagulation factors), which target NP–protein complexes to specific cells and tissues, thereby ultimately determining their pharmacokinetic profile. There is also species specificity as to which RES cells (e.g., spleen vs. liver) will clear the same NP, a topic nicely reviewed by Moghimi et al.[7] This is not at all surprising, yet it has not been taken into consideration when extrapolating ADME data across different species. There is a great deal of biological diversity in plasma protein content as a function of species, as well as from genetics, age, diet, disease, and environment. If NP–protein interactions are a crucial determinant of biodistribution, individual differences in the plasma proteome will modulate pharmacokinetic markers of biodistribution. Walkey et al.[16] has defined the *adsorbome* as the family of proteins that have been associated with forming NP biocoronas with a wide variety of NPs, the dominant proteins are listed in Table 8.1.

There is a need for a robust characterization of protein interaction patterns using biologically relevant metrics. This knowledge is a prerequisite for defining robust PBNPK models that would apply to a wide variety of nanomaterials.

TABLE 8.1

Major Proteins Comprising the Adsorbome of a Nanoparticle Protein Corona

Albumin	Antithrombin III
Apolipoproteins (A-I, A-IV, B-100, C-II, C-III, E)	Clusterin
Complement C3	Fibrinogen
Haptoglobin	Hemoglobin
Histidine-rich glycoprotein	Immunoglobulin G (specifically μ, γ, and light chains)
Inter-(α-trypsin inhibitor H1	Mannose-binding protein C
Paraoxonase-1	Transferrin

Source: Data from Walkey CD and Chan WCW, *Chem. Soc. Rev.*, 41, 2780–99, 2012.

As discussed earlier in this section, it is well known that coating of NPs with PEG, block copolymers, and hyaluronic acid will suppress RES uptake and prolong circulation half-life.[5–8] In addition, shape of surface adducts may change the nature of protein interaction and subsequent deposition.[7] Particle shape including surface area and curvature are also important parameters.[17,18] Size has long been known to play a major role in determining the ability of a nanoscale particulate body (usually a liposome) to circulate in the blood without passage through so-called *leaky* fenestrated or discontinuous capillary endothelium and uptake by RES or phagocytic cells.[6,19,20] Splenic filtration cells tend to capture particles greater than 100 nm in contrast to hepatic Kupffer cells that engulf smaller particles. In addition, larger but flexible micrometer-long filamentous particles may *stream through* RES lined splenic capillaries, which normally would exclude rigid spheres greater than 150 nm in diameter.[21] An aspect of biodistribution that has received attention is the movement of a therapeutic nanomaterial out of the systemic circulation to reach a targeted tumor. Optimal targeting of a cancer chemotherapeutic NP to a tumor would require a size that would evade liver uptake but not be too large to be instead captured by the spleen; this range being somewhere between 100 and 200 nm.

Movement of any NP in the viscous medium of blood introduces rheological factors that further dictate ability of an NP to exit a capillary bed.[22] The geometry, porosity, and surface chemistry of a series of silica NPs were critical attributes for determining the pattern of biodistribution after intravenous injection to mice.[23] For similar-sized particles, mesoporous silica and NPs with a high geometric aspect ratio (rods vs. spheres) tended to preferentially cause deposition into pulmonary capillaries, suggesting these factors are also important determinants of biodistribution. In this 96-hour study, up to 85% of nanomaterial was not excreted from the body and remained in the animal.

Once an NP is circulating in the systemic circulation, some of the same physical and chemical properties that modulate biocorona formation may also modulate cellular uptake. For the RES system, neutral liposomes are taken up less than are negatively charged particles,[24] as are many neutral iron oxide NPs.[25] This is similar to that seen with quantum dot (QD) studies in human epidermal keratinocyte cultures that showed neutral QD uptake < QD-NH$_2$ < QD-COOH.[26] Of particular significance to pharmacokinetic model development is the observation that the time course of cellular uptake also differed between QDs with different surface coatings, indicating different rate constants would be required. Cellular uptake ultimately determines tissue deposition and the extent of biodistribution in a pharmacokinetic model. Studies are lacking using NPs that have a biocorona similar to what would be found under in vivo conditions, or using particles that are capable of actually crossing the systemic capillary barrier to gain exposure to cell surfaces. What is the nature of the NP–biocorona complex that would migrate out of a capillary to be available for cellular uptake? In vitro experiments are conducted using standard cell culture techniques that include the use of media that have protein and other biomolecule compositions optimized for cell growth. What impact do these media constituents have on biocorona formation around the tested particles? Recent work has demonstrated that NPs incubated in cell culture medium have protein coronas reflecting the proteins in the medium.[27] Our group has shown that preincubation of two types (citrate or silica coated) and two sizes (20–40 and 110–120 nm) of silver NP with albumin, transferrin, or immunoglobulin G resulted in significantly different uptake profiles into cultured human keratinocyte than did pristine NP, supporting the hypothesis that protein corona composition impacts cellular uptake.[28] Lipophilic particulate matter often transits the body via the lymphatic system, a pathway not normally accounted for in classic pharmacokinetic models. Research has shown that intramuscular and subcutaneous NP injection,[29] as well as inhalational[9] and intradermal administration,[30] may result in NP accumulation within the lymphatic system. In fact, NP QDs can be used to image lymphatic drainage to lymph nodes during surgery,[31] a study that demonstrated QDs between 15 and 20 nm tend to be retained in the first lymph node encountered.

8.2.3 METABOLISM

There are no published studies on the in vivo metabolism of NPs.

8.2.4 ELIMINATION

A small molecule's elimination and clearance from the body is principally determined via diffusion or by enzymatic active transport systems. As with biodistribution and cellular uptake processes, NPs are dealt with as particles and undergo different mechanisms of removal, when elimination actually occurs.

Studies using QDs of different sizes have suggested that only particles smaller than 5.5 nm are capable of being cleared by the kidney,[32] the primary route that most small molecules or their metabolites are excreted from the body. This is consistent with the kidney's normal function of not filtering plasma proteins or other formed elements in the blood. Lack of robust urinary or biliary secretion pathways for nanomaterials is a major attribute that must be taken into consideration when constructing PBNPK models. This lack of excretion for stable nonbiodegradable materials, which cannot be degraded within cells, results in their persistence in the body. In fact, excretion for such materials is normally by degradation in lysosomes of the RES cells. This is highly dependent on the particle's composition and its susceptibility to metabolism.

Surface coatings and coronas may get the NP targeted to a cell, but its ultimate stability is a function of the particle's composition. Hence so-called biologically labile polymers (e.g., albumin, chitosan, cellulose, milk protein, and poly(lactic-*co*-glycolic) acid), some iron oxide particles, and nanostructured colloid liquids are easily degraded and thus effectively eliminated from the body. Silver particles often undergo pH-mediated dissolution within lysosomes.[28] With iron oxide NPs, once surface coatings degrade, the iron oxide core could be incorporated into the body's iron pool.[25] Substances such as styrene and metallic structures including QDs may persist due to resistance to degradation. Recently, it has been demonstrated that certain carbon nanotubes can be degraded in vitro by neutrophil myeloperoxidases,[33] suggesting that if this occurs in vivo, even relatively *hard* nanostructures may ultimately be cleared from the body. This mechanism of degradation suggests that unlike most small molecules where the organs of elimination are the liver or kidney, for nanomaterials the elimination organs may be dispersed throughout the body.

8.3 PHARMACOKINETIC MODELS

With this background in mind, how has pharmacokinetics been applied in general to the ADME parameters of nanomaterial disposition? It is well beyond the scope of this chapter to review the science and art of pharmacokinetics that is adequately covered in many textbooks.[34–36] Pharmacokinetics allows quantitation of nanomaterial biodistribution parameters across different studies that facilitate cross study comparisons. Half-life ($T_{1/2}$) of disappearance of material in the systemic circulation is the parameter most often calculated for nanomaterials. This is the time it takes for the concentration in blood or plasma to decrease by half, and is calculated as $[(\ln 2)/K]$, where K is the elimination rate constant determined from the slope of the ln (concentration) versus time profile of a substance's decay in blood. For nanomaterials, a longer $T_{1/2}$ correlates to persistence of the particle in the circulation. However, unlike the interpretation for small molecules, a short $T_{1/2}$ does not necessarily imply increased clearance from the body, but rather may reflect opsonization and removal of particles by the RES system. Because in many cases NP biodistribution is an irreversible process, unlike chemical distribution, which is often driven by reversible diffusion that reaches an equilibrium, removal of NPs from the vascular system does not imply clearance from the body. These limitations have been discussed previously in detail elsewhere.[8]

In another modeling approach, noncompartmental or stochastic models may be used where sojourn of a particle in the body is assessed by calculating mean residence times (MRTs), whose interpretation is similar to $T_{1/2}$. In addition, $T_{1/2}$ or MRT for absorption and distribution can be calculated if enough data over proper time frames are collected to allow statistical definition of multiple slopes in the concentration versus time profile. There are very few classic pharmacokinetic modeling studies reported for NPs, especially for manufactured hard NPs not designed for biological applications.

In classic pharmacokinetic analysis, the body is modeled as being composed of compartments that consist of those body regions where particle movement into and out of the vasculature (reference point for most pharmacokinetic models) is similar. These compartments have no anatomical or physiological reality. When compartmental models are constructed, a simple exponential equation relating concentration (C) at any time (t) can be written as $C_t = C_0 e^{-Kt}$, where K is the elimination rate constant and C0 is the initial concentration at $t = 0$. Because this is an equation of a first-order rate process (driving force of diffusion is a first-order process), K is the slope of the ln C versus t plot. From this equation, pharmacokinetic parameters of volume of distribution ($V_d = \text{Dose}/C_0$) and clearance ($Cl = K\,V_d$) may be calculated. Some of these descriptive parameters have been occasionally calculated for a number of nanomaterials.[8,25] If more complex distribution patterns occur (e.g., the body is not homogenous but slow and rapid distribution phases occur), bi- or triexponential models may be needed and these parameters can be determined using different equations. Additional compartments can be added if multiple rates can be discerned.

8.3.1 Physiologically Based Pharmacokinetic Models

A particularly useful approach to pharmacokinetic modeling takes into account the body's actual physiology and anatomical structure. These are termed physiologically based pharmacokinetic (PBPK) models, an area well reviewed in a recent text.[37] They quantitate a molecule's time course of biodistribution and elimination in actual tissues and organs linked together by blood flow through the vascular system, an example of which is illustrated in Figure 8.1. As will be seen in Section 8.3.1.1, they also allow introduction of mechanisms of biodistribution and cell incorporation, as well as being more amenable to incorporation of in vitro data and ultimately interspecies extrapolations.

8.3.1.1 In Vitro Perfused Tissue Biodistribution Studies

Briefly, it is worth mentioning application of simple pharmacokinetic models to isolated perfused tissue preparations to model nanomaterial biodistribution from the infused arterial circulation to the perfused tissue bed. Our laboratory has used the previously developed isolated perfused porcine skin flap (IPPSF) model developed for studying transdermal chemical absorption to study biodistribution by infusing nanomaterial into the arterial circulation perfusing the IPPSF in vitro, and measuring NP concentrations in arterial and venous drainage samples.[38,39] Experiments are designed with an infusion phase, where NP is included in the arterial media, followed by a washout phase where clean media is then infused. This allows accumulation and washout phases to be separately modeled. Arterial/venous extraction is then determined and simple pharmacokinetic models (Figure 8.2) can be developed to describe NP movement from the arterial to venous vasculature, reflecting biodistribution. Alternatively, a model similar to an organ block in the PBPK model above may be used. These models have been used to describe tissue biodistribution of QD 621,[40] $n\text{C}_{60}$ fullerene,[41] and silver NPs.[42] The studies allow tissue extraction to be directly calculated, tissue volume of distributions as well as the time course of particle movement into a defined tissue bed. These parameters differed with NP

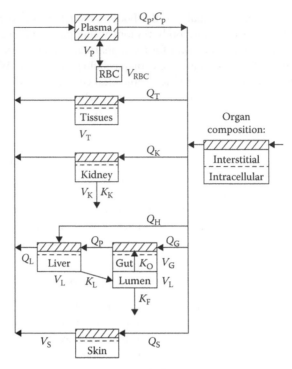

FIGURE 8.1 Physiologically based pharmacokinetic model. Tissue blocks are connected by the vascular system with organ blood flows (Q), volumes (V), elimination rate constants (K) when an organ is a route of excretion, and absorption rates (K_0) from the gut.

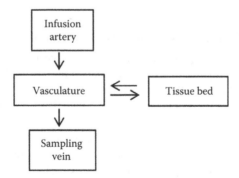

FIGURE 8.2 Perfused tissue pharmacokinetic model. Samples are collected from the perfusing artery and venous drainage.

composition and coating for QD infusions (anionic QD uptake > neutral QD). An interesting finding with QD infusions for both coatings was a periodicity of approximately 100 min in the arterial to venous extraction. If such a phenomenon occurs in vivo, tissue redistribution between vascular effects could be observed. The structure of these models allows easy incorporation into whole-animal PBPK model.

An intriguing finding in some IPPSF infusion studies was the induction of vascular toxicity after infusion of certain NPs.[43] Compared to a large control series of small-molecule infusions and nC_{60} in the IPPSF, infusion with larger silica, iron oxide, silver, and QD NPs caused vascular congestion evidenced by perfused flap weight gain and a gradual increase in arterial perfusion pressure in QD flaps. The kinetics of this gradual change is different than the immediate rise seen after arterial infusion of a vasoconstrictive drug.[44] This finding of what could be considered vascular congestion is consistent with recent findings in mice in vivo where silica NPs caused acute toxicity due to mechanical obstruction of the vasculature in multiple organs.[45] This vascular toxicity, most evident in the kidney, was modulated by surface geometry and chemical characteristics, with mesoporous and high aspect ratio NPs being more toxic. This phenomenon has implications to the construction of PBNPK models because direct vascular effects will impact both clearance and distribution parameters.

8.3.2 WHOLE-ANIMAL IN VIVO PHYSIOLOGICALLY BASED NANOPARTICLE PHARMACOKINETIC MODELS TO DATE

How have PBPK models been applied to nanomaterials? To clarify this discussion, as defined in Section 8.1, PBPK models of nanomaterials will be termed PBNPK. An excellent review of specific PBNPK models was published by Li et al.[46] There have been few attempts to construct PBNPK models due to the scarcity of available animal data. As already discussed, the mechanism describing NP movement in vivo is often lacking. For example, only a few types of NPs have been modeled and none have included or considered lymphatic transport. No attempts were made to monitor biocorona formation or in situ NP aggregation states in any biophase included in the models. In a phenomenon very similar to so-called *pH partitioning*, where a weak acid or base may accumulate in an organ where pH gradients favoring the ionized form of the drug exist across certain membranes (e.g., milk, prostate), a similar phenomenon could occur with NPs where the driving force would now be aggregation or agglomeration in the tissue favoring accumulation. This has not been addressed in any models developed.

Lin et al.[47] described a QD 705 model in mice based on a complete biodistribution study reported earlier.[48] This dataset had shown only partial and gradual QD excretion over 6 months, with tissue redistribution occurring, a phenomenon consistent with the QD IPPSF data discussed in Section 8.3.1.1. These workers proposed a one-way (blood to tissue) distribution coefficient as they argued a true partition coefficient does not exist for NPs. Our group[49] reported on a single PBNPK model using the data from Yang et al.[48] and published studies from other QDs (705, 525, 621, 800, conjugated) ranging in size from 7 to 80 nm in mice and rats. A single model could not describe all data and the work reached a similar conclusion to Li relative to the unsuitability of partition coefficients and flow-limited models. In contrast, in the isolated tissue perfusion studies described in Section 8.3.1.1 using carbon and silver NPs, equilibrium was achieved in the perfused skin tissue bed allowing for the concept of a partition coefficient to be used.[41,42] The periodicity in arterial–venous extraction observed with QD infusion[40] is more consistent with the QD mice data[48] and strongly suggests that even in the same tissue bed, the approach used to model tissue distribution may be particle dependent. Finally, direct interaction with NPs on modulating vascular function as discussed in Section 8.3.1.1[43] also makes constructing tissue compartments in a PBNPK model problematic.

Lankveld et al.[50] developed a PBNPK model in rats over 16 days to describe the disposition of 20-, 80-, and 110-nm silver NPs. This study showed rapid clearance of NPs from the blood into all tissues studied independent of particle size, although smaller particles tended to accumulate in the liver whereas larger particles tended toward the spleen, a finding consistent with RES distribution data discussed in Section 8.2.2 for other nanomaterials. Accumulation was observed in all organs after multiple dose administration. The largest particles persisted in all tissues at the end of the experiment. These studies clearly suggest that disposition of 20-nm particles were different than the larger ones studied. A *quasi-irreversible* tissue incorporation parameter was used to account for 1.3%, 9.9%, and 6.5% of particles across all sampled tissues for 20-, 80-, and 110-nm particles, respectively. The tissue–blood partition value obtained in this study for the reversible tissue concentrations (which would include skin) was similar as that reported in the IPPSF studies discussed in Section 8.3.1.1 for 20- and 40-nm silver NPs.[42]

This model also illustrates a number of other characteristics of a PBNPK model relative to where *elimination* occurs. As discussed earlier,[8] irreversible tissue sequestration, as seen with the silver particles in this model and QD in the models above, becomes a primary mechanism of clearance. Thus unlike many small-molecule PBPK models, the organ of elimination may be a misnomer as elimination via sequestration and degradation occurs in multiple tissues. In the Lankveld model, blood was defined as the elimination organ. This phenomenon is similar to pharmacokinetic models describing protein and peptide disposition (e.g., kinins)[35] where enzymatic degradation by pulmonary enzymes results in effective elimination.

For NPs containing endogenous materials such as iron oxide, models may get more complex as a PBNPK model may be needed to describe initial biodistribution of the NP itself and then a traditional PBPK model used to quantitate incorporation of iron into the body pool. The same could be postulated for any metallic particle (e.g., silver) that would leach out elemental metal. In fact for silver NPs, NPs may undergo dissolution and reassembly across membranes depending on local pH and ionic strengths.[51] For many particles such as QD and silica, particles often persist in the body at the end of an experiment. A limitation of these existing models and studies is that data are not available from long-term trials to accurately estimate the final fate of distributed materials.

8.4 NEED FOR BIOLOGICALLY FRIENDLY CHARACTERIZATION INDICES

As can be seen from this analysis of data needs for developing PBNPK models, there is also a need for including parameters that quantitate interactions of an NP with biological molecules and structures, a phenomenon that seems central to a complete understanding of nanomaterial disposition. How do you take into account the nature of the protein corona, and what specific cellular uptake processes will be involved in a particle's deposition?

Computational approaches to describe nanomaterials using biologically relevant descriptors were recently reviewed[5] as were approaches to predict biocorona formation and composition.[16] Physical and chemical properties important in understanding unique effects of nanomaterials include

- Particle size and its distribution
- Agglomeration state
- Particle shape including aspect ratio
- Crystal structure
- Chemical composition
- Surface area, chemistry, charge, and porosity
- Electronic properties

What differentiates NPs from other molecules is that quantum effects may also be present in true nanosized structures. Thus, as some materials approach the nano scale, properties such as conductivity, reactivity, color, strength, and solubility change. In other cases, physical and chemical properties of the underlying material do not change; only the size effects and particulate structure become important. For all nanomaterials, the surface to volume ratio dramatically increases, and this may be the single attribute that distinguished nano- from larger-sized particles. These changes make extrapolation of properties and biological activity difficult from bulk chemicals of the same composition. When quantitative structure–activity relationship is determined for small molecules, descriptors can be easily calculated based on atomic and molecular properties. Size and surface charge can easily be measured using electron microscopy or methods based on dynamic light scattering. However for nanomaterials, there are significant challenges in using theory to compute accurate properties. Variation in the actual materials manufactured, except for all but very-defined structures (e.g., C_{60} fullerenes, defined carbon nanotubes, and graphene sheets), makes calculation of molecular properties nearly impossible. Similarly, nanomaterials that are essentially stable colloids (e.g., QDs) also make computational approaches difficult. Indirect approaches to probe complex nanostructure properties relevant to biological interactions have been studied. Experimental methods that indirectly probe such properties show promise.

One such approach developed in our laboratory is the multidimensional biological surface adsorption index (BSAI) developed to provide a surrogate metric describing NP interactions to substitute for the partition coefficient used for small molecules.[52,53] The BSAI is a novel experimental-based approach to characterize the adsorption properties of nanomaterials governing behaviors such as aggregation, biocorona formation, and other interactions that dictate biological deposition. The BSAI is composed of five nanodescriptors that represent the surface adsorption forces (hydrophobicity, hydrogen bonding, polarity/polarizability, and lone-pair electrons) that dictate NP interactions under conditions encountered in aqueous biological systems. This results in a five-parameter fingerprint that describes the molecular forces that occur across the NP–water interface. The five descriptors R, P, A, B, and V are strength coefficients representing the strength of the R (excess molar fraction describing lone-pair electrons), π (effective solute dipolarity or polarizability), α (hydrogen bond acidity), β (hydrogen bond basicity), and V (McGowan characteristic that reflects London dispersion or hydrophobicity) molecular forces, respectively. A comparison of three BSAI fingerprints is depicted in Figure 8.3 for three different types of carbon-based nanostructures.

Positive values of these coefficients represent a molecule's tendency to adsorb to the NP surface, whereas negative coefficients represent a tendency to stay in aqueous solutions. These carbon-based NPs all have large V values that tend to indicate they would bind hydrophobic molecules, such as amino acid residues on protein components of the biocorona, or a small molecule encountered in the aqueous environment. It is anticipated that such signatures could be used to describe association constants for the protein–surface NP interaction. The index was successful in predicting log octanol–water partition coefficients as well as adsorption of a variety of small molecules onto multiwalled carbon nanotubes.[52] When this five-dimensional nanodescriptor fingerprint is reduced to two dimensions using principal component analysis, a series of 16 diverse nanomaterials could be classified into distinct clusters based on these surface adsorption properties.[53]

Assuming that the biocorona is a primary determinant of NP biodistribution, and biocorona formation is a dynamic and changing process, any PBNPK model would also have to incorporate the kinetics of NP biocorona formation to be able to accurately predict behavior. Such modeling has begun to be described as the composition of NP biocoronas becomes better studied.[54–56] To date, such models are constructed based on defining association constants between the surface of the NPs and interacting proteins. The models predict rapid association of proteins to immediately form an NP–protein *soft* corona on exposure to proteins that slowly evolve by exchanging with different proteins to form a more stable *hard* biocorona. These observations suggest that depending on the environment of the NP (pH, ionic strength, protein

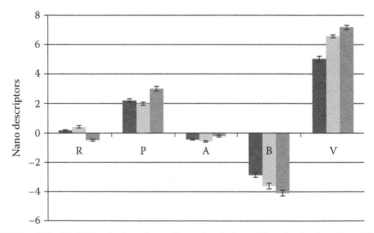

FIGURE 8.3 Relative biological surface adsorption index molecular interactions (R, P, A, B, V) for three types of carbon nanomaterials: single-walled carbon nanotube ■, fullerene ■, and carbon hollow sphere ■. The five descriptors are strength coefficients representing R (lone-pair electrons), P (solute dipolarity or polarizability), A (hydrogen bond acidity), B (hydrogen bond basicity), and V (London dispersion) molecular forces.

makeup, and concentrations), its composition will evolve during its sojourn through the body. The makeup of these particles will also have to be described using distributions because the associations are not deterministic like an organic chemical reaction would be, but rather are probabilistic and determined by properties of the individual NPs as well as variations in the local protein environment.

We postulate that knowledge of the surface adsorption properties embedded in an index such as the BSAI, coupled with classic characterization of the physical (particle size and its distribution, shape, surface area, etc.) and biochemical properties (antibody–antigen interactions, chemical reactivity—oxidation, reduction, complexion, etc.) of the NPs, along with NP–biocorona dynamic models would be required if truly robust PBNPK and toxicodynamic models are to be developed. These parameters would then be embedded into the model to govern uptake or elimination for particles with these characteristics. For example, if biocorona formation targets an NP primarily to the spleen, then a spleen compartment would be added to the scheme in Figure 8.1. If in vitro studies define uptake of particles via vesicular transport mechanisms, then instead of assuming a reversible transport process between the interstitial and intercellular compartments within a tissue, a specific one-way rate process could be included based on parameters obtained from in vitro studies. If a particle traffics in the lymphatic system, lymphatic flow would be added to the basic PBNPK model structure. Another approach is to use in vitro characterization studies to classify the type of NP interactions with biomolecules are expected, based on physical and chemical properties of the NP surface, organ selection for the final model could be estimated. Finally, embedded into the vascular compartment would be an NP–biocorona association model describing the state of the biocorona that would determine which tissues the NP is distributed to.

8.5 CONCLUSION

There remains a serious lack of well-characterized in vivo data to precisely define the structure of PBNPK models needed to predict in vivo ADME parameters across a reasonable number of NPs. The initial PBNPK models developed have been limited to nanosilver and QDs in laboratory rodents. Unfortunately, neither human nor large laboratory animal data are available for comparison. As discussed in Section 8.2.2, the RES shows pronounced species differences that impact on the pattern of nanomaterial biodistribution. However, one of the advantages of developing PBNPK models is that they are directly amenable to making interspecies comparisons should appropriate experiments be conducted.

On the basis of the few PBNPK models published and the unique aspects of biodistribution and elimination presented in the chapter, there are several characteristics of NPs that must be considered when formulating PBNPK models. These are tabulated in Table 8.2.

The fields of nanomedicine and nanotoxicology are rapidly developing and entering a more mature phase where quantitative in vivo studies must be conducted with the eye toward incorporating the latest characterization metrics and mechanisms of NP distribution and elimination, with a goal of predicting their behavior in humans. The use of PBNPK models is a tool that would greatly facilitate this goal.

TABLE 8.2
Factors That Differentiate Nanoparticle and Small Molecule Pharmacokinetic Properties

	Small Molecule	Nanoparticle
Size	Small molecules are normally of a single molecular weight	Most NPs are best described by a size distribution (mean size, size range)
Structural identifiability	The entity being modeled is normally a single molecular structure that could be modified by classical chemical or enzymatic metabolic processes	For NPs, an aggregate of subparticles could change in size and surface characteristics depending on the biological environment in which it is located
Metrics	Solubility and partition coefficients are often embedded in small-molecule kinetic models to predict tissue deposition by estimating a partition coefficient	NPs have different metrics that describe different properties potentially affecting biodistribution (size, shape, complete surface chemistry, aggregation) and thus a simple surrogate metric difficult
Vascular movement	Concept of flow- and permeability-limited compartments used in small chemical models	Blood rheological considerations may be operative in nanoparticle movement through capillary beds
Protein binding	Small molecules may undergo nonspecific protein binding primarily with albumin by an equilibrium process described using classic protein ligand binding models (e.g., Hill model)	NPs may bind to multiple proteins and other macromolecules in a dynamic manner to form a biocorona that could be a primary determinant of distribution and elimination. Models have not yet been generalized in a mathematical presentation useful for describing this
Circulation through body	For the majority of small molecules studied, systemic distribution occurs primarily through the circulatory system	For some NPs, trafficking through the lymphatic system may predominate and thus dictate the structure of pharmacokinetic model used
Tissue trapping	Differential tissue distribution may occur for small molecules that are weak acids or bases due to pH partitioning across membranes because only the noncharged moiety can diffuse across the membrane. This ratio is calculated using the Henderson–Hasselbalch equation	NPs have colloidal properties that result in aggregation or agglomeration depending on local microenvironment (pH, ionic strength) across a membrane. NPs may thus change size and surface properties when they enter tissue sites or cellular compartments (e.g., lysosomes) that influence their movement (e.g., trapping on side of membrane promoting aggregation)

TABLE 8.2 (*Continued*)
Factors That Differentiate Nanoparticle and Small Molecule Pharmacokinetic Properties

	Small Molecule	Nanoparticle
Cellular uptake	In most cases, small molecules are taken up into cells by either diffusion down a concentration gradient or by classic molecular transporter systems (organic acid transporter system, P-glycoprotein, etc.) that have been well described using saturable but reversible models (e.g., Michaelis–Menten)	NPs enter cells via slower and capacity-limited vesicular transport systems with charge and size specificity (e.g., micro- and macro-pinocytosis, membrane rafts) whose kinetic properties have not been well described. Cell egress via exocytosis has not often been modeled
Elimination	Small-molecule elimination and clearance is normally via the liver and kidney through well-described pathways (biliary excretion, hepatic metabolism, tubular transporters, glomerular filtration, etc.)	NPs or drugs embedded or encapsulated in NPs, may be protected from normal clearance mechanisms. NP clearance may be accomplished by degradation in RES cells or if evading this, they must undergo some other particle-dependent degradation (e.g., metal ion dissolution) or not undergo elimination
First-pass effects	The pathway of absorption of a chemical into the body may have a major effect of its subsequent deposition as classically illustrated by first-pass hepatic metabolism	If an absorbed particle forms a tight association with a biomolecule (e.g., surfactant) in the process of absorption, its subsequent deposition could be affected

REFERENCES

1. Riviere JE, Tran CL. Pharmacokinetics of nanomaterials. In Monteiro-Riviere NA, Tran CL, eds. *Nanotoxicology: Characterization, Dosing and Health Effects.* New York: Informa, 2007: 127–40.
2. Lynch I, Cedervall T, Lundqvist M, Cabaleiro-Lago C, Linse S, Dawson KA. The nanoparticle-protein complex as a biological entity; a complex fluids and surface science challenge for the 21st century. *Adv. Colloid Interface Sci.* 2007; 134–135: 167–74.
3. Nel AE, Madler L, Velego D, Xia T, Hoek EMV, Somasundaran P, Klaessig F, Castranova V, Thompson M. Understanding biophysiochemical interactions at the nano-bio interface. *Nat. Mater.* 2009; 8: 543–57.
4. Monopoli MP, Åberg C, Salvati A, Dawson KA. Biomolecular coronas provide the biological identity of nanosized materials. *Nat. Nanotechnol.* 2012; 7: 779–86.
5. Gajewicz A, Rasulev B, Dinadayalane TC, Urbaszek P, Puzyn T, Leszczynska D, Leszczynska J. Advancing risk assessment of engineered nanomaterials: Application of computational approaches. *Adv. Drug Deliv. Rev.* 2012; 64: 1663–93. doi:10.1016/j.addr.2012.05.014.

6. Li SD, Huang L. Pharmacokinetics and biodistribution of nanoparticles. *Mol. Pharm.* 2008; 5: 496–504.
7. Moghimi SM, Hunter AC, Andresen TL. Factors controlling nanoparticle pharmacokinetics: An integrated analysis and perspective. *Annu. Rev. Pharmacol. Toxicol.* 2012; 52: 481–503.
8. Riviere JE. Pharmacokinetics of nanomaterials: An overview of carbon nanotubes, fullerenes and quantum dots. *Wiley Interdiscip. Rev. Nanomed. Nanobiotechnol.* 2009; 1: 26–34.
9. Yang W, Peters JI, Williams RO. Inhaled nanoparticles—A current review. *Int. J. Pharm.* 2008; 356: 239–47.
10. Adair JH, Parette MP, Altinoglu EI, Kester M. Nanoparticulate alternatives for drug delivery. *ACS Nano.* 2010; 4: 4967–70.
11. Des Rieux A, Fieves V, Garinot M, Schneider YJ, Preat V. Nanoparticles as potential oral delivery systems of proteins and vaccines: A mechanistic approach. *J. Control Release.* 2006; 116: 1–27.
12. Monteiro-Riviere NA, Riviere JE. Interactions of nanomaterials with the skin: Aspects of absorption and biodistribution. *Nanotoxicology.* 2009; 3: 188–93.
13. Monteiro-Riviere NA, Wiench K, Landsiedel R, Schulte S, Inman AO, Riviere JE. Safety evaluation of sunscreen formulations containing titanium dioxide and zinc oxide nanoparticles in UVB-exposed skin: An in vitro and in vivo study. *Toxicol. Sci.* 2011; 123: 264–80.
14. Kreyling WG, Semmler M, Erbe F, Mayer P, Takenaka S, Schultz H, Oberdörster G, Ziesenis A. Translocation of ultrafine insoluble iridium particles from the lung epithelium to extrapulmonary organs is size dependent but very low. *J. Toxicol. Environ. Health. A.* 2002; 65: 1513–30.
15. Cevc G, Vierl U. Nanotechnology and the transdermal route: A state of the art review and critical appraisal. *J. Control Release* 2010; 141: 277–99.
16. Walkey CD, Chan WCW. Understanding and controlling the interaction of nanomaterials with proteins in a physiological environment. *Chem. Soc. Rev.* 2012; 41: 2780–99.
17. Cedervall T, Lynch I, Foy M, Berggård T, Donnelly SC, Cagney G, Linse S, Dawson KA. Detailed identification of plasma proteins adsorbed on copolymer nanoparticles. *Angew. Chem. Int. Ed. Engl.* 2007; 46: 5754–6.
18. Pedersen MB, Zhou X, Larsen EKU, Sorensen US, Kjems J. Curvature of synthetic and natural surfaces is an important target in classical pathway complement activation. *J. Immunol.* 2010; 184: 1931–45.
19. Liu D, Mori A, Huang L. Role of liposome size and RES blockade in controlling biodistribution and tissue uptake of GMI-containing liposomes. *Biochim. Biophys. Acta.* 1992; 1104: 95–101.
20. Moreira JN, Gaspar R, Allen TM. Targeting stealth liposomes in a murine model of human small cell lung cancer. *Biochim. Biophys. Acta.* 2001; 1515: 167–76.
21. Geng Y, Dalhaimer P, Cai S, Tsai R, Tewari M, Minko T, Discher DE. Shape effects of filaments versus spherical particles in flow and drug delivery. *Nat. Nanotechnol.* 2007; 2: 249–55.
22. Gentile F, Ferrari M, Decuzzi P. The transport of nanoparticles in blood vessels: The effect of vessel permeability and blood rheology. *Ann. Biomed. Eng.* 2008; 36: 254–61.
23. Yu T, Hubbard D, Ray A, Ghandehari H. In vivo biodistribution and pharmacokinetics of silica nanoparticles as a function of geometry, porosity and surface characteristics. *J. Control Release.* 2012; 163: 46–54.
24. Levchenko TS, Rammohan R, Lukyanov AN, Whiteman KR, Torchilin VP. Liposome clearance in mice: The effect of a separate and combined presence of surface charge and polymer coating. *Int. J. Pharm.* 2002; 240: 95–102.

25. Almeida JPM, Chen AL, Foster A, Drezek R. In vivo biodistribution of nanoparticles. *Nanomedicine (Lond).* 2011; 6: 815–35.
26. Ryman-Rasmussen JP, Riviere JE, Monteiro-Riviere NA. Variables influencing interactions of untargeted quantum dot, nanoparticles with skin cells and identification of biochemical modulators. *Nano Lett.* 2007; 7:1344–8.
27. Shannahan JH, Brown JM, Chen R, Ke PC, Lai X, Mitra S, Witzmann FA. Comparison of nanotube protein corona composition in cell culture media. *Small.* 2013; 9: 2171–81. doi:10.1002/smll.201202243.
28. Monteiro-Riviere NA, Samberg ME, Oldenburg SJ, Riviere JE. Protein binding modulates the cellular uptake of silver nanoparticles into human cells: Implications for in vitro to in vivo extrapolations. *Toxicol. Lett.* 2013; 220: 286–93.
29. Yim YS, Choi JS, Jang SB, Kim GT, Park K, Kim CH, Cheon J, Kim DG. Pharmacokinetic properties and tissue storage of FITC conjugated SA-MnMEIO nanoparticles in mice. *Curr. Appl. Physiol.* 2009; 9: E304–7.
30. Gopee NV, Roberts DW, Webb P, Cozart CR, Siitonen PH, Warbritton, AR, Yu WW, Colvin VL, Walker NJ, Howard PC. Migration of intradermally injected quantum dots to sentinel organs in mice. *Toxicol. Sci.* 2007; 98: 249–57.
31. Soltesz EG, Kim S, Laurence RG, DeGrand AM, Parungo CP, Dor DM, Cohn LH, Bawendi MG, Frangioni JV, Mihaljevic T. Interoperative sentinel lymph node mapping of the lung using near-infrared fluorescent quantum dots. *Ann. Thorac. Surg.* 2005; 79: 269–77.
32. Choi HS, Liu W, Misra P, Tanaka E, Zimmer JP, Ipe BI, Bawendi MG, Frangioni JV. Renal clearance of quantum dots. *Nat. Biotechnol.* 2007; 25: 1165–70.
33. Kagan VE, Konduru NV, Feng W, Allen BL, Conroy J, Volkov Y, Vlasova II et al. Carbon nanotubes degraded by neutrophil myeloperoxidase induces less pulmonary inflammation. *Nat Nanotechnol.* 2010; 5: 354–59.
34. Ette EI, Williams PJ. *Pharmacometrics: The Science of Quantitative Pharmacology.* Hoboken: Wiley, 2007.
35. Riviere JE. *Comparative Pharmacokinetics: Principles, Techniques and Applications,* 2nd ed. Ames: Wiley-Blackwell, 2011.
36. Shargel L, Wu-Pong S, Yu ABC. *Applied Biopharmaceutics and Pharmacokinetics,* 5th ed. New York: McGraw Hill, 2005.
37. Reddy MA, Yang RSH, Clewell HJ, Andersen ME. *Physiologically Based Pharmacokinetic Modeling.* Hoboken: Wiley, 2005.
38. Riviere JE, Bowman KF, Monteiro-Riviere NA, Dix LP, Carver MP. The isolated perfused porcine skin flap (IPPSF). I. A novel in vitro model for percutaneous absorption and cutaneous toxicology studies. *Fundam. Appl. Toxicol.* 1986; 7: 444–53.
39. Riviere JE, Monteiro-Riviere NA. The isolated perfused porcine skin flap as an in vitro model for percutaneous absorption and cutaneous toxicology. *Crit. Rev. Toxicol.* 1991; 21: 329–44.
40. Lee HA, Imran M, Monteiro-Riviere NA, Colvin VL, Wu W, Riviere JE. Biodistribution of quantum dot nanoparticles in perfused skin: Evidence of coating dependency and periodicity in arterial extraction. *Nano Lett.* 2007; 7: 2865–70.
41. Leavens TL, Xia XR, Lee HY, Monteiro-Riviere NA, Brooks JD, Riviere JE. Evaluation of perfused porcine skin as a model system to quantitate tissue distribution of fullerene nanoparticles. *Toxicol. Lett.* 2010; 197: 1–6.
42. Leavens TL, Monteiro-Riviere NA, Inman AO, Brooks JD, Oldenburg SJ, Riviere, JE. In vitro biodistribution of silver nanomaterial in isolated perfused porcine skin flaps. *J. Appl. Toxicol.* 2012; 32: 913–9.
43. Riviere JE, Leavens TL, Brooks JD, Monteiro-Riviere NA. Acute vascular effects of nanoparticle infusion in isolated perfused skin. *Nanomedicine.* 2012; 8: 428–31.

44. Rogers RA, Riviere JE. Pharmacologic modulation of the cutaneous vasculature in the isolated perfused porcine skin flap (IPPSF). *J. Pharm. Sci.* 1994; 83: 1682–89.
45. Yu T, Greish K, McGill LD, Ray A, Ghandehari H. Influence of geometry, porosity, and surface characteristics of silica nanoparticles on acute toxicity: Their vasculature effects and tolerance threshold. *ACS Nano.* 2012; 6: 2289–301.
46. Li M, Al-Jamal K, Kostarelos K, Reineke J. Physiologically based pharmacokinetic modeling of nanoparticles. *ACS Nano.* 2010; 11: 6303–17.
47. Lin P, Chen JW, Chang LW, Wu JP, Redding L, Chang H, Yeh, TK et al. Computational and ultrastructural toxicology of a nanoparticle, Quantum Dot 705, in mice. *Environ. Sci. Technol.* 2008; 42: 6264–70.
48. Yang RS, Chang LW, Wu JP, Tsai MH, Wang HJ, Kuo YC, Yeh TK, Yang CS, Lin P. Persistent tissue kinetics and redistribution of nanoparticles, Quantum Dot 705, in Mice: ICP-MS Quantitative Assessment. *Environ. Health Perspect.* 2007; 115: 1339–43.
49. Lee HA, Leavens T, Mason SE, Monteiro-Riviere NA, Riviere JE. Comparison of quantum dot biodistribution with blood-flow limited physiologically based pharmacokinetic model. *Nano Lett.* 2009; 9: 794–9.
50. Lankveld DPK, Ooman AG, Krystek P, Neigh A, Troost de Jong A, Noorlander CW, Van Eijkeren JCH, Geertsma RE, de Jong WH. The kinetics of the tissue distribution of silver nanoparticles of different sizes. *Biomaterials.* 2010; 31: 8350–61.
51. Liu J, Wang Z, Liu FD, Kanes AB, Hurt RH. Chemical transformation of nanosilver in biological environments. *ACS Nano.* 2012; 6: 9887–99.
52. Xia XR, Monteiro-Riviere NA, Mathur S, Oldenberg S, Fadell B, Riviere JE. Mapping the surface adsorption forces of nanomaterials in biological systems using the biological surface adsorption index (BSAI) approach. *ACS Nano.* 2011; 5: 9074–81.
53. Xia XR, Monteiro-Riviere NA, Riviere JE. An index for characterization of nanomaterials in biological systems. *Nat. Nanotechnol.* 2010; 5: 671–5.
54. Dell'Orco D, Lundqvist M, Oslakovic C, Cedervall T, Linse S. Modeling the time evolution of the nanoparticle-protein corona in a body fluid. *PLoS One.* 2010; 5: e10949.
55. Dell'Orco D, Lundqvist M, Cedervall T, Linse S. Delivery success rate of engineered nanoparticles in the presence of the protein corona: A systems-level screening. *Nanomedicine.* 2012; 8: 1271–81.
56. Sahneh FD, Scoglio C, Riviere JE. Dynamics of nanoparticle-protein corona complex formation: Analytical results from population balance equations. *PLoS One.* 2013; 8: e64690.

Section IV

Methodologies and Techniques

9 Advanced Methodologies and Techniques for Assessing Nanomaterial Toxicity
From Manufacturing to Nanomedicine Screening

Adriele Prina-Mello, Bashir M. Mohamed, Navin K. Verma, Namrata Jain, and Yuri Volkov

CONTENTS

9.1   INTRODUCTION

Future innovation in nanomedicine is expected, in the next 10 years, to deliver solutions to many challenging problems faced by modern medicine in diagnostics and medical imaging of cancer,[1] theranostics,[2,3] pharmaceuticals,[4] and regenerative medicine.[5,6] To accelerate such process, there is the need to overcome some bottlenecks such as toxicity,[7] biocompatibility,[8] pharmacoefficacy,[4,9] and life-cycle assessment[10,11] of the engineered nanomaterials (ENMs) or engineered nanoparticles (ENPs) used as nanocarriers, nanovectors, or probes. Therefore, effective translation from research into industrial, clinical, marketable products is not going to happen until essential effort is invested into the comprehensive environmental, health, and safety characterization and assessment of the ENMs/ENPs to be used.[10]

Based on the definition of the term *nanomaterial* recently adopted by the European Commission (EC),[12] the patent applications falling under the remit of developing ENP applications are expected to be in significant mass production by 2020,[13] with about one-third of patents and start-up companies falling in the biomedical sector.[14] To put this in the present economical context, the invention of nanotechnology products that are publicly available has grown by nearly 621% since 2006 (from 212 to 1317 products), with the largest increase allocated in health and fitness products (from roughly 150 to 738 products).

Leading European organizations involved in this area (e.g., European Technology Platform on Nanomedicine [ETP Nanomedicine]) and consortia (NanoSafety Cluster) have highlighted the insufficient availability of *decision-making tools* for a planned and designed translation of the most promising and innovative ENPs from the lab bench to stakeholder groups (manufacturers, hospitals, national healthcare systems, and consumers) and industries (pharmaceutical and medical devices industries) in the European market. Recently, both the ETP Nanomedicine and the NanoSafety Cluster have identified that the fundamental hurdle of this insufficient translation is linked to the potential adverse unintended effects of ENPs at any stage of their development, manufacturing, and use, where the lack of decision-making tools is mainly associated with incomplete and/or conflicting information on the properties that determine the ENP impact on human health. As compared to the safety of other pharmaceutical products, ENPs have the added complications of their *nano* size that can influence their properties and activities. This concern has been fueled in the recent years by the consideration that ENPs in their final and/or intermediate formulations may significantly differ from their bulk (unmodified and *ex-synthesis*) states.[15–17]

Large efforts have geared toward establishing consistent risk assessment approaches for ENPs and ENMs.[9,18–20] Currently, nanotechnology policies and regulations are encouraging the development and validation of new test models, the implementation of in vitro high-throughput (HTP) screening tools,[21–25] and the assessment of the physicochemical properties of the tested ENPs within the experimental test systems used for the ENP risk assessment.[26–28] For instance, solutions such as HTP screening and high-content screening (HCS) for in vitro assessment envisaged for quantitative testing of different toxicity and biocompatibility endpoints (e.g., cell viability, morphology, cell cycle, reactive oxidative stress, cytotoxicity, genotoxicity, and immunotoxicity) are representing the most suitable in vitro methodologies for predictive assessment, decision-making, and regulatory acceptance of the nanomedicine ENPs and ENMs.

In recent years, several paradigms have emerged in nanotoxicology: (1) the bio–nano interface/protein corona paradigm,[25] (2) the oxidative stress paradigm,[20] and (3) the pathogenic fiber paradigm,[29] to name but a few. Such theoretical paradigms are helpful as platforms from which to interpret a large range of in vitro data in relation to physicochemical properties, thus accelerating the rate of gathering toxicological data on ENPs by predicting their hazard potential based on their structure.

Yet, given the very high uncertainties associated with the ENP impact on human health, research resources should be directed toward creating decision-making tools based on HTP screening, HCS, and nanoscience-based technologies (e.g., atomic force microscopy [AFM] and others) to comprehensively evaluate large panel of ENPs/ENMs in a systematic way following the most stringent pharmaceutical requirements and standards. Furthermore, to comply with the latest EC directives in animal testing and its reduction,[26] future technologies should be based on suitably tested and accepted in vitro models and validated assays for biological monitoring,[17] for assessing, and possibly for tuning the toxicity, drug delivery, and systemic behavior of ENPs in their tailored nanomedical applications.

This chapter describes advanced methodologies based on the extensive experience accumulated in our Center with in vitro HCS and HTP screening, and AFM tools for assessing and potentially tuning the impact of nanomedicine ENPs on human health. This strategy drives the product development decisions and meets the goals of the recent regulations of the Registration, Evaluation, Authorisation and restriction of Chemicals (REACH); Organisation for Economic Co-operation and Development (OECD); European Medicines Agency (EMA); and EC directives and policies, decreasing development costs and failures and leading to targeted and limited animal testing only if absolutely necessary. Outcome of such approach will result in the translation of fully characterized academic research outcome into industrially relevant and transformable/commutable clinical marketable products.[30]

9.2 HIGH-CONTENT SCREENING IN NANOMEDICINE: BACKGROUND

HCS has been widely used in the pharmaceutical industry for large screening drug formulation and lead compound identification to bring forward drug efficacy investigation. HCS represents a major breakthrough in bringing together quantitative fluorescence microscopy, automated cell-based imaging, and computerized image analysis, providing for the first time a fast and convenient means of conducting multiparametric characterization of biological responses through simultaneous assessment of a multiplicity of molecular and cellular targets.[31–35] This technology built around a decade ago has now reached academia with considerable investments in making the technique accessible for smaller scale or pilot projects. The proliferation of HCS systems available in the market is therefore a function of the multitude of primary requirements desired by industry and academia.

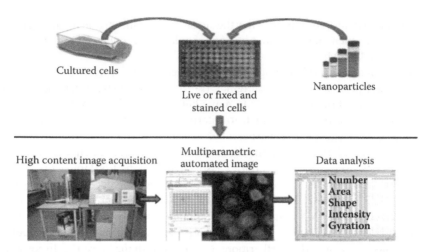

FIGURE 9.1   High-content screening workflow and critical steps for multiparametric analysis.

HCS may include monitoring subcellular localization and redistribution of individual proteins within complex cellular structures such as organelles, thus offering a clear advantage over traditional biochemical or genetic analysis, in order to monitor and characterize physiological responses within the context of the structural and functional networks of cells in both normal and diseased states.[36,37] It presents the capability for precise subcellular imaging, allowing detection of particles in submicrometer range (e.g., quantum dots [QDs] and silica nanoparticles),[38,39] identification of single organelles, tracking and localizing molecular targets[38,40] within live cells providing a high dynamic range. HCS optical operation lies in the range of 350–700 nm wavelengths (excitation wavelengths), which have allowed the commercial development of large number of multiplexing and multicolored assays for diagnostic, screening, and monitoring.

The opportunities offered by such powerful automated image acquisition technology, associated bioinformatics, algorithm for pattern recognition identification, and data mining analysis (Figure 9.1) are extremely powerful in a number of diverse assays, including (1) assay/marker-detection development, (2) lead–ENP candidate screening, (3) toxicity assessment, (4) biocompatibility, (5) cellular uptake and mechanistic investigation, (6) safety assessment, (7) regulatory compliance, (8) batch-to-batch variability, (9) quality control, (10) environmental assessment, and (11) life-cycle assessment.

All these aspects are mandatory in the process of future innovative nanomedicine development.

9.2.1   HIGH-CONTENT SCREENING IN NANOMEDICINE: ENABLING PRODUCT TRANSLATION

Qualitative and quantitative assessment of ENPs by gold standard techniques are often carried out manually or by labor-intensive spectroscopy or image analysis techniques, which are indirectly increasing the overall cost per well plate. HCS and analysis, as opposed to high throughput, is an approach to gain maximum information on many different levels.[31,32,36] This results in a dedicated platform for the assessment of performance of fixed cell, live cell, and kinetic cellular analysis within controlled environmental conditions (temperature, CO_2, and humidity control). Furthermore, multiplex assays can be performed, which is ideal for mapping sequences of events, with minimally automated autofocusing and thereby reduced photodamage and thus phototoxicity.

Automating of cell analysis reduces the probability to introduce systemic errors, operator-dependent problems, or bias problems. Moreover, multiple recording, scanning, or analysis can introduce single-cell targeting or well-plate analysis as a quantitative statistical dataset to be then postprocessed with pattern recognition bio-algorithms (Figure 9.2).

To study nanoparticle–cell interactions, HCS approach offers an advanced eligibility to set up a well-documented library of the key biological indicators of cellular functions, including cell population density, cellular morphology, membrane permeability, lysosomal mass/pH, activation of transcription factor, mitochondrial membrane potential changes, oxidative stress monitoring, and posttranslational modification as also reported in previous works.[38,40–42]

Furthermore, using pharmaceutical methodological approaches, commonly known as quantitative structure–activity relationship (QSAR) or quantitative structure–property relationship (QSPR), allows the determination of the relationships and

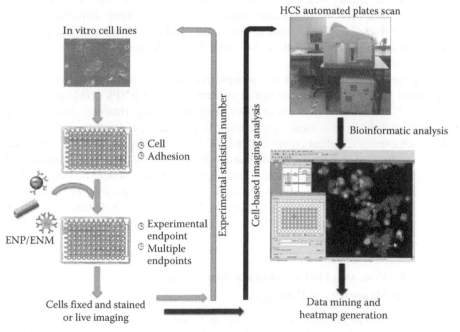

FIGURE 9.2 High-content screening (HCS) and analysis: experimental workflow divided into building statistical significance (light gray arrow path) and achieving robust dataset for data mining (black arrow path).

endpoint responses associated with the key structural properties of the ENPs and ENMs for biomedical use. However, when assessing the translation of these ENPs/ENMs as end products to clinical markets, their endpoints and efficacy should be systematically investigated on a relationship between surface properties and ENP/ENM toxicity behavior by using an approach based on quantitative nanostructure–toxicity relationship (QNTR).[17] Endpoints should be investigated by incremental modification of the toxic behavior of well-characterized ENPs or ENMs when compared to relevant controls.

Categorization and grouping of the data is the next step that should take into account the following:

ENPs/ENMs: their structural properties such as their composition, size, surface coating and functionalization, and surface charge.

Specific cell model adopted: the response of each biological indicator chosen when associated to the relevant structural properties under examination.

Experimental design: the minimal requirement is to have positive and negative controls, incubation endpoints, and incubation doses.

9.2.2 IN VITRO HIGH-CONTENT SCREENING ASSESSMENT OF NANOMEDICAL PRODUCTS

ENPs, such as silica oxide (SiO_2NP), iron oxide (Fe_3O_4NP), and QDs, have been applied in nanomedicine in disease diagnostics, imaging, drug delivery, and biosensors development. Thus, a mechanistic and QNTR evaluation of the potential biological and toxic effects of these and other ENPs becomes crucial to assess their safe applicability limits.

To generate and understand the nanoparticle and cell interactions, several cell lines need to be treated with a range of concentrations of ENPs/ENMs at different intervals (from minutes to hours or days). Appropriate experimental design is then essential to achieve statistical significance. For instance, if working with 96-well plate each experimental data point will need to be designed to have at least triplicates and been repeated in at least three different batches. To ensure reliable identification of toxic events at the cellular level, at least four parameters should be investigated, two qualitative and two quantitative.[43] HCS is particularly suitable for the in vitro investigation of toxicity related to QD uptake,[44] SiO_2 subthreshold toxicity,[38] Fe_3O_4 assessment as theranostic ENPs,[45] single-walled carbon nanotubes as diagnostic probe (Figure 9.3).[40,46]

9.2.3 EX VIVO AND HISTOPATHOLOGY ASSESSMENT OF NANOMEDICAL PRODUCTS

The importance of detecting ENPs/ENMs in ex vivo or tissue samples of different nature is also fundamental when exploring the potential impact of nanomedicine clinical products. Defining the quantitative relationship between postexposure or tissue sample and the ENP/ENM structure–activity relationship creates the basis for standard operating procedures for the human health, risk assessment, reducing

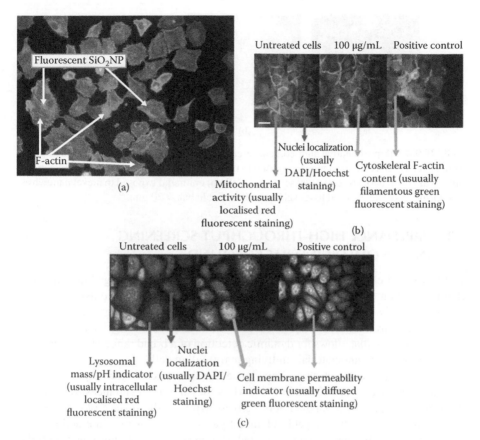

FIGURE 9.3 Representative high-content screening (HCS) images and multiparametric assessment of in vitro cell model exposed to SiO_2 engineered nanoparticles (ENPs) used as nanomedicine carrier. Images are (a) TRITC-labeled SiO_2NP internalization into AGS cell culture model, localization against cytoskeletal F-actin staining. Multiparametric cytotoxicity for untreated cells, 100 µg/mL dose of ENPs and positive control shows by (b) cytoskeletal F-actin and mitochondrial activity and (c) lysosomal mass/pH and localized cellular permeability.

screening costs, and animal experiments. In adherence with the EC directives focused on animal's alternatives, high-content and HTP screening represent a suitable solution for multicellular assessment.[26,47]

Thus, the ability to optically distinguish, threshold, and quantify, in automated way, biologically relevant tissues, such as biopsy, creates the opportunity to comply with histopathological gold standards. It also allows for multiplexed approaches such as conventional tissue staining (e.g., hematoxylin and eosin stains) combined to cellular and subcellular labeling with specific pathway targets (e.g., antibodies and fragment markers) (Figure 9.4). The comprehensive automated analysis of cell morphology and intracellular structures of individual cells in tissue sections enables the development of marketable ENP-enabled products and devices for early-stage clinical testing and further applications.

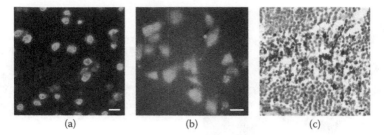

(a)   (b)   (c)

FIGURE 9.4 Representative high-content screening (HCS) images for postexposure assessment of cells and tissue from mouse exposed to nanomaterials for nanomedicine applications. Images are (a) phagocytic cells exposed to SiO$_2$NP, (b) epithelial exposed to nickel nanowires, and (c) spleen tissue exposed to nickel nanowires. Scale bar = 20 μm.

9.3 IMPEDANCE HIGH-THROUGHPUT SCREENING IN NANOMEDICINE: BACKGROUND

The cell-based electrical impedance sensing technology has emerged as a valuable characterization tool for cytotoxicity detection in safety and hazard assessments of ENMs and early-stage efficacy testing of nanoparticle-enabled drugs. In this in vitro label-free noninvasive HTP biophysical assay, cell responses to toxins can be monitored in real-time that allows for dynamic detection of a broad range of physiological responses to toxic nanomaterials in living mammalian cells.

The electrical cell–substrate impedance sensing (ECIS) technique was first described by Giaever and Keese in 1984, when they observed and measured fluctuations in impedance by a population of cells growing on the surface of electrodes.[48] The technology, which is capable of investigating cell–substrate interactions in a quantitative way, has now been recognized as a powerful tool to investigate cellular properties and physiological functions of the cells. To date, impedance measurements have successfully been used to study attachment and spreading of cultured cells,[49] to quantify effects of chemical compounds including cytotoxicity[50] and apoptosis,[51] and to analyze cell migration in wound-healing assays.[52,53] Of particular interest, the use of ECIS technology to access cytotoxic impact on ENPs/ENMs and nano-enabled therapeutic agents is becoming increasingly important. For example, Verma et al.[54–56] have shown the cytotoxicity of various types of nanoclays and nanoparticles in cultured human cells. Additional studies have also reported the applicability of this technique in toxicity testing of metal oxide nanoparticles, inorganic nanomaterials, and therapeutic compounds.[57,58]

9.3.1 ELECTRICAL CELL–SUBSTRATE IMPEDANCE SENSING IN NANOMEDICINE: REAL-TIME SCREENING PRINCIPLES

The basic principle of the impedance sensing technique is to monitor the changes in electric impedance induced by the interaction between test cells and the electrode. Electric impedance of an electrode is primarily determined by the ion environment both at the electrode–solution interface and in the bulk solution or tissue culture medium.[59–61] Upon the application of an electrical field, ions undergo field-directed movement and concentration gradient-driven diffusion, leading to frequency-dependent impedance dispersion. Cells behave as dielectric particles due to the insulating properties of their membranes. Thus, the presence of cells increases the electrical current flow between the electrodes when the cells attach and spread on them. This leads to measureable changes in impedance that increases until cells are fully spread. In other word, the presence of the cells will affect the local ionic environment at the electrode–solution interface, leading to an increase in the electrode impedance. The more the cells are attached to the culture plate (or electrode), the higher the impedance that would result. Changes in cell shape, motion, or both cause corresponding changes in the current flow between the cells and the substratum (i.e., surface of the tissue culture plate over which a cell moves or upon which a cell grows). As the shape of mammalian cells is very sensitive to metabolic alterations as well as to chemical, biological, or physical stimuli including toxic nanomaterials resulting changes in impedance, this technique is applied to quantify cell responses in different experimental settings. Furthermore, the impedance also depends on the extent to which cells attach to the electrodes. For example, if cells spread, there will be a greater cell/electrode contact area, resulting in larger impedance. Thus, the cell biological status including cell viability, cell number, cell morphology, and cell adhesion all affect the measurement of electrode impedance that is reflected by cell index (CI) (described later) on the impedance sensing device. Therefore, a dynamic pattern of a given CI curve may indicate sophisticated physiological and pathological responses of the living cells to a given toxic compound.

The new generation impedance sensing devices such as xCELLigence System (Hoffmann-La Roche, Switzerland) use a circle-on-line microelectrode array that occupies about 80% of the bottom surface area in the wells of a 96-well plate. This allows a large percentage of the cells added into the wells to be monitored and minimizes the well-to-well variation in the detected impedance signals. As this technique monitors cellular activities through weak electrical signals without any need for extra cell labeling, it is considered to be a noninvasive method and be ideal for continuous and label-free cell assays. Thus, impedance measurements can be used to track nanoparticle toxicity over time as well as to identify critical periods following cellular exposure to nano-enabled drugs as shown in its experimental implementation in Figure 9.5.

9.3.2 ELECTRICAL CELL–SUBSTRATE IMPEDANCE SENSING IN NANOMEDICINE: CELL INDEX AS DECISION-MAKING PARAMETER

The CI is an arbitrary dimensionless parameter derived to represent cell status based on the relative change in measured electrical impedance. It is calculated by the following equation:

$$CI = \max_{i=1...N} \left[\frac{R_{cell}(f_i)}{R_b(f_i)} - 1 \right] \tag{9.1}$$

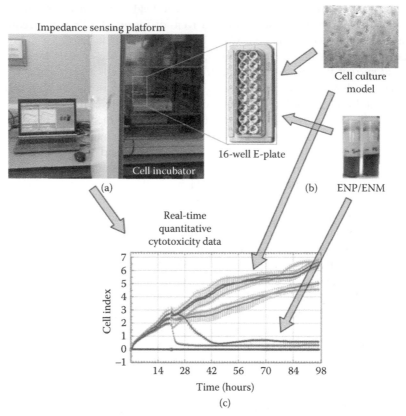

FIGURE 9.5 Electrical cell–substrate impedance sensing (ECIS) experimental workflow: (a) technological platform with sensing unit located inside a tissue culture incubator, (b) a 16-well format impedance plate, and (c) automatic measurement and analysis of cell index (CI) in real-time displayed on the software user interface. Data illustrate the cellular behavior and toxicity linked to engineered nanomaterials or engineered nanoparticles (ENMs/ENPs) exposure.

where $R_{cell}(f)$ is frequency-dependent electrode impedance with cells present in a well; $R_b(f)$ frequency-dependent electrode impedance without cells present in a well; and N the number of the frequency points at which the impedance is measured.

1. When cells are not present on the electrodes or are not well adhered onto the electrodes, then $R_{cell}(f)$ is the same as $R_b(f)$, leading to CI = 0.
2. Under the same physiological conditions, when more cells attach onto the electrodes, then the impedance values are larger. A large $R_{cell}(f)$ value leads to a larger CI, that is, CI > 0.

A schematic diagram of the system and its working principle are shown in Figure 9.6.

For the same number of cells attached to the electrodes, change in a cell status, such as cell morphology, cell adhesion, or cell viability, will lead to a change in CI. For example, an increase in cell adhesion or spread leading to a larger cell/electrode

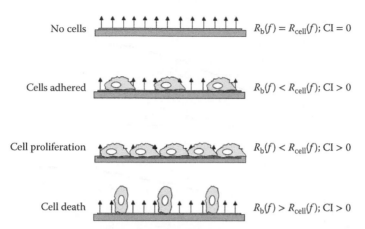

FIGURE 9.6 Electrical cell–substrate impedance sensing (ECIS). Cell index relationship linked to cellular adhesion, proliferation, morphology, and viability.

contact area will lead to an increase in $R_{cell}(f)$ and a larger CI. On the other hand, cell death or toxicity-induced cell detachment or cells rounding up will lead to a smaller $R_{cell}(f)$ and hence smaller CI. Thus, CI is a quantitative measurement that reflects the number of cells attached to the microelectrode surface as well as the nature of cells (e.g., cell adhesion, spreading, and morphology). The applicability of ECIS system and CI measurements to investigate potential toxicity of nanoparticle in question is shown in Figure 9.7. The time-course of CI detected by a typical ECIS system when cells are seeded into the measuring arrays will show an increasing trend (Figure 9.7, shown as light gray line). Following cell treatment with nanoparticles or drug-loaded nanocarriers (Figure 9.7, shown as dark gray line), the CI values will change depending on their toxicity giving a profile of Figure 9.7a for nontoxic, Figure 9.7b for partially toxic, and Figure 9.7c for highly toxic materials.

9.4 ATOMIC FORCE MICROSCOPY IN NANOMEDICINE

As an important aspect in nanomedicine, understanding nanomaterial–cell interactions is imperative to fully explore the efficiency of functional ENPs acting as nanocarriers or drug delivery system.[62] AFM provides a versatile noninvasive platform in nanomedicine with the combination of nanometer-scale resolution and a unique capacity to visualize single biomolecules in their native environment or ENPs within their exposure environmental conditions.[63] This enables addressing key biological questions at nanoscale level such as single-cell-based assays and ultrasensitive biological response to ENP or ENP biomarker surface coating, which could be targeted to quantify cell adhesion and signaling, embryonic and tissue development, cell division and shape, and microbial pathogenesis. AFM is based on detection of attractive or repulsive forces acting between a sharp probe (in order of a few nanometers), known as AFM tip, and the sample's surface (Figure 9.8a and b). The tip is mounted at the very end of a flexible, microscale cantilever. Repulsive force results from the overlapping of electron orbitals between the atoms of the tip and the sample, where attractive force (van der

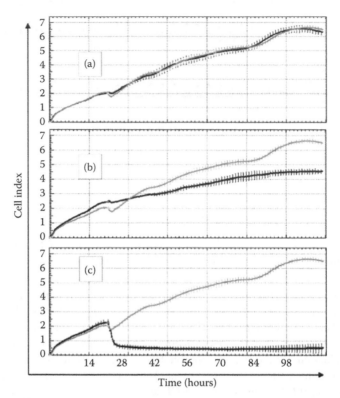

FIGURE 9.7 Electrical cell–substrate impedance sensing (ECIS) cell index plotted against time. Quantitative response of cells exposed to (a) nontoxic engineered nanoparticle (ENP)-coated drug (dark gray line), (b) partially toxic ENP coated drug (dark gray line), and (c) toxic ENP-coated drug (dark gray line) is plotted against control (light gray line).

Waals interaction) is primarily due to dipole–dipole interactions. These forces result in the deflection of a cantilever that is detected by means of a laser spot on a quadrant photodiode. A feedback control maintains the tip–sample separation via cantilever deflection to maintain the constant force experienced by the tip. The deflection signal is recorded digitally and can be visualized on a computer in real-time.

As an imaging tool, AFM determines supramolecular architecture of biological surfaces and their individual constituents. Hence, it acts as a powerful technique for in situ biological detection, an example being the identification and monitoring of oncogenically transformed cell in terms of their growth, morphology, cell–cell interaction, organization of cytoskeleton, and interaction with extracellular matrix. Real-time imaging of cells can provide novel insight into dynamic processes, such as structural changes that are caused by growth or nanoparticle interactions. A recent example revealed that shear stress (SS)-induced cytoskeletal reorganization (Figure 9.9b) aids endocytosis of QDs in human umbilical vein endothelial cells (HUVECs).[64] HUVECs that were not subjected to SS showed a flattened morphology with a smooth surface (Figure 9.9a), whereas their exposure to SS induced the formation of membrane ruffles and stress fibers (Figure 9.9b).

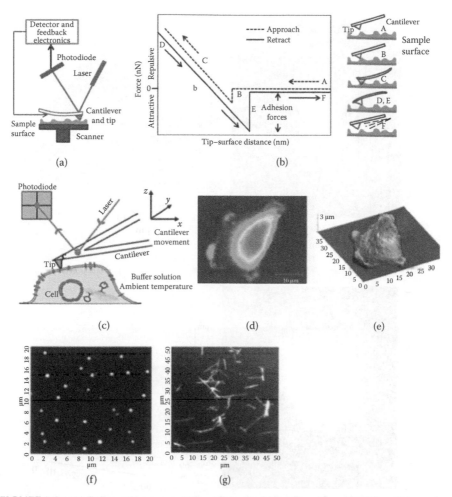

FIGURE 9.8 (a) Schematic representation of atomic force microscopy (AFM) working principle, (b) ideal force–distance curve describing a single approach and retract cycle between the tip and the sample. AFM-based nanoscopy (c) AFM works by sensing piconewton forces between a sharp tip and the cell membrane surface. (d) AFM image of A549 cell line, lung cancer cell line. AFM is also used for (e) three-dimensional profiling of engineered nanoparticles (ENPs) or surface-coated ENPs, for instance (f) SiO$_2$ ENPs ~600 nm, and (g) SiO$_2$ nanowires ~500 nm diameter.

In contrast to the application of AFM for structural studies on nucleic acids, DNA complexes, and various cellular structures, another important feature is the possibility to use it as a force measuring device. Although classic fluorescence microscopy and spectroscopy methods can be used to detect real-time position, distance, distribution, and dynamics of single molecules, force spectroscopy allows manipulation and characterization of mechanical properties, functional state, confirmations, and interactions of biological system to molecular recognition. Furthermore, using AFM also for nanoindentation and nanomechanical analysis of cells or molecular structures at subcellular

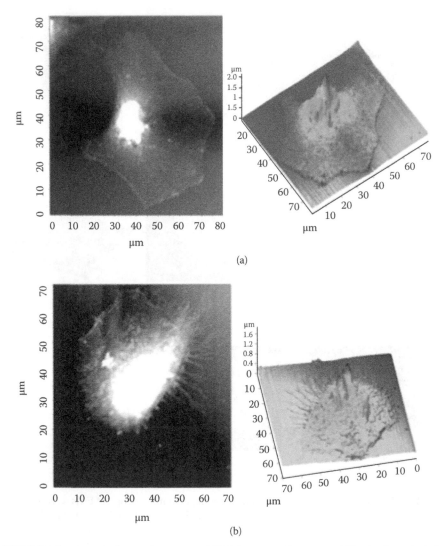

FIGURE 9.9 Atomic force microscopy (AFM) images of human umbilical vein endothelial cells (HUVEC) and their corresponding three-dimensional profile (a) without shear stress and (b) with shear stress.

level is an emerging tool in nanomedicine, particularly where the quantification of the atomic mechanical properties is important for the development of ultrasensitive sensor or highly specific molecular target such as siRNA, DNA, and aptamers of fragment antibodies.[65,66] AFM cantilevers serve as soft nanoindentors allowing for local testing of small and inhomogeneous materials such as cells, tissues, DNA strand, or molecular markers (e.g., antibodies, peptides, and aptamers).[67,68] Cantilevers are exploited as quantitative nanosensors for several applications due to their extreme sensitivity and functionalization opportunities that make it a versatile tool for diverse applications including detection of complementary structures (Figure 9.10), for instance, linking of

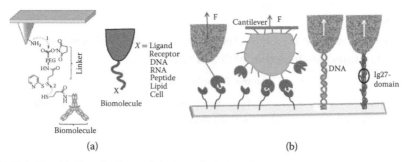

(a)   (b)

FIGURE 9.10 (a) Chemically modified atomic force microscopy (AFM) probe. (b) Applying AFM to probe interaction forces of single biomolecules. These examples measure ligand–receptor interactions in their isolated form (left) and embedded in their cellular environment (probe replaced by a biological cell); stretching of a DNA molecule; unfolding of Ig27-titin.

specific antibodies to the tip via a spacer allows the recognition of its antigen on the surface of a specimen.[69]

Therefore, the development of force spectroscopy can quantitatively measure the interaction of nanomaterials with cell membrane providing insight into the role of these interactions on cellular uptake and intracellular disposition of nanomaterials generating a basis for new diagnostic and therapeutic approaches. Prominent applications of these AFM-based force spectroscopy include measuring interaction forces and dynamics between individual pairs of ligands and receptors, quantifying cell adhesion force from single molecule to the cellular level, deciphering pathways of protein folding and unfolding, approaching molecular mechanisms regulating cell mechanism, measuring single molecules binding to a target protein or nucleic acid, and mapping the spatial distribution of cell-surface receptors.

Thus, AFM represents a rapidly developing technique allowing to go beyond morphology and investigate mechanical (such as elastic and adhesive), electrical, and magnetic properties of a material, in addition to its structure as indicated in Table 9.1.

9.5 NANOMEDICINE METHODOLOGY FOR TRANSLATION: ADVANTAGES AND LIMITATIONS

The measurement of cell viability, proliferation, and changes in cellular physiology for nanotoxicity studies for nanomedicine product translation to the market by HCS, ECIS, and AFM approach has several advantages over conventional methods. Automatic monitoring of cell processes in real-time, such as cell attachment, proliferation, and death, is one of the advantages of the ECIS system. Conventional methods, such as the MTT assay and lactate dehydrogenase (LDH) tests, use the mitochondrial reduction of tetrazolium salts into an insoluble dye and enzyme LDH release, respectively, as markers for cell viability. Although these methods provide a general sense of cytotoxicity, they show results only at a final time point.[70,71] As a result, the kinetic model ADME (absorption, distribution, metabolism and excretion) of the nanoparticle uptake is not usually achievable by the conventional methods. Following biological exposure, the ENMs/ENPs may be transported across cell membranes, especially into mitochondria, causing internal damage that may affect cell behavior and over time, lead to cell

TABLE 9.1

Key Attributes of AFM for Characterization of Engineered Nano Particles and Single-Molecule Detection with Advantages

Atomic Force Microscopy		
Imaging Tool	**Force Spectroscopy**	**Advantages**
Physical Properties:	*Single-Molecule Interactions:*	Easy sample preparation
Size		
Morphology		
Roughness	Antigen–antibody	Image both conductive and nonconductive samples
Surface texture	Receptor–ligand	
	Protein–protein	
Statistical Information:	Protein–DNA	Nanometer scale resolution
Particle counts	Measure interactions forces	
Size distribution		Works in air, vacuum, special air, or fluid environment
Surface area distribution	Protein folding	
Volume distribution	Mechanical properties of cells and its organelles	
Range:		Provides three-dimensional surface profile
Particle size: 1 nm to 8 μm		
Scan range: up to 100 μm		

death. The real-time monitoring of a dynamic cell response to a given toxicant by ECIS methodology provides advantages, which are difficult to achieve by the endpoint assays unless multiparametric analysis is used, as for the HCS methodology. This provides high information content in addition to the information about cell viability change.

For HTP cytotoxicity assessment applications, the speed, simplicity, reproducibility, and reliability are highly desirable. The impedance sensing approach has a much simpler assay protocol than other cell-based assays, such as the microscopy-based HCS, which requires a more scrupulous and time-consuming process, for example, staining, distaining, detection, and subsequent analysis. Thus, the impedance sensing technology is best suited for the continuous detection of possible toxicity caused by ENMs/ENPs, whereas HCS provides a robust platform for large, cost-effective, and multiplexed monitoring of possible ENM/ENP toxicity in nanomedicine research for health and safety assessments.

The ability of AFM to simultaneously detect morphological, topographical, roughness, and micromechanical changes demonstrates the utility for capturing multiple parameter measurements in a single analysis. AFM comes as detailed mapping for fine-tuning of the properties when the system is dynamic and need to be analyzed in their physiological conditions. Being a highly sensitive tool, single-molecule detection becomes possible using force spectroscopy, which allows measuring various mechanical properties of material, such as elasticity and adhesion.

There is virtually no sample preparation or special treatment (such as metal/carbon coatings), apart from one requirement; the object to study should be attached to the sample surface. In addition, AFM provides true three-dimensional surface topographical information. Although most AFM systems are affordable and user-friendly, accurate data collection and interpretation are not trivial undertakings and require extensive expertise and a great deal of patience, especially when dealing with soft samples. Another problem with AFM is rather limited image resolution (around 50–100 nm), typically achieved on soft biological samples in liquid environment. Also, being a surface-sensitive technique, interior of any material cannot be accessed by conventional AFM. Finally, time resolution is a crucial factor that currently limits imaging studies using AFM. The open architecture and invariance to operating environment allows integration of AFM with other techniques for simultaneous structure–function correlation studies, such as the integration of AFM with a fluorescence microscope, including confocal microscope, total internal reflection fluorescence microscope, fluorescence resonance energy transfer microscope, and optoelectronic manipulators. This flexibility offers a unique opportunity to develop AFM into a powerful multidimensional platform of technologies to advance nanomedicine in its every sphere allowing investigation of nanoscale structure and function in complex systems.

9.6 MULTIPARAMETRIC STATISTICAL ANALYSIS AND GRAPHICAL DISPLAY: HEATMAP TABLES

On the basis of the experience derived from the pharmaceutical industry where large compound screening or lead compound identification generates large volume of data, the mining and handling of such information is important in high content, high throughput, or AFM to be able to carry out multiparametric statistical analysis. This is necessary for the identification of experimental leads and also for significant and comprehensive graphical output representation of the analyzed data.

Multiparametric statistical analysis is often carried out on independent grouped dataset after pattern recognition bioinformatics on independent and normally distributed dataset. Principal component analysis (PCA) could be a suitable alternative as the best to explain the variance of multivariate dataset. This allows for the reduction of multidimensional dataset into lower-dimensional spaces by taking only the first principal components to achieve simplified graphical plots.

Histogram data plot, with associated error bars, is also the most common way to display data; however, this suffers from visual-impact immediacy and also, for the case of multiparametric dataset, it will require a large number of graphical plots to comprehensively show the whole data and dependencies. Another way to represent dataset is by scatter plot with interpolated trend lines; this graphical output is very useful when the dataset to be presented is limited.

The generation of heatmap tables is by far the most effective and intuitive way to graphically display multiparametric dataset by means of a colorimetric gradient (Figure 9.11). For instance, in the case of nanotoxicology screening based on three cellular parameter evaluations, the colorimetric changes in cell count, lysosomal mass/pH, and cell membrane permeability can range from dark green (values lower than 15% change from the

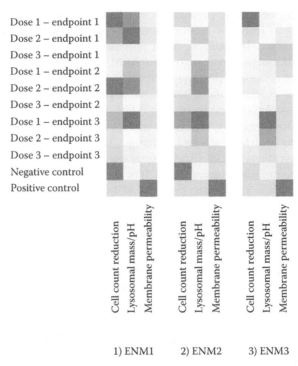

FIGURE 9.11 Representative HCS heatmap graphical table. Each data point is represented by calculated z-score for each associated multiparametric dataset. In this example, three engineered nanomaterials or engineered nanoparticles (ENPs/ENMs), three doses, three time endpoints, one cell line, and three physiological parameters (cell count, lysosomal mass/pH, and cell membrane permeability acquired in different fluorescence channels) are shown. Equivalent histogram representation would have required a minimal of 12 plots each made of 12 histogram bars.

maximum value measured) to bright green (30%), yellow (50%), bright orange (60%), dark orange (75%), and finally to red (>75% change from maximum value).

Clustering, robustness, z-score, and covariance calculations are more sophisticated tools to use for the optimization, identification of correlation, or normalization of multiparametric dataset.

The proliferation of open-source data mining, PCA software, and codes has allowed for the proliferation of several other solutions, among many a bioinformatics initiative supported by Trinity College Dublin–IMM has brought to the development of screening module HiTS based on KNIME modular open-source data manipulation and visualization program (http://KNIME.org).

9.7 CONCLUSIONS AND FUTURE DIRECTIONS

The use of pharmaceutically driven methodology for the assessment of ENPs/ENMs for marketable products has a clear potential to introduce a quicker translation strategy into pre- and post-clinically approved nanomedicine tools. In conclusion, progress in nanotoxicology toward an integrated nanomedicine testing strategy will enable us to define the relationship between ENPs/ENMs and the biological models used to create the basis for the robust tuning of ENP/ENM impact on human health, reducing screening costs, animal refinement or replacement, and life-cycle assessment. Successful implications of such strategy will deliver commercially viable tools with direct impact on the patenting/licensing phase that will allow categorization and labeling of promising ENM-based nanomedical marketable products.

REFERENCES

1. Liu, Y.; Solomon, M.; Achilefu, S. Perspectives and potential applications of nanomedicine in breast and prostate cancer. *Medicinal Research Reviews* 2013, 33, 3–32.
2. Chen, X.; Gambhir, S. S.; Cheon, J. Theranostic nanomedicine. *Accounts of Chemical Research* 2011, 44, 841.
3. Lammers, T.; Aime, S.; Hennink, W. E.; Storm, G.; Kiessling, F. Theranostic nanomedicine. *Accounts of Chemical Research* 2011, 44, 1029–1038.
4. Vizirianakis, I. S. Nanomedicine and personalized medicine toward the application of pharmacotyping in clinical practice to improve drug-delivery outcomes. *Nanomedicine: Nanotechnology, Biology, and Medicine* 2011, 7, 11–17.
5. Grattoni, A.; Tasciotti, E.; Fine, D.; Fernandez-Moure, J. S.; Sakamoto, J.; Hu, Y.; Weiner, B.; Ferrari, M.; Parazynski, S. Nanotechnologies and regenerative medical approaches for space and terrestrial medicine. *Aviation, Space, and Environmental Medicine* 2012, 83, 1025–1036.
6. Prescott, C. Regenerative nanomedicines: An emerging investment prospective? *Journal of the Royal Society, Interface* 2010, 7 Suppl 6, S783–S787.
7. Seaton, A.; Tran, L.; Aitken, R.; Donaldson, K. Nanoparticles, human health hazard and regulation. *Journal of the Royal Society, Interface* 2010, 7 Suppl 1, S119–S129.
8. Naahidi, S.; Jafari, M.; Edalat, F.; Raymond, K.; Khademhosseini, A.; Chen, P., Biocompatibility of engineered nanoparticles for drug delivery. *Journal of Controlled Release: Official Journal of the Controlled Release Society* 2013, 166, 182–194.
9. Eaton, M. A. W. How do we develop nanopharmaceuticals under open innovation? *Nanomedicine: Nanotechnology, Biology, and Medicine* 2011, 7, 371–375.
10. Linkov, I.; Bates, M. E.; Canis, L. J.; Seager, T. P.; Keisler, J. M. A decision-directed approach for prioritizing research into the impact of nanomaterials on the environment and human health. *Nature Nanotechnology* 2011, 6, 784–787.
11. Linkov, I.; Satterstrom, F. K.; Corey, L. M. Nanotoxicology and nanomedicine: Making hard decisions. *Nanomedicine: Nanotechnology, Biology, and Medicine* 2008, 4, 167–171.
12. Potocnik, J. European Commission recommendation on the definition of nanomaterial. *Official Journal of the European Union* 2011, L275/38–L275/40, 3.
13. Dang, Y.; Zhang, Y.; Fan, L.; Chen, H.; Roco, M. C. Trends in worldwide nanotechnology patent applications: 1991 to 2008. *Journal of Nanoparticle Research* 2010, 12, 687–706.
14. Marshall, J., Draft guidelines for nanomedicine unveiled. *Nature* 2011, doi:10.1038/news.2011.562. Available at: http://www.nature.com/news/2011/110928/full/news.2011.562.html.
15. Reich, E. S. Nano rules fall foul of data gap. *Nature* 2011, 480, 160–161.
16. Schütz, C. A.; Juillerat-Jeanneret, L.; Mueller, H.; Lynch, I.; Riediker, M. Therapeutic nanoparticles in clinics and under clinical evaluation. *Nanomedicine* 2013, 8, 449–467.
17. Movia, D.; Poland, C.; Tran, L.; Volkov, Y.; Prina-Mello, A. Multilayered nanoparticles for personalized medicine: Translation into clinical markets. In , Bawa, R.; Audette, G. F.; Rubinstein, I. eds. *Handbook of Clinical Nanomedicine: From Bench to Bedside.* Pan Stanford Publishing: Singapore, 2013; Vol. 1, p 17.

18. Join the dialogue. *Nature Nanotechnology* 2012, 7, 545. Available at: http://www.nature.com/nnano/journal/v7/n9/full/nnano.2012.150.html.
19. Schrurs, F.; Lison, D. Focusing the research efforts. *Nature Nanotechnology* 2012, 7, 546–548.
20. Hartung, T. Toxicology for the twenty-first century. *Nature* 2009, 460, 208–212.
21. Kievit, F. M.; Zhang, M. Surface engineering of iron oxide nanoparticles for targeted cancer therapy. *Accounts of Chemical Research* 2011, 44, 853–862.
22. Ventola, C. L. The nanomedicine revolution: Part 1: Emerging concepts. *P & T: A Peer-Reviewed Journal for Formulary Management* 2012, 37, 512–525.
23. Hoffmann, S.; Cole, T.; Hartung, T., Skin irritation: Prevalence, variability, and regulatory classification of existing in vivo data from industrial chemicals. *Regulatory Toxicology and Pharmacology Journal* 2005, 41, 159–166.
24. Service, R. F. Nanotechnology. Can high-speed tests sort out which nanomaterials are safe? *Science* 2008, 321, 1036–1037.
25. Meng, H.; Xia, T.; George, S.; Nel, A. E. A predictive toxicological paradigm for the safety assessment of nanomaterials. *ACS Nano* 2009, 3, 1620–1627.
26. Höfer, T.; Gerner, I.; Gundert-Remy, U.; Liebsch, M.; Schulte, A.; Spielmann, H.; Vogel, R.; Wettig, K. Animal testing and alternative approaches for the human health risk assessment under the proposed new European chemicals regulation. *Archives of Toxicology* 2004, 78, 549–564.
27. Stone, V.; Johnston, H.; Schins, R. P. F. Development of in vitro systems for nanotoxicology: Methodological considerations. *Critical Reviews in Toxicology* 2009, 39, 613–626.
28. Jones, C. F.; Grainger, D. W. In vitro assessments of nanomaterial toxicity. *Advanced Drug Delivery Reviews* 2009, 61, 438–456.
29. Nel, A.; Xia, T.; Madler, L.; Li, N., Toxic potential of materials at the nanolevel. *Science* 2006, 311, 622–627.
30. Nanomedicine, E. T. P. O. Nanomedicine 2020 Contribution of Nanomedicine to Horizon 2020—White Paper to the Horizon 2020 framework programme for research and innovation—recommendation from the Nanomedicine community. 2013, 20.
31. Abraham, V. C.; Taylor, D. L.; Haskins, J. R. High content screening applied to large-scale cell biology. *Trends in Biotechnology* 2004, 22, 15–22.
32. Abraham, V. C.; Towne, D. L.; Waring, J. F.; Warrior, U.; Burns, D. J. Application of a high-content multiparameter cytotoxicity assay to prioritize compounds based on toxicity potential in humans. *Journal of Biomolecular Screening* 2008, 13, 527–537.
33. Towne, D. L.; Nicholl, E. E.; Comess, K. M.; Galasinski, S. C.; Hajduk, P. J.; Abraham, V. C. Development of a high-content screening assay panel to accelerate mechanism of action studies for oncology research. *Journal of Biomolecular Screening* 2012, 17, 1005–1017.
34. Ghosh, R. N.; Grove, L.; Lapets, O. A quantitative cell-based high-content screening assay for the epidermal growth factor receptor-specific activation of mitogen-activated protein kinase. *Assay and Drug Development Technologies* 2004, 2, 473–481.
35. Ghosh, R. N.; Chen, Y. T.; DeBiasio, R.; DeBiasio, R. L.; Conway, B. R.; Minor, L. K.; Demarest, K. T. Cell-based, high-content screen for receptor internalization, recycling and intracellular trafficking. *Bio Techniques* 2000, 29, 170–175.
36. Ghosh, R. N.; Lapets, O.; Haskins, J. R. Characteristics and value of directed algorithms in high content screening. *Methods in Molecular Biology (Clifton, N.J.)* 2007, 356, 63–81.
37. Mohamed, B. M.; Feighery, C.; Williams, Y.; Davies, A.; Kelleher, D.; Volkov, Y.; Kelly, J.; Abuzakouk, M. The use of cellomics to study enterocyte cytoskeletal proteins in coeliac disease patients. *Central European Journal of Biology* 2008, 3, 258–267.
38. Mohamed, B.; Verma, N.; Prina-Mello, A.; Williams, Y.; Davies, A.; Bakos, G.; Tormey, L. et al. Activation of stress-related signalling pathway in human cells upon SiO$_2$ nanoparticles exposure as an early indicator of cytotoxicity. *Journal of Nanobiotechnology* 2011, 9, 29.

39. Williams, Y.; Sukhanova, A.; Nowostawska, M.; Davies, A. M.; Mitchell, S.; Oleinikov, V.; Gun'ko, Y.; Nabiev, I.; Kelleher, D.; Volkov, Y. Probing cell-type-specific intracellular nanoscale barriers using size-tuned quantum dots. *Small* 2009, 5, 2581–2588.
40. Mohamed, B. M.; Movia, D.; Knyazev, A.; Langevin, D.; Davies, A. M.; Prina-Mello, A.; Volkov, Y. Citrullination as early-stage indicator of cell response to single-walled carbon nanotubes. *Scientific Reports* 2013, 3, doi:10.1038/srep01124.
41. Mohamed, B. M.; Verma, N. K.; Davies, A. M.; McGowan, A.; Crosbie-Staunton, K.; Prina-Mello, A.; Kelleher, D. et al. Citrullination of proteins: A common post-translational modification pathway induced by different nanoparticles in vitro and in vivo. *Nanomedicine* 2012, 7, 1181–1195.
42. Cooper, S. E. J.; Kennedy, N. P.; Mohamed, B. M.; Abuzakouk, M.; Dunne, J.; Byrne, G.; McDonald, G. et al. Immunological indicators of coeliac disease activity are not altered by long-term oats challenge. *Clinical and Experimental Immunology* 2013, 171, 313–318.
43. Healthy challenges. *Nature Nanotechnology* 2007, 2, 451, doi:10.1038/nnano.2007.258. Available at: URL: http://www.nature.com/nnano/journal/v2/n8/full/nnano.2007.258.html.
44. Jan, E.; Byrne, S. J.; Cuddihy, M.; Davies, A. M.; Volkov, Y.; Gun'ko, Y. K.; Kotov, N. A. High-content screening as a universal tool for fingerprinting of cytotoxicity of nanoparticles. *ACS Nano* 2008, 2, 928–938.
45. Prina-Mello, A.; Crosbie-Staunton, K.; Salas, G.; Del Puerto Morales, M.; Volkov, Y. Multiparametric toxicity evaluation of SPIONs by high content screening technique: Identification of biocompatible multifunctional nanoparticles for nanomedicine. *IEEE Transactions on Magnetics* 2013, 49, 377–382.
46. Murphy, F. A.; Poland, C. A.; Duffin, R.; Al-Jamal, K. T.; Ali-Boucetta, H.; Nunes, A.; Byrne, F. et al. Length-dependent retention of carbon nanotubes in the pleural space of mice initiates sustained inflammation and progressive fibrosis on the parietal pleura. *American Journal of Pathology* 2011, 178, 2587–2600.
47. Wagner, K.; Fach, B.; Kolar, R. Inconsistencies in data requirements of EU legislation involving tests on animals. *Altex* 2012, 29, 302–332.
48. Giaever, I.; Keese, C. R. Monitoring fibroblast behavior in tissue culture with an applied electric field. *Proceedings of the National Academy of Sciences of the United States of America* 1984, 81, 3761–3764.
49. Wegener, J.; Keese, C. R.; Giaever, I. Electric cell-substrate impedance sensing (ECIS) as a noninvasive means to monitor the kinetics of cell spreading to artificial surfaces. *Experimental Cell Research* 2000, 259, 158–166.
50. Luong, J. H.; Habibi-Rezaei, M.; Meghrous, J.; Xiao, C.; Male, K. B.; Kamen, A. Monitoring motility, spreading, and mortality of adherent insect cells using an impedance sensor. *Analytical Chemistry* 2001, 73, 1844–1848.
51. Arndt, S.; Seebach, J.; Psathaki, K.; Galla, H. J.; Wegener, J. Bioelectrical impedance assay to monitor changes in cell shape during apoptosis. *Biosensors and Bioelectronics* 2004, 19, 583–94.
52. Ghosh, P. M.; Keese, C. R.; Giaever, I. Monitoring electropermeabilization in the plasma membrane of adherent mammalian cells. *Biophysical Journal* 1993, 64, 1602–1609.
53. Giaever, I.; Keese, C. R. A morphological biosensor for mammalian cells. *Nature* 1993, 366, 591–592.
54. Verma, N. K.; Conroy, J.; Lyons, P. E.; Coleman, J.; O'Sullivan, M. P.; Kornfeld, H.; Kelleher, D.; Volkov, Y. Autophagy induction by silver nanowires: A new aspect in the biocompatibility assessment of nanocomposite thin films. *Toxicology and Applied Pharmacology* 2012, 264, 451–461.
55. Verma, N. K.; Moore, E.; Blau, W.; Volkov, Y.; Babu, P. R. Cytotoxicity evaluation of nanoclays in human epithelial cell line A549 using high content screening and real-time impedance analysis. *Journal of Nanoparticle Research* 2012, 14.

56. Verma, N. K.; Crosbie-Staunton, K.; Satti, A.; Gallagher, S.; Ryan, K. B.; Doody, T.; McAtamney, C. et al. Magnetic core-shell nanoparticles for drug delivery by nebulization. *Journal of Nanobiotechnology* 2013, 11, 1.

57. Lysaght, J.; Verma, N. K.; Maginn, E. N.; Ryan, J. M.; Campiani, G.; Zisterer, D. M.; Williams, D. C.; Browne, P. V.; Lawler, M. P.; McElligott, A. M. The microtubule targeting agent PBOX-15 inhibits integrin-mediated cell adhesion and induces apoptosis in acute lymphoblastic leukaemia cells. *International Journal of Oncology* 2013, 42, 239–246.

58. Hondroulis, E.; Liu, C.; Li, C. Z. Whole cell based electrical impedance sensing approach for a rapid nanotoxicity assay. *Nanotechnology* 2010, 21, 315103.

59. Xiao, C.; Lachance, B.; Sunahara, G.; Luong, J. Assessment of cytotoxicity using electric cell-substrate impedance sensing: Concentration and time response function approach. *Analytical Chemistry* 2002, 74, 5748–5753.

60. Xiao, C.; Luong, J. H. T. Assessment of cytotoxicity by emerging impedance spectroscopy. *Toxicology and Applied Pharmacology* 2005, 206, 102–112.

61. Hug, T. S. Biophysical methods for monitoring cell–substrate interactions in drug discovery. *Assay and Drug Development Technologies* 2003, 1, 479–488.

62. Zhang, X. Q.; Xu, X.; Bertrand, N.; Pridgen, E.; Swami, A.; Farokhzad, O. C. Interactions of nanomaterials and biological systems: Implications to personalized nanomedicine. *Advanced Drug Delivery Reviews* 2012, 64, 1363–1384.

63. Ramachandran, S.; Lal, R. Scope of atomic force microscopy in the advancement of nanomedicine. *Indian Journal of Experimental Biology* 2010, 48, 1020–1036.

64. Samuel, S. P.; Jain, N.; O'Dowd, F.; Paul, T.; Kashanin, D.; Gerard, V. A.; Gun'ko, Y. K.; Prina-Mello, A.; Volkov, Y. Multifactorial determinants that govern nanoparticle uptake by human endothelial cells under flow. *International Journal of Nanomedicine* 2012, 7, 2943–2956.

65. Ramachandran, S.; Teran Arce, F.; Lal, R. Potential role of atomic force microscopy in systems biology. *Wiley Interdisciplinary Reviews. Systems Biology and Medicine* 2011, 3, 702–716.

66. Hinterdorfer, P.; Dufrene, Y. F. Detection and localization of single molecular recognition events using atomic force microscopy. *Nature Methods* 2006, 3, 347–355.

67. Dammer, U.; Hegner, M.; Anselmetti, D.; Wagner, P.; Dreier, M.; Huber, W.; Guntherodt, H. J. Specific antigen/antibody interactions measured by force microscopy. *Biophysical Journal* 1996, 70, 2437–2441.

68. McKendry, R.; Zhang, J.; Arntz, Y.; Strunz, T.; Hegner, M.; Lang, H. P.; Baller, M. K. et al. Multiple label-free biodetection and quantitative DNA-binding assays on a nanomechanical cantilever array. *Proceedings of the National Academy of Sciences of the United States of America* 2002, 99, 9783–9788.

69. Almqvist, N.; Bhatia, R.; Primbs, G.; Desai, N.; Banerjee, S.; Lal, R. Elasticity and adhesion force mapping reveals real-time clustering of growth factor receptors and associated changes in local cellular rheological properties. *Biophysical Journal* 2004, 86, 1753–1762.

70. Hussain, S. M.; Hess, K. L.; Gearhart, J. M.; Geiss, K. T.; Schlager, J. J. In vitro toxicity of nanoparticles in BRL 3A rat liver cells. *Toxicology in Vitro* 2005, 19, 975–983.

71. Schrand, A. M.; Lin, J. B.; Hussain, S. M. Assessment of cytotoxicity of carbon nanoparticles using 3-(4,5-dimethylthiazol-2-yl)-5-(3-carboxymethoxyphenyl)-2-(4-sulfophenyl)-2H-tetrazolium (MTS) cell viability assay. *Methods in Molecular Biology (Clifton, N.J.)* 2012, 906, 395–402.

10 Detection Methods for the In Vivo Biodistribution of Iron Oxide and Silica Nanoparticles
Effects of Size, Surface Chemistry, and Shape

Heather A. Enright and Michael A. Malfatti

CONTENTS

10.1 INTRODUCTION

Nanoparticles are being developed and utilized for many biological applications: drug delivery,[1-4] as imaging agents,[5-8] and for diagnostic purposes.[9,10] In addition to biomedical applications, the risk of serious health effects and toxicity on environmental or occupational exposure for nanoparticles remains.[11-13] Although some studies evaluating the toxicity and biological fate of nanoparticles have been reported, a comprehensive understanding on how size, shape, composition, and surface properties influence nanoparticle fate in biological systems is still lacking.

The in vivo biodistribution of nanoparticles has been investigated with an array of detection methods, including highly sensitive ex vivo methods and noninvasive imaging. The methods mentioned in this chapter are outlined in Table 10.1.[14-21] Analysis using ex vivo methods (i.e., inductively coupled plasma mass spectrometry [ICP-MS]), and gamma counting provides quantification of an element of interest with high sensitivity and high throughput.[16,22] Noninvasive imaging, such as magnetic resonance imaging (MRI) and positron emission tomography (PET), can provide real-time in vivo mapping of materials in the same animal over time. This is especially of value for agents delivered to the lungs, where initial deposition influences clearance and transport.[23]

Just recently, accelerator mass spectrometry (AMS) was implemented for nanoparticle pharmacokinetics and biodistribution.[24] AMS is a measurement technique that can quantify rare, long-lived isotopes with high sensitivity in the range of attomoles.[15,25] The mass ratio of the isotope of interest is measured with respect to a stable isotope of the element. For biological applications, ^{14}C ($t_{1/2} = 5730$ years) is a common detection radioisotope because most biological samples contain endogenous levels of carbon. The long half-life and ultrahigh sensitivity of AMS allow for long-term analysis of very small amounts of compounds; therefore, their therapeutic or environmental levels can be analyzed with AMS unlike with many other techniques, in which a higher amount of administered compound is needed for detection. Additionally, trace amounts of material can be quantified; this is tremendously useful in assessing nanoparticle fate and toxicity where miniscule amounts of nanoparticle accumulation may be missed by other less sensitive modes of detection.

TABLE 10.1
Summary of the Detection Methods Described in This Chapter

Detection Method	Sensitivity	Sample Analysis	Label/Detection
ICP-MS	ppb–ppt	Ex vivo	Element
ESR	nM	Ex vivo	Paramagnetic element
ICP-OES/ICP-AES	ppb	Ex vivo	Element
AAS	ppb	Ex vivo	Element
AMS	pmol–amol	Ex vivo	Radioisotope
Gamma counting	<1 nCi	Ex vivo	Radioisotope
MRI	nM–μM	In vivo	Paramagnetic element
Optical imaging	pM	Ex vivo and in vivo	Fluorescent tag
PET	pM	In vivo	Radioisotope

Although each method has its advantages, limitations do exist. For most ex vivo methods, samples must be homogenized for measurement, thereby destroying any regional specific information within the samples. Furthermore, changes in distribution cannot be visualized in the same animal, increasing the number of animals needed for analysis. Although imaging can provide a real-time measurement of material in the same animal, highly sensitive methods such as PET suffer from low resolution, which makes organ distinction difficult unless they are coupled with high-resolution anatomical methods (MRI and computed tomography).[26] Additionally, imaging (i.e., PET) has a lower throughput and is much higher in cost.[27]

In this chapter, we review the in vivo biodistribution of nanoparticles with a specific focus on iron oxide and silica nanoparticles. The effects of size, surface chemistry, and shape are discussed. The findings are summarized in Tables 10.2 and 10.3 for iron oxide and silica nanoparticles, respectively.

10.2 IRON OXIDE NANOPARTICLES

Iron oxide nanoparticles are becoming more popular for use as imaging agents and also as drug delivery vehicles. The magnetic properties of the iron oxide core make these nanoparticles desirable for use as contrast agents for MRI. Imaging applications for iron oxide nanoparticles range from magnetic targeting of brain tumors[28] and tumor imaging[10] to the mapping of vascular inflammation.[5,7] Iron oxide nanoparticles have also been investigated for drug delivery, including sustained release of anticancer agents in tumors[4,29] and gene therapy.[30,31]

The in vivo pharmacokinetics and biodistribution of iron oxide nanoparticles has been reported with both ex vivo and noninvasive methods. In the following, findings from the literature are discussed for each detection modality. The studies discussed in this section are outlined in Table 10.2.

10.2.1 SPECTROMETRY AND SPECTROSCOPY METHODS

The biodistribution of iron oxide has been evaluated with numerous spectrometry and spectroscopy techniques including ICP-MS, electron spin resonance (ESR), inductively coupled plasma optical emission spectrometry (ICP-OES), and atomic absorption spectroscopy (AAS). For ICP-MS, only a few studies have been reported for short- and long-term distributions of iron oxide. In 2009, Tate et al.[32] determined a preliminary short-term biodistribution of 110 and 20 nm hydroxylethyl starch-coated magnetite nanoparticles in a tumor-bearing mouse model over 72 hours after an intravenous injection (80 mg Fe/kg). Iron oxide nanoparticles are desirable for tumor therapy because they can be visualized with imaging methods noninvasively and also targeted to a specific region with magnetic targeting methods. In this study, blood concentrations were found to be higher for smaller sized nanoparticles; at 4 hours post injection, 0.56 mg Fe/g was measured compared to 0.46 mg Fe/g for larger nanoparticles. These observations are in agreement with the hypothesis that larger nanoparticles are more readily taken up by the reticuloendothelial system (RES), therefore decreasing their circulation time.[33,34] As expected, major organs accumulated higher amounts of the larger nanoparticles over the 72 hours. Larger nanoparticles had a

TABLE 10.2

Summary of the Iron Oxide Biodistribution Studies Described in This Chapter

Analysis Methods	Nanoparticle	Size (nm)	Species	Route of Delivery	Time	Plasma Half-Life ($t_{1/2}$)	Major Organs of Accumulation	Reference
ICP-MS	Starch-Fe$_3$O$_4$	20	Mouse	Intravenous	72 hours	ND	Liver, spleen, tumor, lungs, kidney	Tate et al. (2011)
ICP-MS	MNP	110 193	Rat	Intravenous	21 days	ND	Liver, spleen, heart, lung, kidney, brain	Jain et al. (2008)
ESR	PEG-MNP MNP	140–170 104	Rat	Intravenous	60 hours	7.3–11.8 hours 0.12 hour	Liver, spleen, lungs, tumor	Cole et al. (2011)
ESR	MNP	110	Rats	Intravenous	50 minutes	ND	Liver, spleen, lungs, kidney, brain, tumor	Chertok et al. (2010)
ICP-OES							Liver, spleen, lungs	
AAS	SPIO	30	Mouse	SC	24 hours	ND	Heart, spleen, liver, lungs	Liu et al. (2008)
AAS	Fe$_3$O$_4$	20	Mouse	Intragastric	10 days	ND	Liver, spleen, heart, marrow, intestine Brain, stomach, kidneys	Wang et al. (2010)
Gamma counting	USPIO	4.3–6.2	Rat	SC Intravenous	72 hours	3.7 hours	Spleen, lymph nodes Liver, spleen, lymph nodes	Bengele et al. (1994)
Gamma counting	OH-iron oxide	12	Rat	Intravenous	18 hours	39 minutes	Liver, spleen, carcass, bone	Portet et al. (2001)
	NH$_2$-iron oxide					25 minutes	Liver, spleen, bone	
	NMe$_2$-iron oxide					32 minutes	Liver, spleen, bone	
	N+Me$_3$-iron oxide					27 minutes	Liver, spleen, bone	
	OH/N+Me$_3$-iron oxide					36 minutes	Liver, spleen, bone	

Method	Material	Size	Animal	Administration			Distribution	Reference
Gamma counting	Fe_2O_3	144	Rat	Intratracheal instillation	50 days	1.9 days	Liver, spleen, heart, kidney, pancreas	Zhu et al. (2009)
MRI	SPIO	15	Rat	Iron oxide patch	6 months	ND	Testicle, brain none	Schlachter et al. (2011)
MRI	USPIO	20–30	Rat	Intratracheal instillation	14 days	ND	Lungs only	Al Faraj et al. (2008)
ICP-MS				Intravenous	14 days		Liver, spleen, kidney	
Fluorescence imaging	Iron oxide	7	Mouse	Intravenous	48 hours	7–8 hours	Liver, spleen, marrow, kidneys, heart	Jeung-Eun Lee et al. (2010)
AMS	^{14}C-iron oxide	100	Mouse	Inhalation	7 days	ND	Liver, spleen, intestine, olfactory bulb	Unpublished, Enright et al.

Note: ND, not determined; SPIO, superparamagnetic iron oxide.

TABLE 10.3

Summary of the Silica Biodistribution Studies Described in This Chapter

Analysis Methods	Nanoparticle	Size (nm)	Species	Route of Delivery	Time	Plasma Half-Life ($t_{1/2}$)	Major Organs of Accumulation	Reference
Fluorescence/PET imaging	^{124}I-core/shell silica	3.3 6	Mouse	Intravenous	96 hours	ND	Kidney, spleen, lungs, liver	Burns et al. (2009)
Fluorescence/PET imaging	^{124}I-silica	20–25	Mouse	Intravenous	24 hours	ND	Liver, spleen Kidney, lungs, intestine, heart Stomach, skin	Kumar et al. (2010)
Fluorescence imaging Elemental analysis	ICG-silica	50–100	Rat Mouse	Intravenous	3 hours	ND	Liver, kidney, lungs Spleen, heart	Lee et al. (2009)
Fluorescence imaging	OH-silica COOH-silica PEG-silica	45	Mouse	Intravenous	24 hours	80 minutes 35 minutes 180 minutes	Liver, bladder, urinary meatus Liver, bladder, urinary meatus Bladder, urinary meatus	He et al. (2008)
Fluorescence	Silica	50 100 200	Mouse	Intravenous	7 days	ND	Liver, spleen, kidney Liver, spleen, kidney Liver, spleen, kidney	Cho et al. (2009)
Fluorescence	MSN-80 PEG-MSN-80 MSN-120 PEG-MSN-120 MSN-200 PEG-MSN-200	80 80 120 120 200 200	Mouse	Intravenous	1 month	ND	Liver, spleen, lungs, kidney, heart	He et al. (2011)
AMS	^{14}C-silica	33	Mouse	Intravenous	8 weeks	22.8 minutes	Liver, spleen, kidney, lungs	Malfatti et al. (2012)

Method	Particle	Size	Animal	Administration	Time	Half-life	Organs	Reference
Gamma counting	^{125}I-silica	20	Mouse	Intravenous	30 days	ND	Liver, spleen, lungs	Xie et al. (2010)
		80						
Gamma counting	Meso S	120	Mouse	Intravenous	72 hours	ND	Liver, spleen, lungs	Yu et al. (2012)
	MA	120					Liver, spleen	
	AR8	136					Liver, spleen, lungs	
	8A	136					Liver, spleen, lungs	
	Stöber	115					Liver, spleen	
	SA	115					Liver, spleen	
ICP-AES	Silica spheres	700	Mouse	Intravenous	2–6 hours	ND	Liver, lungs, spleen	Decuzzi et al. (2010)
		1000						
		1500						
		2530						
		5000						
	Hemispherical silica	1600					Liver, spleen	
	Discoidal silica	1600					Liver, spleen, lungs, heart	
	Cylindrical	1000					Liver, spleen	
ICP-MS	Silica-coated CdSeS	20	Mouse	Intravenous	5 days	19.8 hours	Liver, spleen, lungs, kidney	Chen et al. (2008)
ICP-AES	MHSN	110	Mouse	Intravenous	4 weeks	ND	Liver, spleen	Liu et al. (2011)
ICP-AES	LR-MSN	185	Mouse	Intravenous	7 days	ND	Liver, spleen, lungs	Huang et al. (2011)
	LR-MSN-PEG	185						
	SR-MSN	720						
	SR-MSN-PEG	720						

Note: ND, not determined. LR, long rod; SR, short rod.

higher liver and spleen uptake with lower tumor accumulation. Tumor accumulation was much higher for the 20 nm nanoparticles (0.05 mg Fe/g); this could be attributed to increased tissue penetration for the smaller nanoparticles through the tumor vasculature. In tumor models, enhanced permeability and retention of nanoparticles in tumors has been observed, especially with smaller nanoparticles.[35,36] Compared to control tissue, lungs and kidney also contained elevated iron concentrations for both sizes, with slightly higher concentrations for the 100 nm nanoparticles.

In a long-term study by Jain et al.,[37] the biodistribution of intravenously administered magnetic nanoparticles (193 nm in diameter) was determined after 3 weeks in rats (10 mg Fe/kg) using ICP-MS. An increase in iron serum levels was observed over the first week, followed by a slow decrease in iron concentration. At 21 days, significant amounts of iron were still detected in serum, suggesting long-term circulation of nanoparticles. Changes in iron levels were detected in major organs, including the liver, lungs, spleen, heart, kidney, and brain. A notable increase in iron levels was observed in the liver, spleen, heart, and brain from 1 week to 3 weeks after injection. At 3 weeks, 50% and 25% of the injected iron were localized in the liver and the spleen, respectively. Lower but detectable levels were also observed at 3 weeks in the other assayed tissues.

Differences in accumulation for polyethylene glycol (PEG)-coated versus non-PEGylated iron oxide were investigated using ESR. PEG-modified, cross-linked starch magnetic iron oxide nanoparticles (PEG-MNPs) were compared to their parent starch-only magnetic iron oxide nanoparticles (MNPs) for magnetic tumor targeting therapy.[38] PEGylation of nanoparticles has been shown to increase their circulation times;[39,40] this is ideal for drug delivery in which increased circulation times are desirable for targeted nanoparticles for evasion of RES. In the work by Cole et al.,[38] biodistributions for both PEG-MNP and parent MNP were investigated in a brain tumor rat model after magnetic targeting (12 mg Fe/kg). At 1 hour post injection, PEG-MNPs showed much higher uptake in 9 L glioma tumors (1%ID/g; %ID stands for the percentage of injected dose), a 15-fold improvement over the parent MNP (0.07%ID/g). Additionally, reductions in liver and spleen concentrations were observed for PEG-MNP compared to MNP at 1 hour. Long-term distribution over 60 hours revealed low amounts of nanoparticles in the lungs and liver; negligible amounts were detected in the kidney. Although a general reduction in RES uptake was observed, the spleen contained the highest concentration of PEG-MNP over 60 hours; >4000 nmol Fe/g tissue was observed at both 12- and 60-hour time points. The increased uptake of PEG-MNP in the spleen and the lower retention in the liver are discussed; a potential mechanism for higher levels in the spleen is that PEGylation of the nanoparticle may sterically inhibit macrophage's ability to phagocytize PEG-MNPs in the liver. Another plausible mechanism is an increase in size of the PEG-MNP once in circulation (>200 nm). Nanoparticles larger than 200 nm are filtered by the spleen more efficiently than by the liver.[2] Reducing liver uptake is an important consideration when designing nanoparticles for drug delivery. Avoidance of long-term accumulation is vital for avoiding potential toxicity and for advancement of candidates to the clinic.[41,42]

In addition to the consideration of surface coating, selecting the best method for detection is necessary; this may vary depending on the type of nanoparticle. An evaluation comparing methods for the accurate analysis of biodistribution of iron oxide nanoparticles (~110 nm) was investigated by Chertok et al.[28] using ESR and ICP-OES. Similar to the work by Cole et al.,[38] MNPs were administered (12–25 mg Fe/kg) intravenously with magnetic targeting. Analysis of tissue for MNP accumulation was performed with both techniques. A strong correlation ($r = 0.94$–0.97) was found between the two methods for organs that contained high levels (>1000 nmol Fe/g tissue) of MNPs (liver, spleen, and lungs). Discrepancies between the two methods were found in organs with lower accumulations (<50 nmol Fe/g tissue) of MNPs (kidney, brain, and tumor); ESR successfully resolved low amounts, whereas ICP-OES did not. The susceptibility for ICP-OES was due to the high levels of background iron, which masked exogenous signal in the tissue. The ESR analysis in this work clearly achieved a greater sensitivity than ICP-OES, illustrating the need for identifying the best suited method when characterizing nanoparticle distribution; endogenous signal present in tissue may confound results, depending on particle composition.

The effects of single and repeated dosing of iron oxide nanoparticles were investigated with AAS.[43,44] Liu et al.[43] compared single and repeated dose biodistributions at 24 hours post subcutaneous (SC) injection of 30 nm iron oxide nanoparticles (100 mg/kg) in mice. For the repeated dose group, doses were given on 10 consecutive days and tissues were harvested 24 hours after the last injection; heart, spleen, liver, lungs, kidney, and brain were analyzed with AAS. Similar concentrations of nanoparticles were observed in both the single and repeated dose groups, and the highest concentrations were found in the heart, spleen, liver, and lungs. The similar trends in biodistribution indicate that the iron oxide nanoparticles were incorporated into the body's iron storage, since no increases in accumulation were observed for the repeated dosing. Additionally, no iron staining was present within macrophages in any of the organs, which suggested nanoparticle breakdown and incorporation into tissue.

Pharmacokinetics and tissue distribution have also been assessed for iron oxide nanoparticles after intragastric administration[44] using AAS. Oral exposure is ideal for drug delivery as it is noninvasive to the patient. Additionally, since silica nanoparticles have been detected in food products, assessing potential absorption and toxicity upon ingestion is essential. MNPs were intragastrically administered to mice (600 mg/kg), and blood and tissue were collected over 10 days. Significant concentrations of MNPs were detected in the blood over 10 days with peaks at 6 hours (439 µg/mL) and 5 days (436 µg/mL), indicating absorption into the bloodstream from the gastrointestinal (GI) system. The main organs of MNP accumulation were liver and spleen. A close relationship between blood and liver and spleen concentrations was observed; increases in liver and spleen concentrations were noted when there was a dip in blood concentration. Significant concentrations of the MNPs were also observed in heart, bone marrow, small intestine, brain, stomach, and kidneys, revealing potential widespread absorption of the nanoparticles over the study period ($p < .05$).

10.2.2 ACCELERATOR MASS SPECTROMETRY

Using the ultrasensitive capabilities of AMS, the pharmacokinetics and biodistribution of ^{14}C-labeled iron oxide nanoparticles after inhalation delivery have recently been evaluated in mice over 7 days in our group at Lawrence Livermore National

Laboratory, California (unpublished work, Enright et al.). In this study, the nanoparticles were delivered using a nose-only nano-aerosol generation system developed by Mikheev et al.[45] at Battelle Memorial Institute, Columbus, Ohio. We observed a 64% reduction in lung retention at 7 days post delivery; extrapulmonary accumulation was mostly found in the GI system, with detectable quantities in the liver and spleen, indicating some blood transport had occurred. Furthermore, detectable concentrations were also observed in the olfactory bulb at 7 days (~1 ng/mg). Clearance of nanoparticles was slow and primarily occurred through the feces ($t_{1/2} = 1.42$ days). The slow excretion and higher concentrations in the GI system indicated mucociliary clearance as the primary mechanism for lung clearance of the iron oxide nanoparticles after deposition.

10.2.3 GAMMA COUNTING

To date, only a few have utilized ^{59}Fe-labeled iron oxide nanoparticles to determine the effects of surface coating and delivery route on pharmacokinetics and biodistribution.[46–48] Bengele et al.[47] compared the effects of SC and intravenous injections of BMS 180549, an ultrasmall paramagnetic iron oxide (USPIO) colloid under clinical development, and their accumulation in lymph nodes after injection in rats. For SC delivery, changes in uptake were observed depending on the site of the injection. SC injections of 10 µmol Fe/kg of agent were administered to the ventral surface of either front paw or back paw, and blood, liver, spleen, and lymph nodes were harvested and analyzed over 72 hours. Using gamma counting, for front paw SC injection axillary and brachial nodes accumulated the highest amount of USPIO; after back paw SC injection, the highest amounts of USPIO were in the popliteal, iliac, and axillary lymph nodes. Both SC routes showed very low concentrations in the blood and liver (<1% Fe/g), indicating mostly lymphatic transport; however, after 4 hours post injection indications of blood transport were observed by accumulated concentrations in the spleen (~20% Fe/g). The accumulation in the spleen is most likely from agent transport from the injection site through the corresponding draining lymph nodes into circulation. Upon entering the vascular compartment, spleen-derived macrophages can uptake the nanoparticles. For the intravenous route, 40 µmol Fe/kg of agent was injected into the femoral vein. In terms of lymphatic uptake, the iliac, mesenteric, and mediastinal lymph nodes accumulated the highest levels of USPIO. Smaller amounts (<5% Fe/g) of USPIO were detected in the blood and liver. The spleen contained much higher concentrations, and accumulation increased over time with the maximum occurring at 72 hours post injection (~52% Fe/g). In this work, the authors clearly illustrated that BMS 180549 could be directed to specific tissues depending on the administration route and site of injection. Additionally, front paw SC injection highlighted the feasibility of using BMS 180549 as a contrast agent for spleen.

As described in Section 10.2.1 with PEG-coated iron oxide, surface properties can influence the uptake and accumulation of nanoparticles in vivo. The effects of varying surface coatings of iron oxide nanoparticles (12 nm) were evaluated by Portet et al.[48] in 2001. The authors compared the impact of different functional groups (hydroxyl or amino) and the saturation degree of the last chemical group (primary amine, tertiary amine, and quaternary ammonium) on plasma half-life and biodistribution. The nanoparticles were radiolabeled with both ^{59}Fe and ^{99m}Tc for detection with gamma counting; dual labeling was used to verify that signal was from intact nanoparticles. Rats were injected intravenously with 1 mg of iron nanoparticles, and organs were harvested and counted at 10 minutes, 90 minutes, and 18 hours post injection. At 18 hours, all five types of nanoparticles showed major accumulation in the liver, followed by the spleen. Hydroxyl-coated nanoparticles also showed high levels of nanoparticles in the carcass (13%ID/g) at 18 hours. In the liver, at 10 minutes post injection accumulation was found to be inversely proportional to the degree of saturation for the aminated nanoparticles, with 79%ID/g, 59%ID/g, and 46%ID/g for the primary, secondary, and tertiary amines, respectively. Blood clearance also varied for each surface coating; primary amine–terminated nanoparticles were cleared the quickest, with only 9%ID at 10 minutes ($t\frac{1}{2} = 25$ minutes). The hydroxyl-coated nanoparticles demonstrated the longest circulation time with a half-life, $t\frac{1}{2}$, of 39 minutes. Bone accumulation of nanoparticles was also observed for all types, with the hydroxyl-terminated nanoparticles showing the highest levels at 18 hours (6.0%ID/g).

Long-term extrapulmonary distribution was assessed after intratracheal instillation of 144 nm ^{59}Fe-labeled iron oxide nanoparticles in rats (4 mg per rat) by Zhu et al.[46] Rapid transport of the nanoparticles from the lungs into circulation was observed; nanoparticles were detected in blood within 10 minutes of instillation. Levels in the blood increased for up to 7 days (~6 µg/mL) and then gradually declined until day 50. The half-life in the blood was 22.8 days. In terms of lung clearance, once deposited the elimination of the nanoparticles from the lungs was slow and consistent with zero-order elimination kinetics with only 3.06 µg of the nanoparticles shown to clear per day. Extrapulmonary distribution showed that the highest levels of accumulation were found in the liver, followed by the spleen, heart, kidney, pancreas, testicle, and brain. Excretion occurred primarily through the feces, consistent with mucociliary clearance of nanoparticles, which is observed for larger particles.[23,49] Considering the high dose (4 mg per rat) administered in this study, larger aggregates may have formed, resulting in deposition in the upper airways, similar to larger particles.[50] In terms of mass balance, less than 5% of the instilled dose was found in the analyzed organs, including the lungs. The majority of the dose was found in feces (73%) and other parts (20%), presumably the carcass.

10.2.4 IMAGING

MRI and fluorescence imaging have both been used to investigate the biodistribution of iron oxide nanoparticles in vivo. Iron oxide nanoparticles have proved to be an effective contrast agent in MRI due to their low toxicity, high magnetic strength, and long-lasting contrast enhancement. MRI allows for noninvasive analysis of iron oxide nanoparticle biodistribution. To date, a few have used MRI to noninvasively assess nanoparticle biodistribution. One study by Schlachter et al.[51] in 2011 assessed the degradation of implanted iron oxide patches in rats over 6 months. Iron oxide nanoparticle–containing patches are currently under investigation to use in electromagnetic tissue soldering; however, the metabolism and long-term effects once implanted have yet to be tested. In the study reported by Schlachter et al., rats were implanted with iron oxide nanoparticle–containing albumin patches subcutaneously in the skull. MRI assessment was conducted over 6 months. With MRI, changes in signal were observed in the patch implantation area throughout the study; however, no changes in any organs of interest (liver, kidney, and nucleus caudatus) were observed, indicating that no systemic absorption of iron oxide nanoparticles from the patch had occurred. These observations were verified with histological methods ex vivo.

A combinatory ^{3}He and proton MRI study by Al Faraj et al.[52] illustrated the feasibility of MRI for combined biodistribution and pulmonary function analysis for iron oxide nanoparticles. Hyperpolarized ^{3}He was used to image pulmonary ventilation after instillation of the iron oxide nanoparticles; in standard MRI, lung tissue does not appear on the images due to its weak proton density and the heterogeneity of the tissue (air–tissue interfaces).[53] In a dose range study, noncoated iron oxide nanoparticles (20–30 nm) were intratracheally instilled into rats, with three different groups receiving 1, 0.5, and 0.1 mg. Imaging was acquired on day 0 and day 3; the image signal intensity was shown to decrease with the quantity of instilled iron nanoparticles, with 1 mg showing the highest loss in signal. For all three doses, no differences in signal in any of the organs of interest (kidneys, spleen, and liver) were observed in the proton imaging. In a longer follow-up study, 0.5 mg of iron oxide nanoparticles were instilled into rats and imaging was performed at days 0, 2, 7, and 14. The ^{3}He ventilation images over 2 weeks showed a gradual increase in signal intensity over time, indicating that some lung clearance had occurred. Similar to the dose-range group, there was no indication of nanoparticle accumulation in any peripheral organs over the 2-week time period. These observations were verified ex vivo with ICP-MS analysis. A comparison was made with an intravenously injected control group of rats (0.5 mg); on day 0, as expected, this group showed an increase in the contrast to noise ratio (CNR) for the liver and spleen, indicating that nanoparticle accumulation had occurred. Over the 2 weeks, the CNR decreased gradually in these organs along with the kidneys, indicating clearance. As a whole, the study validated the combination of the MRI techniques (^{3}He and proton) for investigating nanoparticle distribution and clearance. Clear differences were noninvasively identified in both lung and peripheral tissue when changes in nanoparticle concentration occurred.

Recently, fluorescence imaging has also been implemented to assess biodistribution of nanoparticles. Jeung-Eun Lee et al. developed an ex vivo fluorescent imaging method using a near-infrared (NIR) fluorophore (Cy5.5) to evaluate serum half-life and biodistribution of targeted and nontargeted iron oxide nanoparticles.[54] NIR dyes allow for fluorescent measurement that is not confounded by autofluorescence generated by tissue such as hair, skin, and GI contents; these tissues are relatively transparent in the NIR range (700–1000 nm).[55] Mice were injected with 0.2 mg of the nanoparticles, and blood and whole organs were collected over 48 hours. Excised organs were imaged using a fluorescent imaging system (IVIS-100), followed by sectioning for slice analysis using an NIR scanner capable of quantitative assessment at 21 μ resolution. For both nanoparticles, accumulation in tissue was similar. Tissues of the RES (liver, spleen, and bone marrow), as well as the kidneys and the heart, showed significant localization of the nanoparticles. Tissue slice analysis revealed specific areas within the major organs of binding for the targeted nanoparticle. Patterns in the liver and spleen indicated macrophage uptake; high levels of nanoparticles were observed in the renal cortex in the kidney, and significant accumulation of nanoparticles was observed in the aorta in heart tissue.

10.2.5 SUMMARY

Given what is reported in the literature, several trends for specific physiochemical properties of iron oxide are evident. For circulation time, smaller nanoparticles tend to remain in the blood longer than their larger counterparts; this effect is even more obvious when comparing bare iron oxide with coated iron oxide (i.e., PEG). Additionally, smaller size nanoparticles tend to accumulate more in tumor tissue, which is important when considering iron oxide nanoparticles for cancer therapy. Macrophage-rich organs are the primary sites of uptake; for most studies, the liver and spleen were the organs of highest uptake, followed by other RES organs, such as the lungs and bone. Finally, surface chemistry clearly influences not only circulation time but also excretion and distribution. PEGylation of iron oxide nanoparticles noticeably increases circulation time, reduces organ uptake, and increases clearance. Plasma half-life and organ accumulation have also been shown to be dependent not only for the chemical groups on the surface (amine, carboxyl, and hydroxyl) but also for their degree of saturation.

10.3 SILICA NANOPARTICLES

The use of silica nanoparticles is increasing, with biomedical applications ranging from targeted drug delivery[56,57] and optical imaging[6,58] to cancer therapy[59,60] and controlled drug release.[57] In addition, silica nanoparticles are used as additives for personal care products and have also been found in some food products. With their widespread use, a comprehensive understanding on the pharmacokinetics and biodistribution of these nanoparticles is needed to determine if potential hazards exist and which properties (i.e., size, surface charge, or shape) may influence potential toxicity. Similar detection modalities used for iron oxide nanoparticle analysis have also been used for silica, with the exception of PET and AMS. These detection modalities have more recently been implemented for nanoparticle pharmacokinetic and biodistribution analysis. Section 10.3.1 summarizes the in vivo biodistribution and pharmacokinetics of silica by the detection method. The literature referenced in this section is shown in Table 10.3; not all data are shown in the table. Detailed findings can be found in the corresponding reference.

10.3.1 IMAGING

A few recent combination optical/PET imaging studies have been reported for short- and long-term distribution using dual labeled silica nanoparticles.[58,61] For PET biodistribution studies, a longer lived positron-emitting radioisotope is needed for detection (i.e., ^{124}I, ^{64}Cu, or ^{89}Zr).[27] Similar to MRI, noninvasive measurement of nanoparticle biodistribution can be performed in the same animal over time.

To date, two multimodal imaging studies have been reported for silica nanoparticles. In 2009, Burns et al.[58] developed a multimodal silica nanoparticle, which consisted of a near-infrared fluorescent (NIRF) core-shell silica-based nanoparticle (~3 nm) radiolabeled with ^{124}I on the surface for PET detection. The authors investigated the clearance and the accumulation of bare silica in nude mice using whole-body NIRF imaging and found significant renal clearance and liver accumulation at 45 minutes post injection. To identify the effects of PEG, silica nanoparticles were coated with PEG; a reduction in liver uptake and an increase in renal clearance were observed. A more detailed study was completed in the same work; two differently sized nanoparticles (3.3 and 6.0 nm) were compared using optical and PET imaging. Absolute quantitation of nanoparticles in tissues and blood were assessed by ex vivo fluorescence measurements. Analysis of tissue for both nanoparticle sizes showed minimal tissue accumulation 48 hours post injection (1%ID, 3.3 nm; and 2.4%ID, 6.0 nm). Over 48 hours, the blood showed the highest levels of nanoparticles followed by the kidney, spleen, lungs, and liver for both sizes. PET imaging over 96 hours verified the findings observed by optical imaging. As expected, the larger nanoparticles demonstrated a longer tissue half-life ($t_{1/2}$ = 350 minutes) than the nanoparticles of size 3.3 nm ($t_{1/2}$ = 190 minutes) and faster excretion in the urine was observed for the smaller nanoparticles ($t_{1/2}$ = 180 minutes, 3.3 nm; and $t_{1/2}$ = 360 minutes, 6.0 nm). The estimated total excreted fractions after 48 hours were 73%ID and 64%ID for the 3.3 and 6.0 nm nanoparticles, respectively.

In 2010, Kumar et al.[61] also used dual labeled silica nanoparticles for biodistribution and clearance analysis using optical and PET imaging. Organically modified silica nanoparticles 20–25 nm in size were conjugated to an NIR fluorophore and were radiolabeled with ^{124}I for PET detection. Biodistribution was interrogated over 24 hours after intravenous injection; with fluorescence imaging, the liver and spleen accounted for 75% of the injected dose and less than 5% was detected in the kidney, lungs, intestine, and heart. Slightly higher levels were observed in the stomach (~8%) and skin (9%). The authors postulated that the conjugated fluorophore (DY776) targeted the nanoparticles to the skin; after injection of free DY776 this hypothesis was verified, and a major percentage of the signal was observed in the skin. PET imaging was also acquired over 24 hours, and the results obtained agreed with the trends observed with fluorescence imaging. For PET, a major percentage accumulated in the liver and spleen; however, differences in absolute amounts were observed between the methods. For PET, 61% was retained in the spleen and 46% in the liver at 24 hours. This was markedly different from fluorescence imaging, which showed approximately 5% in the spleen and 75% in the liver at 24 hours. These variances were attributed to a change in surface charge for the ^{124}I labeling on the surface. Once conjugated, the overall surface charge on the nanoparticle may have been neutralized, thereby changing the zeta potential, which can affect the distribution.[62]

Other dual detection methods have also been used for the evaluation of silica nanoparticles in vivo. Optical imaging in combination with elemental analysis for silicon was used to characterize an NIR silica nanoparticle in vivo. In the work by Lee et al.,[6] silica nanoparticles 50–100 nm in size were tagged with indocyanine green (ICG) for the generation of an efficient NIR contrast agent for noninvasive optical imaging. In the rat with a dose of 16 mg/kg, major organs such as the liver, spleen, lungs, kidney, and heart were identified for nanoparticle accumulation using optical imaging ex vivo at 3 hours. These results were verified using inductively coupled plasma atomic-emission spectrometry (ICP-AES) for silicon analysis, with the liver containing the highest amount of the dose (35%), followed by the kidney (9.0%), lungs (8.3%), spleen (8.0%), and heart (4.5%). Noninvasive in vivo imaging capabilities of the ICG-silica nanoparticle were also tested in nude mice. Accumulation in the liver was successfully visualized noninvasively at 3 hours post injection, demonstrating the potential use of ICG-silica nanoparticle as an optical imaging agent.

Variations for both size and surface charge for silica nanoparticles on biodistribution have been investigated by a few groups using optical imaging methods.[63–65] In 2008, He et al.[65] examined the changes in biodistribution and excretion of three different surface-modified silica nanoparticles (45 nm), hydroxyl (OH-SiNPs), carboxyl (COOH-SiNPs), and PEGylated (PEG-SiNPs), by doping each with RuBPY dye for imaging purposes.[65] In nude mice, 0.03 mg of each agent was intravenously injected and biodistribution was followed with optical imaging over 24 hours. PEG-SiNPs remained in circulation longer ($t_{1/2}$ = 180 minutes) than OH-SiNPs ($t_{1/2}$ = 80 minutes) and COOH-SiNPs ($t_{1/2}$ = 35 minutes); fluorescent signal for PEG-SiNPs in the circulation was visible up to 5 hours post injection. Liver accumulation of both OH-SiNPs and COOH-SiNPs was evident by 3 hours, whereas for PEG-SiNPs signal no liver signal was present over the 24-hour time frame, indicating efficient evasion by the RES. All three types of SiNPs showed urinary excretion by a fluorescent signal in the bladder and urinary meatus; this excretion was also verified in the rat model and by fluorescence and optical measurements. Transmission electron microscopy (TEM) and energy-dispersed x-ray spectrum were also used to verify nanoparticles in urine.

The impact of size of silica nanoparticles on tissue distribution and elimination was investigated by Cho et al.[63] using fluorescent dye–labeled silica particles 50, 100, and 200 nm in size. Each size was intravenously administered to mice (50 mg/kg), and distribution and excretion were followed over 7 days using confocal laser scanning microscopy and a fluorescence microplate reader. Using confocal microscopy, all three sizes of nanoparticles were visualized in the kidney. The largest nanoparticles (200 nm) were in higher concentrations in the liver and spleen compared to the smaller sized nanoparticles. For all sizes, no indications of particle accumulation were observed in the brain and lungs. Within the liver, nanoparticles were associated with macrophages; splenic uptake of nanoparticles was primarily found in the red pulp. Although trends in organ uptake were consistent for all three sizes, excretion differed. The smallest nanoparticles (50 nm) peaked in urine earlier (12 hours) and excreted faster than the larger sized nanoparticles; fecal excretion peaked at 24 hours. The largest nanoparticles (200 nm) did not peak in urine until 24 hours and were excreted at lower concentrations overall than the 50 and 100 nm particles.

Particle size and coating were varied in a recent study in 2011.[64] Four different sizes of mesoporous silica nanoparticles (MSNs) were tested with and without PEGylation in mice over a month (MSNs-80, PEG-MSNs-80, MSNs-120, PEG-MSNs-120, MSNs-200, PEG-MSNs-200, MSNs-360, and PEG-MSNs-360). Using fluorescence detection, all particles were primarily found in the liver and spleen, with decreasing amounts in the lungs, kidney, and heart. Overall, PEG-MSNs accumulated less in organs than MSNs of similar size, as early as 30 minutes post injection, suggesting that PEGylation decreased uptake by the liver and spleen phagocytes in circulation. For particle size, tissue concentrations increased with an increase in particle size with the exception of PEG-MSNs-360, which showed much lower concentrations in the spleen. As expected, the nanoparticles of smaller sizes had a longer blood-circulation lifetime than larger particles. In terms of excretion, larger particles cleared faster than smaller ones, presumably due to faster biodegradation by the liver and spleen upon capture. Tissue toxicity was also examined over a month for all particles; for both MSNs and PEG-MSNs, no significant toxicity or inflammation was observed.

10.3.2 Accelerator Mass Spectrometry

Recently, we investigated the pharmacokinetics and biodistribution of silica nanoparticles using AMS.[24] In the study, 33 nm silica nanoparticles (^{14}C-SiNPs) were investigated after a single intravenous injection (0.5 mg) over 8 weeks. Over the first 48 hours, ^{14}C-SiNPs were rapidly cleared from the plasma with a distribution half-life of 22.8 minutes and an elimination half-life of 78.4 hours. The long elimination half-life is indicative of recirculating nanoparticles from either enterohepatic circulation or release from other peripheral organs back into the bloodstream. Over the first 24 hours, uptake of ^{14}C-SiNP primarily occurred in the liver, spleen, kidney, and lungs, with 62%, 6.4%, 5.1%, and 1.5% of the injected dose at 2 hours, respectively. Other peripheral tissues such as the lymph nodes, bone marrow, fat, and muscle also contained detectable levels of ^{14}C-SiNP (<1%). Long-term biodistribution showed a gradual decline of ^{14}C-SiNP in tissues over time. At 8 weeks, significant levels of ^{14}C-SiNP were observed in the liver, spleen, kidney, lungs, and cervical lymph nodes. Clearance of ^{14}C-SiNP occurred through both urine and feces, with urine excreting 69.4% of the total amount and feces 30.6%. Clearance through urine was also more rapid ($t_{1/2} = 1.9$ hours) compared to that through feces ($t_{1/2} = 22.1$ hours). The results reported demonstrate the utility of AMS for long-term pharmacokinetic and biodistribution analysis of nanomaterials; even very small amounts of tissue including lymph nodes and bone marrow can be quantified with great precision over extended periods of time.

10.3.3 Gamma Counting

Gamma counting has been used to investigate the effects of size, geometry, porosity, and surface charge for silica nanoparticles. One study investigating the influence of size using gamma counting was reported by Xie et al.[42] in 2009. Two sizes of ^{125}I-labeled amine-terminated silica nanoparticles (20 and 80 nm) were intravenously injected into mice (10 mg/kg), and long-term tissue distribution was evaluated over 30 days. The authors found the pattern of distribution of both sizes to be similar; accumulation primarily occurred in the liver, spleen, and lungs. Much smaller quantities (<1%ID/g) were found in other tissues except for bone, in which concentrations ranging from approximately 1%ID/g to 2%ID/g were observed for the 20 nm nanoparticles over 30 days. Additionally, the 20 nm nanoparticles accumulated at a higher percentage in organs, indicating that the smaller size of the nanoparticles allowed for higher levels of diffusion in organs such as the liver and spleen. These findings were verified with TEM of the splenic and liver tissue; an enhanced uptake of the 20 nm nanoparticles in macrophages was observed, which could be attributed to their larger surface area. Smaller nanoparticles have been shown to bind more readily to surface receptors for uptake into the cells.[66,67] At 1 week post injection, histological analysis also showed pathological changes in the liver, indicating potential toxicity effects.

The effects of geometry, porosity, and surface characteristics on silica nanoparticle biodistribution were recently evaluated by Yu et al.[68] in mice over 72 hours. While all nanoparticles tested showed accumulation primarily in the liver and spleen, changes in the lungs accumulation were also observed. When comparing different geometries, both spherical and rod-shaped nanoparticles accumulated to a high degree in the lungs; however, compared to their amine-modified counterparts, almost a complete reduction in lung accumulation was observed for the spherical nanoparticles. Only slightly lower levels were observed for the amine-modified rod-shaped silica. Porosity was also compared; nonporous silica showed very low levels of accumulation in the lungs, whereas mesoporous silica primarily accumulated in the lung compared with the liver and spleen.

10.3.4 Spectrometry and Spectroscopy Methods

The differences in shape and size in biodistribution were also investigated in a tumor model using ICP-AES by Decuzzi.[69] Silica spheres (0.7, 1, 1.5, 2.53, and 5 µm) were injected into mice at two different doses (10^7 [low] or 10^8 [high]), and major organs were harvested along with tumors 2–6 hours after injection. The liver, lungs, and spleen showed a higher uptake of silica with the higher dose of particles for all sizes, except for the 1.0 and 2.5 µm spheres in the spleen. This trend was not observed in the other organs analyzed: brain, kidneys, tumor, and heart. In terms of total percentage of dose absorbed, an increase in overall percentage ID was higher for particles of smaller size; this was mostly evident in the smaller organs, including the brain, heart, tumor, and kidneys. Statistically, the correlation in the liver, spleen, and lungs was not as clear. Histological analysis of the tissues showed adherence of particles to the blood vessel walls in the heart, kidneys, liver, and spleen. Vascular adhesion was also observed in the lungs for smaller sized particles in conjunction with partial or total obstruction of the capillaries.

The authors also clearly demonstrated that the shape could influence distribution. Silicon-based particles with hemispherical, discoidal, and cylindrical shapes were investigated and compared to spherical silica of similar size (1 µm). In lung tissue, discoidal particles accumulated four times higher than spherical particles and eight times higher than cylindrical and quasi-hemispherical particles. A similar trend for discoidal particle accumulation was also observed in the heart. Cylindrical particles accumulated at a higher amount in the liver than spherical (2×), quasi-hemispherical (2×), and discoidal (5×) particles. For spleen, quasi-hemispherical and discoidal particles accumulated at the same level but at a higher amount than spherical and cylindrical particles. For tumor, brain, and kidneys, no clear differences in uptake were observed for any of the particle shapes.

Metabolic pathway identification for silica-coated nanoparticles was investigated with ICP-MS. Chen et al.[70] utilized a silica-coated quantum dot (CdSeS) approximately 20 nm in size for single dose studies (5 nmol per mouse) over 5 days. For pharmacokinetic analysis, the authors found a long plasma half-life for the nanoparticles ($t_{1/2} = 19.8$ hours) and a high volume of distribution ($V_d = 1611$ mL/kg), indicating an extensive accumulation of the nanoparticles in tissues. Tissue distribution using ICP-MS revealed accumulation primarily localized to the liver, spleen, lungs, and kidney; levels close to background levels were observed in other peripheral tissues (heart, brain, bones, muscles, and spermary). Peak concentrations of nanoparticles in tissue were observed at 6 hours post injection. When normalized to dose, liver and kidney had the highest levels at 6 hours, approximately 50%ID and approximately 20%ID, respectively, followed by the spleen (~2.2%ID) and lungs (~2.0%ID). Elimination occurred through both feces (33.3%) and urine (23.8%), with quicker excretion taking place through feces. Peak excretion through the fecal route was noted at 6–12 hours post exposure, whereas urine was more delayed and did not peak until 24–36 hours. Three metabolic pathways were identified in the study reported: (1) excretion by the kidneys through urine was from nanoparticles that retained their original size, (2) protein-bound nanoparticles were translocated to the liver and excreted through the fecal route, and (3) a very small fraction of the nanoparticles aggregated to a larger size and was retained in the liver tissue. Within 48 hours after injection, most of the organs eliminated the nanoparticles completely with the exception of the liver, in which 8.6% of the dose remained at 5 days post exposure.

In a single and repeated dose study, Liu et al.[71] utilized mesoporous hollow silica nanoparticles (MHSNs) 110 nm in size to investigate toxicity, biodistribution, and clearance after intravenous injection in mice. For single-dose toxicity, lethal dose 50 was found to be higher than 1000 mg/kg. Repeated dose studies for 20, 40, and 80 mg/kg of nanoparticles that were injected daily for 14 days resulted in no deaths. Taken together, the nanoparticles demonstrated low toxicity for both dosing regimens. ICP-OES and TEM were used to identify sites of nanoparticle accumulation. At 24 hours post injection, silicon levels in the spleen and liver peaked and then declined over 4 weeks with the spleen containing higher concentrations of MHSNs. Of the total dose administered, 85% of MHSNs localized in the liver and spleen combined at 24 hours, which declined to 60% after 1 week. At 4 weeks, 41% was still retained in the spleen and 7% in the liver; only approximately 50% of the dose was excreted. In other tissues, much smaller but detectable amounts were found in the lungs, kidney, brain, and testicle (≤5%), which decreased in concentration over the 4-week study. TEM analysis provided visualization of the MHSNs in the lysosomal compartment within macrophages in the liver and spleen at 24 hours.

The effects of nanoparticle shape have also been investigated for silica in a combined study by Huang et al.[72] using fluorescence imaging and ICP-AES. Bare and PEGylated long- and short-rod fluorescent MSNs were tested; after intravenous administration to mice, all MSNs primarily accumulated in the liver, spleen, and lungs in amounts greater than 80% of the administered dose. Clear particle shape effects were observed with short-rod MSNs accumulating at higher levels in the liver, whereas long-rod MSNs distributed mostly in the spleen. The concentrations of MSNs in the lungs increased with the PEGylated MSNs compared to their bare counterparts. For excretion, short-rod MSNs cleared at a faster rate through both urine and feces routes compared to long-rod MSNs.

10.3.5 Summary

Similar to iron oxide nanoparticles, a clear trend in uptake for silica nanoparticles was observed with localization mainly in the liver and spleen. For physiochemical properties, the shape of silica was shown to influence uptake by other organs, such as the heart and lungs; shape and porosity also affected the level of uptake for these target tissues. In line with iron oxide, PEGylation of the nanoparticle also increased both circulation and excretion times and reduced the degree of uptake by RES organs. Finally, varying the functional groups on the surface was shown to influence both circulation time and uptake by tissue.

10.4 COMPARISON OF SILICA AND IRON OXIDE NANOPARTICLES

When comparing the main findings for each nanoparticle type, some agreement exists for main organs of accumulation; however, for half-life, surface charge, and coating some findings did not agree. For iron oxide nanoparticles, in the Bengele work the plasma half-life for nanoparticles 4.3–6.2 nm in size was $t_{1/2} = 3.7$ hours. Conversely, Jeung-Eun Lee et al. observed a much longer half-life ($t_{1/2} = 7$ to 8 hours) for a similar sized iron oxide nanoparticle (7 nm). This could be due to the differences in animal models used for the studies. Additionally, for the work by Jeung-Eun Lee a fluorophore was conjugated to the surface for detection, whereas for the Bengele work the radioisotope was incorporated into the nanoparticle for detection. These changes in surface chemistry could also explain the differences in half-life for the two studies. For silica, plasma half-life was only determined for larger nanoparticles $\geq$ 20 nm. In the study by Chen et al., for silica-coated nanoparticles 20 nm in size a $t_{1/2} = 19.8$ hours was observed, which is much longer than what was found for smaller sizes of iron oxide. While experimental details such as species and route of delivery were similar, the behavior of the nanoparticles once injected may have been different (i.e., aggregation state). Changes in the coating of the two types of nanoparticles once in the bloodstream could have changed the size or aggregation state, which could have led to differences in plasma half-life and organ uptake.

For surface coating, PEG was used for both silica and iron oxide nanoparticles. Although a direct comparison cannot be made due to the differences in the size of nanoparticles used, in the works by both He et al. and Cole et al. longer circulation half-lives were observed for the PEG-coated nanoparticles when compared with their uncoated counterparts. Cole et al. observed a $t_{1/2} = 7.3$–11.8 hours half-life for PEG-coated iron oxide nanoparticles (140–170 nm) in the rat; uncoated nanoparticles (104 nm) were cleared out of the bloodstream at a much quicker rate ($t_{1/2} = 0.12$ hour). For silica, He et al. compared silica nanoparticles (45 nm) with three different surface coatings: hydroxyl, carboxyl, and PEG. The un-PEGylated nanoparticles showed a much shorter half-life ($t_{1/2}$ = 80 minutes for hydroxyl and $t_{1/2}$ = 35 minutes for carboxyl), whereas PEGylated nanoparticles were capable of longer circulation ($t_{1/2}$ = 180 minutes). Extending the work by He et al., comparison of similar surface functional groups of silica (hydroxyl) with iron oxide showed some differences in biodistribution. For silica, uptake was primarily localized to the liver, bladder, and urinary meatus over 24 hours. In a slightly shorter study with a smaller nanoparticle size (12 nm), Portet et al. observed iron oxide accumulation in the liver, spleen, carcass, and bone. Whereas a small amount of signal was observed ex vivo in the spleen by He et al., there was no obvious signal in the bone.

Variations in organ uptake were also observed for nanoparticles of the same size. In the works of Kumar et al.[61] and Al Faraj et al.[52], nanoparticles 20 nm in size were investigated. For iron oxide, accumulation was observed only in the liver, spleen, and kidney; for silica, distribution was more widespread, with accumulation extending to the lungs, intestine, heart, stomach, and skin in addition to the liver, spleen, and kidney. For silica, the authors demonstrated that conjugation of the dye to the nanoparticle increased the affinity of the nanoparticle for skin. Therefore, the discrepancies between the two may be due to the change in the surface chemistry of the nanoparticles. For the iron oxide MRI and ICP-MS were used for detection and solely relied on iron content in the nanoparticle, whereas the silica detection was reliant on the conjugated dye.

10.5   CONCLUDING REMARKS

In this chapter, we summarize some of the major observations for the biodistribution of silica and iron oxide nanoparticles. This is by no means an exhaustive report of the published data; there are many other studies in the literature for these materials including core/shell hybrid nanoparticles for multidetection applications. The experiments we have outlined clearly demonstrate the complexity of nanomaterials and that each aspect (size, shape, composition, and surface chemistry) can dramatically influence their behavior in vivo. The trends identified for iron oxide and silica can be applied to both biomedical and environmental cases of nanoparticle distribution and toxicity. When considering nanoparticles for drug delivery or diagnostics, the size and surface coating/chemistry should be designed to facilitate efficient delivery of the nanoparticles to the target, while minimizing toxicity from extensive organ accumulation. For cases of environmental exposure, both size and shape can provide insight into probable sites of accumulation and potential toxicity, whereas chemical information can provide knowledge on surface chemistry and how it may affect clearance and uptake.

Additionally, when investigating nanoparticles it is important to consider both detection method and label used for the study; these can influence the results obtained and the behavior of the nanoparticles, respectively. Although some growth has been made in investigating which properties dictate nanoparticle distribution, more progress remains to be made. Ultimately, the combination of several methods to acquire a multiscale perspective and analysis will remain imperative for investigating the nanoparticle distribution in vivo.

REFERENCES

1. Papasani, M. R., Wang, G. & Hill, R. A. Gold nanoparticles: The importance of physiological principles to devise strategies for targeted drug delivery. *Nanomedicine: Nanotechnology, Biology, and Medicine* **8**, 804–14 (2012).
2. Gupta, A. K. & Wells, S. Surface-modified superparamagnetic nanoparticles for drug delivery: Preparation, characterization, and cytotoxicity studies. *IEEE Transactions on Nanobioscience* **3**, 66–73 (2004).
3. Cho, K., Wang, X., Nie, S., Chen, Z. G. & Shin, D. M. Therapeutic nanoparticles for drug delivery in cancer. *Clinical Cancer Research: An Official Journal of the American Association for Cancer Research* **14**, 1310–6 (2008).
4. Sun, C., Fang, C., Stephen, Z., Veiseh, O., Hansen, S., Lee, D., Ellenbogen R., Olson, J. & Zhang, M. Tumor-targeted drug delivery and MRI contrast enhancement by chlorotoxin-conjugated iron oxide nanoparticles. *Nanomedicine* **3**, 495–505 (2008).
5. Nahrendorf, M., Zhang, H., Hembrador, S., Panizzi, P., Sosnovik, D., Aikawa, E., Libby, P., Swirski, F. & Weissleder, R. Nanoparticle PET-CT imaging of macrophages in inflammatory atherosclerosis. *Circulation* **117**, 379–87 (2008).
6. Lee, C. H., Cheng, S., Wang, Y., Chen, Y., Chen, N., Souris, J., Chen, C., Mou, C., Yang, C. & Lo, L. Near-infrared mesoporous silica nanoparticles for optical imaging: Characterization and in vivo biodistribution. *Advanced Functional Materials* **19**, 215–22 (2009).
7. Jarrett, B. R., Correa, C., Ma, K. L. & Louie, A. Y. In vivo mapping of vascular inflammation using multimodal imaging. *PloS One* **5**, e13254 (2010).
8. Longmire, M., Choyke, P. L. & Kobayashi, H. Clearance properties of nano-sized particles and molecules as imaging agents: Considerations and caveats. *Nanomedicine* **3**, 703–17 (2008).
9. Tassa, C., Shaw, S. Y. & Weissleder, R. Dextran-coated iron oxide nanoparticles: A versatile platform for targeted molecular imaging, molecular diagnostics, and therapy. *Accounts of Chemical Research* **44**, 842–52 (2011).
10. Rosen, J. E., Chan, L., Shieh, D. & Gu, F. X. Iron oxide nanoparticles for targeted cancer imaging and diagnostics. *Nanomedicine: Nanotechnology, Biology, and Medicine* **8**, 275–90 (2012).
11. Merget, R., Bauer, T., Kupper, H., Philippou, S., Bauer, H., Breitstadt, R. & Bruening, T. Health hazards due to the inhalation of amorphous silica. *Archives of Toxicology* **75**, 625–34 (2002).
12. Simkó, M. & Mattsson, M. O. Risks from accidental exposures to engineered nanoparticles and neurological health effects: A critical review. *Particle and Fibre Toxicology* **7**, 42 (2010).
13. Card, J. W., Zeldin, D. C., Bonner, J. C. & Nestmann, E. R. Pulmonary applications and toxicity of engineered nanoparticles. *American Journal of Physiology. Lung Cellular and Molecular Physiology* **295**, L400–11 (2008).
14. Fifield, L. Accelerator mass spectrometry and its applications. *Reports on Progress in Physics* **62**, 1223–74 (1999).
15. Vogel, J. S., Turteltaub, K. W., Finkel, R. & Nelson, D. E. Accelerator mass spectrometry. *Analytical Chemistry* **67**, 353A–9A (1995).
16. Nageswara Rao, R. & Talluri, M. V. N. K. An overview of recent applications of inductively coupled plasma-mass spectrometry (ICP-MS) in determination of inorganic impurities in drugs and pharmaceuticals. *Journal of Pharmaceutical and Biomedical Analysis* **43**, 1–13 (2007).
17. Lappin, G. & Garner, R. C. Big physics, small doses: The use of AMS and PET in human microdosing of development drugs. *Nature Reviews. Drug Discovery* **2**, 233–40 (2003).
18. Millhauser, G. L. Selective placement of electron spin resonance spin labels: New structural methods for peptides and proteins. *Trends in Biochemical Sciences* **17**, 448–52 (1992).

19. Manning, T. J. & Grow, W. R. Inductively coupled plasma–atomic emission spectrometry. *The Chemical Educator* **2**, 1–19 (1997).
20. Morris, A. C. A diagnostic-level whole-body counter. *Journal of Nuclear Medicine* **6**, 481–8 (1965).
21. Luker, G. D. & Luker, K. E. Optical imaging: Current applications and future directions. *Journal of Nuclear Medicine: Official Publication, Society of Nuclear Medicine* **49**, 1–4 (2008).
22. Scheffer, A., Engelhard, C., Sperling, M. & Buscher, W. ICP-MS as a new tool for the determination of gold nanoparticles in bioanalytical applications. *Analytical and Bioanalytical Chemistry* **390**, 249–52 (2008).
23. Oberdörster, G., Oberdörster, E. & Oberdörster, J. Nanotoxicology: An emerging discipline evolving from studies of ultrafine particles. *Environmental Health Perspectives* **113**, 823–39 (2005).
24. Malfatti, M. A., Palko, H. A., Kuhn, E. A. & Turteltaub, K. W. Determining the pharmacokinetics and long-term biodistribution of SiO_2 nanoparticles in vivo using accelerator mass spectrometry. *Nano Letters* **12**, 5532–8 (2012).
25. Turteltaub, K. W. & Dingley, K. H. Application of accelerated mass spectrometry (AMS) in DNA adduct quantification and identification. *Toxicology Letters* **102–103**, 435–9 (1998).
26. Rudin, M. & Weissleder, R. Molecular imaging in drug discovery and development. *Nature Reviews. Drug Discovery* **2**, 123–31 (2003).
27. Cherry, S. R. & Gambhir, S. S. Use of positron emission tomography in animal research. *ILAR Journal/National Research Council, Institute of Laboratory Animal Resources* **42**, 219–32 (2001).
28. Chertok, B., Cole, A. J., David, A. E. & Yang, V. C. Articles comparison of electron spin resonance spectroscopy and inductively-coupled plasma optical emission spectroscopy for biodistribution analysis of iron-oxide nanoparticles. *Molecular Pharmaceutics* **7**, 375–85 (2010).
29. Jain, T. K., Morales, M. A., Sahoo, S. K., Leslie-pelecky, D. L. & Labhasetwar, V. Iron oxide nanoparticles for sustained delivery of anticancer agents. *Molecular Pharmaceutics* **2**, 194–205 (2005).
30. Pan, B., Cui, D., Sheng, Y., Ozkan, C., Gao, F., He, R., Li, Q., Xu, P. & Huang, T. Dendrimer-modified magnetic nanoparticles enhance efficiency of gene delivery system. *Cancer Research* **67**, 8156–63 (2007).
31. Ryoo, S. R., Jang, H., Kim, K., Lee, B., Bo Kim, K., Kim, Y., Yeo, W., Lee, Y., Kim, D. & Min, D. Functional delivery of DNAzyme with iron oxide nanoparticles for hepatitis C virus gene knockdown. *Biomaterials* **33**, 2754–61 (2012).
32. Tate, J. A., Petryk, A. A., Giustini, A. J. & Hoopes, P. J. In vivo biodistribution of iron oxide nanoparticles: An overview. *SPIE Proceedings* **7901**, 790117–9 (2011).
33. Wahajuddin & Arora, S. Superparamagnetic iron oxide nanoparticles: Magnetic nanoplatforms as drug carriers. *International Journal of Nanomedicine* **7**, 3445–71 (2012).
34. Seijo, B., Fattal, E., Roblot-Treupel, L. & Couvreur, P. Design of nanoparticles of less than 50 nm diameter: Preparation, characterization and drug loading. *International Journal of Pharmaceutics* **62**, 1–7 (1990).
35. Loomis, K., McNeeley, K. & Bellamkonda, R. V. Nanoparticles with targeting, triggered release, and imaging functionality for cancer applications. *Soft Matter* **7**, 839 (2011).
36. Schluep, T., Hwang, J., Hildebrandt, I., Czernin, J., Choi, C., Alabi, C., Mack, B. & Davis, M. Pharmacokinetics and tumor dynamics of the nanoparticle IT-101 from PET imaging and tumor histological measurements. *Proceedings of the National Academy of Sciences of the United States of America* **106**, 11394–9 (2009).
37. Jain, T. K., Reddy, M. K., Morales, M. A., Leslie-pelecky, D. L. & Labhasetwar, V. Articles biodistribution, clearance, and biocompatibility of iron oxide magnetic nanoparticles in rats. *Molecular Pharmaceutics* **5**, 316–27 (2008).

38. Cole, A. J., David, A. E., Wang, J., Galban, C. J. & Yang, V. C. Magnetic brain tumor targeting and biodistribution of long-circulating PEG-modified, cross-linked starch coated iron oxide nanoparticles. *Biomaterials* **32**, 6291–301 (2011).
39. Park, J. Y., Daksha, P., Lee, G. H., Woo, S. & Chang, Y. Highly water-dispersible PEG surface modified ultra small superparamagnetic iron oxide nanoparticles useful for target-specific biomedical applications. *Nanotechnology* **19**, 365603 (2008).
40. Varna, M. In vivo distribution of inorganic nanoparticles in preclinical models. *Journal of Biomaterials and Nanobiotechnology* **03**, 269–79 (2012).
41. Lee, W. M. & Senior, J. R. Recognizing drug-induced liver injury: Current problems, possible solutions. *Toxicologic Pathology* **33**, 155–64 (2005).
42. Xie, G., Sun, J., Zhong, G., Shi, L. & Zhang, D. Biodistribution and toxicity of intravenously administered silica nanoparticles in mice. *Archives of Toxicology* **84**, 183–90 (2010).
43. Liu, S. Y., Han, Y., Yin, L. P., Long, L. & Liu, R. Toxicology studies of a superparamagnetic iron oxide nanoparticle in vivo. *Advanced Materials Research* **47–50**, 1097–100 (2008).
44. Wang, J., Chen, Y., Chen, B., Ding, J., Xia, G., Gao, C., Cheng, J. et al. Pharmacokinetic parameters and tissue distribution of magnetic Fe(3)O(4) nanoparticles in mice. *International Journal of Nanomedicine* **5**, 861–6 (2010).
45. Mikheev, V. B., Forsythe, W. C. & Swita, B. N. *Nano-Aerosol Generation System and Methods*. At <http://www.freepatentsonline.com/WO2012058246A1.html.> (2012).
46. Zhu, M. T., Feng, W., Wang, Y., Wang, B., Wang, M., Ouyang, H., Zhao, Y. & Chai, Z. Particokinetics and extrapulmonary translocation of intratracheally instilled ferric oxide nanoparticles in rats and the potential health risk assessment. *Toxicological Sciences: An Official Journal of the Society of Toxicology* **107**, 342–51 (2009).
47. Bengele, H., Palmacci, S., Rogers, J., Jung, C., Crenshaw, J. & Josephson, L. Biodistribution of an ultrasmall superparamagnetic iron oxide colloid, BMS 180549, by different routes of administration. *Magnetic Resonance Imaging* **12**, 433–42 (1994).
48. Portet, D., Denizot, B., Rump, E., Hindre, F., Le Jeune, J. & Jallet, P. Comparative biodistribution of thin-coated iron oxide nanoparticles TCION: Effect of different bisphosphonate coatings. *Drug Development Research* **54**, 173–81 (2001).
49. Mistry, A., Stolnik, S. & Illum, L. Nanoparticles for direct nose-to-brain delivery of drugs. *International Journal of Pharmaceutics* **379**, 146–57 (2009).
50. Yang, W., Peters, J. I. & Williams, R. O. Inhaled nanoparticles—a current review. *International Journal of Pharmaceutics* **356**, 239–47 (2008).
51. Schlachter, E. K., Widmer, H., Bregy, A., Lonnfors-Weitzel, T., Vajtai, I., Corazza, N., Bernau, V. et al. Metabolic pathway and distribution of superparamagnetic iron oxide nanoparticles: In vivo study. *International Journal of Nanomedicine* **6**, 1793–800 (2011).
52. Al Faraj, A., Lacroix, G., Alsaid, H., Elgrabi, D., Stupar, V., Robidel, F., Gaillard, S., Canet-Soulas, E. & Cremillieux, Y. Longitudinal ^{3}He and proton imaging of magnetite biodistribution in a rat model of instilled nanoparticles. *Magnetic Resonance in Medicine: Official Journal of the Society of Magnetic Resonance in Medicine/Society of Magnetic Resonance in Medicine* **59**, 1298–303 (2008).
53. Kauczor, H. & Kreitner, K. MRI of the pulmonary parenchyma. *European Radiology* 1755–64 (1999).
54. Lee, M. J., Veiseh, O., Bhattarai, N., Sun, C., Hansen, S.J., Ditzler, S., Knoblaugh, S., Lee, D., Ellenbogen, R., Zhang, M. & Olson J. M. Rapid pharmacokinetics and biodistribution studies using chlorotoxin-conjugated iron oxide nanoparticles: A novel nonradioactive method. *PLoS one* **5**, e9536. doi:10.1371/journal.pone.0009536.
55. Ntziachristos, V. Fluorescence molecular imaging. *Annual Review of Biomedical Engineering* **8**, 1–33 (2006).

56. Mamaeva, V., Sahlgren, C. & Lindén, M. Mesoporous silica nanoparticles in medicine—recent advances. *Advanced Drug Delivery Reviews* (2012). doi:10.1016/j.addr.2012.07.018.
57. Yang, P., Gai, S. & Lin, J. Functionalized mesoporous silica materials for controlled drug delivery. *Chemical Society Reviews* **41**, 3679–98 (2012).
58. Burns, A., Vider, J., Ow, H., Herz, E., Penate-Medina, O., Baumgart, M., Larson, S., Wiesner, U. & Bradbury, M. Fluorescent silica nanoparticles with efficient urinary excretion for nanomedicine. *Nano Letters* **9**, 442–8 (2009).
59. Gary-Bobo, M., Hocine, O., Brevet, D., Maynadier, M., Raehm, L., Richeter, S., Charasson, V. et al. Cancer therapy improvement with mesoporous silica nanoparticles combining targeting, drug delivery and PDT. *International Journal of Pharmaceutics* **423**, 509–15 (2012).
60. Lu, J., Liong, M., Li, Z., Zink, J. I. & Tamanoi, F. Biocompatibility, biodistribution, and drug-delivery efficiency of mesoporous silica nanoparticles for cancer therapy in animals. *Small (Weinheim an Der Bergstrasse, Germany)* **6**, 1794–805 (2010).
61. Kumar, R., Roy, I., Ohulchanskky, T., Vathy, L., Bergey, E., Sajjad, M. & Prasad, P. In vivo biodistribution and clearance studies using multimodal organically modified silica nanoparticles. *ACS Nano* **4**, 699–708 (2010).
62. Duan, X. & Li, Y. Physicochemical characteristics of nanoparticles affect circulation, biodistribution, cellular internalization, and trafficking. *Small* **9**, 1521–32 (2013). doi:10.1002/smll.201201390.
63. Cho, M., Cho, W., Choi, M., Kim, S., Han, B., Kim, S., Kim, H., Sheen, Y. & Jeong, J. The impact of size on tissue distribution and elimination by single intravenous injection of silica nanoparticles. *Toxicology Letters* **189**, 177–83 (2009).
64. He, Q., Zhang, Z., Gao, F., Li, Y. & Shi, J. In vivo biodistribution and urinary excretion of mesoporous silica nanoparticles: Effects of particle size and PEGylation. *Small (Weinheim an Der Bergstrasse, Germany)* **7**, 271–80 (2011).
65. He, X., Nie, H., Wang, K., Tan, W., Wu, X. & Zhang, P. In vivo study of biodistribution and urinary excretion of surface-modified silica nanoparticles. *Analytical Chemistry* **80**, 9597–603 (2008).
66. Gao, H., Shi, W. & Freund, L. B. Mechanics of receptor-mediated endocytosis. *Proceedings of the National Academy of Sciences of the United States of America* **102**, 9469–74 (2005).
67. Bao, G. & Bao, R. Shedding light on the dynamics of endocytosis and viral budding. *Proceedings of the National Academy of Sciences* **102**, 9997–8 (2005).
68. Yu, T., Hubbard, D., Ray, A. & Ghandehari, H. In vivo biodistribution and pharmacokinetics of silica nanoparticles as a function of geometry, porosity and surface characteristics. *Journal of Controlled Release: Official Journal of the Controlled Release Society* **163**, 46–54 (2012).
69. Decuzzi, P., Godin, B., Tanaka, T., Lee, S., Chiappini C., Liu, X. & Ferrari, M. Size and shape effects in the biodistribution of intravascularly injected particles. *Journal of Controlled Release: Official Journal of the Controlled Release Society* **141**, 320–7 (2010).
70. Chen, Z., Chen, H., Meng, H., Xing, G., Gao, X., Sun, B., Shi, X. et al. Bio-distribution and metabolic paths of silica coated CdSeS quantum dots. *Toxicology and Applied Pharmacology* **230**, 364–71 (2008).
71. Liu, T., Li, L., Teng, X., Huang, X., Liu, H., Chen, D., Ren, J., He, J. & Tang, F. Single and repeated dose toxicity of mesoporous hollow silica nanoparticles in intravenously exposed mice. *Biomaterials* **32**, 1657–68 (2011).
72. Huang, X., Li, L., Liu, T., Hao, N., Liu, H., Chen, D. & Tang, F. The shape effect of mesoporous silica nanoparticles on biodistribution, clearance, and biocompatibility in vivo. *ACS Nano* **5**, 5390–9 (2011).

11 Quantitative Single-Cell Approaches to Assessing Nanotoxicity in Nanomedicine Systems

James F. Leary

CONTENTS

11.1 INTRODUCTION

Nanomedical approaches to advanced diagnostics and drug delivery involve the use of nanomaterials, image contrast agents, targeting molecules, and therapeutic drugs. It is important to assess therapeutic efficacy in terms of intended (targeted) versus unintended (bystander) cell toxicity. Design of these systems requires quantitative assessment of nanoparticle biodistribution and measures of nanotoxicity at the single cell and tissue level. In vitro nanotoxicity studies should be carried out in a way that has some prediction of in vivo nanotoxicity at the tissue and organ levels. The overall problem of toxicity of nanomaterials is discussed in the following sections.

11.1.1 Why Nanomaterials Are Potentially a Problem (Large Surface to Volume Ratio)

While there are several reasons why nanomaterials may potentially be a problem in terms of nanotoxicity, one of the most important reasons is its very large surface area to volume ratio. Since many chemical reactions are aided by surfaces that can bring together different molecules for potential interactions and aid in their reorientations for potential interactions, small nanoparticles can act as catalysts and like enzymes to amplify the effects of interactions between molecules. The reactivity of nanomaterials can vary widely with size. For this reason their nanotoxicity must be measured using nanoparticles of specific size ranges. The take-home message should be that the same nanomaterials can have different nanotoxicity depending on their size.

11.1.2 Why Nanomaterial Toxicity Is Hard to Measure: Hydrophobic Nanoparticles Covered with Hydrophilic Coatings That Can Mask True Toxicity

Nanomaterial toxicity is particularly difficult to measure because most nanomaterials are hydrophobic. For this reason, they are typically treated with hydrophilic or biological surface coatings to enable them to go into aqueous solutions. If the biocoating is stable and relatively permanent in its environment, then the nanotoxicity may be measured to be very low. Such is the case for cadmium-derived quantum dots that have biocoatings. With the biocoatings intact one is measuring the toxicity of the biocoating, not the nanomaterials underneath. But if the biocoating is removed, the nanomaterial may become much more toxic. Different biological cell types may cause disruption of these biocoatings. The take-home message is that the nanotoxicity evaluations must take into account the stability of the biocoatings in a given environment and that the measured nanotoxicity may be highly dependent on the cell type. To avoid mistakes in assessing nanotoxicity, the measurements should be carried out in the full range of environments and cell types likely to be exposed.

11.1.3 SIZE, SHAPE, AND ELECTRICAL CHARGE MATTER

As mentioned earlier, the size of nanomaterials matters because it relates to the surface to volume ratio that can affect overall reactivity. But it is not only the size that matters but also its shape. Charge properties and interaction surfaces vary with the local radius of curvature. So nanomaterials may have different toxicity than spherically shaped nanoparticles of the same material.[1] The shape of the nanoparticle affects the local charge density of the counter-ions in the aqueous solvent. Since most nanotoxicity measurements on living cells must be carried out in these aqueous solutions that have the proper composition to be suitable, or at least possible for living cells, the interaction of the solution with the nanomaterials must be taken into account. Needless to say, while the toxicity of the nanomaterials must be deduced by elimination of the toxicity of the solvent alone (e.g., many cells find phosphate buffered saline at least mildly cytotoxic), it should not be assumed that the two toxicities are entirely independent. The nanotoxicity of the net system may be either enhanced or decreased by the interactions between nanomaterials and the solutions in which the cells are exposed.

11.1.4 SOME NANOMATERIALS CONTAIN VERY TOXIC ELEMENTAL COMPONENTS

Nanomaterials may contain atoms that, while embedded in the nanostructures, are not very toxic. But changes in environmental conditions may release those atoms from their initial nanostructures rendering them more toxic. An example is cadmium semiconductor-derived *quantum dots*, which, while cadmium is tightly linked to the nanostructure, have lower toxicity. But when the nanostructure starts breaking down, for example, with the exposure to light, free elemental cadmium is released. Elemental cadmium is much more toxic and can lead to increased nanotoxicity over alternative forms of quantum dots not containing cadmium.[2]

11.1.5 WHY MEASURING BIODISTRIBUTION OF NANOMATERIALS IS CHALLENGING

One of the most challenging aspects of measuring the toxicity of nanoparticles is that the nanotoxicity is at least partially dependent on the relative distribution within tissues and within individual cells. This makes for the very challenging experimental testing situation of trying to enumerate the actual number of nanoparticles over vast areas of tissues and organs and determining if the nanoparticles are inside or outside cells. There are very little data on this problem because of the extreme challenges of making these measurements. By their nature nanoparticles are suboptical, meaning they are below the diffraction limit, and cannot be measured individually by conventional light or fluorescence microscopy. Since they are very small it is difficult or impossible to tag them with enough fluorescent molecules per nanoparticle so that they can be individually seen. While single molecule detection scheme can indeed measure individual nanoparticles, that measurement process cannot currently be done at high enough speeds to permit high-speed image analysis of cellular tissues. Atomic force microscopy has the same problem. Super-resolution microscopy may, or may not, be able to simultaneously measure suboptical objects while being fast enough to provide realistic measurements across large areas of tissues. One attempt to deal with these issues using *nanobarcoding* and in situ PCR amplification of oligonucleotides attached to nanoparticles[3–5] tries to solve the problem by finding the location of nanoparticles across large areas of tissues using optically sized spots that can then be used with scanning cytometry to measure individual cells containing nanoparticles. Since the nanoparticles can be labeled with different barcodes, this raises the interesting possibility of performing multiple experiments simultaneously in animals, thereby reducing both the number of animals needed and inter-animal variations.

11.1.6   WHY MEASURING IN SINGLE CELLS RATHER THAN ORGANS IS IMPORTANT: BUT NOT SUFFICIENT

There is a temptation to measure what we can measure quickly and easily rather than what we should measure, which may be much more difficult. Most measures of nanotoxicity with cells involve measuring two-dimensional (2D) monolayers of tissue culture cells of one type rather than the actual three-dimensional (3D) structures of multiple cell types of non-immortalized cells. Actual thin slices of real tissue are perhaps a better model but lack the complex functions of thick sections of normal tissue. Tissue culture cells (immortalized and altered in ways that typically make them cancer cells lines) are not the equivalent of actual cells as they exist in the body. Cells of different types send signals to one another to maintain the integrity of morphology and function that constitute tissues. Different tissues come together to constitute organs that have overall functions. Making rapid measurements on 3D tissues and organs is extremely difficult and the amount of information acquired is huge. These very large datasets are then very difficult to analyze.

Even if all of the cells of a tissue or organ could be measured, the question is whether the measurement of the collection of single cells will provide an adequate predictor of the function of an organ. Structural measurements of individual cells at single-cell level are frequently less complex to make than functional tests of a tissue or organ. Cells in tissues and organs interact with each other and the overall behavior of the tissue or organ may, or may not, be predictable based on the collection of single-cell measurements. Nonetheless, high-speed measurements of large numbers of single cells are possible and can provide at least some prediction of overall nanotoxicity. Single-cell measurements can be quite sophisticated and need not be crude. Multiple measurements per cell can be made using multicolor fluorescence techniques by both flow cytometry and scanning image cytometry. The important thing to remember is that science is an iterative process. It is never perfect. Likewise nanotoxicity measurements are always going to be deficient in some ways. But we can make progress on the overall problem of nanotoxicity by making progressively better measurements.

11.2 SINGLE-CELL MEASURES OF TOXICITY

Direct or indirect measurements of toxicity can be made in a wide variety of ways at the single-cell level. Typically those measurements are made either on isolated single cells in suspension using flow cytometry or in attached cell monolayers or tissues using scanning image cytometry. When cells from monolayers or tissues are dissociated into single cell suspensions for flow cytometric analysis, we lose information about the 2D and 3D architecture and cell–cell associations and morphology. Sometimes that information loss is crucial—sometimes it is not. We also should not forget that one of the body's largest tissues is a complex liquid suspension of cells, namely blood.

Analysis of blood cells for toxicity is perhaps best made by flow cytometry, which can analyze multiple, as many as 17 colors simultaneously at many thousands of cells per second,[6] colors of fluorescence. The level of detailed analysis at the single-cell level as well as the vast numbers of cells that can be quickly measured enable us to study the toxicity of nanomaterials simultaneously on many different blood cell types, including some that might be extremely rare but important.

11.2.1 WHY SINGLE-CELL MEASURES ARE IMPORTANT

At this time it is perhaps important to show why single-cell measurements are so important. Most realistic biological situations involve the interaction of a number of different cell types, each of which can have a different toxicity when exposed to nanomaterials. Making *bulk cell* measures of toxicity of mixtures of cell types can be very misleading and sometimes completely incorrect and irrelevant. Figure 11.1 shows the importance of single-cell *molecular* measurements for nanotoxicity assays. The upper left panel of Figure 11.1 shows a hypothetical collection of cells that all respond similarly in terms of toxicity to nanomaterials. A bulk cell measurement (e.g., a 3-(4,5-dimethylthiazol-2-yl)-2,5-diphenyltetrazolium bromide (MTT) test) and a single-cell test (e.g., a propidium iodide [PI] assay by flow cytometry) will yield a similar answer. But in the lower left-hand panel of Figure 11.1 we see a hypothetical case (which is much more typical than case 1A) where there are two cell subpopulations of different cell types 1 and 2, each of which responds very differently in terms of cytotoxicity. The bulk cell assay will yield a result, which is the weighted (by relative cell number or frequency) average of the two or more cell types. That answer may prove to be poor or even irrelevant depending on the relative frequencies of cell types and their relative cytotoxicities. A more extreme (but still common and relevant) example is shown in the hypothetical situation outlined in the right-hand panel of Figure 11.1. In this case, the nanomaterials are cytotoxic differently to four cell subpopulations. But the bulk cell measurement completely misses the correct result.

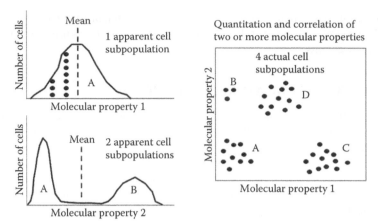

FIGURE 11.1   The power of single-cell measurements is that it is not just the frequency-weighted average of multiple cell subpopulations. If only one molecular property is separately measured there may appear to be a single distribution of cells with a biological variability (the error of flow cytometric measurements is so small that the variation shown is true biological variation, not an error in measurement). When a second molecular property is separately measured there may now appear to be two distinct cell subpopulations. However, when both molecular measurements are made on the same cells and correlated, we see that there are four distinct cell subpopulations in this example. As the number of correlated measurements per cell increases, finer and finer distinctions can be made of more and more cell subpopulations.

The take-home message is that if measurements can be performed at single-cell level and can look at several different cell types separately, then the measurements should be made at single-cell level. This paradigm is true for all measurements regardless of the technology used to acquire the data. The data shown in Figure 11.1 could be from flow cytometry, scanning image cytometry, or some completely different technology. The specific technology used to make the measurements does not matter in this regard.

11.2.2   MEASURES OF CELL DEATH

Most researchers try to measure cell death directly. While measures of cell death may seem obvious and simple, they are not. Death is a complex, multistep process and, as a consequence, simple measures of it are difficult. An analogy might be *how do we measure the death of a person?* Currently, we would not declare a person *dead* if they stopped breathing or their heart stopped beating. The reason is that these are not good measures of a person's death because the person can sometimes be revived. Usually we consider a person *dead* when there is absence of measurable brain electrical activity because a person rarely, if ever, revives after total lack of electrical brain activity for more than a brief time. So whatever single-cell assay we use for measuring cell death should be an absolute predictor of that cell's death. If we take a *late stage* measure of cell death, we can be reasonably certain that the cell will not recover. But what if a cell, exposed to nanomaterials, is in an early stage of the death process and may, or may not, recover? A good measure of toxicity should also know about the cells earlier in the death process, if it is not to underestimate the true final toxicity. There may also be a certain probability of a cell dying or recovering at a given point in the death process.

11.2.2.1 Simple Measures of Cell Death

One of the simplest measures of *cell death* is lack of cell membrane activity. Living cells must maintain an intact cell membrane to survive. When that membrane integrity is compromised, the cell is unable to control the inflow or outflow of molecules essential to life. For this reason a number of assays have been developed involving the exclusion of dyes, fluorescent or nonfluorescent. A *simple assay* involves exclusion of Trypan Blue. This is one of the most commonly used assays of cell death. Unfortunately, it is a poorly understood assay that is highly susceptible to serious errors. As it turns out, the correct concentration of Trypan Blue is highly dependent on cell type. If a too high concentration of Trypan Blue is used for a given cell type, the dye will penetrate even intact cell membranes giving a false-positive measure of cell death. Recall the earlier descriptions of the variabilities of different cell types. A cell mixture containing different cell types may contain cells of widely varying membrane integrity. Devising a single concentration of Trypan Blue that gives an accurate measure of all cell types is problematical. Another simple measure of cell death is the PI exclusion assay that allows fluorescence measurements of cell death to be made. PI has very low quantum efficiency unless it is bound to DNA or RNA whereupon its quantum efficiency increases by orders of magnitude. So cells with leaky membranes allow PI to get through the cell membrane and bind to DNA or RNA within the cell. The result is usually a brightly stained red fluorescent cell that can be readily measured by human eye looking through a fluorescence microscope, and more rapidly measured by flow cytometry or scanning image cytometry. Simple measurement? Not really. Some cell types are inherently pretty leaky and yet survive and function. Other cell types have normally very *tight* membrane structure so that they measure death at different stages and sometimes can recover.

All of this variation with simple one-color fluorescence measures of cell death have led to development of slightly more sophisticated measures not only of cell death requiring some signs of cell life. Examples of the latter include the addition of a fluorogenic substrate (meaning it is a molecule that is nonfluorescent unless it is cleaved by another molecule). An early example of this was the use of fluorescein diacetate with PI. If the nonfluorescent, but otherwise membrane permeable, fluorescein diacetate penetrated the cell membrane it would have its diacetate group cleaved by intracellular esterases present in virtually all living cells leading to a green fluorescence because of the free fluorescein after the diacetate group is cleaved by the intracellular esterases. A more recent version of the preceding assay is that of *calcein AM* (an acetoxymethyl ester of calcein) where the *AM* (acetomethoxy-) portion of the molecule confers easier permeability into cells with intact membranes and the calcein portion of the molecule is less pH sensitive than fluorescein and has a very high quantum efficiency leading to intense green fluorescence if the AM group is cleaved by intracellular esterases present in the cytoplasm of nearly all living cells.

11.2.2.2   More Sophisticated Measures of Cell Death

We now have more sophisticated understanding of the cell death process. Some of the interesting and important measures of cell death are measures of *programmed cell death*. While the earlier descriptions of cell death are due to cell injury (*necrosis*), we now understand that every normal cell (not necessarily true for cancer cells or some cell lines) has a complicated and sophisticated programmed cell death agenda. These programmed cell death signal transduction pathways contain multistep processes for the cell to go through its death process in an orderly manner. A chief characteristic of these programmed cell death pathways (sometimes referred to as apoptosis) is that in early apoptosis the cell membrane remains intact and is able to exclude dyes such as PI and Trypan Blue. During apoptosis the cell is shutting down in a controlled manner and is preparing all of its molecular contents for recycling into basic molecular subcomponents by neighboring cells or the body. Only near the end of the process is the cell membrane permeable to PI, Trypan Blue, and other such dyes. There are a number of assays for measuring the different stages of apoptosis. One early-apoptosis assay involves the externalization of phosphatidylserine (PS) from the inner portion of the cell membrane to the outer portion.[7] When PS is located in the internal side of the membrane it is inaccessible to its ligand annexin-V, which can be labeled with a wide variety of fluorescent probes. So a cell with an intact membrane and whose PS is on the inner side of the cell membrane will be negative for binding of annexin-V. But early in the process of apoptosis, while the membrane is still impermeable to dyes such as Trypan Blue and PI, the PS relocates on the outer side of the cell membrane and can bind annexin-V. An example of this single-cell assay for toxicity, which has caused the cell to start to go into apoptosis, is shown in Figure 11.2.

Later on in the process of apoptosis, the DNA of the cell is starting to be degraded. At this point there are nicks and fragmentation in the DNA that can be labeled by reagents in the so-called TUNEL (terminal deoxynucleotidyl transferase dUTP nick end labeling) assay.[8] Terminal deoxynucleotidyl transferase (TdT) is an enzyme that catalyzes the addition of dUTPs (deoxyuridine triphosphates), to the end of nicked DNA occurring in late apoptotic cells. The interesting aspect of dUTPs is that they chemically compete with the normal dTTP DNA precursor and will effectively incorporate into the DNA sequence. But dUTPs can be recognized with a monoclonal antibody that can differentiate between dUTPs and dTTPs. Therefore, nicked DNA within late apoptotic stage cells can be labeled with a fluorescently labeled antibody of a specific color. An example of late stage apoptotic cells identified by flow cytometry is shown in Figure 11.3.

Interestingly there is a less well-known, but an even simpler, assay for detecting nicks in DNA during oxidative stress toxicity to cells. A superoxide indicator *dihydroethidium*, also called hydroethidine, exhibits blue fluorescence in the cytosol of a cell until oxidized, where it intercalates within the cell's DNA, staining its nucleus a bright fluorescent red (Figure 11.4). This test is simple but very effective at detecting early oxidative stress in cells even prior to detection by Trypan Blue or PI exclusion, or annexin-V (early apoptosis) or TUNEL (late apoptosis) assays. For some very detailed single-cell protocols of some of these assays by both flow and scanning image cytometry, please refer to the recently published book by Eustaquio and Leary.[9]

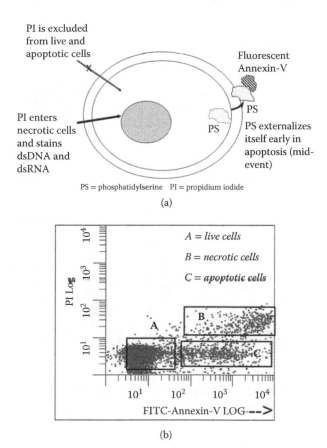

FIGURE 11.2 (a) The annexin-V concept, (b) annexin-V flow cytometry data.

11.2.3 MORE SUBTLE CHANGES NOT RESULTING IN DIRECT CELL DEATH

All of the preceding discussion concerns cells either killed by injury (necrosis) or induced into programmed cell death (apoptosis).[10] If we truly want to consider all of the potentially bad effects of nanomaterials on cells, we must study how they might change the patterns of cell proliferation and differentiation and should be considered as measures of toxicity.

11.2.3.1 Changes in Cell Proliferation

Changes in cell proliferation can be measured in multiple ways. First, and most obvious, is studying how fast, or not, the cells are replicating by measuring absolute cell numbers. The cells may not be dying but they may be changing in their rate of proliferation. Usually when a cell is stressed, but not killed or induced into apoptosis by cytotoxic agents, the proliferation rate slows. In longer term this can be studied by measuring the total cell numbers at longer time intervals allowing the cell to go into cell divisions. But this type of measurement is difficult and time consuming. An easier

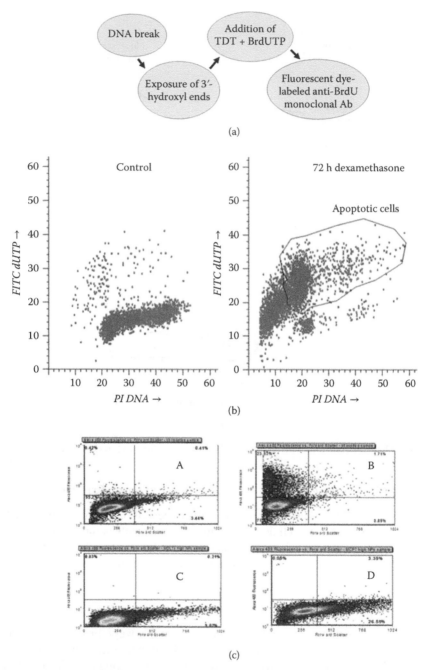

FIGURE 11.3 (a) Annexin-V concept, (b) flow cytometric data of a campothecin-treated positive control cells for terminal deoxynucleotidyl transferase dUTP nick end labeling (TUNEL) apoptosis, (c) superparamagnetic iron oxide nanoparticle-treated cells show little TUNEL toxicity.[25]

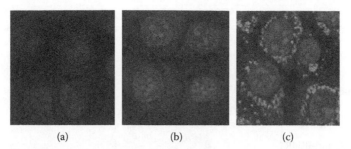

(a)   (b)   (c)

FIGURE 11.4 Grayscale images of fluorescent staining of cells with dihydroethidium (DHE). (a) Live cells without oxidative stress show little nuclear fluorescence (usually red), (b) a hydrogen peroxide-treated positive control for oxidative stress shows brightly labeled nuclei, (c) cells treated with supposedly nontoxic Q-Tracker quantum dots (bright white dots that are actually green) are negative for standard dye exclusion, annexin-V, and terminal deoxynucleotidyl transferase dUTP nick end labeling (TUNEL) assays but show extensive oxidative stress with this DHE assay.

way is to measure the DNA synthesis process within single cells. This can be measured in a variety of ways. The simplest way is to treat the cells with a DNA-specific fluorescent dye. Some of these dyes such as Hoechst 33342 and recently available new dyes such as the Vybrant® DyeCycle™ family of dyes (Invitrogen, Carlsbad, CA) are actually cell permeable and can be used to label live cells with excitation at different non-UV wavelengths. DAPI (4′,6-diamidino-2-phenylindole) is a fluorescent stain that binds strongly to A-T-rich regions in DNA. DAPI can pass through an intact cell membrane therefore it can be used to stain both live and fixed cells, though it passes through the membrane less efficiently in live cells and therefore the effectiveness of the stain is lower so that if done at the proper concentrations live and dead cells differentially stain (dead cells stain more brightly with DAPI). Among other dyes DRAQ5™, a far-red fluorescent DNA dye, is a novel cellular imaging reagent for use in live cells, dead cells, or fixed cells in combination with other common fluorophores (e.g., FITC [fluorescein isothiocyanate], PE [phycoerythrin], GFP [green fluorescent protein]) in the visible part of the spectrum. DRAQ7™, a recently available dye, also emits in the far-red part of the spectrum and labels only dead or permeant (e.g., fixed) cells.

As cells undergo replication they must synthesize enough DNA to be able to divide into two daughter cells. People used to perform laborious tritiated thymidine (³TdR) uptake experiments. Now that data can be much more easily obtained in minutes by either flow cytometry or scanning image cytometry using BUdR (a metabolic analog of ³TdR) that can be easily recognized by a fluorescently labeled antibody. When combined with PI labeling of the DNA the data appear as in Figure 11.5.

Sometimes the situation is a bit more complicated and can measure DNA mutagenic effects such as induced aneuploidy, meaning changes in chromosome numbers or DNA translocations in single cells. There might be other measurable cytotoxic changes, including apoptosis, that can be measured on large numbers of cells treated with a potentially cytotoxic or mutagenic agent. A more complicated example, showing the appearance of aneuploid cell subpopulations with aberrant chromosome numbers as measured by G0/G1 peaks beyond the normal euploid peak, is shown in Figure 11.6. The simplest assay is to measure the amount of DNA per cell and then

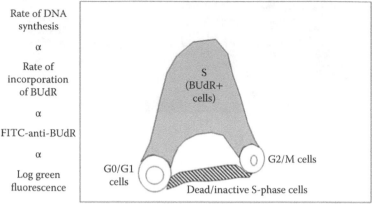

Red fluorescence α propidium iodide α DNA/cell

(a)

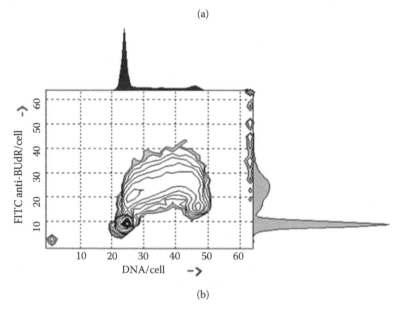

(b)

FIGURE 11.5 A rapid anti-BUdR assay by flow cytometry (a) concept is shown whereby flow cytometric measurement of anti-BUdR fluorescence is proportional to BUdR uptake during S-phase DNA synthesis, (b) actual flow cytometric data are shown in a bivariate display with contour mapping showing the relative number of cells in the Z-direction coming out of the page toward the reader.

to use *DNA cell cycle* curve/model fitting data analysis software (e.g., ModfitLite™; Verity Software House, Topsham, Maine) to extract the relative numbers of G1, S, and G2 phase cells.

A final caveat—DNA measurements alone do not always indicate that proliferation compartment assignments (e.g., G1, S, G2, M) are correct. In particular, cells highly perturbed or synchronized are not always equivalent to normal cells in those same subcompartments. For a more detailed analysis, the cyclin expression pattern should be studied.[11]

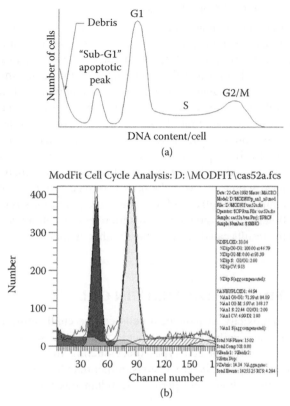

FIGURE 11.6 The toxic effects of materials on cells cannot only cause perturbations of the normal cell cycle but can, after some growth time in culture, potentially cause mutagenic aberrations in the numbers of chromosomes per cell. This aberrant number of chromosomes (usually an increase) appears as a hyperdiploid peak (aneuploidy) in a DNA content flow cytometry histogram, as shown conceptually in (a). In addition to the number of cells in each cell cycle compartment, the number of diploid and aneuploidy cells can be determined by appropriate models and curve-fitting data analyses, as shown in (b).

11.2.3.2 Changes in Cell Differentiation

Proliferation and differentiation are usually opposing forces in a cell's life. But if we are to measure the true toxic effects of agents on cells we must also check to see if the normal differentiation processes are still proceeding or if a different differentiation pathway has been induced. The most common way we can rapidly test for evidence of altered differentiation is to use monoclonal antibodies of different fluorescent colors to label one or more cell surface molecules. These cell surface molecules are usually referred to as *biomarkers*, which allow us to determine the differentiation state of a cell. Again, single-cell analysis by flow cytometry or scanning image cytometry is essential because only a subpopulation of cells may have changed their differentiation states and bulk cell measurements, as shown in Figure 11.1, will reflect a frequency-weighted distribution of differentiation states.

11.2.3.3 Changes in Cell Gene Expression

The preceding discussion assumed that the changes in the cell could be measured by cell surface biomarkers to measure changes in the cell's differentiation state. But sometimes the changes are more subtle and cannot be easily measured with fluorescent monoclonal antibodies and flow cytometry or image scanning cytometry. There may be more subtle changes in the gene expression profile of the cells. While gene expression profiles can be measured on many cells, this general approach is yet another example of the potential pitfalls of a bulk cell assay approach. If there are subpopulations of cells present, they may be missed as shown earlier in this chapter. The relative purity of cells required to accurately measure gene expression profiles has been previously studied.[12] The number of cells required and their relative purity isolated from other cell types to avoid the pitfalls of measuring cell mixtures varies with the smallest number of cells and lowest purity required for strongly expressing genes. Single-cell gene expression analysis is more complicated and expensive and the reliability of the results can depend on the number of copies of an expressed gene per cell.

11.3   MEASURING NANOTOXICITY IN CELLS

Measuring the nanotoxicity of biological cells can be challenging because of the variations within single cell types (e.g., specific cell cycle sensitivities), cell–cell interactions between different cell types (necessitating the ability to distinguish between different cell types in heterogeneous, rather than homogeneous, cell types), the need for high-throughput (and perhaps simultaneous high-content) screening methods.

11.3.1   IN VITRO MEASURES

In vitro measurements are by far the most common because of the availability of methodologies and technologies for making rapid, quantitative single-cell measurements. In vitro measurements also provide a simpler, closed system laboratory to make measurements in a controlled and reproducible manner. That said, it should be understood that while the measurements may be faster, easier, and reproducible, the results may, or may not, be relevant.

11.3.1.1   Cell Lines

The simplest types of in vitro measurements are made on a single cell type, usually a *cell line*. Cell lines are immortalized cells, usually of cancerous origin, that grow rapidly and easily in cell culture. Cell lines for virtually all human organs and cell types have been constructed and are commercially available from a variety of sources that characterize and maintain the purity and availability of each cell line. Cell lines for specific organs (e.g., liver cells, myocardial cells, neuronal cells) provide researchers a rapid means of assessing the toxicity of specific organ cell types to various chemicals and agents. One caveat is that cell lines are, in fact, abnormal and sometimes unrealistic cell models for predicting the toxicities of actual cells of human organs. Care should be taken to note the possible differences.

11.3.1.2   Primary Cell Strains

Primary cell strains are not cell lines and are closer to the actual cells in human organs. They are more difficult and expensive to grow and are available in a more limited manner than cell lines. Although they are closer to clinical reality, it should be kept in mind that these cells are difficult to maintain in their original form and may change their characteristics and relative sensitivity to nanotoxic agents, over time and after a number of cell divisions. For this reason it is important to freeze down aliquots of primary cells before many cell divisions and to frequently go back to those frozen cells to avoid large number of cell divisions that tend to produce variations and distortions in the results.

11.3.1.3   Two-Dimensional versus Three-Dimensional Cultures

Most nanotoxicity assays are performed in 2D cultures on specially treated micro-well plates. This aids in making high-throughput assays of nanotoxicity by reducing or eliminating the need to autofocus in multiple planes. However, cells grown in 2D frequently change their gene expression profiles and behave differently than their 3D counterparts. 3D measurements are much more challenging to make because of the need to focus properly in 3D and the immense scale-up of the amount and complexity of the data. For this reason, 3D in vitro measurements are still relatively rare in the literature.

11.3.2   High-Throughput Methods

Both to obtain statistically robust data and to adequately sample outlier cell subpopulations, it is important to make measurements rapidly. It should be kept in mind that live cells are changing, including dying, during long measurement processes. Therefore, an obvious, but sometimes neglected requirement is that the measurements must take place much faster than the rate at which the cells are changing over time. Luckily, there are a number of technologies available to make these measurements. It is important to weigh both the suitability and the advantages/disadvantages of each technology before determining the best choice.

11.3.2.1   Flow Cytometry

Flow cytometry is a relatively mature technology that is perhaps the most suitable technology for making measurements on cell within blood or other fluid organ systems. In addition to being very rapid, multiple highly quantitative measurements can be made on each single cell. If desired, the cell subpopulations can be simultaneously isolated to very high purity by cell sorting, usually based on inkjet technology. Up to 17 colors of fluorescence have been simultaneously measured at rates of many thousands of cells per second. If cells are either naturally in suspension, or can be put into a cell suspension without changing important characteristics, flow cytometry is probably the preferred technology for making nanotoxicity measurements. For simplicity, a flow cytometer can be thought of as a type of fluorescence microscope with a virtual flowing (rather than static) microscope slide as shown in Figure 11.7a.

11.3.2.2 Scanning Image Cytometry

When cells must remain attached to a substrate, flow cytometry is not the preferred technology. Instead, one can use scanning image cytometry, meaning quantitative measurements are carried out with a type of modified microscope that can capture images of cells at high speed, which can then be individually identified and classified in a manner similar to that of flow cytometry. While great advances have been made in scanning image cytometry, the technology is inherently slower and more complex, particularly in terms of the data analysis, than flow cytometry. If cell morphology and cell–cell architecture are important, then scanning image cytometry must be used, since this information is lost when dissociating cells into a single cell suspension for flow cytometry. Most scanning image cytometry systems analyze 2D cell cultures or thin sections of tissues, which is both their advantage and their disadvantage. A basic configuration for a typical scanning image cytometer is an inverted fluorescence microscope that may, or may not, have the ability to optically section a cell for 3D fluorescence content as shown in the confocal scanning cytometer of Figure 11.7b.

11.3.2.3 High-Speed Three-Dimensional Confocal Scanning

As discussed above most scanning image cytometers are 2D systems. But as previously mentioned, the most realistic cell systems to examine are 3D. Obtaining 3D measurements rapidly is extremely challenging for multiple reasons. First, there is an inherent need to rapidly autofocus cells occurring in different focal planes. Second, it is very difficult to measure fluorescence quantitatively in three dimensions without photobleaching cells in different focal planes during the measuring process. For live cells, it is virtually impossible to make good quantitative measurements in 3D. If the 3D cell systems are chemically fixed, then a substantial part of the photobleaching can be reduced or eliminated by use of a variety of commercially available antifading solutions that can be used on fixed, but not live, cells.

11.3.3 In Vivo Measures

Obviously the most biologically relevant cell systems to measure are actual human or animal organ systems. But these systems are open systems, meaning that they are inherently changing with time and are designed to be in continual communication with other organ systems. They also represent even more challenging 3D systems to image quantitatively. Since they require use of actual organs, human studies are challenging and expensive. Animals may, or may not, represent relevant and reasonable alternatives to use of actual human organs.

11.3.3.1 Measures in Animals: What Is a Good Animal Model?

Many studies are made in mice for a variety of reasons. First, they are relatively inexpensive systems to use. Second, the variations from animal to animal are small. Third, these are small organs systems and require only small or modest amount of

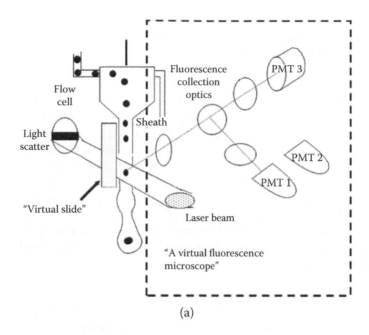

(a)

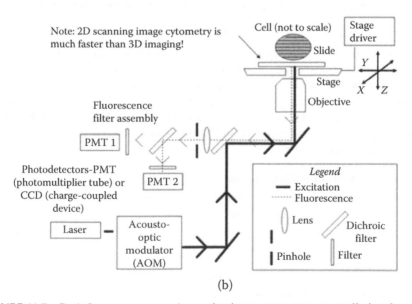

(b)

FIGURE 11.7 Both flow cytometer and scanning image cytometers are really just forms of fluorescence microscopes. (a) A flow cytometer has cells flowing single file individually past one or more laser light sources in a *virtual moving slide* that allows it to make fluorescence measurements rapidly on thousands of cells per second. (b) A scanning image cytometer is typically a form of inverted fluorescence microscope that moves the stage to perform image analysis on cells in culture or in thin sections of tissue. Data analysis algorithms separate the cells from each other using image analysis software segmentation algorithms so that single-cell analysis can be achieved.

reagents to test (scale-up to pigs or dogs can be extremely challenging). However, the most important thing is that the animal used must represent a good animal model for predicting human toxicities.

11.3.3.1.1   Nude Mice

The most common type of animal initially used in toxicity studies are *nude mice*. The mice are indeed free of hair (hence nude), which can create a high background fluorescence in whole-animal measurements. But the other, and perhaps more important reason, is that the nude mice are genetically bred to have a deficient immune system. This allows us to separate out the immunological from the toxicological components of the measurement. While some people may wonder why it is important to separate these two variables, a simple example is the case of immune-compromised cancer patients undergoing chemotherapy. Since these patients do not have an active immune system effectively reducing part of the dose, they may have a much higher toxicity reaction to a given dose. Interestingly, they may behave more like nude mice in terms of toxicity.

11.3.3.1.2   Immune Competent Mice

Since most people have an active immune system, immune competent mice would perhaps be a better predictor of human toxicity than nude mice. That said, there may be very different immunological reactions of mice to reagents than of humans.

11.3.3.1.3   Other Animals: Problems of Scale-up

The obvious problem in using mice, immune competent or not, is that people are not mice. In that sense mice are usually not a very good model animal system for predicting human toxicities. While it varies with the substance being tested, better animal models for predicting human toxicities are pigs and dogs. However, the complexity and expense of doing these studies, and the fact that most dogs in the United States are pet animals and therefore are similar to using humans in terms of ethical practices, make such studies rare.

11.3.3.2   Organ-on-a-Chip Approaches

An interesting new alternative to the use of actual humans or animals for nanotoxicity studies is the use of human *organs-on-a-chip*. These new systems represent a lower-cost alternative to human or animal studies and can be specifically constructed for easier high-throughput quantitative measurements.

11.3.3.2.1   What Is an Organ-on-a-Chip and Why Is It Important?

In essence an organ-on-a-chip is an artificial 3D engineered tissue grown on some kind of scaffolding structure to try to mimic as closely as possible the structure and function of an actual human organ. The term *chip* came from the fact that many of the original organs-on-a-chip were grown on or around a microfluidic chip scaffold. Usually these structures are much smaller (e.g., 1–10 mm) than a real human organ. This allows their use within formats such as 96-well plates allowing high-throughput screening assays. While the 3D tissue structure provides a more realistic situation, it does make the screening process more challenging if assays are performed (e.g., by image analysis) in

3D rather than 2D. While still early in development, it is possible to circumvent some of these limitations by designing sensors directly into the 3D structures.

11.3.3.2.2 Examples of Organs-on-a-Chip

Organs-on-a-chip have been constructed for more than 20 human organ systems.[13] For a rapid overview of the technology, one can visit the organs-on-a-chip section of the excellent website of Harvard University's Wyss Institute (http://wyss.harvard .edu/viewpage/293/). There is even a large ongoing *human-on-a-chip* project at the Wyss Institute funded by Defense Advanced Research Projects Agency that links 10 organs-on-a-chip to study the interactions between different organs. One example is a nanomedical system being developed on an organ-on-a-chip designed to mimic human ductal breast cancer.[14,15]

11.3.3.2.3 Ways to Measure Toxicity in Three-Dimensional Organs-on-a-Chip

It is still very early in the process for determining the best ways to measure toxicity in these 3D structures. The most conventional way is to analyze 3D organs-on-a-chip by whole organ in vivo imaging techniques similar to what would be done in small animals such as nude mice. If fluorescence probes are chosen (e.g., near-infrared fluorescent probes) to allow deep-body imaging, the entire 3D structure can be rapidly scanned. Quantitative high-resolution 3D imaging is much more challenging because of the immense amount of data generated by single or multiphoton confocal microscopy. In addition to the very large number of imaging voxels required, there are many challenges because of laser photobleaching issues with conventional confocal microscopy and the need to prevent fluorescent probe destruction by too intense localized spots from multiphoton confocal systems.

11.4 OVERVIEW OF NANOMEDICAL DRUG/ GENE DELIVERY SYSTEMS

One important, relatively new application of some significance is the use of nanoparticles for drug/gene delivery for medicine (*nanomedicine*). For a free introduction and review of nanomedicine see http://nanohub.org/resources/11877, a 16-lecture biomedical engineering course Engineering Nanomedical Systems. Nanomedicine will have profound impacts on human health care[16] and will inevitably lead to ethical issues that should be considered (for a perspective on some of the ethical issues of nanotechnology in general, and nanomedicine in particular, see Leary, 2013).[17] Interestingly, nanotechnology and nanomedicine will probably be first applied to humans in the field of ophthalmology for treatment of retinal and optic nerve diseases.[18–20]

An important issue is the potential nanotoxic effects of the nanoparticle carrier vehicle itself in addition to the toxicity of the drugs it contains. An important assumption to test is whether the collection of subcomponent parts can be separately tested for toxicity. That is important because, as indicated earlier in this chapter, the total nanovehicle may, or may not, be well characterized in terms of its toxicity or in terms of the individual toxicities of the subcomponent parts. This is an important question to answer because otherwise the amount of testing of slightly different amounts of subcomponents and their order of nanoassembly may prove prohibitively expensive to pharmaceutical companies. Worse yet, if these small differences lead to large differences in toxicity, the economic and the U.S. Food and Drug Administration (FDA) approval processes may quickly stifle the creative development of new nanomedicine strategies.

11.4.1 WHY ARE NANOMEDICAL SYSTEMS IMPORTANT TO DRUG/GENE DELIVERY?

Nanomedical systems are a vital next step in drug delivery. Untargeted drugs can result in excessive patient dose with accompanying side effects, some of which can lead to individual patients deciding to discontinue the use of the drug and more serious side effects leading to withdrawal of the drug from the market or at least very focused and limited use based on a risk/benefit analysis. Drugs need to be targeted to the diseased cells and tissues of relevance.

They also need to be protected as much as possible from immune rejection or excretion from the body. The addition of so-called *stealth factors* (the simplest being pegylation) allows much lower doses to be given to the patient. The combination of targeting and stealth factors may solve or at least mitigate the huge costs and problems of bringing new drugs to market. They may also allow re-packaging and use of already FDA-approved drugs along with accompanying patent extensions as *combo devices* (most current combo devices are drug-eluting stents, but nanomedical systems with targeting and stealth layers would certainly qualify as combo devices in the FDA definition).

11.4.2 EXAMPLES OF SIMPLE NANOMEDICAL SYSTEMS

A simple *nanomedical system* currently in testing is that of folate-targeted drugs.[21] These drugs make use of the fact that many cancers have significantly elevated levels of folate receptors. In addition to the benefits of differential targeting through the folate receptor, these systems can also provide for real-time fluorescence-guided surgery of tumor metastases readily visible to the surgeon when fluorescence probes attached to the folate-targeted drug are excited during patient surgery.

11.4.3 MORE SOPHISTICATED NANOMEDICAL SYSTEMS

Much more sophisticated, multilayered nanomedical systems have been designed.[22–25] Some of these have even involved sophisticated tethering of transcribable DNA sequences under control of upstream molecular biosensors for initial feasibility tests of potential feedback-controlled drug–gene delivery nanomedical systems.[26–28]

11.4.4 FDA AND NANOTOXICITY CONCERNS

The FDA is in the process of reviewing many new nanomedical systems. But this process is being slowed by the nanotoxicity issues mentioned previously. There are concerns about the nanotoxicity of the nanoparticle drug delivery systems that must be satisfied by systematic appraisals of the toxicity of the nanoparticle vehicles with and without the encapsulated drug to insure that the use of these nanomedical systems is safe. That said, the concerns in many cases are overblown because of the *nano* factor.

However, when the overall situation is put in perspective, these nanomedical systems will expose the patient to perhaps 10% as much drug in targeted form as in untargeted doses. Furthermore, if the nanoparticle delivery systems are constructed entirely of highly biocompatible and biodegradable subcomponents, the probability of these nanoparticle vehicles contributing to overall toxicity is minimal. This does assume that the toxicity of the total structure is close to the sum of the toxicity of the separate parts, but this is probably the case in most instances. The concern about high reactivity of nanoparticles because of their very high surface to volume factors is a valid concern, but this is a testable rather than theoretical concept and just needs to be tested to avoid nanotoxicity concerns based on speculation rather than actual data. In the experience of this author, most nanomedical systems designed with biocompatible and biodegradable subcomponents have insignificant toxicity levels even when measured with the most sensitive single-cell assays.

11.4.5 MEASURING THESE SYSTEMS IN VITRO AND IN VIVO

While in vitro measurements (including some on miniature 3D organs-on-a-chip) can be at least semiquantitatively made, in vivo measurements of nanoparticle distributions have been primarily made on very small animals such as nude and immune-competent mice where the use of infrared fluorescence probes can penetrate the relatively small tissue distances. For human testing other techniques will need to be used. Radioisotope tracers linked to nanoparticle systems is the most obvious, and perhaps most sensitive, choice. But those experiments on volunteer human subjects will need to be done very carefully to avoid risky exposure levels. A less sensitive, but perhaps safer, method would be to use magnetic resonance imaging (MRI) contrast agents in the nanoparticles. This would only allow concentrations of nanoparticles to be seen under conventional 3 Tesla level MRI scans. This works on animals and can also be done in dual imaging modality mode fluorescence and MRI,[29-33] but is still relatively untested on humans. Macrocyclic gadolinium T1 contrast agents or superparamagnetic iron oxide T2 contrast agents could be used initially to insure that no large concentrations of these nanoparticle systems accumulate anywhere other than in targeted tumors.

11.5 CONCLUSIONS

The reader should not despair at the potential complexity of good toxicity measurements. While I have attempted to warn the reader of oversimplification of the toxicity assays, I also believe that measurements need not be perfect, but just good enough, to make decisions and to get a reasonable assessment of the overall toxic nature of assayed materials. The key point is to appreciate that good toxicity measurements need to consider more than simple necrotic cell death. In this chapter, I have shown the increasingly sophisticated measurements that can be used to avoid missing more subtle, but perhaps important, effects of materials on living cells.

REFERENCES

1. Wani, M.Y., Hashim, M.A., Nabi, F., Malik, M.A. 2011. Nanotoxicity: Dimensional and morphological concerns. *Advances in Physical Chemistry* 450912–450927.
2. Pons, T., Pic, E., Lequeux, N., Cassette, E., Bezdetnaya, L., Guillemin, F., Marchal, F., Dubertret, B. 2010. Cadmium-free CuInS$_2$/ZnS quantum dots for sentinel lymph node imaging with reduced toxicity. *ACS Nano* 4(5): 2531–2538.
3. Eustaquio, T., Leary, J.F. 2011. Nanobarcoding: A novel method of single nanoparticle detection in cells and tissues for nanomedical biodistribution studies. *Proceedings of SPIE* 8099: 80990V-1–80990V-13.
4. Eustaquio, T., Leary, J.F. 2012. Nanobarcoding: Detecting nanoparticles in biological samples using in situ polymerase chain reaction. *International Journal of Nanomedicine* 7: 5625–5639.
5. Eustaquio, T., Cooper, C.L., Leary, J.F. 2011. Single-cell imaging detection of nanobarcoded nanoparticle biodistributions in tissues for nanomedicine. *Proceedings of SPIE* 7910: 791000-1–791000-11.
6. Perfetto, S.P., Chattopadhyay, P.K., Roederer, M. 2004. Perspectives. Innovation: Seventeen-colour flow cytometry: Unravelling the immune system. *Nature Reviews Immunology* 4: 648–655.
7. van Engeland, M., Nieland, L.J., Ramaekers, F.C., Schutte, B., Reutelingsperger, C.P. 1998. Annexin V-affinity assay: A review on an apoptosis detection system based on phosphatidylserine exposure. *Cytometry* 31: 1–9.
8. Loo, D.T. 2011. In situ detection of apoptosis by the TUNEL assay: An overview of techniques. *Methods Molecular Biology* 682: 3–13.
9. Eustaquio, T., Leary, J.F. 2012. Single-cell nanotoxicity assays of superparamagnetic iron oxide nanoparticles. *Methods in Molecular Biology* 926: 69–85.
10. Darzynkiewicz, Z., Juan, G., Li, X., Gorczyca, W., Murakami, T., Traganos, F. 1997. Cytometry in cell necrobiology: Analysis of apoptosis and accidental cell death (necrosis). *Cytometry* 27: 1–20.
11. Gong, J., Traganos, F., Darzynkiewicz, Z. 1995. Growth imbalance and altered expression of cyclins B1, A, E, and D3 in MOLT-4 cells synchronized in the cell cycle by inhibitors of DNA replication. *Cell Growth and Differentiation: The Molecular Biology Journal of the American Association for Cancer Research* 6(11): 1485–1493.
12. Szaniszlo, P., Wang, N., Sinha, M., Reece, L.M., Van Hook, J.W., Luxon, B.A., Leary, J.F. 2004. Getting the right cells to the array: Gene expression microarray analysis of cell mixtures and sorted cells. *Cytometry* 59A: 191–202.
13. van der Meer, A.D., van den Berg, A. 2012. Critical review: Organs-on-chips: Breaking the in vitro impasse. *Integrative Biology* 4: 461–470.
14. Grafton, M., Wang, L., Vidi, P.-A., Leary, J.F., Lelievre, S. 2011. Breast on-a-chip: Mimicry of the channeling system of the breast for development of theranostics. *Integrative Biology* 3: 451–459.
15. Leary, J.F., Key, J., Vidi, P.-A., Cooper, C.L., Kole, A., Reece, L.M., Lelièvre, S.A. 2013. Human organ-on-a-chip BioMEMS devices for testing new diagnostic and therapeutic strategies. *Proceedings of SPIE* 8615: 8615-1–8615-10.
16. Leary, J.F. 2010. Nanotechnology: What is it and why is small so big. *Canadian Journal of Ophthalmology* 45(5): 449–456.
17. Leary, J.F. 2013. Nanotechnologies: Science and public policy. In: *Perspectives in Bioethics, Science, and Public Policy*, Editor: J. Beever, N. Morar, Published in collaboration with the Global Policy Research Institute by Purdue University Press, West Lafayette, Indiana.

18. Zarbin, M.A., Montemagno, C., Leary, J.F., Ritch, R. 2010a. Nanomedicine in ophthalmology: The new frontier. *American Journal of Ophthalmology* 150(2):144–162.
19. Zarbin, M.A., Montemagno, C., Leary, J.F., Ritch, R. 2010b. Nanotechnology in ophthalmology. *Canadian Journal of Ophthalmology* 45: 457–476.
20. Zarbin, M.A., Montemagno, C., Leary, J.F., Ritch, R. 2012. Regenerative nanomedicine and the treatment of degenerative retinal diseases. *WIREs Nanomedicine Nanobiotechnology* 4: 113–137.
21. van Dam, G.M., Themelis, G., Crane, L.M., Harlaar, N.J., Pleijhuis, R.G., Kelder, W., Sarantopoulos, A. et al. 2011. Intraoperative tumor-specific fluorescence imaging in ovarian cancer by folate receptor-[alpha] targeting: First in-human results. *Nature Medicine* 17(10): 1315–1319.
22. Seale, M.-M., Zemlyanov, D., Haglund, E., Prow, T.W., Cooper, C.L., Reece, L.M., Leary, J.F. 2007. Multifunctional nanoparticles for drug/gene delivery in nanomedicine. *Proceedings of SPIE* 6447: 64470E-E1–64470E-E9.
23. Seale, M., Haglund, E., Cooper, C.L., Reece, L.M., Leary, J.F. 2007. Design of programmable multilayered nanoparticles with in situ manufacture of therapeutic genes for nanomedicine. *Proceedings of SPIE* 6430: 6430-1–6430-7.
24. Seale, M.-M., Leary, J.F. 2009. Nanobiosystems. In: *Wiley Interdisciplinary Reviews: Nanomedicine and Nanobiotechnology*, Editor: J.R. Baker, Wiley Press, New York: NY *Nanomedicine and Nanobiotechnology* 1: 553–567 (2009).
25. Haglund, E., Seale, M.-M., Leary, J.F. 2009. Design of multifunctional nanomedical systems. *Annals of Biomedical Engineering* 37(10): 2048–2063.
26. Prow, T.W., Rose, W.A., Wang, N.A., Reece, L.M., Lvov, Y., Leary, J.F. 2005. Biosensor-controlled gene therapy/drug delivery with nanoparticles for nanomedicine. *Proceedings of SPIE* 5692: 199–208.
27. Prow, T.W., Smith, J.N., Grebe, R., Salazar, J.H., Wang, N., Kotov, N., Lutty, G., Leary, J.F. 2006a. Construction, gene delivery, and expression of DNA tethered nanoparticles. *Molecular Vision* 12: 606–615.
28. Prow, T.W., Grebe, R., Merges, C., Smith, J.N., McLeod, D.S., Leary, J.F., Gerard A., Lutty, G.A. 2006b. Novel therapeutic gene regulation by genetic biosensor tethered to magnetic nanoparticles for the detection and treatment of retinopathy of prematurity. *Molecular Vision* 12: 616–625.
29. Nam, T., Park, S., Lee, S.-Y., Park, K., Choi, K., Song, I.C., Han, M.H. et al. 2010. Tumor targeting chitosan nanoparticles for dual-modality optical/MR cancer imaging. *Bioconjugate Chemistry* 21: 578–582.
30. Key, J., Kim, K., Dhawan, D., Knapp, D.W., Kwon, I.C., Choi, K., Leary, J.F. 2011. Dual-modality in vivo imaging for MRI detection of tumors and NIRF-guided surgery using multi-component nanoparticle. *Proceedings of SPIE* 7908: 7908-1–7908-8.
31. Key, J., Dhawan, D., Knapp, D.K., Kim, K., Kwon, I.C., Choi, K., Leary, J.F. 2012a. Multimodal in-vivo MRI and NIRF imaging of bladder tumors using peptide conjugated glycol chitosan nanoparticles. *Proceedings of SPIE* 8225: 82251F–82258F.
32. Key, J., Dhawan, D., Knapp, D.K., Kim, K., Kwon, I.C., Choi, K., Leary, J.F. 2012b. Design of peptide-conjugated glycol chitosan nanoparticles for near-infrared fluorescent (NIRF) in vivo imaging of bladder tumors. *Proceedings of SPIE* 8233: 82330R1–82330R10.
33. Key, J., Cooper, C., Kim, A.Y., Dhawan, D., Knapp, D.W., Kim, K., Park, J.H. et al. 2012c. In vivo NIRF and MR dual modality imaging using glycolchitosan nanoparticles. *Journal of Controlled Release* 163: 249–255.

12 Imaging Techniques for Nanoparticles in Skin

Tarl W. Prow

CONTENTS

12.1 INTRODUCTION

The skin is exposed to environmental and applied nanoparticles. This continual exposure to natural and man-made nanomaterials is usually harmless because the skin is an excellent barrier to these materials. The skin is remarkably good at protecting against the penetration of virtually all nanomaterials that we come in contact with. This is primarily because of the physical properties of the skin as a stratified epithelium. The skin is composed of an outer layer of stratum corneum, viable epithelium, and dermis (Figure 12.1a). These layers can expand (Figure 12.1b) or contract dramatically due to pathological processes.

The skin is also an exceptional barrier because the keratinocytes of the viable epithelium are continuously dividing, differentiating, and sloughing off. This natural turnover continuously pushes invading nanoparticulates out and off of the skin. Therefore, if particulates penetrate into the viable epidermis, then they will be removed via the natural turnover of the skin in less than a week. This property is unique to the epithelium and must be taken into consideration when interpreting nanotoxicology data in context with dermatopathology.

Another, less tangible barrier within skin is the cutaneous immune system. This protective apparatus is composed of many different cell types including immune, pigment, keratinocytes, and fibroblasts. The cutaneous immune system is relevant to

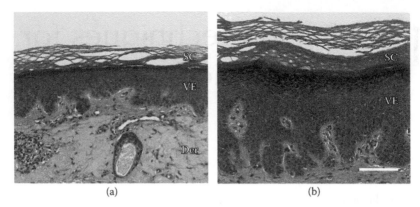

(a)   (b)

FIGURE 12.1   The skin is composed of three primary layers: the stratum corneum (SC), viable epidermis (VE), and dermis (Der). The VE is separated from the vascularized dermis by the dermal–epidermal junction. Both panels were obtained from different regions of the same histopathological specimen to illustrate the dynamic nature of human skin. (a) A relatively normal region of skin, (b) a hyperkeratotic region exemplified by increased thickness of the SC and VE.

nanotoxicology in terms of hypersensitivity, but is largely being investigated as an easily accessible site for minimally invasive immunization with antigen-displaying nanoparticles.

The accessibility of skin makes it a likely site for exposure, drug delivery, and research in the context of nanotoxicology. The skin's regenerative capacity enables low-risk volunteer studies. Together, all of these factors make skin one of the most suitable sites for imaging in the field of nanotoxicology.

12.2   SKIN IMAGING APPROACHES

The majority of skin imaging is done in dermatology clinics. An amazing array of skin pathology is assessed on a daily basis in these clinics with very little concern for nanotoxicology. The three major categories that give rise to dermatologist intervention are cosmetic concerns, skin cancer, and inflammatory rashes. The primary imaging tool that dermatologists use is the eye coupled with clinical experience. Clinical photography is also used to document skin lesions and treatment for follow-up. In the past 20 years dermoscopy, also known as dermatoscopy, has become a staple for many dermatologists. A general description of the dermatoscope device is a 10X lens that uses nonpolarized or polarized light and some of these devices also have a transparent window that flattens the skin being imaged. Dermatoscopes began as stand-alone devices (Figure 12.2a) that have since been integrated into cases attaching to smart phones for digital photography and teledermatology (Figure 12.2b). Interestingly, despite the widespread use of dermoscopy, there are only a couple of nanotoxicology publications showing dermoscopy images of nanoparticle-treated skin. The reason for this is likely a lack of collaborative work between dermatologists and researchers studying nanotoxicology in skin.

Optical coherence tomography (OCT) is an emerging technology being put to use in dermatology clinics primarily to define lesion borders.[1] The key feature for OCT is the capacity to generate micrometer-resolved images up to several millimeters deep. This is possible because the device uses infrared light that can penetrate deeply into and out of the skin.

Another emerging skin imaging technology in dermatology is reflectance confocal microscopy (RCM) (Figure 12.2c). RCM uses light scattering to visualize skin structures at the micrometer scale. An expert can diagnose skin lesions based on morphological cues found at different depths within the skin brought on by a

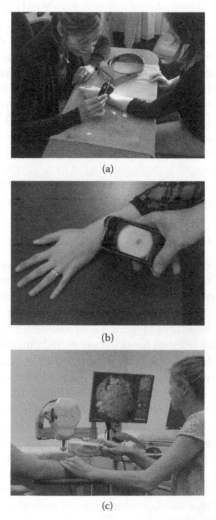

(a)

(b)

(c)

FIGURE 12.2 Routine clinical skin imaging is done by eye, dermatoscope (a and b), and reflectance confocal microscopy (c). Dermatoscope technology is improving, costs are declining, and it can now be coupled to smart phones for automated analysis, telemedicine, and documentation (b). These devices are now beginning to be used by nanotoxicology researchers with an interest in clinical outcomes.

pathological condition, for example, melanoma or photoaging. It stands to reason that toxic nanoparticles could induce morphological changes. However, RCM is rarely used in research publications, with a few exceptions, but is almost exclusively used in clinical dermatology reports, with respect to confocal microscopy-focused studies. Research studies almost exclusively use confocal laser scanning microscopy, commonly called fluorescence confocal microscopy or single-photon confocal microscopy. This additional disparity in clinical and research imaging is another indication of a disconnect between basic and clinical nanotoxicology research in skin.

Finally, there are a handful of clinical dermatology research facilities in the world that have access to clinical multiphoton microscopes (MPM). These super-specialized pieces of equipment require experts to run and analyze data, although having a significant up-front and maintenance costs. A fraction of these researchers have more highly specialized detectors capable of time-correlated single-photon counting (TCSPC). This capacity enables fluorescence lifetime measurements to be made. These data can be interpreted to yield quantitative information on molecular interactions and help to separate endogenous from nanoparticle signals.

In summary, the disconnect between clinical imaging and basic science imaging is apparent in the literature, but this gap is closing. There is a global push to integrate benchtop research with clinical research termed "translational research." This movement is directly relevant to those working in the area where nanotoxicology and nanodermatology meet. Imaging data are expected by grant/manuscript/regulatory assessors and never has clinical relevance been higher on the respective agendas.

12.2.1 Dermoscopy

Handheld dermatoscopes are used to diagnose skin lesions noninvasively. Using this relatively simple imaging device reveals morphological features that cannot be seen by the naked eye. Dermatoscopes represent a link between clinical observation and microscopic evaluation. In particular, dermoscopy enables the visualization of skin pigment and vascular structures (Figure 12.3a and b); for a review see Zalaudek et al.[2] These features are relevant for skin cancer, inflammatory and infectious diseases, all of which could be downstream consequences of particulate or solvent toxicity. We have used dermoscopy to examine gold nanoparticle-treated skin with and without a penetration enhancing solvent, toluene (Figure 12.3c through f).[3]

In this study we treated viable excised human skin with gold nanoparticles (90 μg/mL) for 4 and 24 hours in Franz diffusion cells. The nanoparticles were suspended in either aqueous or toluene vehicles. We observed gold nanoparticle staining in the skin furrows and variations in the color of the nanoparticle staining, indicating precipitation in the aqueous vehicle and not in the toluene vehicle. The dermoscopy images in Figure 12.2 show that the skin treated with the aqueous gold nanoparticle solution originally had blue staining, which was different from the stock solution that was lighter in color. The color change of this nature is indicative of gold nanoparticle precipitation. In contrast, the dermoscopy images of the skin treated with gold nanoparticles in toluene had lighter staining as in the stock solution. This lack of a color change suggests that nanoparticles were still dispersed within the skin. These observations were confirmed with subsequent analysis discussed later. This appears to

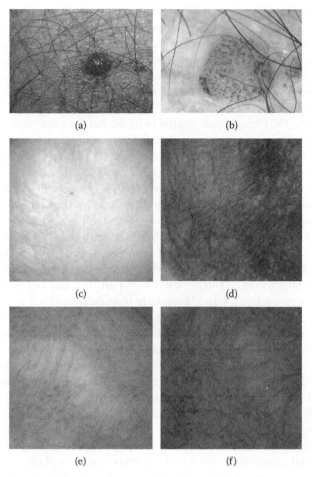

FIGURE 12.3 Dermoscopy is used to visualize features of the skin that are not visible to the naked eye. (a) A clinical photo of a skin lesion and (b) the same lesion with dermoscopy where the unique blood vessel signature of amelanotic melanoma is visible. Dermoscopy photos of aqueous vehicle (c), gold nanoparticles in aqueous media (d), toluene (e), and gold nanoparticles in toluene-treated (f) skin.

be the only report utilizing dermoscopy in the area of nanotoxicology despite the widespread use of dermoscopy in the dermatology clinic and relative ease of use. The output of dermatoscopy imaging is image data that should be interpreted by an experienced dermatopathologist. These images can be useful in both animal and volunteer studies.

12.2.2 OPTICAL COHERENCE TOMOGRAPHY

Optical coherence tomography uses low-coherent interferometry in the near-infrared wavelengths to image relatively deep into skin >2 mm at micrometer resolution. See Gambichler et al.[1] for a comprehensive review on OCT in the context of dermatopathology. The relatively long wavelength light used in OCT enables deep skin penetration. The technology uses interferometry to gather data from within turbid structures like skin. There are two light paths used in OCT, one goes down a reference arm to a mirror and the other to the sample. Importantly, the light goes the same optical distance. The reflectivity from the mirror path generates a reference profile to which the sample path is compared. Areas scanned in the sample path that have high light-scattering properties will generate increased interference than low light-scattering areas (Figure 12.4).

Conceptually, OCT may appear similar to RCM, but there are some critical differences. OCT uses interferometry to quantify differences in wave propagation information, whereas RCM detects backscattered light levels. These are two fundamentally different analytical techniques. Another critical difference is that OCT uses much longer wavelength light (e.g., 1300 nm) than RCM (785 nm). Practically, OCT is a technology that is significantly more technically complex than the confocal microscopy techniques. One important similarity between OCT and RCM is that both techniques allow for morphological characterization without the need for contrast agents or fluorescent dyes. OCT has a history of both gold and silver nanoparticles being used as contrast agents.[4–6] Kim et al.[6] report a phase-sensitive OCT system capable of differentiating 60 nm gold and 40 nm silver nanoparticles. The key was to separately excite the two nanoparticles with 532 nm light modulated with a 1 kHz square wave for the larger gold nanoparticles and 405 nm light modulated with a 200 Hz square wave for the smaller silver nanoparticles. This resulted in unique frequency and time domain signals for each nanoparticle. Finally, the amplitude in the Fourier domain increased linearly with the concentration of each nanoparticle suggesting that this approach could also be used to estimate the concentration of these nanoparticles. The authors tested gold nanoparticle concentrations from 1.17 to 4.5×10^{11} particles/mL ($2.0–7.9 \times 10^3$ µg/mL for comparison) and silver nanoparticle concentrations from 0.50 to 1.70×10^{11} particles/mL ($3.3–12.7 \times 10^3$ µg/mL for comparison). Although these are quite narrow windows with relatively high concentrations, OCT appears to be a relevant technology for future nanotoxicology studies in skin.

12.2.3 REFLECTANCE CONFOCAL MICROSCOPY

RCM is a noninvasive real-time imaging technology that gives the user quasi-histological resolution images of the skin in an *en face* orientation. For a full description see Nehal et al.[7] RCM is gaining ground in the dermatology clinic for use as a diagnostic aid. An easy way to think about RCM is that the device operates with the same equipment as a conventional fluorescence confocal microscope with the exception that the backscattered laser light is detected instead of fluorescence light. The state-of-the-art system is a handheld device that uses an 830 nm Class I diode laser powered from 1 to 25 mW at the skin surface. The resolution is <1.25 µm in the horizontal plane and <5.0 µm in the vertical plane with a field of view of 1 × 1 mm. The objective lens has a numerical aperture of 0.9 and is infinity corrected. Detailing these specifications is only possible because there is only one manufacturer, Lucid (Rochester, NY), at this time.

In 2011, we were the first to investigate topically applied nanoparticles with RCM.[8] The focus of the study was to determine whether topically applied silver

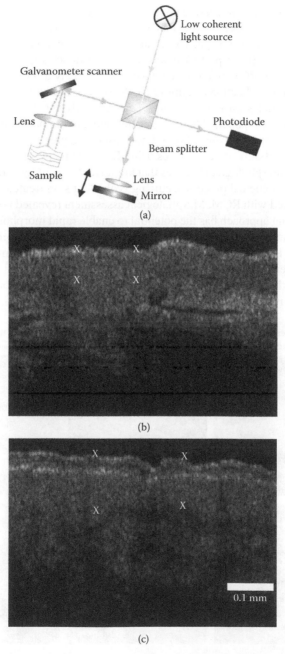

FIGURE 12.4 (a) A schematic of a simplified optical coherence tomography imaging system, (b) example images of baseline, and (c) a corresponding skin image 3 days following exposure to three minimal erythema doses of ultraviolet B. In the damaged skin, there is a disruption of the entrance signal with a signal-poor center is observed. The borders (x) between the skin surface and the dermal–epidermal junction are shown indicating the usual increase of epidermal thickness after UVB exposure.

nanoparticles penetrated intact or damaged skin. A secondary goal was to determine how long that nanoparticles remained on the skin surface. RCM was used to address this secondary issue of nanoparticle clearance. A commercial silver nanoparticle containing spray (<50 nm particles) was applied to intact and tape-stripped (20X tape stripping) skin. RCM images were taken immediately after and at intervals for up to 20 days after application (Figure 12.5). We observed relatively large aggregates of silver nanoparticles on the skin surface, particularly on the tape-stripped skin (Figure 12.5, top right panel). These aggregates appeared to persist up to 10 days on the skin surface and in furrows. We did not observe aggregates below the stratum corneum and could only see aggregates using RCM. This imaging tool is clinically used to observe morphological features that are related to healthy skin or skin where there is some pathological process occurring. Volunteer skin treated with nanoparticles was examined with RCM. Morphological assessment revealed no abnormalities. This skin imaging approach has the potential to enable rapid morphological imaging in human and animal nanotoxicology experiments with morphological assessment as an outcome. RCM is also capable of detecting nanoparticle aggregates that can efficiently scatter light, for example, gold and silver nanoparticles. This capacity, combined with the growing clinical use of RCM, makes this technology relevant to nanotoxicology research in skin.

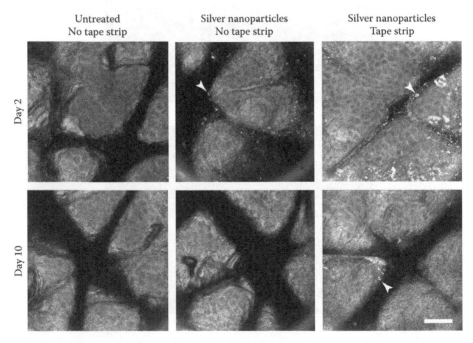

FIGURE 12.5 In vivo reflectance confocal microscopy images of untreated and silver nanoparticle-treated skin that was intact (no tape strip) or tape stripped. Silver nanoparticle aggregates (arrowheads) were seen 2 and 6 days (data not shown) after application but not at 10 days after application in intact skin. In contrast, silver nanoparticle aggregates were still seen in tape-stripped skin 10 days after application, but only in discrete areas. Each image is 0.5×0.5 mm^2.

12.2.4 Single-Photon Confocal Microscopy

Single-photon confocal microscopy, usually called laser scanning confocal microscopy (LSCM), uses singe wavelength laser light to excite a fluorescent molecule or nanomaterial and detects the fluorescence emission of that entity using optical filters to select the relevant light wavelengths (Figure 12.6). For a thorough background on this technology see Pawley.[9] For an in-context review see Zhang and Monteiro-Riviere.[10] Instruments range from the common Zeiss LSM 510 to exotic custom machines tailored to specific purposes. The most common configuration is a 488-nm air-cooled laser light source to illuminate a derivative of fluorescein with a 40X oil immersion objective and 1.3 numerical aperture objective. Table 12.1 shows a snapshot of recent reports that used LSCM imaging to investigate nanoparticles in skin. Emission light for this setup would likely be filtered by a 488 nm band reject optical filter to remove backscattered excitation light and a 505–530 nm band-pass optical filter to capture the green fluorescence emission light.

LSCM is used in the nanotoxicology context for several reasons. LSCM imaging generates images using samples such as cells, tissue sections, and even in vivo skin imaging. Conventional fluorescence microscopy has lower resolution and higher levels of noise compared with LSCM. This is primarily due to the capacity for LSCM to take optical sections of samples. This is the defining feature of LSCM and is the primary result of using a pinhole in front of the detector. The pinhole excludes out of focus light, thus enabling images to be taken from a narrow (<5 µm in many cases) depth plane. Further, the excitation source is single wavelength laser light that can penetrate skin up to a few hundred micrometers. Experience in this area suggests that the best quality LSCM images of skin come from the upper 200 µm. LSCM performed in vivo eliminates the common artifacts from tissue processing that occur with fixation and sectioning.

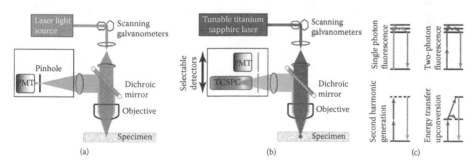

FIGURE 12.6 Schematic of single (a) and multiphoton microscope (MPM) (b) and energy transfer diagrams. MPMs generally have tunable 80 MHz titanium:sapphire laser sources for infrared excitation light. Scanning galvanometers are used to raster scan the excitation beam over the imaging area. The excitation beam passes through a dichroic mirror and an objective before reaching the specimen. Emission light then passes back through the objective and is reflected to one or more detectors by the dichroic mirror. Photomultiplier tube and time-correlated single-photon counting (PMT and TCSPC, respectively) detectors can be stationary or be selected depending on the application. These components are shown in schematic form within panels a and b. (c) Energy transfer diagrams for single-photon fluorescence, two-photon fluorescence, second-harmonic generation, and energy transfer upconversion.

TABLE 12.1

Recent Laser Scanning Confocal Microscopy Nanoparticle Imaging Configuration

Nanoparticle Material	Diameter	Dye	Laser Scanning Confocal Microscopy System	Excitation	Emission
Dendrimers[11]	10.6–18.3 nm	Fluorescein isothiocyanate	Olympus FluoView FV300	488 nm	Not reported
Dendrimers[12]	3–10 nm	Rhodamine isothiocyanate	Zeiss LSM 510 META	543 nm	465–495 nm
Ethosomes[13]	74.2 ± 5.9 nm	Rhodamine 6GO	Zeiss LSM 510	543 nm	>560 nm
Ethyl cellulose[14]	90–118 nm	Nile red	Nikon Eclipse Ti	488 and 543 nm	Not reported
Microemulsion[15]	74 ± 7.07 nm	Rhodamine B	Leica TCS SP2	554 nm	575 nm
Multiwalled carbon nanotubes[16]	50–300 nm × 3–10 μm	None	Leica TCS SP1	488 nm	510–540 nm
Nanoemulsions[17]	18–125 nm	Endogenous protoporphyrin IX	Leica TCS SP2	488 nm	590 nm
Nanotransfersomes[18]	72–190 nm	Rhodamine Red-X	Olympus FluoView FV1000	488 nm	590 nm
PLGA[19]	100 nm	Coumarin-6	Olympus FluoView FV1000	488 nm	519 nm
Polystyrene[20]	20 and 100 nm	FluoSpheres	Zeiss LSM 510 META; Invitrogen	488 nm	505–530 nm
Quantum dot[21]	18 nm	None	Nikon Eclipse TE2000 CIsi	Not reported	Not reported
Quantum dot[22]	18 nm	None	Leica TCS SP2	488 nm	500–535 nm
Quantum dot[23]	Not reported	None	Lucid VivaScope (modified)	600 nm	>664 nm
Quantum dot[24]	13–29 nm	None	Zeiss LSM 510 META	4888 nm	560–615 nm
SiO_2[25]	91–103 nm	Rhodamine	Not reported	569 nm	585 nm

The output from LSCM instruments are gray-scaled images where each pixel ranges from 0 to 255, black to white. These images can be analyzed in a number of ways, but the most common analysis is subjective penetration assessments done by morphological assessment. Table 12.2 shows a summary of recent reports in this context. Basic image analysis includes using software, for example ImageJ, to extract integrated density information from the images.[11,13] This basic approach can be extended to normalization using positive or negative controls.[24] More advanced image analysis techniques include modeling and curve fitting for semiquantitative analysis.[20,23] The majority of recent studies use LSCM data for determining nanoparticle penetration into skin and secondarily to study the localization of nanomaterials with skin structures.

One example of LSCM data used to determine nanoparticle penetration into skin was recently reported by Campbell et al.[20] The focus of this report was to objectively assess nanoparticle-disposition-intact and barrier-compromised skin. One of the driving forces in the field of nanodermatology is the persistent question of topical nanoparticle toxicity. At the end of the day, it is impossible to show that a particular nanoparticle is not toxic. One of the key features of toxicity is the concept of exposure. Thus, there are many groups around the globe assessing nanoparticle exposure in skin. This report by Campbell et al. feeds into this niche. The authors used 20–200 nm particles and treated porcine skin from 5 minutes to 16 hours. Modest barrier disruption was achieved by taking four tape strips just before treatment in Franz cells. Imaging was carried out on randomly selected fields (143×143 μm²) from 10 to 60 μm deep as 512×512 pixel images. These images were down-sampled to 128×128 pixel images to improve the signal-to-noise ratio. These data were then processed with a custom model designed to incorporate autofluorescence signals to detect the skin surface and calculate the nanoparticle signal intensity (Figure 12.7). These data were then interpreted as supporting evidence that the nanoparticles cannot penetrate beyond the outer layers of the skin. This observation is echoed throughout most publications in this nanoparticle-focused area.

Although multiwalled carbon nanotubes (MWCNT) are not nanoparticles, they are nanomaterials that have the potential for topical contact in the industrial setting. Zhang and Monteiro-Riviere[16] used LSCM to investigate MWCNT interactions with cultured human epidermal keratinocytes. MWCNT have many potential uses in biomedical technologies including use as sensors and transdermal drug delivery vectors.[26,27] Zhang and Monteiro-Riviere were focused on elucidating mechanisms of MWCNT uptake in keratinocytes with an emphasis on lectin-mediated uptake inhibition. During this process they developed a marker-free technique for visualizing MWCNT by using the LSCM to image scattered excitation light in a manner similar to RCM. This enabled them to generate some very high quality images of nanomaterial interactions with living and labeled keratinocytes (Figure 12.8). Human epidermal keratinocytes were grown to 70% confluence and then treated with 0.1 mg/mL MWCNT. The cells were treated for 24 hours for LSCM and up to 48 hours for toxicity experiments. The outcome of this study was the observation that all of the lectin cocktails tested reduced MWCNT uptake and that MWCNT could transverse from one keratinocyte to another through the cytoskeleton and plasma membranes. This information could be used to develop integrated biosensors or long-term drug delivery platforms in the future.

TABLE 12.2

Recent Laser Scanning Confocal Microscopy Skin Models, Imaging Approaches, and Analysis Strategies

Nanoparticle Material	Model	Imaging Approach	Analysis	Interpretation
Dendrimers[11]	Porcine skin	*En face*, cross section shown	Total integrated pixels	Penetration
Dendrimers[12]	Porcine skin	Sections	Photos	Penetration and localization
Ethosomes[13]	Human hypertrophic scar	Sections	Mean fluorescence intensity and fluorescent area	Penetration and delivery
Ethyl cellulose[14]	Mouse skin	*En face*, cross section shown	Photos	Penetration, autofluorescence, and localization
Microemulsion[15]	Mouse fibroblasts	In vitro	Photos	Cell uptake
Multiwalled carbon nanotubes[16]	Human keratinocytes	In vitro	Photos	Localization
Nanoemulsions[17]	Porcine and mouse skin	*En face*	Photos	Drug function
Nanotransfersomes[18]	Rat skin	Sections	Photos	Penetration
PLGA[19]	Rat skin	Sections	Photos	Penetration and localization
Polystyrene[20]	Porcine skin	*En face*	Signal intensity and nonlinear mixed-effects modeling	Penetration
Quantum dot[21]	Mouse skin	*En face*	Photos	Localization
Quantum dot[22]	Mouse skin	*En face*	Photos	Penetration
Quantum dot[23]	Excised human skin	*En face*	Relative fluorescent intensity	Penetration
Quantum dot[24]	Excised human skin	*En face*, cross section shown	Normalized integrated density	Penetration
SiO_2[25]	Excised human skin	Cut skin in cross section	Photos	Penetration

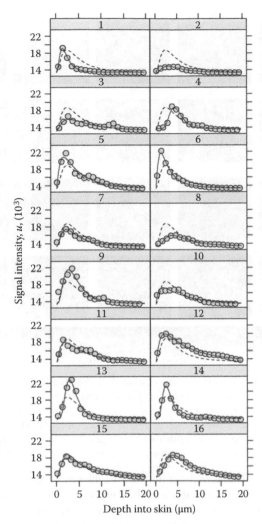

FIGURE 12.7 Calculated and observed distribution of particle signal intensity (u) in one $143 \times 143\ \mu m^2$ region of the skin following application of 200 nm FluoSpheres® for 16 hours. A log-normal distribution was fitted to the data for each of the sixteen $36 \times 36\ \mu m^2$ quadrants using nonlinear mixed-effects modeling with R software. Dashed lines show the population mean profile; solid lines the local mean for each quadrant.

12.2.5 MULTIPHOTON MICROSCOPY AND TIME-CORRELATED SINGLE-PHOTON COUNTING

Multiphoton microscopy uses femtosecond pulsed lasers to excite fluorescent molecules, endogenous fluorophores, and some specific nanomaterials for high-resolution imaging. For a recent review related to MPM imaging of nanoparticles in skin see Prow.[28] The two key technical differences between single-photon microscopy and MPM are the laser source (pulsed lasers are used for MPM) and that there is no pinhole needed for MPM

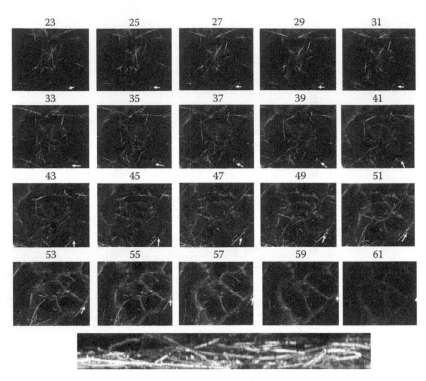

FIGURE 12.8 Z-series optical sections of multiwalled carbon nanotube (MWCNT)-dosed human epidermal keratinocytes stained with F-actin. The interaction of MWCNT with cytoskeleton was visualized by optical sectioning cells into 70 sections. Here 20 sections (layers 23–61) are presented. An arrow in each section depicts a single MWCNT extending sharply to the lower layers, suggesting its protrusion into the cell. The bottom image shows a side view and the above sections show the strong interaction between MWCNT and F-actins.

(Figure 12.6a and b). The use of the pulsed laser enables more than one photon to interact with a single molecule or nanoparticle at the same time. This phenomena result in the emission of a single photon within the focal volume, a tiny football-shaped region at the excitation beam convergence point. Excitation and emission occur only in this 3D space, so all of the emission light is in the focal plane, thus negating the need for a pinhole to exclude out-of-plane emission light. The practical follow-on effect is that imaging in skin strata does not result in out-of-plane fluorescence before or after beam convergence and therefore there is much lower background noise with MPM compared to LSCM. This property of MPM is efficiently shown when imaging endogenous fluorophore-like NAD(P)H, a major metabolic molecule whose fluorescence decreases with toxicity.[3]

Another beneficial aspect of MPM imaging is the capacity to image second-harmonic generation (SHG) and multiphoton-enhanced photoluminescence (MEP) (Figure 12.6c). Dermal collagen is a vital structural skin component with a specific morphology that changes with pathology and can be directly visualized with MPM by SHG.[29] Metal nanoparticles can also be visualized with a similar process called MEP. Prow and others have successfully used this approach with gold, silver, and zinc oxide nanoparticles in skin. MEP results in easily detectable signals from metal and metal oxide nanoparticles. This powerful approach has been used for estimating the nanoparticles per pixel based on thresholding to isolate pixels with nanoparticle signal and then using nanoparticle Rayleigh scattering and the size of a single pixel to estimate the minimal number of nanoparticles that could be present in the defined area.[30]

Estimating the number of nanoparticles present at various depths in skin is a complex and relevant endeavor in the field of skin nanotoxicology. This information would feed back into the earlier discussed issue of topical nanoparticle toxicity and the question of penetration. Nanoparticle penetration studies are also directly relevant to drug delivery applications. The MPM is suited for these types of studies, but alone is only capable of generating gray scale image data. TCSPC detectors and signal processing enable MPM to be used for quantitative studies. For in-context reviews of this technology see References 28 and 31.

In 2011, our group published the first report detailing an MPM-TCSPC-based approach to quantify metal nanoparticles in skin.[32] We treated volunteers that had psoriasis and atopic dermatitis with ZinClear-S 60 in a caprylic/capric triglyceride vehicle at 60% (w/w) to 2 cm² area for 3 hours. We also treated healthy volunteers with the same nanoparticle formulation after tape stripping for 4 and 24 hours. These studies were designed to evaluate the penetration potential for zinc oxide nanoparticles in diseased and barrier-disrupted skin. The key to this approach is that MPM-TCSPC could be used to identify pixels with nanoparticle signal signatures. These pixels were isolated by using the proportion of the MEP photons that return to the detector nearly instantaneously to separate it from the proportion of photons generated from the process of fluorescence emission that have both short and long lifetimes in a given pixel (Figure 12.9). This enabled us to identify nanoparticle-containing pixels and exclude non-nanoparticle-containing pixels. We then used a standard curve-based approach to quantify the nanoparticle signal.

This method and the one described earlier have obvious benefits and critical limitations. These approaches have not been validated and modeled in phantoms or in skin at depth in different strata or in diseased skin. Likewise, there has not been a systematic validation with different metal nanoparticles of different sizes, not to mention exploring the effects of nanoparticle coating or vehicle effects. Suffice to say, there is much work ahead before we can confidently say how many nanoparticles are present in a particular place within the complex media of skin without resorting to invasive techniques such as electron microscopy (EM).

12.2.6 ELECTRON MICROSCOPY

Electron microscopy is simply the best way to image single and small numbers of nanoparticles. Our laboratory has previously published pilot data that compared MPM and transmission electron microscopy (TEM) to process and image nanoparticle-containing sunscreen formulations.[33] We found that the two imaging modalities were comparable at MPM level magnification, but of course MPM cannot achieve EM levels of magnification. This is because MPM uses photons to image whereas EM uses electrons as the name suggests. This has several follow-on consequences. First, using electron beams with electromagnetic lenses enables such high magnification that individual nanoparticles can be examined in detail. Second, contrast within an image is

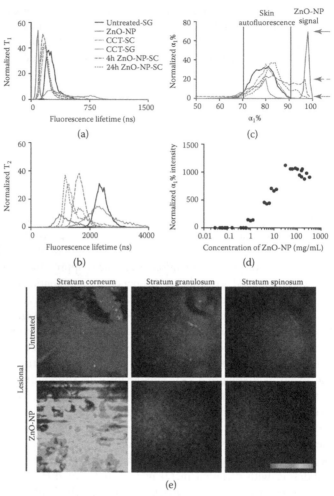

(e)

FIGURE 12.9 MPM-TCSPC-derived photonic characteristics of skin and ZnO-NP and pseudo-colored images. (a–c) τ_1, τ_2, and $\alpha_1\%$ histograms of the untreated stratum granulosum (Untreated-SG), caprylic/capric triglyceride (CCT)-treated stratum corneum after 24 hour (CCT-SC), CCT-treated stratum granulosum after 24 hour (CCT-SG), ZnO-NP alone (ZnO-NP), ZnO-NP on the stratum corneum after 4 hours (4h ZnO-NP-SC), and ZnO-NP on the stratum corneum after 24 hours (4h ZnO-NP-SC). Panel c shows distinct α_1 signals from skin autofluorescence and ZnO-NP multiphoton-enhanced photoluminescence. The arrows in panel c indicate the maximal ZnO-NP signals. (d) The generation of the in vitro ZnO-NP standard curve. Normalized $\alpha_1\%$ 90–100 data were obtained from ZnO-NP solutions in vitro. The equation derived from the linear region (1.0–50.0 mg/mL) of this standard curve was $y = 0.6573e^{0.004X}$. The dotted line represents the mean value of normalized $\alpha_1\%$ 45–85 from vehicle-treated SG. The laser optical power was decreased to image ZnO-NP at 150.0 and 200.0 mg/mL. (e) In vivo multiphoton images of lesional volunteer skin of different depths after 2 hour treatment with ZnO-NP. Each image is $214 \times 214 \times 1$ µm³. These color images depict the autofluorescence as blue ($\alpha_1\%$ 0–85) and ZnO-NP as yellow/red ($\alpha_1\%$ 90–100) in volunteer skin. All bars indicate 100 µm. The color scale bar represents $\alpha_1\%$ 85–100, blue to red.

generated by electron density. This means that if gold nanoparticles are being imaged in skin there is significant contrast, but not if ethosomes or dendrimers are used.

TEM and scanning electron microscopy (SEM) are the two relevant EM approaches. TEM is commonly used to characterize nanomaterials, including the following recent reports from Tables 12.1 and 12.2.[13,15,16,18] Monteiro-Riviere et al.[34] published a report on the penetration potential of TiO_2 and ZnO nanoparticles in sunscreen formulations using TEM and SEM as the primary endpoints, in both excised and in vivo porcine skin. Nanoparticle penetration was tested in ultraviolet (UV)-exposed skin to mimic the real-world condition. Skin was dosed with 50 µL per 0.64 cm^2 of each of the four sunscreens (10% TiO_2 coated with hydrated and aluminum hydroxide in o/w; 10% TiO_2 coated with and aluminum hydroxide in w/o; 5% ZnO coated with in o/w; 5% uncoated ZnO in o/w). In vivo skin treated with UV and the same nanoparticles was processed for conventional histopathology and imaged in cross section with TEM. These data show that the sunscreen nanoparticles were able to penetrate between the corneocyte layers of the stratum corneum but were not detected beyond this barrier, even after UV exposure. This effective use of skin imaging technology was further strengthened by using SEM to image these same nanoparticles on the surface of skin and hair. SEM has the added benefit of energy-dispersive X-ray spectroscopy detectors. This analytical technique can yield elemental analysis data that are registered to a particular location within the field of view. In Monteiro-Riviere et al.[34] this approach yields information about the location of nanoparticle aggregates on the surface of the skin. Together, these techniques provide convincing localization and penetration information. One major limitation of the TEM-based approach is that only a limited amount of analysis can be carried out due to the high magnification and very thin and small sections used. Another limitation of TEM is the processing time is lengthy and can be prone to artifacts. That said, it is possible to find and collaborate with one of the many experienced EM experts to achieve high-quality outcomes.

12.3 LIMITATIONS AND CONSIDERATIONS FOR IMAGING NANOPARTICLES IN SKIN

There are many inherent limitations and considerations that need to be taken into account when imaging nanoparticles in skin. Nanoparticles have special properties because of the nature of their physical size and shape. Indeed a 100 nm diameter particle is 500–1000 times smaller than the width of a hair. This creates a sort of paradox where we are looking for objects within objects at two vastly different size scales. If we can see the layers of the skin, we cannot see the nanoparticles and vice versa. Therefore, two basic approaches can be imagined. One approach is to use dermoscopy, OCT, RCM, LSCM, and MPM to visualize the skin strata and accept error in terms of nanoparticle number/position. The other approach uses EM to directly examine a handful of nanoparticles and their interactions at the subcellular level at the expense of only analyzing a tiny area and increasing the likelihood of overinterpreting artifacts. In the context of nanotoxicology, perhaps a combination of low- and high-magnification approaches is best in a given study.

The incorporation of functional assays and real-world experimental designs is the real key to determining toxicity. For example, Raphael et al.[35] noted a ZnO nanoparticle signal in the viable epidermis indicating potential for toxicity to occur, but this signal disappeared with simple soap and water washing. This illustrated a real-world scenario where the apparent risk of nanoparticle exposure can be easily overcome. The risk is "apparent" because using imaging technologies to identify nanoparticles in complex media like skin is prone to significant error depending on the experimental design, instrument settings, and methodology. It is all too easy to increase the laser power or gain a little too much with LSCM and MPM and record artifacts. This is where controls are key, that is more than one control. Generally, we have untreated skin, vehicle-treated skin, nanoparticles in vehicle, and the experimental group with nanoparticle-treated skin. The risk from not having comprehensive controls is that of false-positive or-negative nanoparticle identification issues.

One of the most difficult aspects of studying nanoparticles in skin is actually identifying what is and what is not a nanoparticle. An example is shown in Figure 12.5 and the use of elemental analyses (see Chapter 13). Identifying where the nanoparticles are is impossible without having done many pilots and control experiments. In nanotoxicology the real-world concentration of nanoparticles in question can be below the detection limit of the imaging equipment available. There are also occasions where a researcher follows a previous report and cannot visualize the nanomaterials in the same way as others. To overcome this detection issue, we routinely blot a high concentration of nanoparticles onto paper and use that to set up equipment for a new experiment. This simple approach helps researchers be sure that they are really looking at nanoparticles and not an artifact of inappropriate system settings.

System settings are critical and the difficulty of using these instruments properly increases with the complexity of the machines. For example, the dermatoscope always works, but reproducing MPM settings is rarely successful without significant tweaking. Each instrument has limitations: OCT has excellent depth information, but is limited by low contrast. RCM is easy to use in the clinic but only resolves highly reflective materials at the skin surface. LSCM has good x/y resolution but poor depth resolution for nanoparticles. MPM requires an expert for use and data analysis, otherwise significant misinterpretation of data can follow. SEM lacks good contrast. TEM is limited to tiny samples and sample preparation requires an expert. That said, multiple imaging approaches can be used to offset the inherent limitations, for example, the approaches taken by Monteiro-Riviere et al.[34] and Labouta et al.[36]

12.4 SUMMARY

More is better when it comes to imaging nanoparticles in skin. Combining imaging modalities and incorporating functional readouts, for example, NAD(P)H, only strengthens the interpretation of imaging data. Another way to improve the power of this type of data is through image analysis. ImageJ is a free tool from National Institutes of Health that can be used to do basic intensity and localization measurements in single images and stacks. More advanced image analysis can be carried out by those skilled in the craft and sometimes the process itself can be published.[37] The capacity to use automated or at least mathematically defined approaches to tease out information from images dramatically improves the credibility of the observation, for example, Campbell et al.[20] In summary, there is no perfect nanoparticle imaging tool for skin, but when image analysis is combined with high-quality imaging a picture can be worth more than a thousand words.

REFERENCES

1. Gambichler T, Jaedicke V, Terras S. Optical coherence tomography in dermatology: Technical and clinical aspects. *Arch Dermatol Res.* 2011;303(7):457–73.
2. Zalaudek I, Argenziano G, Di Stefani A, Ferrara G, Marghoob AA, Hofmann-Wellenhof R, Soyer HP, Braun R, Kerl H. Dermoscopy in general dermatology. *Dermatology.* 2006;212(1):7–18.
3. Labouta HI, Liu DC, Lin LL, Butler MK, Grice JE, Raphael AP, Kraus T et al. Gold nanoparticle penetration and reduced metabolism in human skin by toluene. *Pharm Res-Dordr.* 2011;28(11):2931–44.
4. Lee TM, Oldenburg AL, Sitafalwalla S, Marks DL, Luo W, Toublan FJ, Suslick KS, Boppart SA. Engineered microsphere contrast agents for optical coherence tomography. *Opt Lett.* 2003;28(17):1546–8.
5. Zagaynova EV, Shirmanova MV, Kirillin MY, Khlebtsov BN, Orlova AG, Balalaeva IV, Sirotkina MA, Bugrova ML, Agrba PD, Kamensky VA. Contrasting properties of gold nanoparticles for optical coherence tomography: Phantom, in vivo studies and Monte Carlo simulation. *Phys Med Biol.* 2008;53(18):4995–5009.
6. Kim S, Rinehart MT, Park H, Zhu YZ, Wax A. Phase-sensitive OCT imaging of multiple nanoparticle species using spectrally multiplexed single pulse photothermal excitation. *Biomed Opt Express.* 2012;3(10):2579–86.
7. Nehal KS, Gareau D, Rajadhyaksha M. Skin imaging with reflectance confocal microscopy. *Semin Cutan Med Surg.* 2008;27(1):37–43.
8. Prow TW, Grice JE, Lin LL, Faye R, Butler M, Becker W, Wurm EM et al. Nanoparticles and microparticles for skin drug delivery. *Adv Drug Deliver Rev.* 2011;63(6):470–91.
9. Pawley JB. *Handbook of Biological Confocal Microscopy.* 3rd ed. New York, NY: Springer; 2006. xxviii, 985 pp.
10. Zhang LW, Monteiro-Riviere NA. Use of confocal microscopy for nanoparticle drug delivery through skin. *J Biomed Opt.* 2013;18(6):061214.
11. Venuganti VV, Sahdev P, Hildreth M, Guan X, Perumal O. Structure–skin permeability relationship of dendrimers. *Pharm Res-Dordr.* 2011;28(9):2246–60.
12. Yang Y, Sunoqrot S, Stowell C, Ji J, Lee CW, Kim JW, Khan SA, Hong S. Effect of size, surface charge, and hydrophobicity of poly(amidoamine) dendrimers on their skin penetration. *Biomacromolecules.* 2012;13(7):2154–62.
13. Zhang Z, Wo Y, Zhang Y, Wang D, He R, Chen H, Cui D. In vitro study of ethosome penetration in human skin and hypertrophic scar tissue. *Nanomedicine.* 2012;8(6):1026–33.
14. Abdel-Mottaleb MM, Moulari B, Beduneau A, Pellequer Y, Lamprecht A. Surface-charge-dependent nanoparticles accumulation in inflamed skin. *J Pharm Sci.* 2012;101(11):4231–9.
15. Zhang YT, Huang ZB, Zhang SJ, Zhao JH, Wang Z, Liu Y, Feng NP. In vitro cellular uptake of evodiamine and rutaecarpine using a microemulsion. *Int J Nanomedicine.* 2012;7:2465–72.
16. Zhang LW, Monteiro-Riviere NA. Lectins modulate multi-walled carbon nanotubes cellular uptake in human epidermal keratinocytes. *Toxicol In Vitro.* 2010;24(2):546–51.
17. Zhang LW, Al-Suwayeh SA, Hung CF, Chen CC, Fang JY. Oil components modulate the skin delivery of 5-aminolevulinic acid and its ester prodrug from oil-in-water and water-in-oil nanoemulsions. *Int J Nanomedicine.* 2011;6:693–704.
18. Ahad A, Aqil M, Kohli K, Sultana Y, Mujeeb M, Ali A. Formulation and optimization of nanotransfersomes using experimental design technique for accentuated transdermal delivery of valsartan. *Nanomed-Nanotechnol.* 2012;8(2):237–49.

19. Tomoda K, Terashima H, Suzuki K, Inagi T, Terada H, Makino K. Enhanced transdermal delivery of indomethacin using combination of PLGA nanoparticles and iontophoresis in vivo. *Colloid Surf B.* 2012;92:50–4.

20. Campbell CS, Contreras-Rojas LR, Delgado-Charro MB, Guy RH. Objective assessment of nanoparticle disposition in mammalian skin after topical exposure. *J Contr Rel.* 2012;162(1):201–7.

21. Kulvietis V, Zurauskas E, Rotomskis R. Distribution of polyethylene glycol coated quantum dots in mice skin. *Exp Dermatol.* 2013;22(2):157–9.

22. Lee WR, Shen SC, Al-Suwayeh SA, Yang HH, Li YC, Fang JY. Skin permeation of small-molecule drugs, macromolecules, and nanoparticles mediated by a fractional carbon dioxide laser: The role of hair follicles. *Pharm Res-Dordr.* 2013;30(3):792–802.

23. Mortensen LJ, Glazowski CE, Zavislan JM, Delouise LA. Near-IR fluorescence and reflectance confocal microscopy for imaging of quantum dots in mammalian skin. *Biomed Opt Express.* 2011;2(6):1610–25.

24. Prow TW, Monteiro-Riviere NA, Inman AO, Grice JE, Chen XF, Zhao X, Sanchez WH et al. Quantum dot penetration into viable human skin. *Nanotoxicology* 2012;6(2): 173–85.

25. Staronova K, Nielsen JB, Roursgaard M, Knudsen LE. Transport of SiO_2 nanoparticles through human skin. *Basic Clin Pharmacol.* 2012;111(2):142–4.

26. Baughman RH, Zakhidov AA, de HWA. Carbon nanotubes—the route toward applications. *Science.* 2002;297(5582):787–92.

27. Ilbasmis S, Ismail T, Degim T. A feasible way to use carbon nanotubes to deliver drug molecules: Transdermal application. *Expert Opin Drug Del.* 2012;9(8):991–9.

28. Prow TW. Multiphoton microscopy applications in nanodermatology. *Wires Nanomed Nanobi.* 2012;4(6):680–90.

29. Wheller L, Lin LL, Chai E, Sinnya S, Soyer HP, Prow TW. Noninvasive methods for the assessment of photoageing. *Australas J Dermatol.* 2013. doi: 10.1111/ajd.12030.

30. Labouta HI, Schaefer UF, Schneider M. Laser scanning microscopy approach for semi-quantitation of in vitro dermal particle penetration. *Methods Mol Biol.* 2013;961:151–64.

31. Konig K, Raphael AP, Lin L, Grice JE, Soyer HP, Breunig HG, Roberts MS, Prow TW. Applications of multiphoton tomographs and femtosecond laser nanoprocessing microscopes in drug delivery research. *Adv Drug Deliver Rev.* 2011;63(4–5):388–404.

32. Lin LL, Grice JE, Butler MK, Zvyagin AV, Becker W, Robertson TA, Soyer HP, Prow TW. Time-correlated single photon counting for simultaneous monitoring of zinc oxide nanoparticles and NAD(P)H in intact and barrier-disrupted volunteer skin. *Pharm Res-Dordr.* 2011;28(11):2920–30.

33. Butler MK, Prow TW, Guo YN, Lin LL, Webb RI, Martin DJ. High-pressure freezing/freeze substitution and transmission electron microscopy for characterization of metal oxide nanoparticles within sunscreens. *Nanomedicine (London)* 2012;7(4):541–51.

34. Monteiro-Riviere NA, Wiench K, Landsiedel R, Schulte S, Inman AO, Riviere JE. Safety evaluation of sunscreen formulations containing titanium dioxide and zinc oxide nanoparticles in UVB sunburned skin: An in vitro and in vivo study. *Toxicol Sci.* 2011;123(1):264–80.

35. Raphael AP, Sundh D, Grice JE, Roberts MS, Soyer HP, Prow TW. Zinc oxide nanoparticle removal from wounded human skin. *Nanomedicine (London)* 2013.

36. Labouta HI, Liu DC, Lin LL, Butler MK, Grice JE, Raphael AP, Kraus T et al. Gold nanoparticle penetration and reduced metabolism in human skin by toluene. *Pharm Res-Dordr.* 2011;28(11):2931–44.

37. Kurugol S, Dy JG, Rajadhyaksha M, Gossage KW, Weissman J, Brooks DH. Semi-automated algorithm for localization of dermal/epidermal junction in reflectance confocal microscopy images of human skin. *Proc SPIE.* 2011;7904:7901A.

Section V

Hazards

13 Safety Implications of Nanomaterial Exposure to Skin

Nancy A. Monteiro-Riviere

CONTENTS

13.1 INTRODUCTION

The field of nanotechnology has drastically grown because of development of engineered nanomaterials (ENMs) that can exhibit a variety of unique and tunable chemical and physical properties. It is these unique physicochemical properties that have made ENMs key components in material science, engineering, and medicine. This has led to an array of emerging technologies and numerous commercial products. Although nanomaterials (NMs) have widespread potential applications, skin penetration and toxicity of these ENMs have not been thoroughly evaluated under likely environmental, occupational, and medicinal exposure scenarios. Many challenges must be overcome before NMs can be used in nanomedicine and consumer products.

Skin, a complex and dynamic organ, is one of the largest organs of the body that serves as a principal portal of entry by which chemicals, environmental toxicants, or ENMs can traverse through to penetrate into the body. Currently, there is a lack of information on whether ENMs or nanoparticles (NPs) can be absorbed across the stratum corneum barrier or whether systemically administered particles can accumulate in dermal tissue. Most studies so far conducted have been short-term toxicity studies, suggesting a great need to conduct long-term in vivo human and animal studies for a realistic risk assessment to be made.

Although a plethora of studies are being reported, as will be seen later, there is still a deficiency in the safety and toxicology data gathered over longer exposure periods (days to months) that characterize the usage of many ENMs in daily life. As with many other studies in nanotoxicology, lack of characterization as well as inconsistency in types of NPs studied (size, shape, surface coatings, vehicles/formulations), use of the wrong model system or species to study dermal absorption/dermal penetration, prevents generalizations to be easily made.

One of the most commonly asked questions is whether ENMs or NPs gain access to the epidermal layers of the skin? Would these particles preferentially locate in the lipids of the stratum corneum layers after topical exposure? Can NPs gain access to tissue spaces, a prerequisite for systemic toxicity? Does the concept of partitioning, central to predicting chemical absorption, apply to NPs? What are the toxicological consequences of *dirty* NPs (catalyst residue) becoming lodged in the upper layers of the skin or in the viable epidermal layers? In fact, it is this relative biological isolation in the lipid domains of the epidermis that has allowed for the delivery of drugs to the skin using lipid NPs and liposomes.

13.2 SKIN ANATOMY AND PHYSIOLOGY

Skin has many functions, a primary one being to act as a barrier to the external environment. It is the primary exposure route for cosmetics and is the preferred route of administration for topically applied drugs targeted either locally for dermatological applications or transdermally for systemic therapy. Recent advances in nanotechnology result in potential exposure of NPs to skin after occupational or consumer product use (e.g., sunscreens).

The tendency for ENMs to traverse the skin is a primary determinant of its dermatotoxic potential. That is, NMs or NPs must penetrate the outermost stratum corneum rate-limiting barrier to exert toxicity in the cell layers below. The quantitative prediction of the rate and extent of percutaneous penetration (into skin) and absorption (through skin) of topically applied NMs is complicated because the processes that drive NPs into skin may be different than those that govern chemicals. ENM or NP absorption may occur through several potential skin routes: the majority of lipid-soluble particles may move through the intercellular lipid pathway between the stratum corneum cells (intercellular), through the cells (transcellular), or through the hair follicle or sweat ducts (transappendageal) openings. For a more complete review of skin anatomy and physiology and comparative species differences see Monteiro-Riviere.[1,2]

13.3 MODEL SYSTEMS USED TO ASSESS NANOPARTICLE ABSORPTION AND PENETRATION

13.3.1 Pig Skin Is an Accepted Animal Model for Human Skin

Percutaneous absorption of chemicals in humans and different animal species[3–9] has been extensively studied. To compare such datasets, we must control and take into consideration many factors such as applied dose, surface area, use of occlusive dressings, and dosing in a vehicle or formulation. Rodents are often used as the primary laboratory animal to assess the safety of drugs and chemicals in dermal absorption studies to make route-to-route extrapolations feasible because the majority of preclinical toxicology studies are conducted in rodents. However, the skin of rodents (e.g., mice, rats) is much more permeable to chemicals than human skin, thereby making rodents suitable for defining worst-case scenario for absorption and toxicological endpoints. If the goal is to predict the rate and extent of chemical absorption in humans, then a species with minimal hair or fur must be used in pharmaceutical development studies for extrapolation to humans. The hair follicle density in rats ($\sim$300/cm^2) or mice ($\sim$650/cm^2) is much greater than that in humans making the resulting interfollicular region of skin thinner, two factors potentially increasing the rate of chemical or NP absorption. The hair follicle density in humans is only 11/cm^2, which is similar to that of pigs.[5] Monteiro-Riviere (unpublished observations) conducted an age comparison study on the hair density of 8-week-old weanling pigs, 14-week-old pigs, and 15-month-old pigs and showed that at 8 weeks the hair follicle density was 68.9 cm^2, at 14 weeks 29.7 cm^2, and at 15 months only 8.4 cm^2. If hair follicles play a role in absorption, the age of the animal may be an important determinant for absorption or penetration studies. Apes are often selected because of their evolutionary closeness to humans; however, only the abdomen with minimal hair should be used. Many other anatomical factors can influence the absorption of chemicals or NPs. Regional and species differences, thickness, hair follicle density, blood flow, age, and disease states may all influence the barrier function of skin.[10,11]

The domestic pig is thus an appropriate and accepted animal model for studying dermal chemical absorption in humans.[3,4,7,8] In addition, if pigs are obtained from an abattoir then one must be sure to harvest skin before scalding occurs in the carcass decontamination process. In addition to their similarities in hair follicle density, number of epidermal cell layers, skin thickness, and cutaneous blood flow, biochemistry and biophysics of the stratum corneum lipids of pigs are comparable to humans.[10,11] The body mass/surface area ratios are similar to humans, thereby making extrapolations of systemic exposure easier without doing complex allometric analyses. This is particularly important when chemical reactions or protein–NP interactions change ENM surface characteristics in a timescale not influenced by biological factors that are weight dependent (e.g., blood circulation times) when extrapolated across species. Body temperatures across species are different. Scaling these interactions from mice to humans may be extremely difficult. Depending on the species of choice, it is important to be cognizant of the fact that regional differences in skin anatomy exist making both species and body site crucial descriptors of experimental protocols.

13.3.2 In Vivo Models

Once the appropriate species has been selected, the decision must be made between using intact animals in vivo and numerous in vitro approaches. The *gold standard* for absorption studies is in vivo; however, the extent of variability inherent to this work often precludes intact animals from being used for detailed probing of mechanisms involved. In addition, in vivo approaches are expensive and require special facilities that are not accepted by European regulatory authorities for cosmetic screening. Both of these approaches are extensively reviewed elsewhere.[9]

The classic in vivo approach used to assess absorption of any compound exposed to the body is by measuring the amount excreted into urine and feces compared with that excreted after intravenous administration of an equivalent dose. The ratio of the total amount excreted after dermal exposure compared to intravenous dosing is termed *bioavailability*, referred to in pharmacokinetic literature as F. This parameter is calculated by measuring the plasma concentrations of the compound after each route of exposure and comparing the ratio of their areas under the concentration–time curve (AUC); that is AUC_{dermal}/AUC_{iv}. This can also be calculated by measuring total amount of drug or NP excreted in the urine and feces. The reason F has to be determined relative to a parenteral route such as intravenous is that differences in metabolism between chemicals may change the rate and/or extent of excretion or the percentage of a drug present in the central plasma compartment. Once the fraction of the dose eliminated in urine or feces is known, then only the excreta needs to be collected to get future estimates of absorption. An extension of this procedure is used when urine is monitored for assessing systemic exposure. If creatinine concentrations are also measured (marker of urine production), and drug mass normalized by creatinine concentration, then monitoring of only chemical concentrations and creatinine is sufficient to make estimates of relative systemic exposure or absorption after different topical chemical treatments. This approach is used in occupational medicine and field studies monitoring dermal exposure to topical NPs or chemicals.[12]

All of these approaches require timed samples and careful analysis. Experiments should be terminated only when the majority of the compound has been eliminated (e.g., ~80%). For dermal application of compounds with very slow rates of absorption, the duration of the study required to insure complete systemic distribution and elimination can be lengthy. However, most studies with NPs have been conducted in short time periods of hours or days. No long-term studies have been conducted with NPs and that is why there is still controversy as to whether NPs can be systemically absorbed. Truncating studies earlier may lead to erroneous conclusions. When conducting such studies, care must be taken to insure the dose is completely covering the marked exposure area, and that this surface area is measured and occluded to prevent contamination and evaporation of the compound or NM.

Additional approaches have been developed, which attempt to assess in vivo absorption by monitoring the drug absorption process from the perspective of nonabsorbed concentrations remaining in stratum corneum tape strips.[13,14] This approach is known as dermatopharmacokinetics, which analyzes the amount of drug in the skin at the application site by sequentially removing and analyzing the drug or NP on the surface (nonabsorbed drug) by gentle washing and swabbing and then tape-stripping.

Various approaches have been used to model the diffusion gradient reflected in the stratum corneum depth profiles.[15] These techniques must first be calibrated to normal human absorption endpoints before reliable predictions can be made.

13.3.3 IN VITRO DIFFUSION CELL MODELS

A primary in vitro technique used to study dermal absorption for both humans and animals is to mount dermatomed skin samples into a diffusion cell and measure the chemical flux into the perfusate bathing the dermal side of the skin sample. Specific protocols have been established relative to membrane preparation and perfusate composition.[6,16] Skin may be either full-thickness, heat-separated epidermal membranes, or dermatomed skin (skin is sectioned to a predetermined thickness) and the size punched out to place on the diffusion cell. The skin is clamped between two chambers, one of which contains a vehicle from which absorbed chemical will be sampled. Static diffusion cells sample this chamber and replace with new perfusate at each time point. Flow-through cells use a pump to pass the perfusate through the receptor chamber and the flux is collected repeatedly in the perfusate.

In some systems, the skin disk is first fully hydrated and allowed to equilibrate with media before dosing. Receptor fluid used is usually saline for pharmaceutical drug studies or a media containing albumin or a surfactant/solvent for studies of organic chemicals where some degree of lipid solubility in the receptor fluid is required. Perfusate should be heated to 32°C or 37°C according to experimental guidelines. If the chemical is dissolved in water and dosed in the donor chamber at a dose that far exceeds the amount absorbed, this experimental condition is termed an *infinite* dose experiment. In contrast, if the dose is applied at typical exposure situations and exposed to ambient air or occluded, the experiment is termed *finite dose*. In this case, relative humidity of the dosing environment is crucial.

It is also important to know if the skin obtained is fresh and not frozen. In human studies, skin may be obtained either fresh from reconstructive surgical procedures (within 1–2 hours) or dead from cadaver sources. In the latter case, skin is often frozen and then thawed before use. In these cases, a membrane integrity test is conducted using tritiated water (³H) absorption to screen skin disks for lack of barrier integrity or permeability. Fresh skin should be used to insure an intact barrier and healthy tissue, which is an extremely important factor for metabolism studies, and to minimize the hydration caused from the ³H integrity studies. These skin samples are perfused with oxygenated media to maintain metabolic functions. Artificial skin grown on air–liquid interface cultures is commercially available, although the permeability through such in vitro models of these artificial membranes is much greater than that of normal human skin and not recommended for skin absorption or penetration studies. Many artificial 3D model systems contain only keratinocytes and do not contain all of the other cell types, appendages, or a lipid barrier that are present in normal skin. In vitro models are the predominant approach used to assess dermal absorption today. These methodologies allow calculation of absorption parameters such as the permeability constant. Numerous protocols have been proposed for specific purposes from assessing absorption of lipophilic pesticides versus transdermal delivery of more hydrophilic drugs. Differences between such protocols often relate to surface decontamination methods (swabs, wash, etc.), perfusate composition (addition of vehicle to penetrate solubility of penetrant), and length of experiment. If penetration is to be assessed, chemical concentrations remaining in the skin disk after surface drug removal must be measured. Finally, specific types of experimental designs must be used to provide data for specific mathematical modeling approaches, considerations that often dictate specific approaches.

The optimal in vitro cutaneous model should possess viable cells and structures similar to intact skin as well as a functional vasculature. Such a model would allow topical chemical absorption to be assessed simultaneously with direct toxicity. Our laboratory has developed the isolated perfused porcine skin flap (IPPSF), an ex vivo perfused skin preparation precisely for this purpose.[17–20] The IPPSF is an anatomically intact alternative animal model that possesses a viable epidermis and dermis and an accessible and functional microcirculation. It is experimentally much closer to in vivo than excised skin and is thus a more appropriate in vitro model for assessing percutaneous absorption. The IPPSF model has been shown to be predictive of in vivo human absorption.[8]

13.4 NANOPARTICLE CONSIDERATIONS USING IN VIVO AND IN VITRO TEST SYSTEMS

In vitro cell systems and differences in animal species provide several limitations for a thorough understanding of NP interaction and penetration through the skin. This will be a major challenge for scientists in understanding the safety of NMs in drugs, cosmetics, clothing, medical devices, or any other items that might come into direct contact with the skin.

One of the major decisions in assessing the skin absorption and toxicity of ENMs is how to properly conduct experiments. Topical application of NPs to pig or human skin is preferable than to rodent skin, which is very permeable and does not accurately predict the human response. To determine whether absorption of NPs has occurred, static or flow-through diffusion cells are commonly used. In addition, in vitro cell culture studies are used to assess the toxicity of NPs; however, these have been shown to have limitations when NPs are assessed.[21–23] NPs may interact and interfere with the dye-based assays used to determine cell viability. Invalid results may be obtained because of interactions between the NP with the dye and adsorption of the dye products generated in the assay system. This is also true for NP absorption of cytokines and essential nutrients.[22,23] These issues are fundamentally different from in vitro chemical testing. Positive and negative controls are required, and often classic viability assays (e.g., MTT (3-[4,5-dimethyl-2-thiazol]-2,5-diphenyl-2H-tetrazolium bromide) cannot be used when ENMs are being studied.

Another concern is the difficulty in obtaining a large quantity and good quality of ENMs to conduct in vivo studies. It is much easier to conduct preliminary screening of NPs using in vitro cell systems to estimate the in vivo starting dose for toxicity studies. ENM or NP absorption may not be similar to chemical absorption because these NPs are made of different materials such as carbon, or heavy metals, and have many different physicochemical properties such as different sizes, shapes, charges, porosity, and surface coatings. Also, they may occur in a range of surface modifications and in different vehicles or excipients. All of these physicochemical properties will influence their toxicological response and how they traverse through the upper stratum corneum layers of the skin.

More long-term data are needed for defining systemic exposure after topical administration as well as cutaneous hazard after topical or systemic exposure, the two essential components of any risk assessment. Should NPs be accidentally modified or if exposure occurs before cleansing, consequences could occur if they gain entry into tissues. This has not been evaluated for any type of ENM. A single study will not definitively answer all of the pertinent questions relative to dermal risk assessment of the myriad of ENMs available, but should be able to start to provide an insight into the nature of their potential hazard as well as an initial estimate of dermal exposure parameters that can be used to design more definitive studies.

Despite this increase in the number of studies being reported, as will be seen later, there is still a paucity of safety and toxicology data gathered over longer exposure periods (days to months). As with many other studies in nanotoxicology, lack of characterization as well as inconsistency in types of materials studied (size, shape, surface coatings, vehicles/formulations) prevents generalizations to be easily made. These general factors and limitations have been adequately discussed in other chapters of this book.

This chapter will focus on depicting how skin can serve as a potential route of exposure to several types of NMs, and will discuss how size, shape, charge, surface properties, and vehicles can be important determinants on the penetration through the rate-limiting lipid barrier of the stratum corneum. There are many factors that can influence the extent of uptake in the skin in addition to the physicochemical properties of the NPs, such as the integrity of the skin barrier, ultraviolet (UV) light, mechanical action (tape-stripping), mechanical flexion, abrasion, chemicals, contaminated surfaces, anatomical structures, species differences (rat, pig, or human), presence of diseases such as allergic and irritant dermatitis, atopic eczema, psoriasis, detergents, surfactants, and solvents may potentially increase the rate of absorption or penetration into skin. Therefore, this chapter will review the current information and provide background information relating to NM toxicity in human epidermal keratinocytes (HEK), penetration, and absorption through skin under many different scenarios.

13.5 NANOPARTICLE PENETRATION STUDIES

The penetration of ENMs into and through the skin may be studied from several different perspectives. A material scientist may be interested in the chemical composition, dimension, shape, and coating whereas an anatomist is very interested in the skin morphology. A toxicologist would be interested on the mechanism of interaction with biological structures/systems and a pharmacologist would appreciate the physicochemical properties of the NM formulation (solid, powder, inorganic, or organic dispersion) and absorption kinetics. All these perspectives should be taken into consideration when conducting ENM experiments.

13.5.1 PENETRATION STUDIES

The transdermal flux of ENMs with systemic absorption would have obvious implications to the field of toxicology, therapeutics, and drug delivery studies. However, the depth and mechanism of how ENMs or NPs penetrate into the stratum corneum rate-limiting barrier are crucial. The avascular epidermis is a potential site where individual particles can become lodged within the stratum corneum layers and then be removed by phagocytosis and still be recognized by the immune system through the interaction of Langerhans cells in the suprabasal layers of the epidermis. This can occur in the absence of any significant transdermal flux. This relative biological isolation of the lipid domains of the epidermis has allowed for the delivery of drugs to skin using lipid NPs and liposomes. Therefore, it is possible that NPs could get through or get lodged within the lipid matrix of skin without being absorbed unto the systemic circulation. Several studies evaluated this hypothesis.[24,25]

13.5.2 Quantum Dot Penetration

Quantum dot (QD) NPs have the potential to be used in diagnostics, drug delivery, and imaging in biomedicine or therapeutic applications because of their optical characteristics that result in strong fluorescence without photobleaching.[26] QDs have intense and photostable fluorescence and are commercially available in various sizes, shapes, and diverse surface coatings, making them useful tools to study cellular uptake in HEK and penetration into skin.

Our laboratory has extensively studied the penetration of QDs of different sizes, shapes, and surface coatings in three species to see if these physicochemical properties or species had an effect on skin penetration. QD NP penetration was investigated through porcine, rat, and human skin.[11,24,25,27] The QD565 and QD655 contain a cadmium selenide (CdSe) core with a zinc sulfide (ZnS) shell. Transmission electron microscopy (TEM) was used to depict their size and shape. QD565 were spherical with a diameter of 4.6 nm, whereas QD655 were ellipsoid with a diameter of 6 nm (minor axis) $\times$ 12 nm (major axis). The hydrodynamic diameter was 35 nm for QD565 polyethylene glycol (PEG) coated with a neutral charge, 14 nm carboxylic acid (COOH) with a negative charge, and 15 nm (PEG-amine [NH_2]) with a positive charge. The hydrodynamic diameters for the elliptical QD655 were 45 nm (PEG), 18 nm (COOH), and 20 nm (NH_2).[27] In comparison, the nail-shaped QD621 coated with PEG had a CdSe core with a CdS (cadmium sulfide) shell coated with PEG polymer coils with the mean width of 5.78 ± 0.97 nm, length of 8.40 ± 1.9 nm, and a hydrodynamic size of 39 ± 1 nm evaluated by size-exclusion chromatography.[28] It was proposed that the heavy metals Cd and Se in the QD may be responsible for the toxic effect on cells or tissues. Also, QDs can degrade in oxidative environments and this degradation could cause the release of Cd in vivo and potentially pose a toxic risk.[29,30]

Is it possible for QDs to penetrate porcine skin? The intercellular lipid space between the corneocytes is the most common route of penetration through skin. Our laboratory showed the diameter of porcine corneocytes to be 32 µm and the vertical and lateral gaps between corneocytes to be 19 nm.[31] Therefore, it is feasible for QD to pass through the lateral intercellular spaces of the corneocytes, because the QD621 has a rigid core length of 8.4 nm and width of 5.8 nm but an overall size of 39–40 nm. It is theoretically possible that the outer PEG coating is *soft* allowing for QD621 to *squeeze* through the intercellular space and remain lodged within the stratum corneum lipid bilayers (Figure 13.1). Due to their large size and irregular nail shape, the QD621-PEG may have great difficulty moving through the epidermal layers.

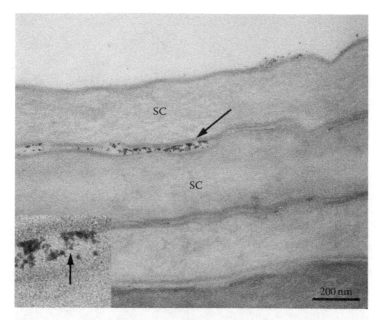

FIGURE 13.1 Transmission electron micrograph depicting the localization of QD621 within the intercellular lipid matrix of the stratum corneum (SC) layers. Arrow is pointing to the quantum dots. (Inset) A higher magnification of QD621 showing their nail-shaped appearance (arrow).

The penetration of QD565 and QD655 with a CdSe core and a ZnS shell dispersed in borate buffer with diverse physicochemical properties was studied in porcine skin flow-through diffusion cells. Spherical QD565 coated with PEG (35 nm; neutral), PEG-NH$_2$ (15 nm; cationic), or COOH (14 nm; anionic) were detected in the stratum corneum layers of the epidermis and localized within the epidermal and dermal layers by 8 hours. The QD655 with PEG (45 nm) and PEG-NH$_2$ (20 nm) coatings were localized within the epidermal layers by 8 hours. The penetration of QD655-COOH (18 nm) into epidermal layers was evident only at 24 hours.[27] Our laboratory also studied QD621 (nail-shaped, 39 ± 1 nm), with a CdSe core and a CdS shell dispersed in water, topically applied to porcine skin in flow-through diffusion cells to assess penetration. After 24 hours, the lower concentration of 1 µM depicted QD621 on the surface of the stratum corneum (Figure 13.2). No fluorescence was detected in the epidermal layers. However, when the concentration was increased to 10 µM, both confocal and TEM depicted QD621 in the upper stratum corneum bi-lipid layers (Figure 13.1) or in between the stratum granulosum and corneum interface, with only a small amount of fluorescence noted in the upper viable epidermal layers of the skin. Occasionally, QD621 were noted in the outer root sheath of the hair follicle.[25]

The penetration seen with QD565, QD655, and QD621 in flow-through diffusion porcine skin cells showed how differences in composition, size, configuration, surface charges, dispersing medium, and other physicochemical parameters could influence penetration. The QD621-PEG were capable of penetrating only the uppermost layers of the stratum corneum after 24 hours of exposure, whereas confocal microscopy

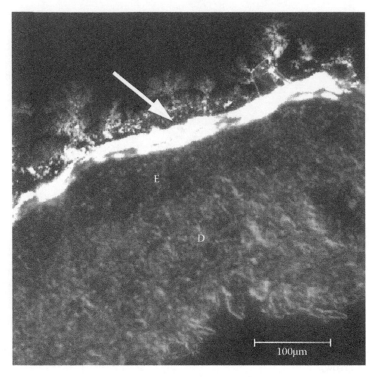

FIGURE 13.2 Laser scanning confocal micrograph depicting a 1 μM dose of QD621 that was topically applied to porcine skin in flow-through diffusion cells for 24 hours. This overlay shows QD621 (arrow, quantum dots appear as bright white) on the surface of the stratum corneum layer of the epidermis (E), (D) dermis.

showed that all three surface coatings of the QD565 penetrated at 8 and 24 hours, but only the QD655-COOH took 24 hours to penetrate the skin. These studies clearly indicate that proper penetration studies must assess both extent and rate of NP penetration. Fluorescence was not detected in any of the perfusate samples, suggesting absorption did not occur with any of these QDs. QDs that are synthesized with the same core/shell and similar surface coatings and hydrodynamic diameters have different shapes and showed different rates of penetration into intact skin.[27] In these studies, spherical QDs penetrated better than the ellipsoid and the latter better than the nail-shaped structures (comparison of neutral QD with similar hydrodynamic dimensions). These results were not unexpected if one considers skin morphology and dimensions of the torturous penetration pathway. It is possible that the vehicle altered the skin, and the penetration effects that we reported remain valid because the dosing solutions are representative of QD as commercially supplied.[27] In fact, studies with the borate vehicles conducted with skin cells showed no effect on viability or toxicity.[32]

Dynamic light scattering dimensions of QDs account for the metallic core, coating (if present) and a double layer of liquid rich in counter-ions (inner layer), and similarly charged ions (outer layer). This ionic cloud surrounding a charged NM constantly changes in composition and thickness, and therefore its hydrodynamic diameter and the effective Z-potential may change as a result of changes in the environment to which NMs are exposed. The spherical QDs penetrated better than the ellipsoid and cationic better than anionic.[27] Although the skin might be affected by the borate buffer to justify the entrance of charged materials (normally not allowed[33–37]), the mechanism facilitating the ingress of cationic QD is difficult to extrapolate to other exposure scenarios because it is not possible to control changes in hydrodynamic dimensions and Z-potential of a material in situ within a biological specimen. Nonetheless, the demonstration that charged NMs entered the skin is extremely important, because the general public may be exposed to a many different types of ENMs and NM formulations. These studies illustrate the complexity of this process and the number of variables that should be either controlled or assessed if generalizations about ENM penetration are to be made.

13.5.3 Do Quantum Dots Penetrate Intact or Tape-Stripped Human Skin?

Another hypothesis was designed to test if water-soluble QD565 of three sizes and three surface coatings could penetrate intact and tape-stripped (30X) human skin. Human skin flow-through diffusion cells were conducted in our laboratory with QD565 having three surface coatings of PEG, PEG amine, or carboxylic acid; with hydrodynamic diameters of 35, 15, and 14 nm, respectively; exposed at 62.5 pmol/cm² for 8 and 24 hours.[38] The perfusate samples were collected hourly and analyzed for fluorescence and Cd by inductively coupled plasma-optical emission spectroscopy (ICP-OES). Laser scanning confocal microscopy of all three surface coatings and sizes did not show penetration through fresh intact human skin after 8 and 24 hours of exposure. Fresh human intact skin exposed to QD565-PEG-amine (Figure 13.3a through c) depicts QD565-PEG-amine remaining on the surface of the stratum corneum layer. When fresh human skin was tape-stripped 30X (to remove some of the rate-limiting stratum corneum layers) (Figure 13.3d through f) and placed in flow-through diffusion cells for 8 hours, the QDs still remained on the surface of the skin even though tape-stripping removed most of the stratum corneum layers. This study, using laser scanning confocal microscopy and TEM, suggested that all three surface coatings of QDs remained on the surface of the skin in intact or tape-stripped skin. If hair follicles were present, they followed the surface contour and dipped down the invagination of the epidermis but the QDs remained within the stratum corneum layers (Figure 13.4) and not within the viable epidermis. There was no absorption into the skin with all three QD surface coatings of different sizes by fluorescence or for Cd by ICP-OES.

Our study showed that diverse QDs did not penetrate through fresh intact or tape-stripped human skin, but if hair follicles were present, they followed the contour of the surface and appeared in the hair follicle invagination in all treatments[38] (Figure 13.4). In contrast, weanling porcine skin showed penetration most probably due to the greater number of hair follicles in the young animal, as discussed earlier in the text. However, another variable relates to the increased elasticity of human skin compared to porcine skin. Human skin tends to shrink when removed from the body because it has great elasticity but porcine skin remains very firm. When dermatomed pig skin is removed from the animal, it remains the same size. In contrast, when

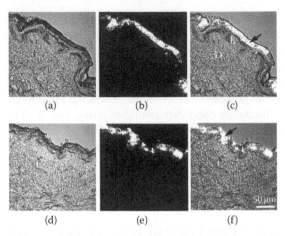

(a)   (b)   (c)

(d)   (e)   (f)

FIGURE 13.3 QD565-PEG-amine was topically applied to surface of fresh intact human skin and tape-stripped 30 times and then placed in flow-through diffusion cells for 8 or 24 hours. Top panel illustrates fresh human intact skin dosed with QD565-PEG-amine: (a) differential interference contrast, (b) fluorescent, and (c) fluorescent overlay. Bottom panel illustrates human skin that was tape stripped 30 times and then QD-PEG amine was topically applied for 8 hours: (d) differential interference contrast, (e) fluorescent, and (f) fluorescent overlay. Arrow depicts quantum dots (QDs) that appear bright white on the surface of the skin. Note that after stratum corneum removal, the QD did not penetrate into the viable epidermis (E) or dermis (D) but remained on the surface of the skin (arrow).

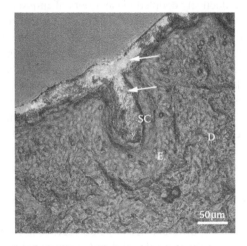

FIGURE 13.4 Quantum dots (QDs) can follow the contour of human skin and embed down into the epidermal invagination of a hair follicle. Notice that the QDs are within or on the surface of the stratum corneum and not penetrating into the viable epidermis (E) or dermis (D).

dermatomed human skin is removed, it shrinks and requires stretching to mount on diffusion cells so the in vivo dimensions may be altered. This change in size could affect the intercellular pathway geometry and is not controlled for in such studies.

Despite the fact that small molecules, chemicals, and drugs have a similar rate and extent of absorption through human and porcine skin, penetration of NPs in porcine and human skin may be different because of the geometrical conformation of stratum corneum lipid packing, which would affect pathway size. Normal human stratum corneum has an orthorhombic lipid organization that may theoretically decrease NP penetration; in contrast to the hexagonal lipid organization of the more permeable porcine stratum corneum. In fact, it is well known that the hexagonal organization confers permeability to the stratum corneum.[39] Terpenes can alter the lipid organization of the stratum corneum from orthorhombic to hexagonal.[40,41] Therefore, to verify this hypothesis, human skin was pretreated for 1 hour with the terpenes; eucalyptol, menthol, and limonene and dosed with QD565-COOH and PEG-amine. Laser scanning confocal microscopy showed no penetration of QD into the viable layers of the skin and QDs remained primarily on the surface of the stratum corneum layers but TEM showed that the QDs penetrated deeper into the terpene-treated stratum corneum layers compared with the water and ethanol vehicle controls.[42] QDs did not reach the viable epidermal layers and were not detected in the perfusate.

When QD621 were intradermally injected in SKH-1 hairless mice, migration occurred from the injection site to the regional lymph nodes through the lymphatic duct system and then to the liver and other organs.[43] The biodistribution of intra-arterially infused QD621 was studied in perfused skin and showed that the QD621 can migrate out of the capillaries into the surrounding tissue.[44] This pathway of QD migration is very difficult to study using in vitro diffusion cell systems and reflects movement of minute quantities of penetrated NPs.

13.5.4 MECHANICAL ACTIONS

Studies in another species have been conducted to assess if mechanical actions could perturb the barrier and affect skin penetration. When QD655 and QD565 coated with carboxylic acid were topically applied for 8 and 24 hours in flow-through diffusion cells with flexed, tape-stripped, and abraded rat skin, no penetration was seen in the nonflexed control, flexed, and tape-stripped rat skin, but minimal penetration occurred in the abraded skin. This was expected because the rate-limiting barrier of the stratum corneum was removed by tape-stripping and abrasion removed most of the epithelial layer. In some instances, there was retention of QDs in hair follicles and in abraded skin. This type of research provided a better understanding of absorption and/or penetration of materials through damaged skin. NP penetration not only occurs on the surface of the stratum corneum or within the stratum corneum layers, but may penetrate deeper with skin flexing. This research suggests that there is risk for potential health effects to medical personnel exposed to QDs during medical applications with damaged or abnormal skin. In addition, this study also provided information on QD NP absorption that could occur only in abraded skin and could also be relevant as a method of drug delivery.[24]

Mechanical stressors could also play a role on how NPs traverses the skin. An occupational environment may require repetitive motions such as wrist bending or flexing that could accentuate naturally occurring biomechanical forces of the skin and thereby altering the structural organization of the skin. This could potentially lead to an increase in penetration of the NP. Penetration was assessed after mechanical stress and tension with a relatively small fullerene amino acid-derivatized peptide

NP of 3.5 nm dosed in flow-through porcine diffusion cells.After 8 hours of perfusion, NP penetrated into the epidermal layers with 60 minutes of flexion and into the dermal layers after 90 minutes of flexion. NP penetration was limited only to the upper epidermal layers in nonflexed control skin. Studies with only 15 and 30 minutes of flexion did not show penetration. TEM localized these derivatized fullerenes within the intercellular space of the stratum granulosum.[45] Studies by other investigators with fluorescein isothiocyanate-conjugated dextran beads of 0.5 µm showed penetration through the stratum corneum layers that reached the viable epidermis after 30 minutes of flexion.[46]

13.5.5 SILVER PENETRATION

Silver (Ag) NPs have unique physicochemical properties that differ from their bulk metallic constituents but provide many functional advantages. Ag in its bulk form is inert and exerts no biocidal action, but ionizes in the presence of water or tissue fluids to release antimicrobial Ag ions. The augmented properties of AgNP are likely because of the increased surface area to volume ratio at the nanoscale. In vivo studies were conducted in our laboratory with AgNP on the backs of weanling pigs. Daily repetitive in vivo skin exposure studies were conducted with AgNP for 14 days with 25 and 35 nm carbon-coated AgNP, as well as freshly synthesized and thoroughly washed 20, 50, and 80 nm AgNP.[47] AgNP penetration occurred only in the superficial layers of the stratum corneum as assessed by TEM and confirmed by energy dispersive x-ray spectrometry (EDS).[47] The free ion is precipitated as Ag sulfide in the superficial layers of the stratum corneum. Skin penetration of 25 nm AgNP was shown in intact and damaged in vitro human skin in static diffusion cells[48] but with low flux compared to the absorption rates for metal powders.[49] TEM confirmed the presence of AgNPs into the stratum corneum and the outermost surface of the epidermis but not into the dermis. These results suggest that a small fraction of the NPs dissolved and diffused through the skin layers as elemental Ag.

13.5.6 GOLD NANOPARTICLE PENETRATION

Gold (Au) NPs were studied in in vitro rat skin Franz static diffusion cells with 15, 102, and 198 nm AuNP.[50] AuNP showed size-dependent permeation through rat skin, and TEM revealed accumulation of small NPs within the deep layers of skin, although large particles were observed primarily in the epidermis and dermis. This is not surprising because the larger particles were not of nanosize (<100 nm). Other investigators confirmed that AuNP of 12.6 nm can permeate intact and damaged human skin in Franz static diffusion cells in a dose-dependent manner. Also, when greater amounts of Au were applied three times, there was also a three time increase in permeation. Evaluation of the Au into skin revealed that Au concentration decreased from the superficial to deeper skin layers and a significant amount was found in damaged skin compared to intact skin. The abraded skin showed an increase in the amount of Au that remained in the skin but not in the receptor compartment. This may be because of a strong interaction between AuNP and the skin and extracellular matrix that might hinder the permeation of particle migration during this short 24-hour study.[51]

13.5.7 TITANIUM DIOXIDE AND ZINC DIOXIDE PENETRATION

Broad-spectrum physical sunblockers and nano-sized particles such as nanoscale titanium dioxide (TiO_2) and micronized zinc oxide (ZnO) NPs less than 100 nm have been incorporated into commercial sunscreens. Their nano-sized transparent form is more aesthetically acceptable to consumers. Emulsions of TiO_2 and ZnO on excised human skin remained on the surface of the stratum corneum[52] and when larger 80–200 nm and up to 1 μm particles were applied to in vitro porcine skin in static diffusion cells, no penetration was observed.[53] Cross et al.[54] applied transparent ZnONPs of 26–30 nm in oil/water (o/w) formulations to human skin in in vitro static cells and showed that they remained on the surface of the stratum corneum at 24 hours. Studies showed that larger particles of ZnO and TiO_2 penetrated the stratum corneum barrier of rabbit skin especially with water and oily vehicles.[55] Other studies described penetration in the disjunctum layer, which comprises its more external layers where lipids are organized hexagonally because of sebum penetration.[39,56–60]

Studies of micronized TiO_2 applied to elderly individuals for 9–31 days detected insignificant levels of Ti in the dermis.[61] Lademann et al.[62] noticed TiO_2 within the follicular openings but most remained on the surface. Other studies have reported that ZnO and TiO_2 NPs do not permeate the skin.[63,64] Absorption of 20 nm TiO_2 in a water/oil (w/o) emulsion for 5 hours with in vitro and in vivo human skin showed no penetration into the viable epidermal layers.[65] Studies conducted on human forearms by Schulz et al.,[64] with different size micronized TiO_2 in o/w emulsions for 6 hours, showed the emulsion to remain on the surface of the stratum corneum and neither particle shape, formulation, nor exposure had a significant impact on penetration. Additional studies with TiO_2 in o/w emulsions applied to in vitro human organotypic cultures for 24 hours showed penetration was greater in vitro than in vivo, but tape-stripping indicated that NP remained within the stratum corneum.[66] Small transparent ZnO NP of 26–30 nm in o/w formulations topically applied to in vitro human skin in static cells for 24 hours also found ZnO NPs on the surface of the stratum corneum. Although Zn was detected in the perfusate of the control and treated diffusion cells, it was attributed to normal levels of Zn present in skin.[54]

A sunscreen containing TiO_2 was applied in vivo to minipigs four times per day for 5 days per week for 4 weeks had high levels of Ti in the epidermis by TEM/ energy dispersive X-ray analysis (EDX), in the stratum corneum layers and near hair follicle openings, with only a few particles of TiO_2 in the dermis that were regarded as contamination.[67]

The European NANODERM project provided data from in vivo and in vitro experiments with TiO_2-based sunscreens on human and porcine skin and with human foreskin transplanted to immunodeficient mice. A small amount of TiO_2 was detected in the deep viable epidermal layers as analyzed by particle-induced X-ray emission and scanning transmission ion microscopy (STIM) which tracked Ti particles to the stratum granulosum layer of the epidermis.[68] Studies conducted for 8–24 hours on in vivo porcine skin suggested hair follicles were not important for penetration but detected four formulations of TiO_2 on the stratum corneum and in the stratum granulosum with high-energy ion probe,[69] whereas others showed no TiO_2 penetration through intact skin of severe combined immunodeficiency models.[70] This European project confirmed the safety of the sunscreens formulation containing TiO_2 NPs, that there was no reported evidence of NP penetration.[71] Previous reports by the European Union (EU) Scientific Committee on Cosmetics and Non-Food Products (SCCNFP)[72] found no evidence of TiO_2 NP penetration either in vitro or in vivo skin. The EU Scientific Committee on Consumer Products in 2007 stated that a safety assessment of nano-sized TiO_2 in cosmetics should also take into account abnormal skin conditions and the impact of mechanical effects on skin penetration.[73]

It has been suggested that penetration can occur when the skin barrier is damaged. Studies on sunscreens containing ZnO NPs exposed to humans for 5 days outdoors found small increases in the Zn tracer in the blood and urine using sensitive methods, but it is not known whether[68] Zn was absorbed as ZnO or soluble Zn.[74]

More recent in vitro and in vivo studies were conducted with TiO_2 and ZnO NP to UVB-damaged porcine skin.[75] Flow-through diffusion cells were treated for 24 hours with four sunscreen formulations: 10% coated-TiO_2 o/w, 10% coated TiO_2 in w/o, 5% coated ZnO in o/w, and 5% uncoated ZnO in o/w. TiO_2 primary particle size was 10×50 nm with mean agglomerates of 200 nm; ZnO mean was 140 nm. Skin was evaluated by light microscopy, scanning electron microscopy (SEM), TEM, and time-of-flight–secondary-ion-mass-spectrometry (TOF–SIMS). UVB-exposed skin showed TiO_2 NPs down to 13–17 layers into the stratum corneum compared to normal skin (Figure 13.5a and b), whereas ZnO remained on the surface. TOF–SIMS analysis showed TiO_2 and ZnO penetration. TEM/EDX of the perfusate or inductively coupled plasma mass spectrometry detected no Ti or Zn, indicating minimal transdermal absorption. The TiO_2 in o/w formulation penetrated 13 layers deep into UVB-damaged stratum corneum, whereas only 7 layers deep in non-UVB-treated skin; TiO_2 in w/o penetrated deeper in the UVB-damaged stratum corneum. Coated and uncoated Zn NPs in o/w were localized to the upper 1–2 superficial stratum corneum in UVB-damaged and normal skin

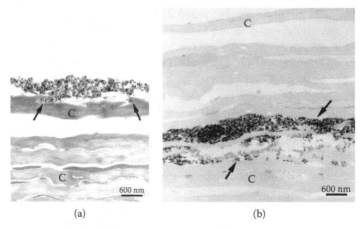

(a)   (b)

FIGURE 13.5 (a) Transmission electron microscopy (TEM) of unexposed in vivo skin treated with TiO_2 sunscreen formulation for 48 hours. Arrows depict TiO_2 on the surface of the stratum corneum (C) layers. (b) TEM of ultraviolet B-exposed in vivo skin treated with TiO_2 sunscreen formulation for 48 hours. Arrows depict TiO_2 localized deep down 13 layers within the stratum corneum (C).

(Figure 13.6a and b). The SEM images of UVB-exposed skin illustrate ZnO NP formulation seen near the base of the hair shafts where the hair emerges from the opening of the skin (Figure 13.7a). The sunscreen formulation seen on the surface of the stratum corneum sheets as a diffuse matrix containing large ZnO NP on UVB-exposed pig skin illustrates how large the NPs are within the formulation (Figure 13.7b), and these Zn crystals were confirmed by EDS. Only by using more sophisticated and sensitive techniques like TOF–SIMS, which involves an ion TOF–SIMS[5] (Physical Electronics Inc., Chanhassen, MN) employing a bismuth (Bi[+]) pulsed primary ion beam was employed to sputter the surface of the skin

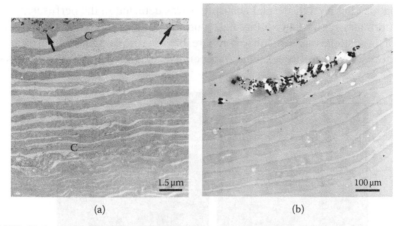

(a)   (b)

FIGURE 13.6 (a) Transmission electron microscopy (TEM) of unexposed in vivo skin treated with ZnO sunscreen formulation for 48 hours. Arrows depict ZnO NPs on the surface of the stratum corneum (C). (b) TEM of ultraviolet B-exposed in vivo skin treated with ZnO for 48 hours. ZnO NP sunscreen formulation was also seen on the surface and between the upper stratum corneum layers.

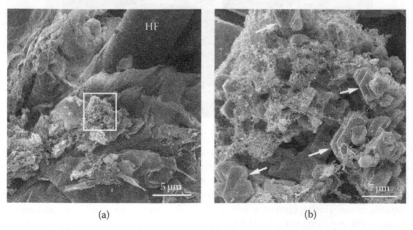

(a)   (b)

FIGURE 13.7 A low-magnification scanning electron microscopy image; (a) image of ZnO sunscreen formulation on ultraviolet B-exposed skin adjacent to a hair follicle (HF) and (b) a higher magnification of the box inset illustrating large Zn crystals (arrows).

sample showed more precise localization of the ions. The ionized species is eroded from the sample surface and extracted into a TOF mass spectrometer and analyzed according to its mass-to-charge ratio by measuring the TOF from the sample surface to the detector to obtain high spatial and mass resolution data. The pulsed Bi^+ beam is rastered across the sample (256×256 pixels) and a mass spectrum is acquired at each pixel. The ion images are obtained retrospectively from the acquired data for selected Ti and Zn isotopes. TOF–SIMS showed Ti within the epidermis and superficial dermis (Figure 13.8a through d,) and Zn was limited to the stratum corneum and upper epidermis (Figure 13.9a through d). In summary, UVB-damaged skin slightly enhanced TiO_2 or ZnO NP penetration in sunscreen formulations but no transdermal absorption was detected in the perfusate.[75]

The primary reason for conflicting reports as to the penetration of TiO_2 is that many of these studies are difficult to compare, because the variables within each study are so diverse. Are Zn NPs or soluble ions being detected? Anatomical variables such

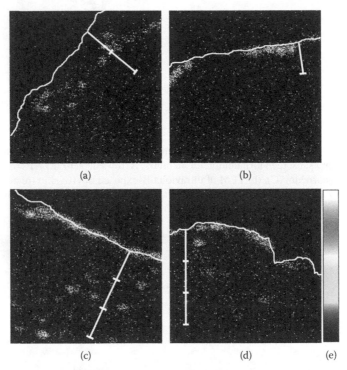

(a)   (b)

(c)   (d)   (e)

FIGURE 13.8   Time-of-flight–secondary-ion-mass-spectrometry of in vivo skin. (a) Unexposed skin treated with the sunscreen TiO_2 formulation. Mapping shows Ti in the epidermis; (b) ultraviolet B (UVB)-exposed skin treated with the TiO_2 formulation. Mapping shows Ti concentrated in the upper epidermis; (c) unexposed skin treated with TiO_2. Mapping shows Ti concentrated in the stratum corneum, with slight diffusion into the epidermis and superficial dermis; (d) UVB-exposed skin treated with TiO_2. Mapping shows Ti concentrated in the stratum corneum, with slight diffusion into the epidermis. Upper white line denotes surface of the skin; (e) intensity scale for Ti, from highest (upper lighter gray zone) to lowest (dark gray zone). Each vertical bar segment represents 100 μm.

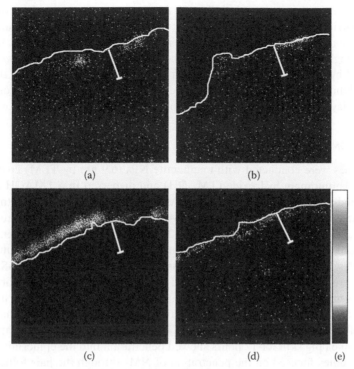

FIGURE 13.9   Time-of-flight–secondary-ion-mass-spectrometry of in vivo pig skin. (a) Unexposed skin treated with the ZnO sunscreen formulation. Mapping shows Zn in the epidermis; (b) ultraviolet B (UVB)-exposed skin treated with the ZnO formulation. Mapping shows Zn in the stratum corneum with slight diffusion into the upper epidermis. (c) Unexposed skin treated with ZnO. Mapping shows Zn on the surface of the skin; (d) UVB-exposed skin treated with the ZnO sunscreen formulation. Mapping shows Zn in the stratum corneum, with slight diffusion into the upper epidermis. Upper white line denotes surface of the stratum corneum. (e) Intensity scale for Zn, from highest (upper light gray zone) to lowest (dark gray zone). Each vertical bar segment represents 100 μm.

as species differences in skin, hair follicle density, regional differences in absorption, vehicle differences, and the length of most of these experiments were of short duration, making a consensus difficult. For a more complete discussion on the aspects of skin penetration with NPs, see the reviews by Monteiro-Riviere et al.[11,76,77]

It is not surprising that most studies showed that TiO_2 and ZnO remained on skin surfaces because they are usually coated with alumina, silica, or silicon oil thus causing their dimensions to be larger. Most of these TiO_2 or ZnO NPs used in sunscreen applications are coated to make them more dispersible in organic media. As these are semiconducting NPs and can be efficient photocatalysts, their coatings help prevent photocatalytic activity from occurring. Their formulations have been designed for NP to stick on the surface of skin so that they can scatter the light and resist removal from swimming. Many of these studies were not conducted in normal environmental conditions where the public will wear sunscreens such as at the beach. In this case, skin is warmed from the sun, hydrated by the seawater, irritated by the sun and sand, and possibly inflamed. All these changes to the skin could alter the skin barrier function allowing for penetration of NPs. It has been assumed that if NPs do not penetrate the skin then they are considered to be safe, not realizing that they might have only been tested in short-term studies and that they could be recognized by the immune system through the interaction of a few NPs that have penetrated through the stratum corneum and interacted with Langerhans cells in the suprabasal layers of the epidermis, which could trigger an immunological response.

13.5.8 IRON OXIDE NANOPARTICLE

Other studies were conducted with maghemite NPs (6.9 nm by TEM) coated with tetramethylammonium hydroxide (TMAOH) and dispersed in a TMAOH aqueous solution, and iron NPs (51%: 4.9 nm by TEM) coated with sodium bis(2-ethylhexyl) sulfosuccinate (AOT) and dispersed in an AOT-rich aqueous solution were applied to human abdominal skin samples for up to 24 hours. Particles were recovered in the stratum corneum, at the interface of stratum corneum–stratum granulosum, in the hair follicles, and in the viable epidermis as assessed by TEM and EDS-SEM. No particles were recovered in the receptor fluid with ICP-OES. Modification of the skin barrier was monitored by measuring skin impedance, and it was found that its degree of modification correlated with penetration depth.[78] Studies in which polymeric NPs coated with a thick PEG block copolymer layer (40 nm) were topically applied to hairless guinea pig skin for 12 hours showed penetration into the epidermis.[79]

Other studies focused on the penetration of NMs through the hair follicle. As it has been discussed elsewhere and in this and other chapters, hair follicles could serve as a shunt route of penetration bypassing the stratum corneum barrier.[2,80] These studies showed that particles, whose dimensions ranged between 7 μm and 20 nm, were almost exclusively found in the hair follicle.[79,81–85] It is noteworthy that penetration to the perifollicular dermis was shown for particles of 40 nm in diameter,[85] demonstrating that the hypothesis of a potential safety issue related with follicular penetration is more than a mere assumption. However, the biological and toxicological implications of NP penetration into hair follicles are not known. From a pharmaceutical perspective, lodging of soluble drug-containing NPs into orifices of hair follicles would be ideal controlled-release drug delivery systems to the surface of the skin.

13.6 POTENTIAL FOR NANOMATERIALS ABSORPTION INTO BLOOD FROM SKIN

The evaluation of ENM absorption into blood is a complex matter, so results from in vitro systems that do not have intact microcirculation should be carefully interpreted. Furthermore, human and porcine skin may react differently with respect to NP penetration as compared to smaller organic chemicals and drugs where human and porcine skin are very similar. Nevertheless, most recent work has shown that absorption into blood would not be predicted following topical application of NMs to skin. For example, QD621 nanocrystals that were applied to porcine skin in flow-through diffusion cells were not found in the perfusate at any time point or concentration.[25] Likewise, studies with QD565 coated with PEG, PEG-amine, or carboxylic acid topically applied to human skin in diffusion cells for 8 or 24 hours showed that all three QD preparations remained on the surface of the stratum corneum or were retained within hair follicle invaginations, but were not detected in the perfusate.[38] Similar observations were made by our group with porcine skin exposed to the same NP.[27] A recent in vivo study, though, showed that nano-sized TiO_2 applied topically to pig skin in sunscreen formulation did not accumulate in lymph node or liver tissue following exposures for 5 days per week for 4 weeks.[67]

In most cases studied to date, topically applied NPs have not been shown to be absorbed into the systemic circulation. However, penetration into the stratum corneum can occur in all animal species studied. This penetration could be significant relative to immunological and carcinogenic endpoints. Extensive studies have not been conducted with very small NPs (e.g., <2–5 nm) that bridge to large organic molecules. We have shown that pristine C_{60} fullerenes exposed topically in industrial solvents could penetrate deep into the stratum corneum, whereas exposure in mineral oil prevented penetration.[86] Vehicles have an important impact. Current findings also suggest that surface coatings as well as NP geometry seem to modulate penetration. All of these factors must be studied further if realistic risk assessments of ENMs are to be made.

13.7 SUMMARY AND DISCUSSION

The skin penetration studies of various types of topical applied NPs reviewed in this chapter illustrate not only the breadth of studies conducted, but also common findings that begin to define properties and experimental parameters that must be considered when studying NP dermal penetration. Most studies have shown that the penetration and distribution into skin for topical administration of NPs are minimal over the short duration of studies reported. Penetration of NPs into the skin is particle dependent and sensitive to the vehicles used to administer them. In almost all cases, transdermal flux has not been detected yet various degrees of stratum corneum and epidermal penetration are common. This is particularly important for immunological endpoints because large amounts of penetrated NPs are not necessary to illicit a biologically and clinically relevant response. Abrogation of the stratum corneum barrier by abrasion or tape-stripping promotes NP penetration into deeper layers, as does mechanical deformation. In contrast, damage to the skin by UVB solar radiation does not significantly increase penetration of topical application of TiO_2 or ZnO sunscreens. Finally, vehicles can modulate NP penetration. Uptake into epidermal cells has also been shown for most types of NPs, with some clearly exhibiting cytotoxicity or inflammatory mediator release. The QD studies clearly showed specific pathways involved and responses that were coating dependent.

From a toxicological perspective where large skin surfaces could be exposed to high exposure levels, skin becomes an important organ for assessment. In most cases, classic irritation might be the only sequela observed and this could be because of solvent or contaminant effects. Large transdermal fluxes are not required for toxicological effects to be initiated if the endpoint is sensitization or neoplastic transformation. This would be especially true if the skin barrier were disrupted by mechanical perturbation or disease.

There is no doubt that dermal exposure to NPs in occupational, cosmetic, and dermatological preparations will continue to occur. It is crucial that experiments be properly conducted and endpoints defined that actually measure intact NPs. Studies of chronic exposure and evaluation for toxicity are lacking, making definitive statements of safety impossible to make. Also, it would be of interest to investigate if very small NPs can penetrate deeper into the skin after repetitive applications and for longer periods of time, for example, several months. Long-term in vivo studies in humans or animals are desperately needed because there are limitations to in vitro models and differences in animal species.

REFERENCES

1. Monteiro-Riviere NA. The integument. In: Eurell J, Frappier B, eds. *Dellmann's Textbook of Veterinary Histology.* 6th ed. Ames, Iowa: Blackwell, 2006; pp. 320–349.
2. Monteiro-Riviere NA. Structure and function of skin. In: Monteiro-Riviere NA, ed. *Toxicology of the Skin*—Target Organ Series 29. New York: Informa Healthcare, 2010; vol 29, Chapter 1, pp. 1–18.
3. Feldmann RJ, Maibach HI. Percutaneous penetration of some pesticides and herbicides in man. *Toxicol Appl Pharmacol* 1974; 28:126–132.
4. Maibach H, Feldmann R. Systemic absorption of pesticides through skin of man. Occupational Exposure to Pesticides: Report to the Federal Working Group on Pest Management for the Task Group, 1974; 120–127.
5. Bronaugh RL, Stewart RF, Congdon ER. Methods for in vitro percutaneous absorption studies. II. Animal models for human skin. *Toxicol Appl Pharmacol* 1982; 62:481–488.
6. Bronaugh RL, Stewart RF. Methods for in vitro percutaneous absorption studies. IV. The flow-through diffusion cell. *J Pharm Sci* 1985; 74:64–67.
7. Scott RC, Corrigan MA, Smith F, Mason H. The influence of skin structure on permeability: An intersite and interspecies comparison with hydrophilic penetrants. *J Invest Dermatol* 1991; 96:921–925.
8. Wester RC, Melendres J, Sedik L, Maibach HI, Riviere JE. Percutaneous absorption of salicylic acid, theophylline, 2,4-dimethylamine, diethylhexyl phthalic acid, and *p*-aminobenzoic acid in the isolated perfused porcine skin flap compared to man. *Toxicol Appl Pharmacol* 1998; 151:159–165.
9. Riviere JE. *Dermal Absorption Models in Toxicology and Pharmacology.* Boca Raton, FL: Taylor and Francis/CRC Press, 2006.
10. Monteiro-Riviere NA, Bristol DG, Manning TO, Riviere JE. Interspecies and interregional analysis of the comparative histological thickness and laser Doppler blood flow measurements at five cutaneous sites in nine species. *J Invest Dermatol* 1990; 95:582–586.
11. Monteiro-Riviere NA, Anatomical factors that affect barrier function. In: Zhai H, Wilhelm KP, Maibach MI, eds. *Dermatotoxicology.* 7th ed. New York: CRC Press, 2008; pp. 39–50.
12. Riviere JE, Monteiro-Riviere NA. Toxicokinetics: Dermal exposure and absorption of chemicals and nanomaterials. In: McQueen C, ed. *Comprehensive Toxicology.* New York: Elsevier, 2010; vol 1, Chapter 1.05, pp. 111–122.
13. Rougier A, Dupuis D, Lotte C, Roguet R. The measurement of stratum corneum reservoir, a predictive method for in vivo percutaneous absorption studies: Influence of application time. *J Invest Dermatol* 1985; 84:66–68.
14. Nylander-French LA. A tape-stripping method for measuring dermal exposure to multifunctional acrylates. *Ann Occup Hyg* 2000; 44:645–651.
15. Reddy MB, Stinchcomb AL, Guy RH, Bunge AL. Determining dermal absorption parameters in vivo from tape strip data. *Pharm Res* 2002; 19:292–298.

16. Bronaugh RL, Stewart RF. Methods for in vitro percutaneous absorption studies. III. Hydrophobic compounds. *J Pharm Sci* 1984; 73:1255–1258.
17. Bowman KF, Monteiro-Riviere NA, Riviere JE. Development of surgical techniques for preparation of in vitro isolated perfused porcine skin flaps for percutaneous absorption studies. *Am J Vet Res* 1991; 52:75–82.
18. Riviere JE, Bowman KF, Monteiro-Riviere NA, Dix LP, Carver MP. The isolated perfused porcine skin flap (IPPSF). I. A novel in vitro model for percutaneous absorption and cutaneous toxicology studies. *Toxicol Sci* 1986; 7:444–453.
19. Monteiro-Riviere NA. Specialized technique: isolated perfused porcine skin flap. In: Kemppainen BW, Reifenrath WG, eds. *Methods for Skin Absorption.* Boca Raton, FL: CRC Press, 1990; Chapter 11, pp.175–189.
20. Riviere JE, Monteiro-Riviere NA. The isolated perfused porcine skin flap as an in vitro model for percutaneous absorption and cutaneous toxicology. *Crit Rev in Toxicol* 1991; 21:329–344.
21. Monteiro-Riviere NA, Inman AO. Challenges for assessing carbon nanomaterial toxicity to the skin. *Carbon* 2006; 44:1070–1078.
22. Monteiro-Riviere NA, Inman AO, Zhang LW. Limitations and relative utility of screening assays to assess engineered nanoparticle toxicity in a human cell line. *Toxicol Appl Pharmacol* 2009; 234:222–235.
23. Zhang LW, Zeng L, Barron AR, Monteiro-Riviere NA. Biological interactions of functionalized single-wall carbon nanotubes in human epidermal keratinocytes. *Int J Toxicol* 2007; 26:103–113.
24. Zhang LW, Monteiro-Riviere NA. Assessment of quantum dot penetration into intact, tape stripped, abraded and flexed rat skin. *Skin Pharmacol Physiol* 2008; 21:166–180.
25. Zhang LW, Yu WW, Colvin VL, Monteiro-Riviere NA. Biological interactions of quantum dot nanoparticles in skin and in human epidermal keratinocytes. *Toxicol Appl Pharmacol* 2008; 228:200–211.
26. Michalet X, Pinaud FF, Bentolila LA, Tsay JM, Doose S, Li JJ, Sundaresan G, Wu AM, Gambhir SS, Weiss S. Quantum dots for live cells, in vivo imaging, and diagnostics. *Science* 2005; 307:538–544.
27. Ryman-Rasmussen J, Riviere JE, Monteiro-Riviere NA. Penetration of intact skin by quantum dots with diverse physicochemical properties. *Toxicol Sci* 2006; 91:159–165.
28. Yu WW, Chang E, Falkner JC, Zhang J, Al-Somali AM, Sayes CM, Johns J, Drezek R, Colvin VL. Forming biocompatible and nonaggregated nanocrystals in water using amphiphilic polymers. *J Am Chem Soc* 2007; 129:2871–2879.
29. Derfus AM, Chan WCW, Bhatia SN. Probing the cytotoxicity of semiconductor quantum dots. *Nano Lett* 2004; 4:11–18.
30. Chang E, Thekkek N, Yu WW, Colvin VL, Drezek R. Evaluation of quantum dot cytotoxicity based on intracellular uptake. *Small* 2006; 12:1412–1417.
31. Van der Merwe D, Brooks JD, Gehring R, Baynes RE, Monteiro-Riviere NA, Riviere JE. A physiologically based pharmacokinetic model of organophosphate dermal absorption. *Toxicol Sci* 2006; 89:188–204.
32. Ryman-Rasmussen JP, Riviere JE, Monteiro-Riviere NA. Surface coatings determine cytotoxicity and irritation potential of quantum dot nanoparticles in epidermal keratinocytes. *J Invest Dermatol* 2007; 127:143–153.
33. Swarbrick J, Lee G, Brom J, Gensmantel NP. Drug permeation through human skin II: Permeability of ionizable compounds. *J Pharm Sci* 1984; 73(10):1352–1355.
34. Wagner H, Kostka KH, Lehr CM, Schaefer UF. pH profiles in human skin: Influence of two in vitro test systems for drug delivery testing. *Eur J Pharm Biopharm* 2003; 55(1):57–65.
35. Elias PM. Stratum corneum defensive functions: an integrated view. *J Invest Dermatol* 2005; 125(2):183–200.

36. Lee WR, Shen SC, Wang KS, Hu CH, Fang JY. Lasers and microdermabrasion enhance and control topical delivery of vitamin C. *J Invest Dermatol* 2003; 121:1118–1125.
37. Schmid-Wendtner MH, Korting HC. The pH of the skin surface and its impact on the barrier function. *Skin Pharm Physiol* 2006; 19:296–302.
38. Monteiro-Riviere NA, Inman AO. Evaluation of quantum dot nanoparticle penetration in human skin. *Toxicologist* 2008; 102(S-1):211.
39. Bouwstra JA, Ponec M. The skin barrier in healthy and diseased state. *Biochim Biophys Acta* 2006; 1758(12):2080–2095.
40. Sapra B, Jain S, Tiwary AK. Percutaneous permeation enhancement by terpenes: Mechanistic view. *AAPS J* 2008; 10(1):120–132.
41. Dos Anjos JLV, Alonso A. Terpenes increase the partitioning and molecular dynamics of an amphipathic spin label in stratum corneum membranes. *Int J Pharm* 2008; 350(1–2):103–112.
42. Monteiro-Riviere NA, Inman AO, Erdmann D, Xia X, Riviere JE. Terpene effects on penetration of nanoparticles in human skin. *Toxicologist* 2011; 120(S2):462, 2154.
43. Gopee NV, Roberts DW, Webb P, Cozart CR, Siitonen PH, Warbritton AR, Yu WW, Colvin VL, Walker NJ, Howard PC. Migration of intradermally injected quantum dots to sentinel organs in mice. *Toxicol Sci* 2007; 98:249–257.
44. Lee HA, Imran M, Monteiro-Riviere NA, Colvin VL, Yu WW, Riviere JE. Biodistribution of quantum dot nanoparticles in perfused skin: Evidence of coating dependency and periodicity in arterial extraction. *Nano Lett* 2007; 9:2865–2870.
45. Rouse JG, Yang J, Ryman-Rasmussen JP, Barron AR, Monteiro-Riviere NA. Effects of mechanical flexion on the penetration of fullerene amino acid-derivatized peptide nanoparticles through skin. *Nano Lett* 2007; 7:155–160.
46. Tinkle SS, Antonini JM, Rich BA, Roberts JR, Salmen R, DePree K, Adkins EJ. Skin as a route of exposure and sensitization in chronic beryllium disease. *Environ Health Perspect* 2003; 111:1202–1208.
47. Samberg ME, Oldenburg SJ, Monteiro-Riviere NA. Evaluation of silver nanoparticle toxicity in skin in vivo and keratinocytes in vitro. *Environ Health Perspect* 2010; 118(3):407–413.
48. Larese Filon F, D'Agostin F, Bovenzi M, Crosera M, Adami G, Romano C, Maina G. Human skin penetration of silver nanoparticles through intact and damaged skin. *Toxicology* 2009; 255:33–37.
49. Larese FF, Adami G, Venier M, Coceani N, Bussani R, Massiccio M, Barbieri P, Spinelli P. In vitro percutaneous absorption of cobalt. *Int Arch Environ Health* 2004; 77:85–89.
50. Sonavane G, Tomoda K, Sano A, Ohshima H, Terada H, Makino K. In vitro permeation of gold nanoparticles through rat skin and rat intestine: Effect of particle size. *Colloids Surf B* 2008; 65:1–10.
51. Larese Filon F, Crosera M, Adami G, Bovenzi M, Rossi F, Maina G. Human skin penetration of gold nanoparticles through intact and damaged skin. *Nanotoxicology* 2011; 5:493–501.
52. Dussert AS, Gooris E. Characterization of the mineral content of a physical sunscreen emulsion and its distribution onto human stratum corneum. *Int J Cosmet Sci* 1997; 19:119–129.
53. Gamer AO, Leibold E, van Ravenzwaay B. The in vitro absorption of microfine zinc oxide and titanium dioxide through porcine skin. *Toxicol In Vitro* 2006; 20(3):301–307.
54. Cross SE, Innes B, Roberts MS, Tsuzuki T, Robertson TA, McCormick P. Human skin penetration of sunscreen nanoparticles: In-vitro assessment of a novel micronized zinc oxide formulation. *Skin Pharmacol Physiol* 2007; 20:148–154.
55. Lansdown AB, Taylor A. Zinc and titanium oxides: Promising UV-absorbers but what influence do they have on the intact skin? *Int J Cosmet Sci* 1997; 19(4):167–172.
56. Lavrijsen APM, Bouwstra JA, Gooris GS, Weerheim A, Boddé HE, Ponec M. Reduced skin barrier function parallels abnormal stratum corneum lipid organization in patients with lamellar ichthyosis. *J Invest Dermatol* 1995; 105:619–624.

57. Landmann L. Epidermal permeability barrier: Transformation of lamellar granule-disks into intercellular sheets by a membrane-fusion process, a freeze-fracture study. *J Invest Dermatol* 1986; 87:202–209.
58. Monteiro-Riviere NA, Inman AO, Mak V, Wertz P, Riviere, JE. Effect of selective lipid extraction from different body regions on epidermal barrier function. *Pharm Res* 2001; 18:992–998.
59. Weerheim A, Ponec M. Determination of stratum corneum lipid profile by tape stripping in combination with high performance thin-layer chromatography. *Arch Dermatol Res* 2001; 293:191–199.
60. Nohynek GJ, Lademann J, Ribaud C, Roberts MS. Grey goo on the skin? Nanotechnology, cosmetic and sunscreen safety. *Crit Rev Toxicol* 2007; 37:251–277.
61. Tan MH, Commens CA, Burnett L, Snitch PJ. A pilot study on the percutaneous absorption of microfine titanium dioxide from sunscreens. *Australas J Dermatol* 1996; 37:185–187.
62. Lademann J, Weigmann HJ, Rickmeyer C, Barthelmes H, Schaefer H, Mueller G, Sterry W. Penetration of titanium dioxide in sunscreen formulation into the horny layer and the follicular orifice. *Skin Pharmacol Appl Skin Physiol* 1999; 12:247–256.
63. Pflücker F, Wendel V, Hohenberg H, Gärtner E, Will T, Pfeiffer S, Wepf R, Gers-Barlag H. The human stratum corneum layer: An effective barrier against dermal uptake of different forms of topically applied micronized titanium dioxide. *Skin Pharmacol Appl Skin Physiol* 2001; 14(S1):92–97.
64. Schulz J, Hohenberg H, Pflücker F, Gartner E, Will T, PfeiVer S, Wepf R, Wendel V, Gers-Barlag H, Wittern KP. Distribution of sunscreens on skin. *Adv Drug Deliv Rev* 2002; 54(S1):S157–S163.
65. Mavon A, Miquel C, Lejeune O, Payre B, Moretto P. In vitro percutaneous absorption and in vivo stratum corneum distribution of an organic and mineral sunscreen. *Skin Pharmacol Physiol* 2007; 20:10–20.
66. Bennat C, Müller-Goymann CC. Skin penetration and stabilization of formulations containing microfine titanium dioxide as physical UV filter. *In J Cosmet Sci* 2000; 22:271–283.
67. Sadrieh N, Wokovich AM, Gopee NV, Zheng J, Haines D, Parmiter D, Siitonen PH et al. Lack of significant dermal penetration of titanium dioxide from sunscreen formulations containing nano- and submicron-size TiO_2 particles. *Toxicol Sci* 2010; 115:156–166.
68. Kertész Zs, Szikszai Z, Gontier E, Moretto P, Surlève-Bazeille JE, Kiss B, Juhász I, Hunyadi J, Kiss AZ. Nuclear microprobe study of TiO_2-penetration in the epidermis of human skin xenografts. *Nucl Instrum Methods Phys Res B* 2005; 231:280–285.
69. Menzel F, Reinert T, Vogt J, Butz T. Investigations of percutaneous uptake of ultrafine TiO_2 particles at the high energy ion nanoprobe LIPSION. *Nucl Instrum Methods Phys Res B* 2004; 219–220:82–86.
70. Kiss B, Bíró T, Czifra G, Tóth BI, Kertész Zs, Szikszai Z, Kiss AZ, Juhász I, Zouboulis CC, Hunyadi J. Investigation of micronized titanium dioxide penetration in human skin xenografts and its effect on cellular functions of human skin-derived cells. *Exp Dermatol* 2008; 17(8):659–667.
71. NANODERM. (2007). Quality of skin as a barrier to ultra-fine particles. Final Report. (Project Number: QLK4-CT-2002-02678) www.uni-leipzig.de/~nanoderm/.
72. SCCNFP. (2000). Opinion of the Scientific Committee on Cosmetic Products and non-food products intended for consumers concerning titanium dioxide. European Commission, Brussels, Belgium. Colipa No S 75.
73. SCCP. Scientific Committee on Consumer Products. (2007). Preliminary opinion on safety of nanomaterials in cosmetic products. European Commission, Brussels, Belgium. http://ec.europa.eu/health/archive/ph_risk/committees/04_sccp/docs/sccp_o_123.pdf. Accessed March 10, 2010.

74. Gulson B, McCall M, Korsch M, Gomez L, Casey P, Oytam Y, Taylor A et al. Small amounts of zinc from zinc oxide particles in sunscreens applied outdoors are absorbed through human skin. *Toxicol Sci* 2010; 118(1):140–149.
75. Monteiro-Riviere NA, Wiench K, Landsiedel R, Schulte S, Inman AO, Riviere JE. Safety evaluation of sunscreen formulations containing titanium dioxide and zinc oxide nanoparticles in UVB-exposed skin: An in vitro and in vivo study. *Toxicol Sci* 2011; 123:264–280.
76. Monteiro-Riviere NA, Baroli B. Nanomaterial penetration. In: Monteiro-Riviere NA, ed. *Toxicology of the Skin*–Target Organ Series. New York: Informa Healthcare, 2010; vol 29 Chapter 22, pp. 333–346.
77. Monteiro-Riviere NA, Riviere JE. Interaction of nanomaterials with skin: Aspects of absorption and biodistribution. *Nanotoxicology* 2009; 3(3):188–193.
78. Baroli B, Ennas MG, Loffredo F, Isola M, Pinna R, Lopez-Quintela A. Penetration of metallic nanoparticles in human full-thickness skin. *J Invest Dermatol* 2007; 127:1701–1712.
79. Shim J, Seok KH, Park WS, Han SH, Kim J, Chang IS. Transdermal delivery of minoxidil with block copolymer nanoparticles. *J Con Rel* 2004; 97:477–484.
80. Patzelt A, Sterry W, Lademann J. Hair follicle delivery. In: Monteiro-Riviere NA, ed. *Toxicology of the Skin*—Target Organ Series. New York: Informa Healthcare, 2010; vol 29 Chapter 9, pp.101–109.
81. Schaefer H, Watts F, Brod J, Illel B. Follicular penetration. In: Scott RC, Guy RH, Hadgraft J, eds. *Prediction of Percutaneous Penetration. Methods, Measurements, Modeling*. London: IBC Technical Service, 1990; pp. 163–732.
82. Lauer AC, Ramachandran C, Lieb LM, Niemiec S, Weiner ND. Targeted delivery to the pilosebaceous unit via liposomes. *Adv Drug Deliv Rev* 1996; 18:311–324.
83. Alvarez-Roman R, Naik A, Kalia YN, Guy RH, Fessi H. Skin penetration and distribution of polymeric nanoparticles. *J Con Rel* 2004; 99:53–62.
84. Meidan VM, Bonner MC, Michniak BB. Transfollicular drug delivery—Is it a reality? *Int J Pharm* 2005; 306:1–14.
85. Vogt A, Combadiere B, Hadam S, Stieler KM, Lademann J, Schaefer H, Autran B, Sterry W, Blume-Peytavi U. 40 nm, but not 750 or 1,500 nm, nanoparticles enter epidermal CD1a+ cells after transcutaneous application on human skin. *J Invest Dermatol* 2006; 126:1316–1322.
86. Xia XR, Monteiro-Riviere NA, Riviere JE. Skin penetration and kinetics of pristine fullerenes (C_{60}) topically exposed in industrial organic solvents. *Toxicol App Pharmacol* 2010; 242:29–37.

14 Interspecies Comparisons of Pulmonary Responses to Fine and/or Nanoscale Particulates
Relevance for Humans of Particle-Overload Responses in the Rat Model

David B. Warheit and Kenneth L. Reed

CONTENTS

14.1   INTRODUCTION

Pulmonary responses to inhaled particles and fibers such as crystalline silica or asbestos have long been considered to be major occupational hazards causing chronic respiratory diseases and death among workers in a variety of industries. Examples of these pathogenic particulates include the various forms of asbestos fibers, which have been associated with the development of pulmonary fibrosis (i.e., asbestosis), lung cancer, and mesothelioma. In addition, the causal relationship is well established between exposures to aerosols of crystalline silica particulates and pulmonary inflammation, and the consequent development of silica-induced pulmonary fibrosis (i.e., silicosis) is well established.[1-3] Although the toxic effects of silica and asbestos are well established, for most other particulates (with a few exceptions—e.g., coal dust, titanium dioxide, and carbon black), the epidemiological database is rather sparse. As a consequence, rodent inhalation bioassays have become a benchmark for evaluating potential health hazards and estimating human health risks from exposures to airborne particulates. However, increasing numbers of inhalation studies in rats have shown that chronic exposures to high concentrations of insoluble particulates result in the development of pulmonary inflammation, fibrosis, and lung tumors. The application of the concept of inhalation bioassays for estimating these health risks for humans is complicated by interspecies differences in dosimetry and pulmonary responses. One issue that is troublesome stems from the fact that lung tumorigenic effects have been produced in rats by test materials ranging from the biologically active (e.g., silica) to those generally considered to be biologically benign such as titanium dioxide (TiO_2) particles. Indeed, with few exceptions most materials of low solubility and low toxicity have produced lung tumors in rats following long-term exposures at high particle-overload concentrations. Alternatively, chronic particle-overload exposures have not produced lung tumors in mice or hamsters.[4] Because of the sensitivity in response, the rat inhalation bioassay has been challenged for its appropriateness as a model to extrapolate to humans.[5] The scope of this chapter is a focus on interspecies differences in pulmonary responses to inhaled fine particles and nanoparticulates. Rats are frequently used for pulmonary hazard assessments. Although the rat model is sensitive and useful for gauging lung responses to low-solubility particles, the pulmonary effects in rats following chronic (i.e., 2-year) particle-overload exposures inevitably lead to lung tumors for low-solubility particulates. This does not occur in other rodent species such as hamsters or mice. Moreover, the few studies that have been conducted in nonhuman primates or retrospectively in dust-exposed workers show a different pattern of particle deposition, pulmonary responses, and lack of lung tumors. Finally, numerous epidemiological studies result in carbon black (Cb) and TiO_2-exposed workers do not correlate with evidence of lung cancer risk or noncancer respiratory disease. Thus it is concluded that the rat model may be viewed as a sensitive animal model for assessing the lung hazard potency for evaluating low-solubility particulate materials, but is inappropriate for drawing conclusions about lung cancer risk for humans.[5] This chapter describes the current database for chronic and subchronic inhalation studies with low-solubility particulates in rats, mice, and hamsters. Subsequently, a section is devoted to the available chronic inhalation studies with dusts comparing rats with nonhuman primates and humans. The interspecies differential lung responses have also been reviewed in three different subchronic 90-day inhalation studies with pigment-grade, nanoscale TiO_2, and Cb studies, wherein rats, mice, and hamsters have been exposed to identical test substances at the same aerosol concentrations. The results of the three subchronic inhalation studies are outlined in this chapter and the general findings of pulmonary responses of each of the three species are consistent throughout the three study results. The take-home message is that although each of the three species—rat, mouse, and hamster was exposed to aerosols of the same dusts, at the same concentrations, progressive cellular pathological changes in the lungs were documented only in the lungs of exposed rats. These pathological events are initiated by the development of particle overload, followed by sustained inflammation and cytotoxicity, and progress to characteristics of fibroproliferative disease; evidenced by cell proliferative effects, septal fibrosis, hyperplasia, and eventually development of lung tumors. In Sections 14.4 and 14.5, we focus on the fundamental differences between pulmonary responses in rats to particle overload and the effects observed in primates and humans exposed to high concentrations of dust particles. Indeed, it has also been reported that particle deposition, retention, and pulmonary inflammatory patterns are different in the lungs of rats when compared to particle effects investigated in the lungs of both nonhuman primates as well as in the respiratory tracts of heavily exposed workers. It is also noteworthy that detailed and numerous epidemiological studies in workers exposed to TiO_2 and Cb particles provide clear evidence of no causal link between particle exposures and lung cancers or other non-neoplastic lung diseases. This chapter focuses on the relevant toxicological database of subchronic and chronic inhalation studies showing the interspecies differences in lung response to particle overload among rodents. Subsequently, the mechanistic differences between pulmonary responses in rats versus humans and nonhuman primates are presented. Accordingly, it is concluded that the rat model presents a uniquely sensitive pulmonary response under conditions of particle overload.

14.2 CHRONIC INHALATION STUDIES IN RATS AND MICE WITH CARBON BLACK PARTICLES, DIESEL PARTICLES, AND TITANIUM DIOXIDE PARTICULATES

Chronic inhalation exposure of rats to very high doses of particles can result in inflammation, fibrosis, and some lung tumors. The tumorigenic response to particle overload observed in the rat appears to be both species specific (i.e., occurs in rats but not in mice) and restricted to doses where there is an overload of the lung clearance mechanisms.

14.2.1 CARBON BLACK AND DIESEL NANOPARTICLES

Two chronic inhalation studies with Cb particles in rats have been reported. One study was designed to investigate the importance of diesel soot–associated organic compounds compared to the carbonaceous core of the soot particulates.[6] In this study, male and female F344 rats were exposed for 24 months to either diluted whole diesel exhaust or Cb particulates at aerosol concentrations of 2.5 or 6.5 mg/m³ for 16 hours/day, 5 days/week. Rats exposed to filtered air served as the control group.

Lung clearance was impaired in both groups. The authors reported that both diesel exhaust and Cb exposures produced significant increases in the incidence and prevalence of lung tumors. However, in a subsequent report of a chronic bioassay inhalation study of CD-1 mice conducted in parallel with a previously reported bioassay of F344 rats, exposures to whole diesel exhaust 7 hours/day, 5 days/week for 24 months at soot concentrations of 0.35, 3.5, or 7.1 mg/m³ caused accumulations of soot in mouse lungs similar to those in mouse lungs of rats.[7] Moreover, in contrast to the dose-related neoplastic response of rats, diesel-exhaust exposures of mice did not increase the incidence of lung neoplasms. The authors concluded that the results are consistent with other data showing that mice, as well as Syrian hamsters, differ from rats in their lung neoplastic and non-neoplastic responses to heavy, chronic inhalation exposure to diesel exhaust soot and several other particles. In another study, female Wistar rats were exposed to high-purity furnace Cb (particle size = 14 nm, specific surface area = 227 m²/g, mass median aerodynamic diameter [MMAD] = 0.64 μm). The extractable organic mass of the furnace black was 0.04%; the content of benzo[a] pyrene was 0.6 pg/mg and that of 1-nitropyrene was <0.5 ng/mg particle mass. Rats were exposed in whole-body exposure chambers for 18 hours/day, 5 days/week to 7.4 mg/m³ Cb for 4 months followed by 12.2 mg/m³ for 20 months. After exposure, the rats were kept in clean air for another 6 months. The incidence of benign and malignant lung tumors was increased in the treated groups after 30 months. However, the rate of lung tumors in carbon-black exposed female NMRI mice was not significantly different from air-exposed control mice.[8]

14.2.2 Pigment-Grade TiO₂ Particles

In 2-year chronic inhalation studies, lung neoplastic responses have been produced in rats under conditions of high dose, particle overload. In a 2-year inhalation study in male and female rats, exposures to TiO₂ particles (rutile type) at concentrations of 10, 50, or 250 mg/m³ resulted in the development of lung tumors only at the highest concentration.[9] With one exception, the originally diagnosed squamous cell carcinoma tumors were reclassified (according to new diagnostic criteria) as primarily benign pulmonary keratin cysts[10] (see Figures 14.1 and 14.2). In a study reported by Muhle et al.,[11] TiO₂ was used as a negative control dust in a 2-year inhalation study with toner particles. Male and female rats were exposed (6 hours/day, 5 days/week) to 5 mg/m³ TiO₂ particles (rutile form) of 1.1 μm MMAD. There were no significant increases in lung neoplastic responses versus control rats exposed for up to 24 months by whole body inhalation to TiO₂ particles in this study.

14.2.3 Ultrafine Grade TiO₂ Particles

As a component of a diesel exhaust chronic inhalation study in rats and mice, Heinrich et al.[8] exposed female rats by whole body inhalation to ultrafine TiO₂ (80% anatase: 20% rutile) at an average aerosol concentration of 10 mg/m³ for 2 years followed by a 6-month postexposure recovery period (i.e., 30-month study). The TiO₂ exposure at a single concentration provided an additional particle control group. The TiO₂ particle size MMAD was determined to be 0.8 μm (agglomerates

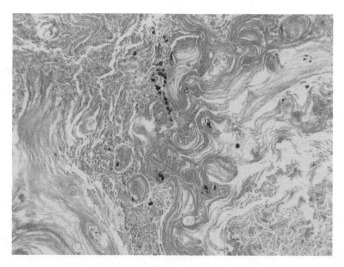

FIGURE 14.1 Light micrograph from lung tissue of a female rat exposed for 24 months to pigment-grade TiO_2 particles at 250 mg/m^3. Micrograph shows presence of keratin cysts. (Magnification × 100.)

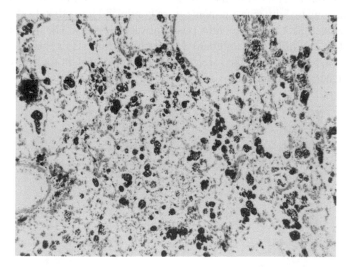

FIGURE 14.2 Light micrograph from lung tissue of a female rat exposed for 24 months to 250 mg/m^3 pigment-grade TiO_2 particles—showing alveoli filled with macrophages containing particles. (Magnification × 100.)

of ultrafine particles). Significant increases in lung tumors versus controls were reported in TiO_2-exposed rats (including benign keratinizing cystic squamous cell tumors, adenocarcinomas, squamous cell carcinomas, and adenomas). However, female mice exposures to ultrafine TiO_2 under the same conditions as for rats (10 mg/m^3) resulted in lung tumor rates, which were not statistically increased over control female mice.[8] These results confirm the species differences between rats and mice in lung responses following chronic exposures to particle-overload concentrations of low-solubility particulates.

14.3 SUBCHRONIC INHALATION STUDIES WITH CARBON BLACK PARTICLES AND TITANIUM DIOXIDE PARTICLES

14.3.1 CARBON BLACK PARTICLES (SEE FIGURE 14.3)

Inhalation exposures to high concentrations of Cb particles have produced lung tumors in rats, but not in mice or hamsters. It has been postulated that the development of lung tumors may be related to a secondary genotoxic mechanisms involving sustained lung inflammation and oxidative/cytotoxic injury. Elder et al.[12] conducted an interspecies, subchronic, 90-day inhalation exposures of two different forms of carbon black, namely high-surface area carbon black (HSCb) and low-surface area carbon black (LSCb) particles. The authors postulated that the pulmonary effects would be more pronounced in rats than in mice and hamsters. Particle retention kinetics, inflammation, and histopathology endpoints were measured in the lungs of female rats, mice, and hamsters exposed for 13 weeks to high-dose HSCb at a range of exposure concentrations, including particle-overload doses, that is, 0, 1, 7, 50 mg/m³. In addition, rats were also exposed to LSCb (50 mg/m³, nominal). Particle retention and lung toxicity measurements were conducted immediately after 13-week exposures as well as 3 and 11 months postexposure recovery. The authors reported that equivalent or similar mass burdens were achieved in rats exposed to high-dose HSCb and LSCb, whereas surface area burdens were equivalent for mid-dose HSCb and LSCb. Particle clearance impairment was measured in rats exposed to mid- and

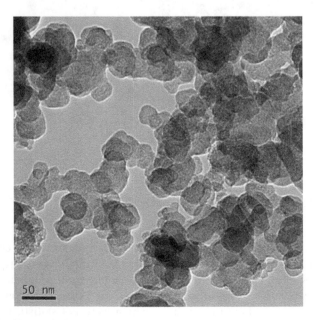

FIGURE 14.3   Transmission electron micrograph of nanoscale carbon black particles.

high-dose HSCb and to LSCb, but LSCb particulates were cleared faster than HSCb. In addition, particle retention was also prolonged in mice exposed to mid- and high-dose HSCb and to LSCb, and in hamsters exposed to high-dose HSCb. The authors concluded that lung inflammation and histopathological effects were more severe and sustained in rats when compared to mice or hamsters; and similar pulmonary effects were observed in rats exposed to mid-dose exposures to the two forms of carbon black, that is, HSCb and LSCb. The findings clearly showed that hamsters have the most efficient pulmonary particle clearance mechanisms and least severe respiratory tract responses of the three species tested.[12]

14.3.2 PIGMENT-GRADE TiO$_2$ PARTICLES (SEE FIGURE 14.4)

Bermudez et al.[13] conducted subchronic, 90-day inhalation exposures of rats, mice, and hamsters to pigmentary TiO$_2$ particles at concentrations likely to induce particle overload and reported a more severe and persistent pulmonary inflammatory response in rats, when compared with either similarly exposed mice or hamsters. Rats were unique among these three rodent species in the development of progressive fibroproliferative lesions and alveolar epithelial metaplasia. In the study, female rats, mice, or hamsters were exposed to 10, 50, or 250 mg/m^3 concentrations of pigmentary (rutile type) TiO$_2$ particles 6 hours/day, 5 days/week for 13 weeks followed by evaluations at several postexposure recovery periods. The investigators reported that lung and associated lymph node burdens of TiO$_2$ particles were increased in a concentration-related manner. It was reported that retained lung burdens were greatest in mice following exposure, with rats and hamsters showing similar lung burdens immediately following 90-day exposures. On the basis of particle burden retention

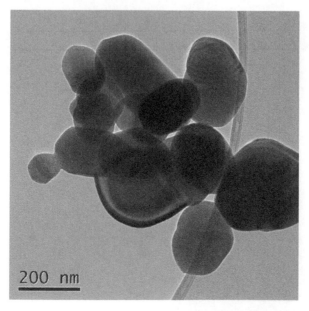

FIGURE 14.4 Transmission electron micrograph of pigmentary TiO$_2$ particles.

data, it was concluded that particle overload in the lungs was reached in both rats and mice at 50 and 250 mg/m^3 concentrations. Pulmonary inflammation was initially measured in all three species at the two highest concentrations. Lung inflammatory responses persisted in rats and mice throughout the postexposure recovery period at 250 mg/m^3. However, in hamsters, lung inflammatory responses were reversible because of the accelerated clearance of particles from the lung. In rats exposed to the highest concentration (250 mg/m^3), histopathological observations revealed that pulmonary lesions were characterized by epithelial and fibroproliferative changes manifested by increased alveolar epithelial cell proliferative labeling indices. Associated with these proliferative changes in the rat were augmented rates of inter-stitial particle accumulations concomitant with the development of alveolar septal fibrotic responses. Although rats exposed to 50 mg/m^3 developed minimal alveolar cell hypertrophy, accumulation of particle-laden macrophages, and inflammation, no alveolar septal fibrosis were observed or relevant cell turnover endpoints were measured at alveolar sites at the lower exposure concentration. One of the major take-home messages of this study: similar pulmonary changes to those seen in rats were not observed in either mice or hamsters exposed to TiO$_2$ particles. The study clearly demonstrated the unique sensitivity of the rat pulmonary response under par-ticle overload concentrations.

14.3.3 ULTRAFINE TiO$_2$ PARTICLES (SEE FIGURE 14.5)

In a subchronic study with ultrafine TiO$_2$ particles (80% anatase: 20% rutile; average primary particle size = 21 nm), female rats, mice, or hamsters were exposed to aero-sol concentrations of 0.5, 2.0, or 10 mg/m^3 TiO$_2$ particles for 6 hours/day, 5 days/week

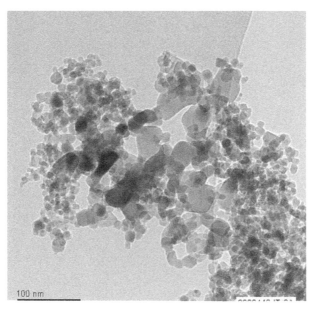

FIGURE 14.5 Transmission electron micrograph of nanoscale TiO$_2$ particles.

for 13 weeks followed by evaluations at several postexposure recovery periods.[14] The experimental design of this ultrafine TiO_2 protocol study was nearly identical to the earlier Bermudez et al.[13] with the exception of differences in the test material and aerosol concentrations. The results showed that mice and rats had similar lung burdens at the end of exposures but hamsters were significantly lower. The significant slowing of particle clearance rates in rats and mice at the highest exposure concentration (10 mg/m³) was consistent with the finding that pulmonary particle overload had been achieved. Similar to the earlier study, pulmonary lesions in rats included sustained inflammation and cytotoxicity (Table 14.1); foci of alveolar epithelial proliferation of metaplastic epithelial cells (alveolar bronchiolization) along with focal areas of heavy, particle-laden macrophages. In rats, these changes were characterized by enhanced alveolar epithelial cell proliferation/labelling indices. Particle overload-related developments were consistent with areas of interstitial particle accumulation and developing alveolar septal fibrosis. These pulmonary lesions measured and observed in rat lungs were sustained and progressive, that is, adverse effects became more pronounced with residence time in the lungs. In contrast mice and hamsters developed less severe inflammatory and cytotoxic responses (see Table 14.2), indeed lacking the progressive epithelial and fibroproliferative changes observed in rats. As described in this section, these findings are consistent with the results of an earlier study with inhaled pigmentary TiO_2 particulates.[13] See Table 14.2 for the differences in rodent interspecies lung responses to particle-overload concentration exposures.

14.4 PULMONARY RESPONSES OF NONHUMAN PRIMATES AND COAL MINERS TO PARTICLE OVERLOAD

The pulmonary responses of rats are unique when compared to other large mammalian species such as nonhuman primates and humans. Nikula et al.,[15,16] have evaluated and compared the pulmonary response patterns of rats relative to nonhuman primates, as well as comparisons of lung responses of rats versus coal miners using lung morphometry methods and have concluded that intrapulmonary particle retention patterns and tissue reactions in rats may not be predictive of lung particle retention patterns and tissue responses in either primates or humans. In one study, male monkeys and rats were exposed to various aerosolized test particle substances including diesel exhaust (2 mg/m³), coal dust (2 mg/m³), or diesel exhaust and coal dust combined (1 mg/m³ each) for a period of 7 hours/day, 5 days/week for 24 months. The respiratory tracts were subsequently evaluated histopathologically.[15] For all of the particle-exposed groups, lung particle retention for the monkeys was similar or greater than similarly exposed rats. However, the disposition and pattern of retained dust within the respiratory tract was different between the two species. Accordingly, rats retained a greater proportion of the particulate material in the alveolar ducts and alveoli, whereas monkeys retained a greater proportion of inhaled dust particles in the pulmonary interstitial compartment (a strong indication of translocation of particles from alveolar epithelial regions to pulmonary interstitium). In addition, rats, but not monkeys, had significant alveolar epithelial hyperplastic responses, lung inflammatory reactions, and development of septal fibrotic responses to the retained particles. In another study comparing rat versus human lung responses (i.e., pulmonary effects

TABLE 14.1
Lactate Dehydrogenase and Total Protein Concentrations in Bronchoalveolar Lavage Fluid from Mice, Rats, and Hamsters

	Mice					Rats					Hamsters				
	Weeks Postexposure					Weeks Postexposure					Weeks Postexposure				
	0	4	13	26	52	0	4	13	26	52	0	4	13	26	49
LDH (U/L [SD])															
Control	53 (35)	38 (10)	37 (16)	35 (10)	28 (9)	24 (2)	29 (1)	29 (3)	34 (16)	30 (8)	26 (5)	25 (3)	26 (5)	18 (5)	6 (5)
10 mg/m³	87 (18)	103* (24)	120* (81)	63 (10)	72 (55)	122* (18)	112* (8)	83* (14)	50 (5)	33 (5)	24 (5)	27 (3)	22 (3)	17 (4)	9 (6)
Protein (µg/ml [SD])															
Control	92 (19)	91 (12)	19 (15)	68 (27)	115 (15)	83 (15)	79 (5)	88 (9)	97 (33)	125 (32)	95 (20)	100 (20)	102 (50)	142 (21)	145 (35)
10 mg/m³	257* (31)	256* (42)	274* (126)	160* (31)	206* (44)	236* (28)	223* (12)	133 (54)	138 (26)	149 (21)	118 (39)	113 (17)	134 (40)	143 (7)	143 (54)

LDH, lactate dehydrogenase.

*Significantly different from concurrent control, p < .05.

TABLE 14.2

Species Comparisons of Rodent Lung Responses to Inhaled Low-Solubility, Fine, and Nanoparticles

Predilection for Developing Particle Overload

1. Rat—high degree
2. Mouse—high degree
3. Hamster—low degree

Alveolar Macrophage Responses to Particle Overload

1. Rat—high but accumulation in alveolar ducts
2. Mouse—high but accumulation in alveolar ducts
3. Hamster—high but favors more rapid lung clearance

Lung Inflammatory Responses

1. Rat—high degree
2. Mouse—high degree
3. Hamster—low degree

Alveolar Epithelial Cell Proliferation

1. Rat—high degree
2. Mouse—medium–low degree
3. Hamster—low degree

Development of Fibroproliferative Effects Including Lung Fibrosis

1. Rat—high and sustained effects
2. Mouse—low degree
3. Hamster—low degree

Location of Retained Particles in the Lung

1. Rat—primarily accumulation of particles in alveolar ducts
2. Mouse—primarily accumulation of particles in alveolar ducts
3. Hamster—less accumulation of particles in alveoli—more rapid clearance

Development of Overload Particle-Related Lung Tumors

1. Rat—Yes
2. Mouse—No
3. Hamster—No

in long-term employed coal miners), differential findings between the two species were similar in nature as those described in the comparative monkey/rat studies. Indeed, nearly identical to the results measured in dust-exposed monkeys, percentages of retained particulate matter (~90%) in the lungs of coal miners were identified in the lung interstitial anatomical compartment.[16] The authors concluded that these fundamental differences in particulate tissue distribution in rats and humans likely bring different lung cells into contact with retained particulates or particle-containing macrophages, and may account for the pulmonary differences in species responses to inhaled particulates. To summarize the careful and fundamental impacts generated by these investigators regarding differential pulmonary responses between rats and nonhuman primates and coal miners when considering lung responses to inhaled particulates, it should be noted that (1) the particle disposition and dosimetry differ between rats and either monkeys or humans (rat = alveolar; monkey/human = interstitial sites), and (2) rats produce significantly enhanced and persistent pulmonary inflammogenic, epithelial, and fibroproliferative responses when compared to respiratory tract responses in either monkeys or humans. The toxicity database also clearly shows that particle overload-exposed rats are significantly more prone and sensitive to develop pathological and progressive lung responses to inhaled particulate exposures when compared to (1) other rodent species, as well as (2) larger mammals such as monkeys and humans. The conclusions reported by Nikula et al. on differential responses between rats and primates are supported by earlier studies comparing the pulmonary responses of rats and primates. For example, Klonne et al.[17] exposed Sprague–Dawley rats and cynomolgus monkeys to 0, 10.2, or 30.7 mg/m³ micronized delayed process petroleum coke for 24 months. The rats exhibited retention of particulate material in macrophages, chronic inflammation, focal fibrosis alveolar–bronchiolar metaplasia, sclerosis, squamous metaplasia of alveolar epithelium, and keratin cysts. In monkeys, the pulmonary histopathology was limited to accumulations of macrophages containing particulate material. In another study, MacFarland et al.[18] exposed F344 rats and cynomolgous monkeys to 0, 10, or 30 mg/m³ respirable raw or processed shale dust for 24 months. All the rats developed proliferative bronchiolitis, alveolitis, and epithelial hyperplasia and most developed chronic inflammation with nonprogressive fibrosis, cholesterol clefts, and microgranulomas. The monkeys accumulated more pigment-laden macrophages in the bronchiolar and alveolar walls than in the alveolar lumens. The majority of monkeys had no reaction to the accumulated dust exposures. A few monkeys had occasional foci of subacute inflammation. The results of comparative studies with rats and monkeys were exposed to identical aerosolized particle test substances indicate that although both rats and monkeys retain particulate material in

---

TABLE 14.3

Comparisons of Rodent Lung Responses versus Human/Primate Responses to Inhaled Low-Solubility, Fine, and Nanoparticles

Likelihood for Developing Particle Overload
1. Rat—high degree
2. Human/Primate—not determined

Alveolar Macrophage Responses to Particle Overload
1. Rat—high but accumulation in alveolar ducts
2. Human/Primate—not extensive due to greater particle translocation to interstitium

Lung Inflammatory Responses
1. Rat—high degree
2. Human/Primate—low degree

Alveolar Epithelial Cell Proliferation
1. Rat—high degree
2. Human/Primate—low degree

TABLE 14.3 (*Continued*)

Comparisons of Rodent Lung Responses versus Human/Primate Responses to Inhaled Low-Solubility, Fine, and Nanoparticles

Development of Fibroproliferative Effects Including Lung Fibrosis
1. Rat—high and sustained effects
2. Human/Primate—low degree

Location of Retained Particles in the Lung
1. Rat—primarily accumulation of particles in alveolar ducts
2. Human/Primate—primarily interstitial

Development of Overload Particle-Related Lung Tumors
1. Rat—Yes
2. Human/Primate—No

their lungs when chronically exposed to high concentrations of dusts or soot, the particle retention patterns and pulmonary reactions to inhaled dusts are fundamentally different. Specifically, rats retain a greater proportion of the material in intralumenal alveolar macrophages, and they respond with a greater degree of epithelial hyperplasia and active inflammation when compared to the response of monkeys (See Table 14.3).

14.5 UNDERSTANDING THE MECHANISMS OF THE UNIQUE RAT LUNG RESPONSES TO PARTICLE OVERLOAD

Studies have been conducted to investigate the mechanisms related to the uniquely sensitive pulmonary responses following particle-overload exposures in rats. Warheit et al.[19] conducted a subchronic inhalation toxicity study that was designed to postulate a potential mechanism for the respiratory tract responses observed in the pivotal rat oncogenicity study.[9] The study showed that the lungs of particle-overload exposed rats are characterized by impaired pulmonary clearance, sustained pulmonary inflammation, cellular hypertrophy, and hyperplasia, and that these effects, following continuous exposure at 250 mg/m^3 (for two years), likely could result in the development of overload-related pulmonary tumors. The Warheit et al. study with pigment-grade TiO_2 in rats used aerosol exposure concentrations similar to those used in the earlier Lee et al.[9] study to simulate the characteristics of "lung overload" in this species, concomitant with an investigation of the rat's ability to recover from this particle-overload challenge. Male rats were exposed to TiO_2 particles 6 hours/day, 5 days/week for 4 weeks at concentrations of 5, 50, or 250 mg/m^3 and the lungs assessed at selected intervals through 6 months postexposure. Results showed that exposures to high dust TiO_2 concentrations produced several interrelated effects including lung inflammation, enhanced pulmonary cell proliferation indices, impairment of particle clearance, deficits in macrophage phagocytic functions, and the appearance of macrophage aggregates at alveolar sites. Rats exposed to 250 mg/m^3 TiO_2 had particle lung burdens of 12 mg/lung. TiO_2 particle exposures induced persistent lung inflammatory responses in animals exposed to 250 mg/m^3. Rats exposed to 250 mg/m^3 showed diminished lung clearance after 1 week through 1 month postexposure. Monoexponential clearance modeling indicated that TiO$_2$ particles were cleared with half-times of approximately 68, 110, and 330 days for the 5, 50, and 250 mg/m^3 test groups, respectively. Lymph node burden data from rats exposed to 250 mg/m^3 TiO$_2$ signaled significant translocation of TiO$_2$ particles to tracheobronchial lymph nodes. In vitro phagocytosis studies showed that alveolar macrophages exposed to 250 mg/m^3 TiO$_2$ were impaired in their phagocytic responses. Rats exposed to 50 mg/m^3 TiO$_2$ had small, sustained inflammatory responses. Rats exposed to 250 mg/m^3 showed diminished lung clearance after 1 week through 1 month postexposure. At the two higher concentrations (50–250 mg/m^3) of TiO$_2$, cellular hypertrophy and hyperplasia were evident at alveolar wall and duct bifurcations that were located adjacent to the phagocytic macrophages (see Figures 14.6 through 14.8). In another study with Cb nanoparticles, Carter et al.,[20] postulated that the uniquely sensitive response of the rat versus other species may relate to an inherent mechanistic inability to generate sufficient anti-inflammatory mediators in the face of continuing pulmonary particulate overload exposures. In addition, it was hypothesized that rats may generate a greater degree of proinflammatory cytokines in response to particle-overload exposures when compared to mice or hamsters, and this might be causally related, in part, to the "exaggerated" pulmonary pathological response documented in the rat model following particle-overload exposures. Accordingly, potential mechanistic differences in the pulmonary responses of rats versus mice or hamsters were investigated following exposures to inhaled Cb particles by assessing the levels of several key pro and anti-inflammatory mediators in the lungs of exposed animals. Thus, the objective of the study was to study proinflammatory and anti-inflammatory mechanisms underlying species specificity in high-dose, Cb particle-induced pulmonary inflammation. For this study, rats, mice, and hamsters were exposed to

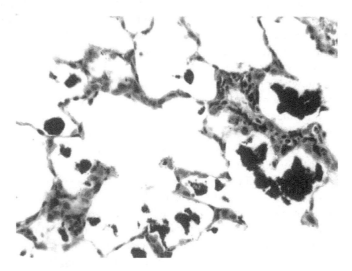

FIGURE 14.6 Lung tissue of a rat exposed for 4 weeks to 250 mg/m^3 TiO$_2$ particles. Macrophages have accumulated at sites of particle deposition 1 week postexposure.

FIGURE 14.7 Transmission electron micrograph showing a macrophage filled with phagocytized TiO_2 particles on the alveolar surface—1 week after a 4-week inhalation exposure.

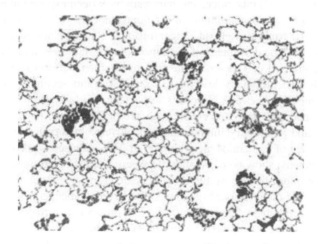

FIGURE 14.8 Lung tissue of a rat exposed for 4 weeks to 250 mg/m³ TiO_2 particles. The micrograph shows focal macrophage accumulation at alveolar sites at 6 months post inhalation exposures.

aerosolized Cb particulates for 13 weeks as described in Section 14.3.1 in the Elder et al.[12] study at three concentrations (1, 7, and 50 mg/m³). Bronchoalveolar lavage studies were conducted to measure relevant inflammatory/cytokine and cytotoxicity indices, concomitant with assessments of reactive oxygen and nitrogen specie endpoints. Moreover, ex vivo mutational analysis of inflammatory cells was implemented by coincubating them with lung epithelial cells. In addition, lung tissue was evaluated for gene expression of selected anti-inflammatory mediators. The investigators reported that rats showed greater tendencies for generating proin-flammatory cytokine responses. In contrast, mice and hamsters showed generated cytokines, which favored increased anti-inflammatory responses. These prelimi-nary findings suggest a potential mechanism for elucidating the differences in rat responses to Cb particle overload-induced lung inflammation when compared to mice or hamster responses.[20]

14.6   CONCLUSIONS AND INTEGRATION OF POINTS

The development of pulmonary tumors in rats after particle exposures to poorly soluble, low-toxicity dusts has been reported only under the conditions of particle overload in the lungs. This finding has been reported following lung exposures to both fine-sized particulates as well as nanoscale particle-types. This pulmonary pathologi-cal response in rats to chronic particulate exposures at high concentrations is unique, as other rodent species such as mice and hamsters; as well as larger mammals such as humans or nonhuman primates do not develop lung tumors or other pulmonary pathological responses under similar conditions of particle lung overload exposures from poorly soluble particulates (PSPs). The evidence to support this position has been addressed in some of the studies listed earlier. To summarize the results of those studies: (1) chronic exposures to PSPs at high concentrations lead to lung tumors in rats, but not in mice; (2) rats, mice, and hamsters have been exposed to identical test substances at identical concentrations of pigment-grade TiO_2, ultrafine TiO_2, as well as Cb particles (at particle-overload concentrations). Rats, but not mice or hamsters, developed a pathological sequelae of sustained lung inflammation and cytotoxicity, followed by increased cell turnover and fibroproliferative effects (e.g., hyperplasia and septal fibrosis) ultimately leading to metaplasia and secondary genotoxicity; (3) studies comparing the lung responses to particle overload of rats versus nonhu-man primates and coal dust workers show different patterns of particle disposition, translocation, and pulmonary inflammatory or adverse histopathological reactions; (4) An abundance of epidemiological studies in TiO_2-exposed workers (the highest exposed individuals) show no correlation between TiO_2 exposures and risks of lung cancer or nontumor-related respiratory effects exposure.[21–26] Accordingly, one can conclude that all of the interspecies studies point to an identical conclusion regarding the unique characteristics of a pathophysiological process operating in the rat, which ultimately leads to the formation of primarily benign lung tumors. The development of pulmonary tumors at particle-overload exposures is triggered by the inability of the rat to effectively clear particles from the respiratory tract, concomitant with the sus-tained development of inflammation and cytotoxic effects. Several lines of evidence suggest that the mechanistic differences could be related to the propensity of rats to undergo a unique pathological sequelae in response to particle-overload conditions. This is initiated by a sustained lung inflammatory and cytotoxicity response, followed sequentially by the development of enhanced cell proliferative activity and corre-sponding fibroproliferative effects, including septal fibrosis, hyperplasia, and second-ary genotoxicity, ultimately leading to the development of lung tumors. Contributing to this pathological effect could be the unique response of rat lung cells to particle-overload concentrations—generating "abnormally high" proinflammatory response, concomitant with a deficiency in anti-inflammatory cytokine responses. This would appear to be a suggested "mechanistic recipe" for enhancing the sensitivity or overriding the protective capacity of the rat lung response to particle overload in rats.

Possible sequence of pathological sequelae in rat lungs following chronic particle overload exposures:

1. Impaired particle clearance → persistent lung inflammation and cytotoxicity
2. → Cell proliferation and fibroproliferative effects—including septal fibrosis, hyperplasia, metaplasia, and (secondary) genotoxicity
3. → Development of lung tumors

REFERENCES

1. Morgan A, Holmes A. Concentrations and dimensions of coated and uncoated as bestosfibres in the human lung. *Br J Ind Med.* 1980;37(1):25–32.
2. Warheit DB. Interspecies comparisons of lung responses to inhaled particles and gases. *Crit Rev Toxicol.* 1989;20(1):1–29.
3. Adamson IY, Letourneau HL, Bowden DH. Comparison of alveolar and interstitial macrophages in fibroblast stimulation after silica and long or short asbestos. *Lab Invest.* 1991;64(3):339–344.
4. ILSI Risk Science Institute Workshop. The relevance of the rat lung response to particle overload for human risk assessment. *Inhal Toxicol.* 2000;12:1–17.
5. Levy LS. The 'particle overload" phenomenon and human risk assessment. *Indoor Built Environ.* 1995;4:254–262.
6. Nikula KJ. Rat lung tumors induced by exposure to selected poorly soluble nonfibrous particles. *Inhal Toxicol.* 2000;12:97–119.
7. Mauderly JL, Banas DA, Griffith WC, Hahn FF, Henderson RF, McClellan RO. Diesel exhaust is not a pulmonary carcinogen in CD-1 mice exposed underconditions carcinogenic to F344 rats. *Fund Appl Toxicol.* 1996;30:233–242.
8. Heinrich U, Fuhst R, Rittinghausen S, Creutzenberg O, Bellman B, Koch W, Levsen K. Chronic inhalation exposure of Wistar rats and two different strains of mice to diesel engine exhaust, carbon black, and titanium dioxide. *Inhal Toxicol.* 1995;7:533–556.
9. Lee KP, Trochimowicz HJ, Reinhart CF. Pulmonary responses of rats exposed to titanium dioxide (TiO$_2$) by inhalation for two years. *Toxicol Appl Pharmacol.* 1985;79:179–192.
10. Warheit DB, Frame SR. Characterization and reclassification of titanium dioxide-related pulmonary lesions. *J Occup Environ Med.* 2006;48(12):1308–1313.
11. Muhle H, Bellmann B, Creutzenberg O, Dasenbrock C, Ernst H, Kilpper R, MacKenzie JC, Morrow P, Mohr U, Takenaka S. Pulmonary response to toner upon chronic inhalation exposure in rats. *Fund Appl Toxicol.* 1991;17:280–299.
12. Elder A, Gelein R, Finkelstein JN, Driscoll KE, Harkema J, Oberdörster G. Effects of subchronically inhaled carbon black in three species. I. Retention kinetics, lung inflammation, and histopathology. *Toxicol Sci.* 2005;88(2):614–629.
13. Bermudez E, Mangum JB, Asgharian B, Wong BA, Reverdy EE, Janszen DB, Hext PM, Warheit DB, Everitt JI. Long-term pulmonary responses of three laboratory rodent species to subchronic inhalation of pigmentary titanium dioxide particles. *Toxicol Sci.* 2002;70(1):86–97.
14. Bermudez E, Mangum JB, Wong BA, Asgharian B, Hext PM, Warheit DB, Everitt JI. Pulmonary responses of mice, rats, and hamsters to subchronic inhalation of ultrafine titanium dioxide particles. *Toxicol Sci.* 2004;77(2):347–357.

15. Nikula KJ, Avila KJ, Griffith WC, Mauderly JL. Lung tissue responses and sites of particle retention differ between rats and cynomolgus monkeys exposed chronically to diesel and coal dust. *Fund Appl Toxicol.* 1997;37:37–53.

16. Nikula KJ, Vallyathan V, Green FH, Hahn FF. Influence of exposure concentration or dose on the distribution of particulate material in rat and human lungs. *Environ Health Perspect.* 2001;109:311–318.

17. Klonne DR, Burns JM, Halder CA, Holdsworth CE, Ulrich CE. Two-year inhalation toxicity study of petroleum coke in rats and monkeys. *Am J Ind Med.* 1987;11:375–389.

18. MacFarland HN, Coate WB, Disbennett DB, Ackerman LJ. Long-term inhalation studies with raw and processed shale dusts. *Ann Occup Hyg.* 1982;26(1–4):213–225.

19. Warheit DB, Hansen JF, Yuen IS, Kelly DP, Snajdr SI, Hartsky MA. Inhalation of high concentrations of low toxicity dusts in rats results in impaired pulmonary clearance mechanisms and persistent inflammation. *Toxicol Appl Pharmacol.* 1997;145:10–22.

20. Carter JM, Corson N, Driscoll KE, Elder A, Finkelstein JN, Harkema JN, Gelein R, Wade-Mercer P, Nguyen K, Oberdorster G. A comparative dose-related response of several key pro- and antiinflammatory mediators in the lungs of rats, mice, and hamsters after subchronic inhalation of carbon black. *J Occup Environ Med.* 2006;48(12):1265–1278.

21. Boffetta P, Gaborieau V, Nadon L, Parent MF, Weiderpass E, Siemiatycki J. Exposure to titanium dioxide and risk of lung cancer in a population-based study from Montreal. *Scand J Work Environ Health.* 2001;27:227–232.

22. Boffetta P, Soutar A, Cherrie J, Granath F, Andersen A, Anttila A, Blettner M et al. Mortality among workers employed in the titanium dioxide industry in Europe. *Cancer Cause Control.* 2004;15:697–706.

23. Fryzek J, Chadda B, Marano D, White K, Schweitzer S, McLaughlin J, Blot W. A cohort mortality study among titanium dioxide manufacturing workers in the United States. *J Occup Environ Med.* 2003;45:400–409.

24. Ellis ED, Watkins J, Tankersley W, Phillips J, Girardi D. Mortality among titanium dioxide workers at three DuPont plants. *J Occup Environ Med.* 2010;52(3):303–309.

25. Ellis ED, Watkins JP, Tankersley WG, Phillips JA, Girardi DJ. Occupational exposure and mortality among workers at three titanium dioxide plants. *Am J Ind Med.* 2013;56:282–291.

26. Chen J, Fayerweather W. Epidemiologic study of workers exposed to titanium dioxide. *J Occup Med.* 1988;30:937–942.

15 Current In Vitro Models for Nanomaterial Testing in Pulmonary Systems

Sonja Boland, Sandra Vranic,
Roel Schins, and Tobias Stöger

CONTENTS

15.1 INTRODUCTION

The pulmonary system is one of the main target organs of nanomaterials (NMs), especially for occupational exposures. The large surface area of the respiratory tract represents a huge surface of deposition and accumulation of airborne particles especially as the clearance mechanisms could be less efficient for particles at the nanoscale. The huge surface area of the respiratory system combined with inadequate clearance mechanisms may also allow nanoparticles (NPs) to pass the epithelial barrier leading to systemic exposures. It is of crucial importance to study not only the direct effect of NMs on resident lung cells but also their uptake and possible transcytosis and effects on the underlying cells and tissues. This capacity of NMs to pass through the epithelial barrier may also be beneficial for nanomedical applications.[1]

For the in vitro testing of NMs on lung cells, several NM-specific issues have to be taken into consideration compared to the toxicity evaluation of solute toxicants. NMs could for instance interfere with assay processes and components due to their specific physicochemical properties. High adsorption capacities, optical properties, hydrophobicity, surface charge, and catalytic activities require the use of appropriate controls and complementary assays.[2-4] Furthermore, several studies have shown that serum or surfactant could modulate the uptake and the effects of NMs by cells and the coating of NMs with biomolecules must thus be considered for in vitro studies. Adsorption of biomolecules present in lung lining fluids or serum could also modify the surface reactivity of the particles, which is particularly important for cytotoxicity studies that often use serum in cell culture or NP dispersions.[5] In contrast to solutes, not only the general chemical composition has to be taken into account but also the surface chemistry as many NM effects are linked, for example, to their surface reactivity, charge, or hydrophobicity. The uptake mechanisms and intracellular fate of NMs are more complex than for solutes. These issues are particularly important in view of biomedical applications of NMs as nanodelivery systems.

15.2 CELL CULTURE MODELS FOR THE PULMONARY SYSTEM

Most studies on NM toxicity are performed on cell lines. Even though these cells are transformed and do not present all physiological features of primary cells, they are easy to maintain in culture. This has the advantage to allow comparisons of large numbers of NMs to determine, for example, the physicochemical features of NMs, which may influence their toxicity. This is facilitated by the development of auto-mated high-throughput and high content screening techniques for NM testing.[6] The use of cell lines also facilitates interlaboratory comparisons to establish and validate protocols for nanotoxicity testing. Nanotoxicological testing is generally performed with cell lines of human origin but murine cell lines are also used, especially when comparing in vitro results to in vivo experiments using mice, or when appropriate human cell lines of specific origin are not available. Many studies have evaluated the effects of NMs on bronchial epithelial cell lines, alveolar epithelial cells, and mac-rophages but only a very few on dendritic cells. Due to the complexity of NM inter-actions with biological systems, the trend of in vitro nanotoxicology is to use more complex cell cultures than cell lines. More realistic physiological culture systems are indeed needed to allow accurate extrapolations to in vivo situations. Cocultures of dendritic cells, macrophages, and epithelial cells have, for example, allowed demon-strating a particle transfer from macrophages to dendritic cells.[7]

Despite the direct effect of NMs on the respiratory epithelium, they may also act on the underlying tissue, either directly after translocation through the epithelial barrier or indirectly by stimulation of the epithelium or resident macrophages (Figure 15.1). For instance, exposure to NM could lead to lung remodeling, and in vitro studies could help understanding the underlying mechanisms. Using pulmonary fibro-blasts it has been shown that titanium dioxide (TiO_2) NPs induce metalloprotease-1 expression and activity through an interleukin (IL)-1beta-dependent mechanism.[8] Epithelial cells may also participate in the lung remodeling. In vitro exposure of air-way epithelial cells to ultrafine atmospheric particles has indeed shown an increase in secretion of epidermal growth factor (EGF)-receptor ligands that could lead to not only a fibroblast proliferation but also autocrine effects such as mucin produc-tion.[9] Epithelial–mesenchymal transition may also contribute to fibrosis induction observed in vivo after exposure to NMs.

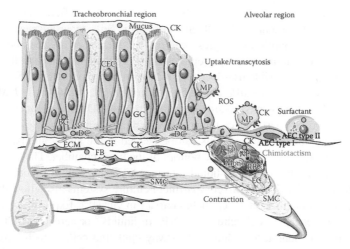

FIGURE 15.1   (See color insert.) Interactions of nanomaterials with different lung cell types studied in monoculture or coculture. Nanoparticles could interact with the different epithelial cell types of the lungs: ciliated cells (CEC), goblet cells (GC) of the tracheobronchial region or alveolar epithelial type I and type II cells (AEC types I and II) of the alveolar region as well as alveolar macrophages (MP). This could lead to nanoparticle uptake, modification of lung lining fluid production (mucus or surfactant), reactive oxygen species (ROS) production, and secretion of inflammatory mediators such as cytokines (CK). These can affect secondary target cells and especially induce chimiotactisme of monocytes (Mono), eosinophils (Eo), and neutrophils (NP). Activation of endothelial cells could lead to pulmonary infiltration of these inflammatory cells. Secretion of CK and growth factors (GF) could also stimulate cells of the underlying tissues such as dendritic cells (DC) and fibroblasts (FB) inducing their proliferation and induction of extracellular matrix production (ECM). Stimulation of smooth muscle cells (SMC) could lead to constriction of bronchi and pulmonary arteries. Nanoparticles could also stimulate directly these secondary target cells of the underlying tissues after translocation through the epithelial barrier, which could also allow nanoparticles entering the bloodstream. These effects of nanoparticles exposure have been investigated in vitro using single-cell cultures or more complex coculture or organotypic culture systems (drawn using Servier medical art).

Inhalation of particles could not only lead to respiratory disorders but may also induce cardiovascular diseases. These effects could be due to either activation of the epithelium, inducing inflammatory reactions, or direct effects on the vasculature after translocation through the epithelial barrier. Studies of pulmonary endothelial cells are thus of high interest to determine possible cardiovascular effects. It has been shown that exposure of pulmonary endothelial cells to silicon dioxide (SiO_2) NPs could increase endothelial adhesiveness.[10] Furthermore, in vitro models of intrapulmonary arteries have allowed demonstrating that exposure to carbon NPs could modify the vasomotor response.[11] Ex vivo studies of the contractile response of pulmonary arteries removed from instilled animals have shown that urban air particles impair vasorelaxation in contrast to TiO_2 NPs.[12] Other ex vivo culture models have been used to account for the complex cellular interactions governing NP responses. Organotypic cultures of precision cut lung slices have been used to evaluate the toxicity of solid lipid NPs used for biomedical applications.[13] To understand the complex interactions between different pulmonary cell types, cocultures of epithelial cells with either dendritic cells, macrophages, endothelial cells, mast cells, neutrophils, monocytes, or fibroblasts have been used to study NM toxicity and uptake. Cocultures of airway epithelial and endothelial cells are, for example, crucial to investigate possible translocation of NMs to the bloodstream. Some culture models using more than two different cell types are even set up to study NM toxicity and uptake.[10] Most of these models are three-dimensional (3D) cultures on porous membranes. These two-compartment chambers also allow studying the transcytosis of NMs through a tight epithelial barrier. Cells can be seeded at the apical and basal sides of the porous membrane as well as on the bottom of the basal chamber. These cell culture systems also allow cultures at air–liquid interface (ALI), which are more realistic to exposure by inhalation. These two compartment culture systems also allow the assessment of translocation through the epithelial barrier in vitro.[14–17] Exposure systems have been developed for these ALI cultures for NPs in liquids[18] or aerosols (Cultex®). ALI cultures also allow long-term cultures of airway epithelial cells, which have already been used to study the effect of repeated exposures to atmospheric particles.[19] ALI cultures allow also the differentiation of primary epithelial cells to obtain a mucociliary epithelium more representative of the airway epithelium. Recently, transgenic models of these differentiated primary cultures have been developed expressing luciferase-based transcription factor reporter functions for high-throughput studies of cell signaling pathway induced by toxicants.[20] Differentiation could also be achieved by spheroid cultures of epithelial cells that could also be used for long-term cultures.[21] Microcarrier cell culture systems could also be used to study long-term effects of anchorage-dependent cells and have been shown to be more sensitive than subculturing of cell lines to study NM toxicity.[22] Three dimensional (3D) tissue mimetic culture models of macrophages were also established to test NMs and compared to classical two-dimensional (2D) cultures.[23] A 3D culture system was developed as a high throughput in vitro lung tumor model to study the effect of various anticancer agents including treatments with NPs.[24] An innovative 3D coculture model of the bronchiole has recently been established using fibroblasts, endothelial cells, smooth muscle cells, and epithelial cells but has not yet been used for NM testing.[25] Even more sophisticated in vitro models for reconstitution of lung functions have been developed. A biomimetic lung-on-a-chip microdevice using compartmentalized micro-channels with flexible membranes allows the imitation of breathing movements.[26] This device was used for cocultures of alveolar epithelial cells at the ALI in proximity to endothelial cells establishing a tight alveolar-capillary barrier that could be submitted to physiological mechanical strain. This system has been used successfully to study the effects and transport of NMs under mechanical strain and neutrophil adhesion upon endothelial activation after ALI exposure of the epithelium to SiO_2 NPs. Even human-on-a-chip devices are nowadays foreseen using microfluidics to connect several organ-on-chip devices that will revolutionize in vitro studies of NMs.

15.2.1 CELLULAR UPTAKE OF NANOPARTICLES

Understanding the mechanisms of NP internalization by lung cells is essential as NP uptake can greatly contribute to NP toxicity especially for insoluble NPs. Furthermore, cellular uptake of NPs could allow them to cross the epithelial barrier leading to their systemic distribution. In recent years, many efforts have been made to decipher the mechanisms underlying the uptake of NPs by cells. A better knowledge of the determinants involved in NP uptake could not only provide clues to create safer NPs but could also improve the development of strategies for NPs used for biomedical applications in drug delivery and imaging. A thorough understanding of the processes involved at the molecular level will, on the one hand, greatly help in the engineering of NPs that do not penetrate cells, which is relevant, for example, for NP-based contrast agents widely used in medical diagnosis. On the other hand, this knowledge is also important for developing NPs designed for the selective uptake by specific cells, for example, for targeted drug delivery. Determining the subcellular localization of NPs is also particularly important for biomedical applications as well as to better understand potential mechanisms of toxicity. Indeed, NPs are developed for transfection of genes or proteins to lung target cells.[1]

Traditionally, NP uptake is studied mostly in in vitro cell cultures grown in 2D environments. Regardless of the cellular model used, internalization is generally studied using light, fluorescence, or electron microscopy and other sophisticated biophysical techniques, such as inductively coupled plasma mass spectroscopy (ICP-MS), particle-induced X-ray emission (PIXE), Raman confocal microscopy, flow cytometry, or imaging flow cytometry. Comparing different available microscopic techniques used to visualize the uptake of NPs by cells, light or fluorescence microscopy can be performed on living cells yielding plenty of data in a relatively short time. However, the optical resolution is limited in standard microscopes, such as wide-field and confocal laser scanning microscopes (CLSM). Ranging in diameter from 1 to 100 nm, NPs do not fall within this limit and thus cannot be optically resolved unless they appear as aggregates or agglomerates, disabling direct absolute quantification of NPs. On the other hand, electron microscopy on fixed cells overcomes this limitation with its superior spatial resolution but brings along its own set of downsides, such as elaborate sample treatment and relatively low throughput. Furthermore, the presence of other cellular and noncellular nano-sized structures in transmission electron microscopy (TEM) cell samples, which may resemble NPs in size, morphology, and electron density, can obstruct the precise intracellular identification of NPs. Therefore, elemental analysis by, for example, energy-dispersive X-ray analysis (EDX) is recommended to confirm the presence of NPs inside the cell. This can also be achieved by energy-filtering TEM that was used to confirm the intracellular NP localization of quantum dots (QDs) for example.[27] Confocal Raman microscopy (CRM) is also a valuable tool to study the uptake of NPs by cells as it combines spontaneous Raman emission with confocal detection. Laser positioning secondary neutral mass spectrometry (laser SNMS) and time-of-flight secondary ion mass spectrometry (TOF-SIMS) have been compared to TEM and confocal Raman microscopy techniques for 3D analysis of NP uptake by macrophages showing that laser SNMS has the highest spatial resolution.[28] PIXE also allows identification of NPs in cells, and micro X-ray fluorescence microscopy (μXRF) and X-ray nano-computed tomography have been compared to study NP uptake in vitro using synchrotron beamlines for high-resolution 3D imaging.[29] As detection of organic NPs in cells and tissues poses a particular challenge, especially for nanomedicine, due to their close similarities to biomolecules, a light microscopic autoradiography (LMA) technique has been established for the tracking of radiolabeled organic

NPs.[30] Radiolabeling of NPs could be achieved by surface attachment, incorporation of radioactive compounds during synthesis, neutron activation, ion beam irradiation by cyclotrons or spark ignition.[31] Scanning electron microscopy (SEM) and atomic force microscopy (AFM) are other high-resolution techniques to study the interactions of NPs at the cell surface.[32] Furthermore, specific physicochemical characteristics of NMs could be used for their detection such as magnetic properties that allow studying the uptake by cell magnetophoresis, for instance, or the light scattering properties of TiO_2 used for their detection by flow cytometry.[33]

A good approach to characterize NP uptake is to use several microscopic techniques to follow membrane interactions, uptake, and subcellular localizations over time.[34,35] The best way to study the uptake of NPs is by combining qualitative approaches that enable visualization of NP internalization with different quantitative techniques that permit the quantification of the uptake.[33] The quantification of NP uptake can be obtained by flow cytometry or using different spectrometric techniques. Flow cytometry has, however, been shown to be more reliable than fluorescence multiplate readers for high-throughput quantification of the uptake of fluorescent NPs.[36] Flow cytometry typically provides good statistical quantitative analysis as a large number of cells can be evaluated very quickly. However, flow cytometry and/or microplate spectrofluorimetry do not yield direct information about the intracellular localization of NPs and thus requires the use of fluorescence quenchers to eliminate the signal of external NPs adsorbed on the cell surface or culture plate to accurately quantify NP uptake. Indeed, most of the studies of NP uptake did not distinguish between adsorbed and internalized NPs. Combining quantitative analysis with localization of NPs can, however, be accomplished using imaging flow cytometry. This technique was used to define the inside of the cell to accurately study the uptake of fluorescently labeled SiO_2 and even nonfluorescent but light-diffracting NPs (TiO_2) in human lung epithelial cells.[33] This high-throughput technique has also allowed quantitative intracellular localization of carbon nanotubes (CNT).[37] One important pitfall of flow cytometric approaches is that the obtained results are not expressed in absolute number of particles, but rather in arbitrary units. To absolutely quantify the uptake of NPs by cells, different spectrometric techniques are available. For example, inductively coupled plasma mass spectrometry (ICP-MS) has been used to determine the quantity of metals associated with the cells treated with NPs. Single-particle ICP-MS is a special adaptation of this method to distinguish signals from NPs and soluble ions.[38] ICP-MS is a destructive method where adsorbed and internalized NPs are considered as a whole leading to a possible overestimation of the quantity of internalized NPs as the spatial information is lost, and NPs that are only adsorbed on the cell surface are also quantified. Determining the number of NPs at the single cell level could be obtained, however, by using laser ablation ICP-MS. Using this technique, Drescher et al. spatially resolved the distribution of silver and gold NPs in fibroblasts.[39] NPs were visualized with respect to cellular substructures and were found to accumulate in the perinuclear region. On the basis of matrix-matched calibration, the authors developed a method for quantification of the number of metal NPs at the single-cell level.

As shown, regardless of the cell type studied, all these techniques used to study the uptake of NPs have different limitations (Table 15.1) and should be chosen depending on NP characteristics (NP size, labeling, physicochemical characteristics, etc.).

TABLE 15.1
Advantages and Pitfalls of Methods for Studying the Internalization of Nanoparticles by Cells

Approach	Method	Advantages	Pitfalls
Qualitative analysis	Light microscopy	• Possibility to analyze living cells • High throughput	• Low optical resolution
	Fluorescence microscopy		• Low optical resolution • Need of NP labeling
	Transmission electron microscopy (TEM)	• Superior spatial resolution	• Low throughput • Need of further confirmation of NPs by elemental analysis
	Scanning electron microscopy (SEM)		
	Atomic force microscopy (AFM)		
	Particle-induced X-ray emission (PIXE)	• Identification of NPs within cells	• Low throughput
	Micro X-ray fluorescence microscopy (µXRF)		
	X-ray nano-computed tomography		
	Light microscopic autoradiography (LMA)		• Low throughput • Need of NP labeling
Quantitative analysis	Inductively coupled plasma mass spectroscopy (ICP-MS)	• Quantification of cell-associated metals	• No direct information about NP localization • Limited to metal NPs
	Laser ablation ICP-MS	• Determination of NP number at the single cell level	• Limited to metal NPs
	Fluorescence microplate readers	• High throughput • Good statistical analysis	• Need of fluorescence quenchers to exclude NPs adsorbed on cellular surface • Relative quantification of NP uptake
	Flow cytometry		
Qualitative and quantitative analysis	Imaging flow cytometry	• High throughput • Distinguishes adsorbed and internalized NPs	• Relative quantification of NP uptake • Need of NP labeling or special physical properties (light scattering/absorption)
	Confocal Raman microscopy	• Distinguishes adsorbed and internalized NPs • Suitable to study nonfluorescent NPs	• Relative quantification of NP uptake

Qualitative approaches (permitting the visualization of NP interactions with cells and their localization), quantitative approaches (enabling relative or absolute quantification of NP uptake by cells), or methods that are combining these two approaches should be used (Table 15.1).

These techniques have allowed showing that the ability of NPs to adhere to and penetrate lung cells depend on their physicochemical properties (Figure 15.2), including size, shape, curvature, surface composition, and surface charge.[32,33,40] These are important factors to be taken into consideration for the design of NPs for biomedical applications for the lungs. Adsorption of biomolecules such as serum proteins and surfactant components on the NP could also modulate their uptake.[33,34,36,41] This is also highly important for the design of NPs for targeted drug delivery, as it could reduce the efficiency of the surface modifications and functionalizations applied to increase their uptake (by grafting receptor ligands on their surface, for example). However, specific surface modifications can also be used to increase mucus penetration of NPs for drug delivery to the lungs as shown, for example, by in vitro studies of functionalized poly(lactic-co-glycolic acid) (PLGA) NPs.[42] Time course studies allowed determination of uptake kinetics and intracellular fate of NPs in lung cells.[33] NP localization and intracellular fate are dependent on NP type and properties, but mostly they move through the endosomal system from early to late endosomes to end up in lysosomes, organelles where they will be digested or stored.[27,33] Rarely NPs are found in other cellular structures such as the endoplasmic reticulum, mitochondria, or nucleus.

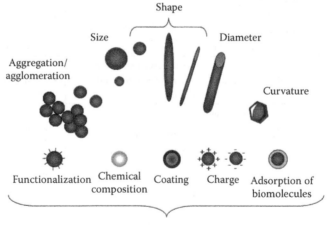

FIGURE 15.2 Nanoparticle (NP) properties influencing their uptake by pulmonary cells. Several physicochemical characteristics of NPs can modulate internalization by cells. Physical properties such as shape, curvature, and size of individual NPs or their agglomerates/aggregates determine NP uptake. Moreover, the surface chemistry including coating, charge, and chemical composition is also essential for internalization. Functionalizations of NPs by grafting antibodies, receptor ligands, or molecules like polyethylene glycol are used in nanomedicine to increase or prevent uptake of NPs. On the other hand, the uncontrolled adsorption of biomolecules such as proteins or lipids present in lung lining fluids could modify NP uptake (drawn using Servier medical art).

These microscopic or analytical techniques also allowed determining the uptake pathways of NPs in different cell types of the lungs. Specialized cells such as macrophages, monocytes, and neutrophils are well described for their ability of NP phagocytosis, a form of endocytosis where the cell engulfs larger particles.[43] On the other hand, almost all cells can internalize NPs by pinocytosis. Four different principal pinocytic mechanisms are currently known: macropinocytosis, clathrin-mediated endocytosis, caveolae-mediated endocytosis, and mechanisms independent of clathrin and/or caveolin.[44] In addition to entering the cells by active processes, NPs may also be internalized by passive penetration through the plasma membrane as NPs are observed within red blood cells or artificial membranes.[32,45,46] Even particles greater than 500 nm in diameter were reported to penetrate cell membranes by inducing strong local membrane deformations.[47]

Since macrophages are "first responders" to all foreign materials, including NPs, mature in vitro murine alveolar macrophage cell lines, such as J774A.1, or human leukemic monocyte macrophage cells, such as RAW264.7 or differentiated THP-1, are commonly used model systems for studying various aspects of NP uptake and potentially related toxicity. These cell lines allow investigation of NP uptake through a variety of possible entry routes, as macrophages take up extracellular material by a wide range of mechanisms. Human lung fibroblasts (e.g., IMR90, WI-38) and different types of epithelial cell lines (e.g., A549, NCI-H292, Calu-3, 16HBE) or primary cells are often used as representative nonphagocytic lung cells for studying the mechanisms of NP uptake. Internalization pathways are often deciphered using pharmacological inhibitors for different endocytic pathways but their specificity and efficiency are often questioned.[33,48] siRNA knockdown approaches have been used to obtain more specific inhibition of one precise pathway.[33,49] In some studies, the conclusion about the internalization pathway is obtained from colocalization studies of NPs with proteins specific for each pathway.[33,50]

Depending on the cell type used, different internalization pathways have been reported for NP uptake by lung cells. These are also highly dependent on the NP characteristics. The size of the NP seems to be the key player in determining the endocytic pathway used by the cells. Larger NPs or aggregates of NPs can be taken up by endocytic processes that lead to the formation of large vesicles (phagocytosis or macropinocytosis). Smaller NPs may also enter the cells by clathrin-mediated endocytosis (which leads to average vesicle sizes between 150 and 200 nm) or caveolae-mediated endocytosis (forming vesicles of less than 100 nm). The uptake is also dependent on other properties of NPs (NP type, shape, surface modifications, etc.). For example, citrate-stabilized gold NPs were shown to be predominantly taken up by macropinocytosis, while polyethylene glycol (PEG)-coated gold NPs were taken up by clathrin- and caveolae-dependent endocytic pathways.[27] Most of the time, macropinocytosis has been identified in nonprofessional phagocytes as the main uptake mechanism of NPs,[27,33,49] but clathrin-mediated endocytosis and even caveolae-mediated endocytosis[27] as well as flotillin-dependent endocytosis[51] have also been reported.

It is evident that for NPs developed for biomedical applications, these issues of characterization of NP uptake, intracellular trafficking, and understanding the endocytic pathways involved are particularly important to optimize clinical strategies for lung disease treatments.

15.2.2 Transcytosis of Nanoparticles through the Respiratory Epithelium

Estimation of internalized NPs is of great interest as NP uptake is the first step leading to the potential translocation of NPs from a target organ to the systemic circulation. Many studies conducted in animals have shown that NPs have the potential to translocate from the site of exposure to the systemic circulation, especially within the lungs. It is essential to completely understand the mechanisms of NP translocation as it could help in the creation of "safer" NPs that do not have the capacity of translocation or it could lead to the production of more efficient NPs for biomedical applications with enhanced capacities of translocation.

In vitro models of mucosal tissues are generally based on epithelial cells cultured as polarized layers on permeable supports in two compartment chambers. This cell culture system allows the quantification of NP translocation to the basal chamber. However, NPs can stick in the pores of the membrane hindering them to reach the basal compartment.[14] This retention of NPs in the porous membrane can thus lead to underestimation of the translocation, which has to be taken into account.

These two compartment chambers also permit to verify the tightness of the epithelial barrier before NP exposure to assess whether this translocation is due to transcytosis through epithelial cells. The functionality of in vitro barriers can be evaluated by measuring the transepithelial electrical resistance (TEER), which indicates efficient tight junctions as well as by analysing the passage of tracer molecules such as the fluorescent dye Lucifer Yellow or radiolabeled mannitol (Figure 15.3). In contrast to immunostaining of tight junctional proteins, these techniques can be used before NP treatment without affecting the cell cultures. Quantification of NPs in the basal compartment can be performed by the same techniques used for NP uptake studies. Depending on the NP nature and labeling, fluorescence measurements, radioactivity counts, or analytical methods such as SP-ICP-MS can be used (Figure 15.3).

The bronchially derived Calu-3 cell line is commonly used as a model of the upper airways in studies of drug absorption across the upper lungs and the nose.[52] In a study of Geys et al., this cell line was compared to other pulmonary epithelial cells (human cell line A549 and primary rat type II pneumocytes) for their capacity to develop a tight epithelium. In contrast to A549 cells, Calu-3 and primary rat type II pneumocytes developed high values of TEER with low passage of sodium fluorescein, which made these cell cultures suitable for studying the potential translocation of NPs.[53] Using different fluorescently labeled polystyrene NPs, the authors observed approximately 6% of translocation of NPs across the Calu-3 epithelium. Using the Calu-3 cell line, George et al. measured the translocation of 50 nm fluorescently labeled SiO_2 NPs across epithelial barriers using confocal microscopy and fluorescence measurements in different culture compartments. After 24 hours of exposure, the authors observed a significant translocation of NPs across the epithelium.[14] Vllasaliu et al. also used this cell line to examine whether surface modification of NPs with the Fc portion of IgG promotes their uptake and transport across airway epithelial cells. This study demonstrated that Fc-modified NPs could shuttle a model therapeutic antibody fragment across the epithelial cell layers.[54] The Calu-3 cell line has also demonstrated good in vitro–in vivo correlation[55] and possesses the capacity to produce mucus,[56] which is advantageous as mucus could influence NP internalization and translocation.

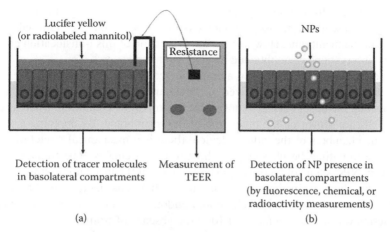

Lucifer yellow
(or radiolabeled mannitol)

Resistance

NPs

Detection of tracer molecules   Measurement of   Detection of NP presence in
in basolateral compartments   TEER   basolateral compartments
(by fluorescence, chemical, or
radioactivity measurements)

(a)   (b)

FIGURE 15.3   Assessment of nanoparticle (NP) transcytosis through the respiratory epithelium. In vitro models of respiratory epithelium are generally based on epithelial cells cultured as polarized layers on permeable supports in two compartment chambers. (a) Two compartment chambers permit to verify the tightness of the epithelial barrier before NP exposure. This is important to assess whether the potential translocation is due to transcytosis through epithelial cells and not because of paracellular passage of NPs. The functionality of in vitro barriers can be evaluated by measuring the transepithelial electrical resistance (TEER), which indicates efficient tight junctions, as well as by studying the passage of tracer molecules such as the fluorescent dye Lucifer Yellow or radiolabeled mannitol from the apical to the basal compartment. (b) Once a tight epithelium is established, cells can be treated with NPs. Quantification of NPs in the basal compartment can be performed by the same techniques described for NP uptake studies. Depending on the NP nature and labeling, fluorescence, radioactivity, or chemical detection methods can be used. (Drawn using Servier Medical Art.)

Primary rat type II pneumocytes[57,58] are also used to study the translocation of NPs through the air–blood barrier. Using this model, no translocation of QD occurred through the intact epithelia regardless of the NP surface charge. A translocation was only observed in the case of disruption of the cell–cell barrier by an oxidant insult.[57] By contrast, trafficking of fluorescently labeled polystyrene NPs was shown across a monolayer of rat alveolar type II cells that was more important for positively charged NPs than negatively charged ones.[58] In addition, this translocation was shown to be transcellular but not through any known major endocytic pathways.

More complex models have been developed for in vitro studies of NP translocation through the air–blood barrier including the most relevant cell types. Triple cell coculture models were established in two compartment cell chambers using human alveolar epithelial cells (A549 cell line or primary human type I cells) or bronchial epithelial cells (16HBE) cocultured with human blood monocyte–derived macrophages on the apical side of the porous membrane and directly exposed to NPs and dendritic cells cultivated on the basal side of the membrane.[17,59] This coculture system enabled to show that dendritic cells extend processes between epithelial cells through the tight junctions to collect NPs in the "luminal space" and to transport them through cytoplasmic processes between epithelial cells across the epithelium or to transmigrate through the epithelium to take up particles on the epithelial surface.

Thus, dendritic cells and macrophages appear to handle in a very effective manner allowing a transcellular, transepithelial transport of NPs from the "luminal side" to the base of the membrane. However, no quantification of this translocation could be made as observations were obtained by electron microscopy.[60]

Huh et al. developed a "lung-on-a-chip" device to study NP translocation.[26] This device is made of compartmentalized microchannels where alveolar epithelial cells are grown on the upper surface of a flexible microporous membrane whereas endothelial cells are grown on the other side of this membrane. By applying a vacuum to the side chambers of the culture device, there is a mechanical stretching of the membrane imitating the physiological breathing movements. The translocation of fluorescent NPs is fourfold more important in case of the mechanical stretching and NPs can be observed in the underlying endothelial cells: more than 70% of cells within mechanically active epithelium and endothelium internalized NPs, whereas this fraction was lower by a factor of 10 in the absence of strain.[26]

Thus, while simple cell culture models are good screening tools to study NP transcytosis through the epithelial cells, more sophisticated coculture systems or lab-on-a-chip devices allow the creation of more physiological conditions. Studies on differentiated bronchial epithelia are also needed for a realistic estimation of translocation at the bronchial level in vivo as ciliary beating and mucus could also prevent the uptake and subsequent transcytosis.

15.2.3 CYTOTOXICITY EVALUATION OF NANOPARTICLES

Various types of cytotoxicity tests have been applied to address potential pulmonary toxicity of NMs. Cytotoxicity methods are usually based on the detection of deficits in the viability of cells and include detection of disturbed plasma membrane integrity or metabolic competence of the cells (e.g., mitochondrial function, cellular ATP content). Loss in membrane integrity can, for instance, be detected by dyes that are selectively taken up by cells with damaged plasma membranes, such as trypan blue and propidium iodide. Alternatively, loss of intracellular constituents may also be used as a measure of membrane integrity loss, such as the enzyme lactate dehydrogenase (LDH). Several assay modifications allowing for the direct or indirect measurement of the amount or activity of this enzyme in cell culture supernatants have been developed. A commonly applied approach to determine the metabolic competence of pulmonary cells in response to NP treatment is the 3-(4,5-dimethylthiazol-2-yl)-2,5-diphenyltetrazolium bromide (MTT) assay. This assay is based on the ability of mitochondrial dehydrogenase enzymes to reduce the yellow tetrazolium salt thiazolyl blue tetrazolium bromide (MTT) into an insoluble purple formazan dye that can be detected spectrophotometrically. Because of the need to dissolve the formazan crystals by dimethylsulfoxide before its detection, an alternative approach has been introduced that is based on the dehydrogenase-based conversion of a related tetrazolium salt, WST-1, into a water-soluble formazan product.

In previous decades, adaptations to these tests and assay principles have been developed by in vitro toxicologists to allow for high-throughput testing approaches. These assays have proven to be highly successful for hazard and risk evaluation of soluble (nonparticulate) chemicals with the uses of multiwell cell culture plates and, for example, spectrophotometry or fluorescence-based detection apparatus. Another technique that has been applied to test NM toxicity toward pulmonary cells is flow cytometry. This approach offers as a further advantage that multiple endpoints can be read simultaneously on the single-cell-based level. As such, for instance, the proportion of apoptotic and necrotic cells can be determined in a cell population with the use of the DNA-staining dyes like propidium iodide or 7-AAD and Annexin V-FITC that specifically binds to phosphatidylserine being presented to the surface of cells that undergo apoptosis.[4] In principle, this method can also be combined with flow cytometry–based uptake assays as described in Section 15.2.1. However, several investigators have highlighted that each of the aforementioned cytotoxicity assays may be prone to artifacts when evaluating the cytotoxicity of NPs, and which consequently can lead to false positive as well as false negative outcomes.[2–4,61–63] For the determination of reliable NM toxicity data, it is therefore important to include positive control samples. Moreover, approaches have been developed and recommended to be included in specific control experiments to address interferences of the tested NM with assay components or readouts. Finally, the concept of using multiple assays in toxicity screening of NM should be encouraged.[3,63]

While cytotoxicity assays were used as major (or even sole) endpoint in toxicity testing and hazard interpretation in "traditional toxicology," it is nowadays widely accepted that pulmonary toxicity, as observed in vivo, can be better explained by in vitro evaluation of cellular responses that take place at subtoxic or nontoxic treatment concentrations. In the field of nanotoxicology, it has emerged that especially the evaluation of the induction of cellular oxidative stress and activation and release of inflammatory mediators can be predictive of the potential toxicity of NMs, as further outlined in Section 15.2.4. Understanding of the mechanisms that drive oxidative stress and proinflammatory properties may also benefit to the development of safer NMs in the field of nanomedicine.

15.2.4 Oxidative Stress Induction by Nanoparticles

Oxidative stress can be defined as the adverse condition resulting from an imbalance in cellular oxidants and antioxidants, with the scale tipping toward the side of the former.[64] There is evidence that oxidative stress induced by particles is the responsible factor involved in the increase of pulmonary inflammation.[65,66] Markedly increased and sustained oxidative stress is implicated in the oxidative attack and damage of crucial cellular (macro)molecules including membrane constituents, proteins, and the genomic DNA.[67] Lower levels of oxidative stress are considered to drive particulate-induced inflammation through activation of redox-sensitive signaling pathways, including nuclear factor-kappa B (NF-κB), activator protein 1 (AP-1), mitogen-activated protein kinases (MAPK), and NF-E2-related factor-2 (Nrf-2), which subsequently drive proinflammatory and proliferative gene expression, or upregulation of antioxidant feedback processes.[67–69]

Oxidative stress by NM in pulmonary cells can originate from several mechanisms. Most importantly, it has been demonstrated that NPs can cause formation of oxidants, which include reactive oxygen species (ROS), such as hydrogen peroxide, hydroxyl radicals, superoxide, and singlet oxygen, as well as reactive nitrogen species including nitric oxide and peroxynitrite. This process of oxidant generation can be principally subdivided into two major mechanisms.[67] On the one hand, specific types of NMs have been shown capable of directly generating oxidants due to their physicochemical properties. On the other hand, it has been demonstrated that specific types of NMs can also trigger oxidant formation from cellular sources upon their interaction with cells. Importantly, this cell-mediated (or cell-dependent) mechanism of oxidant generation can be driven by NMs that do not possess any marked "intrinsic" (i.e., cell-free or cell-independent) oxidant-generating properties. Major cellular sources and compartments that are considered to be involved in cell-dependent oxidant generation include NADPH-oxidase enzyme complexes, mitochondria, and interaction with intracellular calcium stores including the endoplasmic reticulum.[67] Apart from causing increased oxidant generation, specific types of NPs may also trigger the reduction of intracellular antioxidants and as such cause oxidative stress in an indirect manner.

A major cell-dependent mechanism whereby NPs can cause oxidant formation is through the recruitment and activation of professional phagocytic cells including macrophages and neutrophils. The inflammogenic potential of inhaled biopersistent particles, including those in the nano-size range, is considered to be of major importance for the development of subsequent pathologies in the lungs. The excessive and persistent formation of oxidants from activated phagocytes during inflammation is considered to cause pulmonary oxidative stress and cell and tissue remodeling implicated in lung fibrosis and carcinogenesis.[65,70] A particular mechanism of action is considered of key importance for high aspect ratio nanomaterials (HARN), whereby toxicological studies with asbestos have revealed that "frustrated phagocytosis" by long, thin, and biopersistent fibers accelerates chronic inflammatory processes. The exaggerated release of inflammatory mediators including oxidants contribute to profibrotic and mutagenic effects, the latter especially also in the pleura.[65,71]

Various assays have been used to evaluate the oxidant formation by NPs. The oxidant-generating properties of NPs in cell free environment have been considered useful for the prediction of their acute in vivo inflammatory potential in the lungs.[72] Assays to investigate cell-free oxidant generation include measurement of the consumption of dithiothreitol (DTT), the oxidation of the fluorescence dye dichlorofluorescein (DCFH), the reduction of nitroblue tetrazolium (NBT), and electron spin resonance (ESR) spectroscopy.[68,73-77] However, as mentioned above, such approaches cannot identify types of NPs that solely cause oxidant generation in the presence of cells.

Commonly applied methods to determine oxidant generation in pulmonary cells include the use of dyes such as dihydroethidium and dichlorofluorescein diacetate (DCFH-DA), which upon loading into cells can detect ROS with different sensitivity and specificity. This approach allows for the sensitive detection of oxidative stress in lung cells and allows for the detection by, for example, fluorescence microscopy, fluorescence multiwell plate reader, or flow cytometry–based approaches.[78,79] Approaches that have been applied for several decades in the field of particle toxicology involve evaluation of the "respiratory burst" from particle-treated macrophages or neutrophils by lucigenin or luminol-enhanced chemiluminescence or measurement of hydrogen peroxide generation.[80,81] Indirect markers of oxidant generation, which have been used for several decades in oxidative stress research, include measurement of the formation of the lipid peroxidation product malondialdehyde or changes in the amount of the activities of intracellular antioxidants including reduced glutathione, superoxide dismutase, or glutathione peroxidase.[82,83] In more recent times, mRNA expression analysis also has been introduced, specifically of genes that are known to be upregulated in response to cellular oxidant attack and changes in intracellular redox status, such as the Nrf-2, heme oxygenase 1 (HO-1), NAD(P)H dehydrogenase quinone 1, or gamma-glutamylcysteine synthetase genes.[68,77,78,84]

Another relevant marker of intracellular oxidant generation is the formation of the oxidative DNA lesion 8-hydroxydeoxyguanosine. Apart from being a premutagenic lesion and hence used to address the genotoxic potential of NMs, it can also be used as a measure of oxidative stress induction. The formation of this oxidative adduct can be measured in many different ways including immunohistochemistry, high-performance liquid chromatography/electrochemical detection, and the Fpg-modified comet assay.[77,85]

In summary, various in vitro methods can be used to evaluate whether NPs can cause oxidative stress in pulmonary cells. Similar to toxicity testing, great care should also be taken here to avoid artifacts. As with specific components of cytotoxicity assays, binding or chemical modification may also occur with probes that are used to evaluate ROS formation.[63] It can also be recommended here to use multiple assays for oxidative stress detection, not only to avoid false interpretation but also to evaluate potential sources and/or subcellular locations of oxidant generation. For example, TiO_2 NPs, unlike fine TiO_2 particles, were found to cause enhanced extracellular ROS generation from NR8383 macrophages as detected by ESR, enhanced mRNA expression of HO-1, and release of the inflammatory cytokine tumor necrosis factor (TNF)-α release while intracellular ROS could be detected by DCFH-DA assay with both types of particles.[78]

15.2.5 PROINFLAMMATORY RESPONSE INDUCED BY NANOPARTICLES IN PULMONARY CELLS

Pulmonary deposition of inhaled, low-toxic NPs such as made from carbon or TiO_2 has been shown to trigger inflammatory reactions in the lungs and in the alveolar region,[86,87] and this inflammatory response is considered to be crucial for adverse health effects related to inhalation of high particle concentrations.[88] In fact, pulmonary inflammatory responses have been identified to be a sensitive measure for particle inhalation, and blocking this reaction by anti-inflammatory drugs can reduce related health effects.[89] For this reason, characterizing the proinflammatory potency of NPs can be used to assess their toxic potential.[83,90] In related classical toxicological studies, rodents are exposed to NPs by inhalation or instillation (delivery by intratracheal injection into the lungs), but animal experiments are costly and always need sufficient ethical justification. For this reason, well-defined exposure of lung cells to particles has evolved as an alternative method to assess the inflammatory and toxicological potential of new materials. In this case, not inflammatory cell recruitment but the release or expression of inflammatory mediators by the exposed, immunocompetent cells is analyzed and used as readout for "inflammation."[91]

Despite the obvious advantages of avoiding expensive animal experiments and the possibility to establish high-throughput screenings for the cellular responsiveness, limitations of in vitro exposure systems have to be taken into account when aiming to model complex pulmonary responses. In general, it has to be noted that relatively high particle doses—given as mass per volume medium (µg/mL) or mass per surface area of the culture plate (µg/cm²)—have to be used to achieve significant proinflammatory responses in the particle-exposed cells. For example, when exposing the frequently used human alveolar epithelial-like cell line, A549 to zinc oxide (ZnO) NPs, a dose of 5 µg/cm² is required to generate a significant (10-fold) increase in proinflammatory IL-8 cytokine gene expression.[18] In vitro however, already much lower doses are sufficient to cause an inflammatory response in the lungs. For ZnO, an intratracheal instillation of 10 µg into the lungs of a mouse, which equals to a dose of only 1 ng/cm² (assuming a surface area of 0.1 m² for a mouse lung), results in severe pulmonary inflammation, characterized by high alveolar chemokine (C-X-C motif) ligand 1 (Cxcl1; a murine IL-8 homologue) cytokine levels and subsequent accumulation of neutrophils. The in vitro dosimetry of NPs is in general a difficult task for cell exposures, in particular when conventional, submerged cultures are used where NP diffusion and sedimentation in the medium supernatant determine the bioavailable dose for the cells. This problem can be mostly bypassed when using an air–liquid exposure scenario, which is more realistic for the lungs anyway.[18]

Another peculiarity to be incorporated in the experimental design of in vitro toxicity investigations of new NMs is the selection of cells to be exposed. Since alveolar deposited NPs will interact with alveolar epithelial cells and alveolar macrophages, both cell types should be employed. In fact some NPs might show cell-specific toxicity that would be overlooked if the wrong cell type was selected. Figure 15.4 shows an example where a murine alveolar macrophage cell line (MHS) as well as a murine alveolar epithelial cell line (LA4) have been exposed to various concentrations of ZnO NPs or CNT. While the cell viability, assessed by the colorimetric WST-1 assay, shows for both cell lines a comparable decrease with increasing ZnO NP concentrations (Figure 15.4a), the macrophages seem much more sensitive to the CNT as compared to the epithelial cells (Figure 15.4b). A reason for this peculiarity could be related to the cellular uptake.

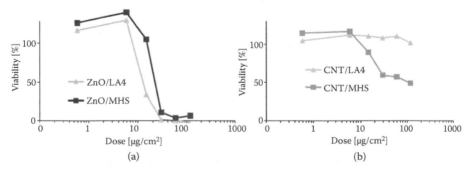

FIGURE 15.4 Cytotoxicity of zinc oxide nanoparticles and carbon nanotubes. Cell viability of murine alveolar epithelial cells (LA4) and macrophages (MHS) exposed for 24 hours to various concentrations of ZnO NPs (a) and carbon nanotubes (b). Cell viability was quantified by the colorimetric WST-1 assay.

Once the size of the particles exceeds the size that can be taken up by endocytosis, this material will not effectively enter the cells. Macrophages as professional phagocytes, however, might internalize individual NMs and their agglomerates by phagocytosis and thus be more susceptible to the toxicity of larger materials. On the molecular level, the crucial role of phagocytic cell activity for the development of the toxicity of needle-shaped inhaled materials has initially been worked out for quartz and asbestos fibers.[92] During this process of failed target engulfment—so-called frustrated phagocytosis—dying macrophages release high levels of inflammatory mediators from their phagolysosomes into the environment. One of these mediators is the master cytokine IL-1 beta, which is known for its pivotal role in initiating the innate immune response in vertebrates. Also for CNT, frustrated phagocytosis has recently been described for rigid and long fibers, and again this mechanism is key to their toxicity.[71] When investigating the inflammatory potential of NPs, cytokine protein release into the medium or cytokine gene expression analysis from exposed cells is commonly performed.

In aiming to choose the dose range for investigation, the viability profile should be considered since unspecific stimulations can be expected at high doses where significant cell death and necrosis contribute to the inflammatory stimulation. Again more than one cell type and several inflammatory markers will have to be analyzed to not overlook specific inflammatory responses. Figure 15.5 shows the gene expression response from two different genes, the mouse IL-8 orthologue Cxcl1 and TNF-α after 24 hours of ZnO or CNT exposure of LA4 epithelial cells and MHS

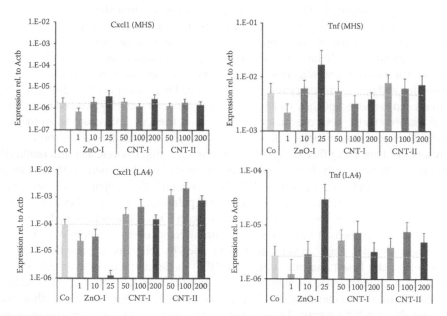

FIGURE 15.5 Proinflammatory response to zinc oxide nanoparticles or carbon nanotubes. Gene expression response of chemokine (C-X-C motif) ligand 1 (Cxcl1) and tumor necrosis factor (Tnf) alpha 24 hours after exposure of murine alveolar epithelial cells (LA4) and macrophages (MHS) to 1–10 µg/cm^2 ZnO NPs or 50–200 µg/cm^2 carbon nanotubes. Gene expression was analyzed by quantitative PCR and normalized to the housekeeping gene beta actin (Actb) and error bars indicate the SEM ($n = 4$).

macrophages. As seen from expression values of the controls, the basal expression of Cxcl1 is almost 100-fold higher in epithelial cells, whereas the inflammatory master cytokine TNF-α is 1000-fold higher expressed in macrophages. For low ZnO NP doses, no inflammatory response could be detected, but high ZnO concentrations of 25 μg/cm², which caused already significant cell death (Figure 15.4a), repressed the epithelial Cxcl1 but stimulated macrophage and epithelial TNF-α expression. CNTs, however, which revealed most toxic to macrophages caused moderate inflammatory gene expression in epithelial cells only. Since both NMs are well known to cause acute inflammatory responses in the lungs, caution may be advised to not overinterpret in vitro data from only few cell types and gene markers.

In summary, many studies have shown that in vitro–based assays can be used to assess the inflammatory potential of engineered NPs, but several other publications also remind about the differences between in vitro and in vivo results and criticize the often used extraordinary high doses necessary to achieve significant responses. For this reason, a confirmation of in vitro result by in vivo exposures is advisable to avoid false negative or positive results as far as possible.

15.3 CONCLUSION

To accurately assess the pulmonary toxicity of NMs in vitro, several endpoints should be studied. The internalization of NMs is an important phenomenon as uptake may not only be a prerequisite for their cytotoxicity but is also the first step for translocation through the epithelial barrier. This transcytosis through the airway epithelium could be evaluated by the use of two compartment cell culture systems. Not only cytotoxicity should be studied as cellular responses that may drive pulmonary disease can already be induced at subtoxic concentrations. Investigations of endpoints such as oxidative stress induction and proinflammatory responses are needed to predict possible adverse health effects and to understand their mechanisms. Each endpoint should be studied using several methods with special attention on possible interferences with assays to avoid artifacts. The use of more complex cell culture models such as cocultures, ALI exposures in presence of lung lining fluids, physiological lung-on-a-chip devices, and differentiated primary cells, which can also be used for long-term studies, allow more realistic in vitro assessments of pulmonary effects of NPs. However, high-throughput screenings using simpler culture models are also needed. Different pulmonary cell types should be investigated to not overlook specific responses, and the in vitro assessment should be based on in vivo observations or be confirmed in animals to not overinterpret in vitro data.

These obtained data can subsequently be used by modeling approaches such as physiologically based pharmacokinetic (PBPK) modeling to allow extrapolation to humans and quantitative structure–activity relationship (QSAR)–like approaches could be used to determine the key physicochemical characteristic(s) that are responsible for NP toxicity. These types of modeling were used, for instance, in the European 7th Framework project ENPRA, which combined in vitro and in vivo dose-response studies of different NPs with varying properties to exposure assessment to develop probabilistic models for risk assessment of NPs.

The determination of NP features responsible for their toxicity is crucially needed to allow their grouping. Indeed, due to the diversity of NPs, their toxic evaluation could not be performed on a single NP base. Numerous NPs are under development for nanomedical applications especially for the lungs[1] and robust and reliable screening approaches of their pulmonary toxicity are needed to design safer NPs. However, some of the observed consequences of NP exposures could also be beneficial for biomedical applications such as their uptake, transcytosis, and bactericidal effects and have given the founding to nanomedicine.

ACKNOWLEDGMENTS

This work has been supported by the ENPRA project, No. 228789, funded by the EC Seventh Framework Programme theme FP7-NMP-2008-1.3-2.

REFERENCES

1. Boland, S.; Guadagnini, R.; Baeza-Squiban, A.; Hussain, S.; Marano, F., Nanoparticles used in medical applications for the lung: hopes for nanomedicine and fears for nanotoxicity. *J Phy: Conference Series* 2011, 304 012031, 9.

2. Guadagnini, R.; Halamoda, B.; Cartwright, L.; Pojana, G.; Magdolenova, Z.; Bilaničová, D.; Saunders et al., Toxicity screenings of nanomaterials: challenges due to interference with assay processes and components of classic in vitro tests. *Nanotoxicol* 2013. Doi:10.3109/17435390.2013.829590.

3. Monteiro-Riviere, N. A.; Inman, A. O.; Zhang, L. W., Limitations and relative utility of screening assays to assess engineered nanoparticle toxicity in a human cell line. *Toxicol Appl Pharmacol* 2009, 234, 222–35.

4. Wilhelmi, V.; Fischer, U.; van Berlo, D.; Schulze-Osthoff, K.; Schins, R. P.; Albrecht, C., Evaluation of apoptosis induced by nanoparticles and fine particles in RAW 264.7 macrophages: facts and artefacts. *Toxicol In Vitro* 2012, 26, 323–34.

5. Val, S.; Hussain, S.; Boland, S.; Hamel, R.; Baeza-Squiban, A.; Marano, F., Carbon black and titanium dioxide nanoparticles induce pro-inflammatory responses in bronchial epithelial cells: need for multiparametric evaluation due to adsorption artifacts. *Inhal Toxicol* 2009, 21 Suppl 1, 115–22.

6. Damoiseaux, R.; George, S.; Li, M.; Pokhrel, S.; Ji, Z.; France, B.; Xia, T. et al., No time to lose—high throughput screening to assess nanomaterial safety. *Nanoscale* 2011, 3, 1345–60.

7. Blank, F.; Wehrli, M.; Lehmann, A.; Baum, O.; Gehr, P.; von Garnier, C.; Rothen-Rutishauser, B. M., Macrophages and dendritic cells express tight junction proteins and exchange particles in an in vitro model of the human airway wall. *Immunobiology* 2011, 216, 86–95.

8. Armand, L.; Dagouassat, M.; Belade, E.; Simon-Deckers, A.; Le Gouvello, S.; Tharabat, C.; Duprez, C. et al., Titanium dioxide nanoparticles induce matrix metalloprotease 1 in human pulmonary fibroblasts partly via an interleukin-1β-dependent mechanism. *Am J Respir Cell Mol Biol* 2013, 48, 354–63.

9. Val, S.; Belade, E.; George, I.; Boczkowski, J.; Baeza-Squiban, A., Fine PM induce airway MUC5AC expression through the autocrine effect of amphiregulin. *Arch Toxicol* 2012, 86, 1851–9.

10. Napierska, D.; Quarck, R.; Thomassen, L. C.; Lison, D.; Martens, J. A.; Delcroix, M.; Nemery, B.; Hoet, P. H., Amorphous silica nanoparticles promote monocyte adhesion to human endothelial cells: size-dependent effect. *Small* 2013, 9, 430–8.

11. Courtois, A.; Andujar, P.; Ladeiro, Y.; Ducret, T.; Rogerieux, F.; Lacroix, G.; Baudrimont, I. et al., Effect of engineered nanoparticles on vasomotor responses in rat intrapulmonary artery. *Toxicol Appl Pharmacol* 2010, 245, 203–10.

12. Courtois, A.; Andujar, P.; Ladeiro, Y.; Baudrimont, I.; Delannoy, E.; Leblais, V.; Begueret, H. et al., Impairment of NO-dependent relaxation in intralobar pulmonary arteries: comparison of urban particulate matter and manufactured nanoparticles. *Environ Health Perspect* 2008, 116, 1294–9.

13. Nassimi, M.; Schleh, C.; Lauenstein, H. D.; Hussein, R.; Hoymann, H. G.; Koch, W.; Pohlmann, G et al., A toxicological evaluation of inhaled solid lipid nanoparticles used as a potential drug delivery system for the lung. *Eur J Pharm Biopharm* 2010, 75, 107–16.

14. George, I.; Vranic, S.; Boland, S.; Borot, M.-C.; Marano, F.; Baeza-Squiban, A., Translocation of SiO2-NPs across in vitro human bronchial epithelial monolayer. *J Phys:Conference Series* 2013, 429, 012022, 1–10.

15. Geys, J.; Nemery, B.; Hoet, P. H., Optimisation of culture conditions to develop an in vitro pulmonary permeability model. *Toxicol In Vitro* 2007, 21, 1215–9.

16. Yacobi, N. R.; Demaio, L.; Xie, J.; Hamm-Alvarez, S. F.; Borok, Z.; Kim, K. J.; Crandall, E. D., Polystyrene nanoparticle trafficking across alveolar epithelium. *Nanomedicine* 2008, 4, 139–45.

17. Rothen-Rutishauser, B. M.; Kiama, S. G.; Gehr, P., A three-dimensional cellular model of the human respiratory tract to study the interaction with particles. *Am J Respir Cell Mol Biol* 2005, 32, 281–9.

18. Lenz, A. G.; Karg, E.; Lentner, B.; Dittrich, V.; Brandenberger, C.; Rothen-Rutishauser, B.; Schulz, H.; Ferron, G. A.; Schmid, O., A dose-controlled system for air-liquid interface cell exposure and application to zinc oxide nanoparticles. *Part Fibre Toxicol* 2009, 6, 32.

19. Boublil, L.; Assémat, E.; Borot, M. C.; Boland, S.; Martinon, L.; Sciare, J.; Baeza-Squiban, A., Development of a repeated exposure protocol of human bronchial epithelium in vitro to study the long-term effects of atmospheric particles. *Toxicol In Vitro* 2013, 27, 533–42.

20. Mankus, C.; Jackson, G.; Bolmarcich, J.; Hayden, P.; Klausner, M., Development of in vitro organotypic 3D epithelial models for high throughput screening of toxicological, immunological and developmental signaling pathways. *Toxicologist* 132S, 485, #2273.

21. Laoukili, J.; Perret, E.; Middendorp, S.; Houcine, O.; Guennou, C.; Marano, F.; Bornens, M.; Tournier, F., Differential expression and cellular distribution of centrin isoforms during human ciliated cell differentiation in vitro. *J Cell Sci* 2000, 113 (Pt 8), 1355–64.

22. Mrakovcic, M.; Absenger, M.; Riedl, R.; Smole, C.; Roblegg, E.; Fröhlich, L. F.; Fröhlich, E., Assessment of long-term effects of nanoparticles in a microcarrier cell culture system. *PLoS One* 2013, 8, e56791.

23. Movia, D.; Prina-Mello, A.; Bazou, D.; Volkov, Y.; Giordani, S., Screening the cytotoxicity of single-walled carbon nanotubes using novel 3D tissue-mimetic models. *ACS Nano* 2011, 5, 9278–90.

24. Godugu, C.; Patel, A. R.; Desai, U.; Andey, T.; Sams, A.; Singh, M., AlgiMatrix™ based 3D cell culture system as an in-vitro tumor model for anticancer studies. *PLoS One* 2013, 8, e53708.

25. Tseng, H.; Gage, J. A.; Raphael, R. M.; Moore, R. H.; Killian, T. C.; Grande-Allen, K. J.; Souza, G. R., Assembly of a three-dimensional multitype bronchiole coculture model using magnetic levitation. *Tissue Eng Part C Methods* 2013, 19(9), 665–75.

26. Huh, D.; Matthews, B. D.; Mammoto, A.; Montoya-Zavala, M.; Hsin, H. Y.; Ingber, D. E., Reconstituting organ-level lung functions on a chip. *Science* 2010, 328, 1662–8.

27. Brandenberger, C.; Mühlfeld, C.; Ali, Z.; Lenz, A. G.; Schmid, O.; Parak, W. J.; Gehr, P.; Rothen-Rutishauser, B., Quantitative evaluation of cellular uptake and trafficking of plain and polyethylene glycol-coated gold nanoparticles. *Small* 2010, 6, 1669–78.

28. Haase, A.; Arlinghaus, H. F.; Tentschert, J.; Jungnickel, H.; Graf, P.; Mantion, A.; Draude, F. et al., Application of laser postionization secondary neutral mass spectrometry/time-of-flight secondary ion mass spectrometry in nanotoxicology: visualization of nanosilver in human macrophages and cellular responses. *ACS Nano* 2011, 5, 3059–68.

29. Astolfo, A.; Arfelli, F.; Schültke, E.; James, S.; Mancini, L.; Menk, R. H., A detailed study of gold-nanoparticle loaded cells using X-ray based techniques for cell-tracking applications with single-cell sensitivity. *Nanoscale* 2013, 5, 3337–45.
30. Holzhausen, C.; Gröger, D.; Mundhenk, L.; Welker, P.; Haag, R.; Gruber, A. D., Tissue and cellular localization of nanoparticles using (35)S labeling and light microscopic autoradiography. *Nanomedicine* 2013, 9, 465–8.
31. Gibson, N.; Holzwarth, U.; Abbas, K.; Simonelli, F.; Kozempel, J.; Cydzik, I.; Cotogno, G. et al., Radiolabelling of engineered nanoparticles for in vitro and in vivo tracing applications using cyclotron accelerators. *Arch Toxicol* 2011, 85, 751–73.
32. Roiter, Y.; Ornatska, M.; Rammohan, A. R.; Balakrishnan, J.; Heine, D. R.; Minko, S., Interaction of lipid membrane with nanostructured surfaces. *Langmuir* 2009, 25, 6287–99.
33. Vranic, S.; Boggetto, N.; Contremoulins, V.; Mornet, S.; Reinhardt, N.; Marano, F.; Baeza-Squiban, A.; Boland, S., Deciphering the mechanisms of cellular uptake of engineered nanoparticles by accurate evaluation of internalization using imaging flow cytometry. *Part Fibre Toxicol* 2013, 10, 2.
34. Lesniak, A.; Fenaroli, F.; Monopoli, M. P.; Åberg, C.; Dawson, K. A.; Salvati, A., Effects of the presence or absence of a protein corona on silica nanoparticle uptake and impact on cells. *ACS Nano* 2012, 6, 5845–57.
35. Bregar, V. B.; Lojk, J.; Suštar, V.; Veranic, P.; Pavlin, M., Visualization of internalization of functionalized cobalt ferrite nanoparticles and their intracellular fate. *Int J Nanomedicine* 2013, 8, 919–31.
36. Vranic, S.; Garcia-Verdugo, I.; Darnis, C.; Sallenave, J. M.; Boggetto, N.; Marano, F.; Boland, S.; Baeza-Squiban, A., Internalization of SiO_2 nanoparticles by alveolar macrophages and lung epithelial cells and its modulation by the lung surfactant substitute Curosurf(®). *Environ Sci Pollut Res Int* 2013, 20, 2761–70.
37. Marangon, I.; Boggetto, N.; Ménard-Moyon, C.; Venturelli, E.; Béoutis, M. L.; Péchoux, C.; Luciani, N.; Wilhelm, C.; Bianco, A.; Gazeau, F., Intercellular carbon nanotube translocation assessed by flow cytometry imaging. *Nano Lett* 2012, 12, 4830–7.
38. Pergantis, S. A.; Jones-Lepp, T. L.; Heithmar, E. M., Hydrodynamic chromatography online with single particle-inductively coupled plasma mass spectrometry for ultratrace detection of metal-containing nanoparticles. *Anal Chem* 2012, 84, 6454–62.
39. Drescher, D.; Giesen, C.; Traub, H.; Panne, U.; Kneipp, J.; Jakubowski, N., Quantitative imaging of gold and silver nanoparticles in single eukaryotic cells by laser ablation ICP-MS. *Anal Chem* 2012, 84, 9684–8.
40. Andersson, P. O.; Lejon, C.; Ekstrand-Hammarström, B.; Akfur, C.; Ahlinder, L.; Bucht, A.; Osterlund, L., Polymorph- and size-dependent uptake and toxicity of TiO_2 nanoparticles in living lung epithelial cells. *Small* 2011, 7, 514–23.
41. Kendall, M.; Ding, P.; Mackay, R. M.; Deb, R.; McKenzie, Z.; Kendall, K.; Madsen, J.; Clark, H., Surfactant protein D (SP-D) alters cellular uptake of particles and nanoparticles. *Nanotoxicology* 2013, 7(5), 963–73.
42. Mura, S.; Hillaireau, H.; Nicolas, J.; Kerdine-Römer, S.; Le Droumaguet, B.; Deloménie, C.; Nicolas, V.; Pallardy, M.; Tsapis, N.; Fattal, E., Biodegradable nanoparticles meet the bronchial airway barrier: how surface properties affect their interaction with mucus and epithelial cells. *Biomacromolecules* 2011, 12, 4136–43.
43. Dobrovolskaia, M. A.; McNeil, S. E., Immunological properties of engineered nanomaterials. *Nat Nanotechnol* 2007, 2, 469–78.
44. Conner, S. D.; Schmid, S. L., Regulated portals of entry into the cell. *Nature* 2003, 422, 37–44.
45. Wang, A. Z.; Langer, R.; Farokhzad, O. C., Nanoparticle delivery of cancer drugs. *Annu Rev Med* 2012, 63, 185–98.

46. Rothen-Rutishauser, B. M.; Schürch, S.; Haenni, B.; Kapp, N.; Gehr, P., Interaction of fine particles and nanoparticles with red blood cells visualized with advanced microscopic techniques. *Environ Sci Technol* 2006, 40, 4353–9.

47. Zhao, Y.; Sun, X.; Zhang, G.; Trewyn, B. G.; Slowing, I. I.; Lin, V. S., Interaction of mesoporous silica nanoparticles with human red blood cell membranes: size and surface effects. *ACS Nano* 2011, 5, 1366–75.

48. dos Santos, T.; Varela, J.; Lynch, I.; Salvati, A.; Dawson, K. A., Effects of transport inhibitors on the cellular uptake of carboxylated polystyrene nanoparticles in different cell lines. *PLoS One* 2011, 6, e24438.

49. Meng, H.; Yang, S.; Li, Z.; Xia, T.; Chen, J.; Ji, Z.; Zhang, H. et al., Aspect ratio determines the quantity of mesoporous silica nanoparticle uptake by a small GTPase-dependent macropinocytosis mechanism. *ACS Nano* 2011, 5, 4434–47.

50. Rejman, J.; Oberle, V.; Zuhorn, I. S.; Hoekstra, D., Size-dependent internalization of particles via the pathways of clathrin- and caveolae-mediated endocytosis. *Biochem J* 2004, 377, 159–69.

51. Kasper, J.; Hermanns, M. I.; Bantz, C.; Koshkina, O.; Lang, T.; Maskos, M.; Pohl, C.; Unger, R. E.; Kirkpatrick, C. J., Interactions of silica nanoparticles with lung epithelial cells and the association to flotillins. *Arch Toxicol* 2013, 87(6), 1053–65.

52. Amoako-Tuffour, M.; Yeung, P. K.; Agu, R. U., Permeation of losartan across human respiratory epithelium: an in vitro study with Calu-3 cells. *Acta Pharm* 2009, 59, 395–405.

53. Geys, J.; Coenegrachts, L.; Vercammen, J.; Engelborghs, Y.; Nemmar, A.; Nemery, B.; Hoet, P. H., In vitro study of the pulmonary translocation of nanoparticles: a preliminary study. *Toxicol Lett* 2006, 160, 218–26.

54. Vllasaliu, D.; Alexander, C.; Garnett, M.; Eaton, M.; Stolnik, S., Fc-mediated transport of nanoparticles across airway epithelial cell layers. *J Control Release* 2012, 158, 479–86.

55. Sakagami, M., In vivo, in vitro and ex vivo models to assess pulmonary absorption and disposition of inhaled therapeutics for systemic delivery. *Adv Drug Deliv Rev* 2006, 58, 1030–60.

56. Grainger, C. I.; Greenwell, L. L.; Lockley, D. J.; Martin, G. P.; Forbes, B., Culture of Calu-3 cells at the air interface provides a representative model of the airway epithelial barrier. *Pharm Res* 2006, 23, 1482–90.

57. Geys, J.; De Vos, R.; Nemery, B.; Hoet, P. H., In vitro translocation of quantum dots and influence of oxidative stress. *Am J Physiol Lung Cell Mol Physiol* 2009, 297, L903–11.

58. Yacobi, N. R.; Malmstadt, N.; Fazlollahi, F.; DeMaio, L.; Marchelletta, R.; Hamm-Alvarez, S. F.; Borok, Z.; Kim, K. J.; Crandall, E. D., Mechanisms of alveolar epithelial translocation of a defined population of nanoparticles. *Am J Respir Cell Mol Biol* 2010, 42, 604–14.

59. Lehmann, A. D.; Daum, N.; Bur, M.; Lehr, C. M.; Gehr, P.; Rothen-Rutishauser, B. M., An in vitro triple cell co-culture model with primary cells mimicking the human alveolar epithelial barrier. *Eur J Pharm Biopharm* 2011, 77, 398–406.

60. Blank, F.; Rothen-Rutishauser, B.; Gehr, P., Dendritic cells and macrophages form a transepithelial network against foreign particulate antigens. *Am J Respir Cell Mol Biol* 2007, 36, 669–77.

61. Wörle-Knirsch, J. M.; Pulskamp, K.; Krug, H. F., Oops they did it again! Carbon nanotubes hoax scientists in viability assays. *Nano Lett* 2006, 6, 1261–8.

62. Samberg, M. E.; Orndorff, P. E.; Monteiro-Riviere, N. A., Antibacterial efficacy of silver nanoparticles of different sizes, surface conditions and synthesis methods. *Nanotoxicology* 2011, 5, 244–53.

63. Stone, V.; Johnston, H.; Schins, R. P., Development of in vitro systems for nanotoxicology: methodological considerations. *Crit Rev Toxicol* 2009, 39, 613–26.

64. Sies, H., Oxidative stress: oxidants and antioxidants. *Exp Physiol* 1997, 82, 291–5.
65. Donaldson, K.; Tran, C. L., Inflammation caused by particles and fibers. *Inhal Toxicol* 2002, 14, 5–27.
66. Xiao, G. G.; Wang, M.; Li, N.; Loo, J. A.; Nel, A. E., Use of proteomics to demonstrate a hierarchical oxidative stress response to diesel exhaust particle chemicals in a macrophage cell line. *J Biol Chem* 2003, 278, 50781–90.
67. Unfried, K.; Albrecht, C.; Klotz, L.-O.; Mikecz, A. V.; Grether-Beck, S.; Schins, R. P. F., Cellular responses to nanoparticles: target structures and mechanisms. *Nanotoxicology* 2007, 1, 52–71.
68. Xia, T.; Kovochich, M.; Liong, M.; Mädler, L.; Gilbert, B.; Shi, H.; Yeh, J. I.; Zink, J. I.; Nel, A. E., Comparison of the mechanism of toxicity of zinc oxide and cerium oxide nanoparticles based on dissolution and oxidative stress properties. *ACS Nano* 2008, 2, 2121–34.
69. Unfried, K.; Sydlik, U.; Bierhals, K.; Weissenberg, A.; Abel, J., Carbon nanoparticle-induced lung epithelial cell proliferation is mediated by receptor-dependent Akt activation. *Am J Physiol Lung Cell Mol Physiol* 2008, 294, L358–67.
70. Driscoll, K. E.; Deyo, L. C.; Carter, J. M.; Howard, B. W.; Hassenbein, D. G.; Bertram, T. A., Effects of particle exposure and particle-elicited inflammatory cells on mutation in rat alveolar epithelial cells. *Carcinogenesis* 1997, 18, 423–30.
71. Donaldson, K.; Murphy, F. A.; Duffin, R.; Poland, C. A., Asbestos, carbon nanotubes and the pleural mesothelium: a review of the hypothesis regarding the role of long fibre retention in the parietal pleura, inflammation and mesothelioma. *Part Fibre Toxicol* 2010, 7, 5.
72. Rushton, E. K.; Jiang, J.; Leonard, S. S.; Eberly, S.; Castranova, V.; Biswas, P.; Elder, A. et al., Concept of assessing nanoparticle hazards considering nanoparticle dosemetric and chemical/biological response metrics. *J Toxicol Environ Health A* 2010, 73, 445–61.
73. Foucaud, L.; Wilson, M. R.; Brown, D. M.; Stone, V., Measurement of reactive species production by nanoparticles prepared in biologically relevant media. *Toxicol Lett* 2007, 174, 1–9.
74. Wardman, P., Fluorescent and luminescent probes for measurement of oxidative and nitrosative species in cells and tissues: progress, pitfalls, and prospects. *Free Radic Biol Med* 2007, 43, 995–1022.
75. Papageorgiou, I.; Brown, C.; Schins, R.; Singh, S.; Newson, R.; Davis, S.; Fisher, J.; Ingham, E.; Case, C. P., The effect of nano- and micron-sized particles of cobalt-chromium alloy on human fibroblasts in vitro. *Biomaterials* 2007, 28, 2946–58.
76. Fenoglio, I.; Greco, G.; Tomatis, M.; Muller, J.; Raymundo-Piñero, E.; Béguin, F.; Fonseca, A.; Nagy, J. B.; Lison, D.; Fubini, B., Structural defects play a major role in the acute lung toxicity of multiwall carbon nanotubes: physicochemical aspects. *Chem Res Toxicol* 2008, 21, 1690–7.
77. Gerloff, K.; Fenoglio, I.; Carella, E.; Kolling, J.; Albrecht, C.; Boots, A. W.; Förster, I.; Schins, R. P., Distinctive toxicity of TiO2 rutile/anatase mixed phase nanoparticles on Caco-2 cells. *Chem Res Toxicol* 2012, 25, 646–55.
78. Scherbart, A. M.; Langer, J.; Bushmelev, A.; van Berlo, D.; Haberzettl, P.; van Schooten, F. J.; Schmidt, A. M.; Rose, C. R.; Schins, R. P.; Albrecht, C., Contrasting macrophage activation by fine and ultrafine titanium dioxide particles is associated with different uptake mechanisms. *Part Fibre Toxicol* 2011, 8, 31.
79. Hussain, S.; Thomassen, L. C.; Ferecatu, I.; Borot, M. C.; Andreau, K.; Martens, J. A.; Fleury, J.; Baeza-Squiban, A.; Marano, F.; Boland, S., Carbon black and titanium dioxide nanoparticles elicit distinct apoptotic pathways in bronchial epithelial cells. *Part Fibre Toxicol* 2010, 7, 10.
80. Faulkner, K.; Fridovich, I., Luminol and lucigenin as detectors for O2.-. *Free Radic Biol Med* 1993, 15, 447–51.

81. Dikalov, S.; Griendling, K. K.; Harrison, D. G., Measurement of reactive oxygen species in cardiovascular studies. *Hypertension* 2007, 49, 717–27.

82. Beck-Speier, I.; Dayal, N.; Karg, E.; Maier, K. L.; Schumann, G.; Schulz, H.; Semmler, M. et al., Oxidative stress and lipid mediators induced in alveolar macrophages by ultrafine particles. *Free Radic Biol Med* 2005, 38, 1080–92.

83. Monteiller, C.; Tran, L.; MacNee, W.; Faux, S.; Jones, A.; Miller, B.; Donaldson, K., The pro-inflammatory effects of low-toxicity low-solubility particles, nanoparticles and fine particles, on epithelial cells in vitro: the role of surface area. *Occup Environ Med* 2007, 64, 609–15.

84. Hussain, S.; Boland, S.; Baeza-Squiban, A.; Hamel, R.; Thomassen, L. C.; Martens, J. A.; Billon-Galland, M. A. et al., Oxidative stress and proinflammatory effects of carbon black and titanium dioxide nanoparticles: role of particle surface area and internalized amount. *Toxicology* 2009, 260, 142–9.

85. Knaapen, A. M.; Borm, P. J.; Albrecht, C.; Schins, R. P., Inhaled particles and lung cancer. Part A: Mechanisms. *Int J Cancer* 2004, 109, 799–809.

86. Oberdorster, G., Significance of particle parameters in the evaluation of exposure-dose-response relationships of inhaled particles. *Inhal Toxicol* 1996, 8 Suppl, 73–89.

87. Stoeger, T.; Reinhard, C.; Takenaka, S.; Schroeppel, A.; Karg, E.; Ritter, B.; Heyder, J.; Schulz, H., Instillation of six different ultrafine carbon particles indicates a surface area threshold dose for acute lung inflammation in mice. *Environ Health Perspect* 2006, 114, 328–33.

88. Donaldson, K.; Mills, N.; MacNee, W.; Robinson, S.; Newby, D., Role of inflammation in cardiopulmonary health effects of PM. *Toxicol Appl Pharmacol* 2005, 207, 483–8.

89. Li, N.; Xia, T.; Nel, A. E., The role of oxidative stress in ambient particulate matter-induced lung diseases and its implications in the toxicity of engineered nanoparticles. *Free Radic Biol Med* 2008, 44, 1689–99.

90. Stoeger, T.; Takenaka, S.; Frankenberger, B.; Ritter, B.; Karg, E.; Maier, K.; Schulz, H.; Schmid, O., Deducing in vivo toxicity of combustion-derived nanoparticles from a cell-free oxidative potency assay and metabolic activation of organic compounds. *Environ Health Perspect* 2009, 117, 54–60.

91. Lu, S.; Duffin, R.; Poland, C.; Daly, P.; Murphy, F.; Drost, E.; Macnee, W.; Stone, V.; Donaldson, K., Efficacy of simple short-term in vitro assays for predicting the potential of metal oxide nanoparticles to cause pulmonary inflammation. *Environ Health Perspect* 2009, 117, 241–7.

92. Dostert, C.; Pétrilli, V.; Van Bruggen, R.; Steele, C.; Mossman, B. T.; Tschopp, J., Innate immune activation through Nalp3 inflammasome sensing of asbestos and silica. *Science* 2008, 320, 674–7.

16 Nanoparticles and the Immune System

Bengt Fadeel and Diana Boraschi

CONTENTS

16.1 INTRODUCTION

The ability to manipulate matter at the nanoscale enables many new properties that are both desirable and exploitable, but the same properties could also give rise to unexpected toxicities that may adversely affect human health; this is, in a nutshell, the *nanomaterial paradox*. However, understanding the physicochemical properties of nanomaterials that determine their toxicity remains a challenge. There is a need for standardization of nanotoxicological approaches including nanomaterial characterization to facilitate the comparison of different studies.[1-3] As discussed in a recent review on the *identity* of the novel discipline of nanotoxicology, it may be instructive to consider the similarities between various nanoscale systems: (1) endogenous (cellular and extracellular) nanostructures or nanomachines, (2) parasites (viruses, bacteria, etc.), and (3) man-made or engineered nanomaterials.[4] In fact, nanotoxicology can be defined as the study of the interference of man-made nanomaterials with endogenous (cellular) nanostructures.[4] Importantly, the immune system has evolved to handle foreign intrusion, including particles and microbes. Thus, important lessons may be learned from immunology in terms of understanding nanomaterial interactions with the human body. For instance, opsonization of microorganisms with antibodies or complement factors to facilitate phagocytosis by professional phagocytes is reminiscent of the process of bio-corona formation on the surface of nanoparticles, which also may affect uptake.[5] Moreover, pathways for biodegradation of microorganisms may be usurped for the degradation of nanomaterials, as exemplified in recent studies of carbon nanotubes (CNTs).[6]

Immunosafety is a key element of the safety assessment of engineered nanomaterials. In this chapter, we discuss recent research on the impact of various classes of nanomaterials on the innate and adaptive arms of the immune system with a view toward safe applications of nanomaterials in the clinical setting.

16.2 INNATE IMMUNITY

16.2.1 INNATE IMMUNE SYSTEM

The innate immune system is a form of generic immune defense that does not rely on specific recognition of *antigen* (see Section 16.4) but instead responds rapidly to offending pathogens. In contrast to the adaptive immune system, innate immune responses do not confer long-lasting or protective immunity. The innate immune system is thought to constitute an evolutionarily old defense strategy and relies on anatomical barriers, on a range of white blood cells including phagocytic cells, and on soluble mediators such as complement factors, but also antibacterial peptides, endogenous antibiotics that are found in all species, from insects to mammals. The major functions of the innate immune system include the recruitment of immune cells to sites of infection, through the production of cytokines and chemokines (the *hormones* of the immune system); activation of the complement cascade to *tag* bacteria and to promote clearance of dead cells or antibody complexes through a process known as opsonization; identification and removal of microbes and other foreign agents present in organs, tissues, the blood, and lymph, by specialized phagocytic cells; and activation of the adaptive immune system through a process known as antigen presentation. There is important cross talk between the innate and adaptive arms of the immune system, and this is relevant for the discussion of nanomaterial effects on the immune system: if nanomaterials exert adverse effects on innate immune cells, such as macrophages or dendritic cells (DCs) (the sentinels of the immune system), then this may have deleterious effects on adaptive immune responses. Macrophages and neutrophils are professional phagocytes and are responsible for the clearance and destruction of bacteria but also of dead (apoptotic) cells, thereby contributing to homeostasis in tissues. The ingestion of bacteria leads to activation of the so-called oxidative burst in phagocytes with the production of vast amounts of reactive oxygen species by the NADPH oxidase and the destruction of the pathogen. Neutrophils can also capture and destroy bacteria in extracellular structures referred to as *neutrophil extracellular traps* consisting of nuclear chromatin studded with antibacterial proteins such as neutrophil elastase and myeloperoxidase (MPO). Neutrophils are by far the most abundant white blood cell and a key player in the inflammatory response and they are the first cells to arrive at a site of infection. Recent studies have disclosed a critical role of the so-called inflammasome complex in phagocytic cells including monocytes-macrophages and DCs for the production of proinflammatory cytokines. The inflammasomes are multiprotein complexes that control the activation of caspase-1 thereby modulating the secretion of the proinflammatory cytokines, IL-1β and IL-18. The activation of several NLR (nucleotide-binding oligomerization domain [NOD]-like receptor) family members drives the formation of inflammasomes.[7] The NLRP3 (or NALP3) inflammasome has been shown to be activated by a range of pathogen-associated molecular patterns (PAMPs) and damage-associated molecular patterns (DAMPs) as well as xenobiotics such as aluminum salts (alum), silica (quartz), asbestos, and ultraviolet B radiation. PAMPs are conserved microbial molecules that are not produced by host cells, such as nucleic acid structures that are unique to microorganisms, bacterial secretion systems and their effector proteins, and microbial cell wall components such as lipoproteins and lipopolysaccharides. In contrast, DAMPs are a set of host-derived molecules that signal cellular stress, damage, or nonphysiological cell death. High-mobility group box 1, uric acid, ATP, and heat-shock proteins Hsp70 and Hsp90 are examples of DAMPs that are thought to play major roles in eliciting inflammation and tissue repair. Toll-like receptors (TLRs) and NLRs are so-called pattern recognition receptors (PRRs) that probe the extracellular milieu and the endosomal compartments, respectively, for PAMPs and DAMPs.[7]

16.2.2 Nanomaterials and Innate Immunity

There are numerous studies on cellular uptake of nanomaterials and some authors have argued that it is the internalized dose of nanomaterials that determines the cytotoxicity. However, it is possible that nanoparticles could also exert indirect effects.[8] Nevertheless, macrophages are equipped with several different classes of receptors including TLRs, scavenger receptors, complement and Fc receptors (for immunoglobulins), and others, and these receptors may play a role in the uptake of nanoparticles, or aggregates of nanoparticles.[9] Scavenger receptors, in particular, have been implicated in macrophage uptake of several types of nanoparticles.[10–12] Importantly, several receptors/pathways may cooperate in uptake of nanoparticles.[13,14] Gao et al.[15] demonstrated that surface chemistry modification on multiwalled carbon nanotubes (MWCNT) changes their preferred binding pattern from mannose receptors to scavenger receptors, leading to reduced immune perturbations in vitro and in an animal model. Several recent studies have shown that the NLRP3 inflammasome is activated in response to a range of engineered nanomaterials including spherical polystyrene nanoparticles[16] and *needle-like* CNTs.[17] Yazdi et al.[18] reported that intranasal administration of titanium dioxide (TiO₂) nanoparticles provoked pulmonary inflammation (neutrophil recruitment), which was suppressed in IL-1 receptor and IL-1α-deficient mice, suggesting that the inflammation is largely driven by IL-1α. Girtsman et al.[19] reported that MWCNT–induced acute inflammation was absent in mice lacking the IL-1 receptor; however, at 28 days postexposure, the inflammatory response was elevated when compared to wild-type mice. Overall, the NLRP3 inflammasome appears to function as a master integrator of diverse stimuli to induce inflammation.[20] Numerous studies have shown that exposure to nanoparticles may induce neutrophil infiltration, a hallmark of the induction of inflammation. However, it is important to note that the purpose of the inflammatory response is to remove or sequester the offending agent (a microorganism, a foreign body), to allow the host to adapt, and ultimately, to restore functionality to the tissues.[21] If the process becomes chronic, the adaptive changes may become detrimental. Thus, it is important to distinguish between transient, protective responses toward nanoparticles versus chronic and maladaptive ones. Shvedova et al.[22] found that pharyngeal aspiration of single-walled carbon nanotubes (SWCNT) in mice resulted in an unusual combination of an acute inflammatory response including early neutrophil accumulation, followed by lymphocyte and macrophage influx, and elevation of proinflammatory cytokines, with early onset yet progressive fibrosis and granuloma formation. Cho et al.[23] reported that metal oxide nanoparticles induce unique inflammatory responses in the lung. Hence, of a comprehensive panel of nanoparticles, only cerium dioxide (CeO_2), nickel oxide (NiO), zinc oxide (ZnO), and copper oxide (CuO) nanoparticles were inflammogenic to the lungs of rats at the high doses used and each of these induced a unique inflammatory *footprint*. Acutely (after 24 h), the patterns of neutrophil and eosinophil infiltrates differed after administration of the different nanoparticles. Chronic inflammatory responses (after 4 weeks) also differed, with neutrophilic, neutrophilic/lymphocytic, eosinophilic/fibrotic/granulomatous, and fibrotic/granulomatous inflammation being caused, respectively, by CeO_2, NiO, ZnO, and CuO nanoparticles.[23] These findings are important as different types of inflammation imply different hazards in terms of pathology and risk to human health. Moreover, in vitro testing could not have differentiated these complex hazard outcomes. While nanomaterials may impact on the immune system and trigger inflammation, with tissue infiltration of neutrophils and/or eosinophils, the immune system may, conversely, affect nanomaterials. Indeed, recent studies have demonstrated the degradation of SWCNT by MPO and eosinophil peroxidase, expressed by neutrophils and eosinophils, respectively,[24,25] and our recent studies have demonstrated that lactoperoxidase, the third major mammalian peroxidase, can also digest SWCNT.[26] Understanding the process of biodegradation of CNTs and its spatiotemporal control will be of considerable importance for biomedical applications of these nanomaterials. Mast cells function in innate immunity including host defense against parasites. Upon activation, the mast cell produces histamine, leukotrienes, proteases, cytokines, chemokines, and other substances that cause immediate airway inflammation, leading to asthma symptoms. Katwa et al.[27] recently reported the crucial involvement of mast cells in pulmonary and cardiovascular responses to MWCNT in mice. TiO_2 nanoparticles were reported to induce histamine secretion in rat RBL-2H3 mast cells in a dose-dependent manner.[28] In contrast, Tahara et al.[29] showed that poly(DL-lactide-co-glycolide) (PLGA) nanoparticles inhibited antigen-induced histamine release from mast cells. The inhibitory effect of the PLGA nanoparticles was recapitulated in vivo using a mouse model for systemic anaphylaxis. Similarly, silicon dioxide (SiO_2) nanoparticles caused a decrease in the release from primary mast cell granules.[30]

16.3 BIO-CORONA ON NANOMATERIALS

It is important to consider whether nanoparticles are sensed as a *danger* signal by the immune system and/or whether the proteins that are adsorbed to the surface of nanoparticles may act as danger signals to elicit immune activation.[31] In fact, nanoparticles may seldom display naked surfaces in a biological system.[5] Furthermore, the bio-corona on nanoparticles may undergo dynamic changes as nanoparticles cross from one biological compartment to another (Figure 16.1). Deng et al.[32] have provided an example of the effects the bio-corona can have in shaping the interactions of nanoparticles with immune-competent cells. Hence, poly(acrylic acid)-coated gold nanoparticles were found to bind fibrinogen in a charge-dependent manner and induce unfolding of the protein. Binding to integrin receptors on the surface of the monocytic cell line, THP-1 led to secretion of the cytokine, TNF-α. Thus, the nanoparticles induced a proinflammatory response as a result of the adsorption of a bio-corona. Walkey et al.[33] investigated the role of size and surface chemistry in mediating serum protein adsorption to gold nanoparticles and their subsequent uptake by macrophages. The authors found that over 70 different serum proteins were adsorbed to the surface of gold nanoparticles. The relative density of each of these adsorbed proteins was found to depend on nanoparticle size and polyethylene glycol (PEG) grafting density, and variations

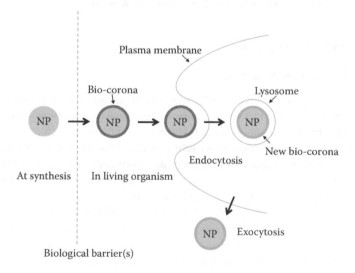

FIGURE 16.1 Nanomaterials quickly acquire a surface *corona* of biomolecules upon introduction into a biological system. Moreover, the composition of the corona may change as the nanoparticles are transported across biological barriers such as the plasma membrane. This may be of particular relevance for nanomaterial interactions with immune-competent cells that express an array of membrane-bound and intracellular *pattern recognition* receptors for pathogen-associated or endogenous molecules. (Pietroiusti, A. et al.: *Small*. 2013;9:1557–1572. Copyright Wiley-WCH Verlag GmbH & Co. KGaA. Reproduced with permission.)

in serum protein adsorption affected the efficiency of nanoparticle uptake by a macrophage cell line. Moreover, Kapralov et al.[34] reported that the lung surfactant *corona* of lipids and proteins enhanced the uptake of SWCNT by macrophage-like RAW 264.7 cells.

The complement system is an ancient and integral part of the innate immune system that helps or *complements* antibodies. The complement system is composed of 30 different plasma and cell-bound proteins that are activated through three different pathways. It is important to note that inadvertent complement activation could lead to anaphylaxis, a serious allergic reaction that is rapid in onset and may cause death. Thus, it is essential to design nanomaterials in such a way that unwanted complement activation (which may be viewed as a special case of corona formation) is avoided, not least in the context of nanomedicine.[35] Hamad et al.[36] showed that polystyrene nanoparticles with surface projected polyethyleneoxide chains in *mushroom-brush* and *brush* configurations activate complement pathways differently. Furthermore, Andersen et al.[37] demonstrated that the coating of SWCNT with methoxypoly(ethylene glycol)-based amphiphiles activates the complement system differently, depending on the amphiphile structure. Notably, an amphiphile with branched PEG architecture activated complement through the lectin pathway but did not lead to the generation of anaphylatoxins. Taken together, these findings show that surface modification of nanomaterials with distinct polymers may significantly improve their immunocompatibility (i.e., *safety by design*).

16.4 ADAPTIVE IMMUNITY

16.4.1 ADAPTIVE IMMUNE SYSTEM

Adaptive immunity is a sophisticated system of defense that has developed in higher vertebrates, but is lacking in lower organisms. Adaptive immunity is highly specific, as opposed to innate immunity, but it takes time to develop (again at variance with innate immunity). Adaptive immunity relies on the capacity of our immune cells to specifically recognize potentially dangerous agents or *antigens*, discriminating them from molecules and structures of the body itself and to mount an antigen-specific response that destroys the foreign agent. Adaptive immunity is accomplished by effector cells, the T and B lymphocytes, and the antigen-presenting cells (APC). When a foreign agent enters the body, the phagocytes immediately come into play with the goal of taking up and degrading the agent (see Section 16.2). While the battle takes place in the tissue under attack, DCs, which are the main APC (together with activated macrophages and other cell types), migrate from the tissue to the nearby lymph node with their cargo of antigens derived from the foreign agent and come in contact with T lymphocytes. These cells recognize the antigen presented by the APC and become activated, exit the lymph node, and recirculate to the tissues to help in destroying the foreign agent. There are several types of T cells, each with its specific function. T helper (Th) cells are necessary to support the production of specific antibodies by B lymphocytes (that become antibody-producing plasma cells upon activation in the presence of Th cells). Th cells can be distinguished into Th1, Th2, and Th17, depending on the type of response they support and on the range of cytokines that they produce. Cytotoxic T cell usually develop (expand) during viral infections and their role is that of killing virus-infected cells, they are also involved in killing malignant cells. Another type of T cell, T regulatory cells, is involved in downregulating immune activation (a mechanism of self-control to avoid tissue damage due to an excessive immune reaction). An important feature of adaptive immunity, which clearly distinguishes it from innate immunity, is memory. Memory is the capacity of immune cells to *remember* the first encounter with an antigen and, upon subsequent encounter, of mounting a quicker and more potent response. Several types of T and B memory cells have been described. Thus, adaptive immunity is a highly regulated system of great specificity and potency that includes recognition of the antigen, presentation to Th cells, activation of a specific response (specific antibodies produced by plasma cells and specific cytotoxic T cell), and memory. The latter phenomenon constitutes the basis for vaccination.

16.4.2 NANOMATERIALS AND ADAPTIVE IMMUNITY

Interaction of nanomaterials with APC has been demonstrated in many studies, showing that nanomaterials have a direct amplification role, mainly by triggering the activation of the inflammasome with subsequent production of IL-1β.[16] Indeed, the efficiency of nanomaterials in activating APC is such that a great effort is being devoted to their use for improving vaccine efficacy and protective immunity[38]. In a recent study, graphene oxide, but not fullerenes, was shown to hamper antigen presentation in bone marrow-derived DC by inhibiting the protein degradation machinery of the immunoproteasome.[39] However, most of the effects of nanomaterials on adaptive immunity that have been reported are indirect, that is, mediated by activation of macrophages and DC to produce cytokines, which then activate and regulate the several types of T and B cells.[40] Direct effects of nanomaterials on other types of immune cells have not been convincing. Indeed, when observing the in vivo distribution of inoculated nanomaterials, it is obvious that most nanomaterials are rapidly taken up by the reticuloendothelial system, composed of the circulating and tissue-resident phagocytes (e.g., in liver and spleen). To obtain a direct interaction with T and B lymphocytes for therapeutic intervention, nanomaterials must be surface modified with lymphocyte-targeting molecules.

The effects of nanomaterials on adaptive immunity can be broadly grouped into immunosuppressive and immunostimulatory effects. Immunosuppression, apart from that induced on purpose by nanomedicines loaded with cytotoxic and/or immunosupressive drugs, has been shown only in a few instances and again appears to be indirect and mediated by macrophages, which release an immunosuppressive cytokine such as TGF-β.[41] Unwanted immunosuppression may pose serious health problems, including increased susceptibility to infections and increased incidence of tumors. Immunostimulation has been reported more widely, as mentioned in the preceding discussion, but in this case the effects are indirect and mediated by macrophages and DC. In a recent study, Zhu et al.[42] showed that in vivo exposure to magnetic iron oxide nanoparticles resulted in significant exosome generation in the alveolar region of Balb/c mice. Through exosome-initiated signals, DCs were found to undergo maturation, while macrophages underwent classical activation to the M1 subtype. These cells released Th1 cytokines driving T cell activation and differentiation. Th1-polarized immune activation is associated with delayed-type hypersensitivity, and the authors suggested that such nanoparticle-triggered Th1-type immune responses might underlie the long-term inflammatory effects associated with nanoparticle exposure. A possible problem related to uncontrolled immuno stimulation, as it may occur with inadvertent exposure to nanomaterials, is that of raising anomalous activation of adaptive immunity against *self* antigens (i.e., the body's own components). The issue here is that the interaction of nanomaterials with host proteins (corona formation) may facilitate structural modifications of the proteins and the consequent erroneous recognition and presentation by APC, thereby causing an autoimmune response (production of specific antibodies against self) and the development of autoimmune diseases.[31] The same considerations hold true for allergy, which is an exaggerated specific immune reaction to antigens that are otherwise harmless. The complexes between nanomaterials and antigens (allergens or host proteins) could exacerbate the immune reaction. However, more studies are required to support the hypothesis that immunostimulation caused by nanomaterials can contribute to autoimmunity or allergy. A recent paper suggests that different types of nanomaterials (SiO_2 nanoparticles, ultrafine carbon black, SWCNT) could induce increased protein citrullination in human tumor cell lines in vitro and in mouse lung in vivo.[43] Citrullination, the deimidation of arginine residues into citrulline by peptidyl arginine deiminases, is the natural posttranslational mechanism for shaping the protein structure (arginine is charged, citrulline is not) and is used for regulating several biological processes including epithelial terminal differentiation and the activity of inflammatory factors, for example, chemokines.[44] It is well known that in rheumatoid arthritis (RA), a common inflammatory/autoimmune disease, there is a significant increase in autoantibodies to citrullinated peptides/proteins, which may imply that citrullinated proteins are more abundant and become immunogenic in RA patients.[45] Indeed, increased protein citrullination is a known consequence of inflammation, and bacterial endotoxin is an excellent inducer of protein citrullination, not least in monocytes. However, a major issue in most studies is that the effects of nanomaterials on macrophages and DC could be attributed to chemical or biological contaminants such as endotoxin[46] and this should be taken into account when evaluating nanomaterial effects on the immune system.

16.4.3 IMMUNOGENICITY OF NANOMATERIALS

An important issue that needs to be addressed, and which is particularly relevant in the case of nanomedicines, is nanomaterial immunogenicity, that is, the ability to elicit a specific adaptive immune response. A specific immune response against nanomaterials, when these are inoculated as diagnostic or therapeutic agents, would imply rapid neutralization and elimination at the second inoculum, with a severe loss in efficacy. Immunogenicity of nanomaterials has been observed only rarely. One case is that of PEG-coated nanoparticles (e.g., liposomes) that have been found able to act as T cell-independent immunogens and elicit a direct B-cell response to the production of anti-PEG IgM antibodies, responsible for accelerated clearance of the nanoparticles upon subsequent administrations.[47,48] While no response against the particles was detected, recognition of PEG represents a problem. PEG is widely used in nanomedicine for particle coating, so as to decrease nonspecific clearance by phagocytes, thus it is obvious that eliciting a specific anti-PEG response is a major drawback for PEGylated nanodrug development. The second case is that of C60 fullerene that, conjugated to a protein (bovine thyroglobulin) and injected repeatedly into mice, were shown to elicit a significant and specific IgG antibody response (an IgG response implies activation of Th cells and generation of memory).[49] Similarly, induction of specific antiparticle antibody responses was reported upon in vivo inoculation of poly(amidoamine) (PAMAM) dendrimers conjugated with a protein (the cytokine, IL-3), constructed to increase the bioavailability and half-life of the protein for therapeutic purposes.[50] Indeed, while the dendrimer was not immunogenic by itself, it became so when conjugated with IL-3, as well as with other proteins, such as bovine serum albumin. This is a well-known phenomenon in immunology, that is, the increased immunogenicity of *haptens* (small molecules that can be antigens, but not immunogens) when conjugated with protein carriers. Thus, immunogenicity of nanomaterials is possible, in particular when the nanomaterials are modified with or adsorbed to proteins, as is often the case in nanomedicine. The observation that hapten–carrier conjugates are strongly immunogenic, and able to induce an adaptive immune response against both components of such conjugates, should raise the attention of scientists on how to design safe and effective nanotherapeutic conjugates. The issue of adsorption of host proteins (corona formation) also needs to be taken into account. Salvati et al.[51] provided evidence that the formation of a bio-corona may obscure targeting ligands (e.g., transferrin) on the surface of nanoparticles. However, contrasting with these studies, transferrin targeting of nanoparticles has been demonstrated in human cancer patients.[52]

16.4.4 IMMUNOLOGICALLY FRAIL POPULATIONS

A great achievement in modern society is the prolongation of life expectancy. By the year 2030, the percentage of the population that will be ≥ 60 years of age is expected to exceed 25% of the world population (Europe and Japan have already achieved this milestone), of which 75% will be living in less developed countries. Consequently, the health of the elderly is rapidly becoming a major public health concern. Specific immune responses in the elderly are less effective than in adults with decreased T-cell and B-cell activation, while innate immune reactions are usually increased due to the constitutive inflammatory microenvironment.[53] Several reports show that macrophages and DC, being constitutively activated in the elderly, are less responsive to activation by external stimuli. Moreover, immunological frailty due to chronic diseases and infection (often concurring with old age) and malnutrition is often associated with a state of enhanced inflammation that induces constitutive macrophage and DC activation. In this respect, despite the lack of any studies in this area, it may be expected that nanomaterials, which interact mainly with phagocytic cells, for example, macrophages and DC, may trigger fewer unwanted reactions in the elderly population. The situation is different in infants and very young children, in which the immune system is immature and therefore less able to cope with external challenges, but not constitutively inflamed. In these young individuals, macrophages and myeloid

DC are also less responsive than in adults, mainly due to a high level of adenosine,[54] while blood plasmacytoid DCs are more abundant than in adults and fully active.[55] Thus, hypothetically, nanomaterials that interact with macrophages and DC may be only partially able to activate them (provided that nanomaterials interact with DC), and consequently pose a reduced risk of detrimental effects. In any case, focused studies are required to dismiss the possibility that nanomaterials may have detrimental effects in immunologically frail populations, with particular consideration of the effects of chronic exposure (highly relevant for the elderly population). It is also important to note that barrier functions of the lungs in infants may differ from those in adults[56] and this could impact on the biodistribution of nanomaterials and, therefore, on the outcome of nanomaterial exposure in infants versus adults.

16.4.5 The Need for Relevant Models

When assessing the effects of nanomaterials on the immune system, scientists face a major problem of the validity of the test system, that is, to what extent does the system reflect what happens in real life in the target population of interest, for instance, healthy adult humans, elderly humans, or chronically ill patients. In many instances, animal models are not valid models of real-life conditions in the human population.[57] On the other hand, the use of human cell lines in vitro has the dual drawback of poor validity both because of the cell behavior (cell lines, most often transformed or of cancerous origin, may behave very differently from primary cells) and because of the artificial culture conditions that do not reproduce the in vivo microenvironment. Human primary cell–based complex tissue–like models (e.g., cocultures of several cell types) may be a solution, at least for identifying and validating relevant markers or end points. Huh et al.[58] have reviewed exciting developments in *organ-on-a-chip* models that could potentially replace animals for toxicity studies and drug development.

16.5 THERAPEUTIC MODULATION OF IMMUNE RESPONSES

The prospect of deploying nanomaterials for therapeutic purposes has bloomed in recent years. Nanomedicines can have different targets and scopes, but their interaction with the immune system is nevertheless of central importance for their efficacy.

16.5.1 Stealth: Avoiding Immune Recognition

Numerous studies have attempted to avoid immune recognition on nanomaterials and their uptake by the reticuloendothelial system, to exploit them for antitumor drug delivery in secluded sites, for example, the brain,[59] attempts that in most cases turned out to be unsatisfactory. Indeed, given all the above considerations, attempts to avoid interaction of nanotherapeutics with the immune system hardly seem possible. Functionalization of nanomaterials with polymers including PEG is a common strategy to prevent nonspecific uptake by phagocytic cells; however, PEGylation does not prevent protein adsorption altogether (see Section 16.3). Cavadas et al.[60] presented examples of alternative strategies with which to achieve stealthy nanocarriers based on pathogen mimetic approaches. Macrophages also engulf apoptotic cells, and numerous receptor–ligand interactions that drive the engulfment process have been described. The phospholipid, phosphatidylserine (PS), is a key *eat-me* signal that is exposed on the surface of dying cells. In a proof-of-principle study, Konduru et al.[61] reported that the adsorption of PS onto SWCNTs resulted in efficient uptake by several classes of macrophages. Conversely, the membrane protein CD47 is a marker of self that prevents the inadvertent phagocytosis of nonapoptotic cells by macrophages. Rodriguez et al.[62] have shown that minimal self-peptides based on human CD47 delayed macrophage-mediated clearance of nanoparticles, which promoted persistent circulation and enhanced drug delivery to tumors in a mouse model. This study underscores that important lessons may be learned from comparative studies of different nanoscale systems, that is, nanoparticles, microorganisms, and cellular structures.

16.5.2 DANGER: ADJUVANTICITY AND VACCINE DELIVERY

It is well known that the immune system recognizes particulate matter more efficiently as opposed to soluble molecules.[63] Immunization strategies in many instances include the conjugation of poorly immunogenic small molecules (haptens) to large and highly immunogenic carriers (large proteins such as albumin, or fragments of microorganisms such as virus-like or bacterium-like particles) to obtain a potent immune response to both hapten and carrier. Effective immunization, in particular for poorly immunogenic vaccine antigens and/or in immunologically frail groups (children, elderly, chronic patients), often requires the presence of a so-called adjuvant, that is, an agent that nonspecifically stimulates the innate/inflammatory reaction at the site of vaccine inoculum, thereby promoting antigen uptake and presentation and consequent initiation of adaptive immunity. Delivery systems and adjuvants often overlap, since good delivery systems should target macrophages and APC and stimulate these cells to promote antigen presentation.[64] Particles have been used since the beginning of vaccinology to increase vaccine efficacy. Alum (a mixture of nanoparticles and microparticles of aluminium hydroxide or aluminum phosphate) is included in licensed vaccines for human use for more than 80 years.[65] Oil-in-water nanoemulsions, such as the preferred influenza vaccine adjuvant MF59, have proved excellent in yielding sustained immunity also in populations at risk.[66] While the mechanism of action of these adjuvants is still not completely understood, there is evidence that the activation of innate immune responses and local inflammation is of central importance, including activation of TLRs and of the NLRP3 inflammasome (see Section 16.2), and induction of chemokine-driven trafficking of leukocytes between the site of reaction and the draining lymph nodes.[38,67]

In this perspective, the design of new nanomaterials with the capacity of selectively inducing the controlled activation of the innate immune response is under intense investigation.[68] The demarcation between physiological innate/inflammatory reactions (those that protect us from danger and that are at the basis of vaccine adjuvants) and pathological inflammation (chronic and persistent reactions) suggests that the design of nanomaterial-based immunostimulants should include the capacity of activating the checkpoints of the inflammatory reaction in a precisely controlled manner.

16.6 CONCLUDING REMARKS

Nanoparticles can stimulate or suppress immune responses and it is our belief—based on the available literature—that most if not all outcomes may be largely driven by the interactions of nanoparticles with macrophages and other phagocytic cells of the innate immune system, constituting the first line of host defense against foreign intrusion. Several factors should be considered in the assessment of the immunological properties of nanomaterials. First, thorough nanomaterial characterization is needed and not only the material intrinsic physicochemical properties but also the acquired biological *identity* should be evaluated, especially when studying interactions with immune-competent cells.[1] Second, relevant and validated model systems are needed, that is, models that are relevant for the human situation and reflective of the complex multicellular immune responses, and validated in terms of being able to capture nanomaterial effects without interference by the nanomaterials in question. Endotoxin contamination of nanomaterials is a nontrivial issue and steps should be taken to control for this. More studies are needed on the impact of nanomaterials in immunologically frail populations (elderly, chronically ill) and individuals with pre-existing conditions including infections or allergy. Chronic exposure studies using realistic doses are needed to understand the real-life impact of engineered nanomaterials. Taken together, understanding how the immune system responds to offending agents (pathogens, particles) is a key to understanding the toxicity and pathogenicity of nanomaterials. The other side of the coin is that nanomaterials could be exploited for immune modulation, for instance, for the delivery of antigen and/or adjuvants for vaccination.

REFERENCES

1. Fadeel B, Feliu N, Vogt C, Abdelmonem AM, Parak WJ. Bridge over troubled waters: understanding the synthetic and biological identities of engineered nanomaterials. *Wiley Interdiscip Rev Nanomed Nanobiotechnol.* 2013 Mar;5(2):111–29.
2. Schrurs F, Lison D. Focusing the research efforts. *Nat Nanotechnol.* 2012 Sep;7(9): 546–8.
3. Fadeel B, Savolainen K. Broaden the discussion. *Nat Nanotechnol.* 2013 Feb;8(2):71.
4. Shvedova AA, Kagan VE, Fadeel B. Close encounters of the small kind: adverse effects of man-made materials interfacing with the nano-cosmos of biological systems. *Annu Rev Pharmacol Toxicol.* 2010;50:63–88.
5. Monopoli MP, Aberg C, Salvati A, Dawson KA. Biomolecular coronas provide the biological identity of nanosized materials. *Nat Nanotechnol.* 2012 Dec;7(12):779–86.
6. Fadeel B, Shvedova AA, Kagan VE. Interactions of carbon nanotubes with the immune system: focus on mechanisms of internalization and biodegradation. In: *Nanomedicine—Basic and Clinical Applications in Diagnostics and Therapy.* Ed. Alexiou C., volume 2, pp. 80–87, Karger, Basel, Switzerland, 2011.
7. Franchi L, Muñoz-Planillo R, Nuñez G. Sensing and reacting to microbes through the inflammasomes. *Nat Immunol.* 2012 Mar 19;13(4):325–32.
8. Bhabra G, Sood A, Fisher B et al. Nanoparticles can cause DNA damage across a cellular barrier. *Nat Nanotechnol.* 2009 Dec;4(12):876–83.
9. Dobrovolskaia MA, McNeil SE. Immunological properties of engineered nanomaterials. *Nat Nanotechnol.* 2007 Aug;2(8):469–78.

10. França A, Aggarwal P, Barsov EV et al. Macrophage scavenger receptor A mediates the uptake of gold colloids by macrophages in vitro. *Nanomedicine* 2011 Sep;6(7):1175–88.

11. Orr GA, Chrisler WB, Cassens KJ et al. Cellular recognition and trafficking of amorphous silica nanoparticles by macrophage scavenger receptor A. *Nanotoxicology* 2011 Sep;5(3):296–311.

12. Wang H, Wu L, Reinhard BM. Scavenger receptor mediated endocytosis of silver nanoparticles into J774A.1 macrophages is heterogeneous. *ACS Nano.* 2012 Aug 28;6(8):7122–32.

13. Zhang LW, Monteiro-Riviere NA. Mechanisms of quantum dot nanoparticle cellular uptake. *Toxicol Sci.* 2009 Jul;110(1):138–55.

14. Zhang LW, Bäumer W, Monteiro-Riviere NA. Cellular uptake mechanisms and toxicity of quantum dots in dendritic cells. *Nanomedicine* 2011 Jul;6(5):777–91.

15. Gao N, Zhang Q, Mu Q et al. Steering carbon nanotubes to scavenger receptor recognition by nanotube surface chemistry modification partially alleviates NFκB activation and reduces its immunotoxicity. *ACS Nano.* 2011 Jun 28;5(6):4581–91.

16. Lunov O, Syrovets T, Loos C et al. Amino-functionalized polystyrene nanoparticles activate the NLRP3 inflammasome in human macrophages. *ACS Nano* 2011;5:9648–57.

17. Palomäki J, Välimäki E, Sund J et al. Long, needle-like carbon nanotubes and asbestos activate the NLRP3 inflammasome through a similar mechanism. *ACS Nano.* 2011 Sep 27;5(9):6861–70.

18. Yazdi AS, Guarda G, Riteau N et al. Nanoparticles activate the NLR pyrin domain containing 3(Nlrp3)inflammasome and cause pulmonary inflammation through release of IL-1α and IL-1β. *Proc Natl Acad Sci USA.* 2010 Nov 9;107(45):19449–54.

19. Girtsman TA, Beamer CA, Wu N, Buford M, Holian A. IL-1R signalling is critical for regulation of multi-walled carbon nanotubes-induced acute lung inflammation in C57Bl/6 mice. *Nanotoxicology* 2012 Nov 14. [Epub ahead of print].

20. Sun B, Wang X, Ji Z, Li R, Xia T. NLRP3 inflammasome activation induced by engineered nanomaterials. 2013 May 27;9(9–10):1595–607.

21. Medzhitov R. Origin and physiological roles of inflammation. *Nature* 2008 Jul 24;454(7203):428–35.

22. Shvedova AA, Kisin ER, Mercer R et al. Unusual inflammatory and fibrogenic pulmonary responses to single-walled carbon nanotubes in mice. *Am J Physiol Lung Cell Mol Physiol.* 2005 Nov;289(5):L698–708.

23. Cho WS, Duffin R, Poland CA et al. Metal oxide nanoparticles induce unique inflammatory footprints in the lung: important implications for nanoparticle testing. *Environ Health Perspect.* 2010 Dec;118(12):1699–706.

24. Shvedova AA, Kapralov AA, Feng WH et al. Impaired clearance and enhanced pulmonary inflammatory/fibrotic response to carbon nanotubes in myeloperoxidase-deficient mice. *PLoS One.* 2012;7(3):e30923.

25. Andón FT, Kapralov AA, Yanamala N et al. Biodegradation of single-walled carbon nanotubes by eosinophil peroxidase. *Small* 2013 Aug;9(16):2721–9.

26. Bhattacharya K, El-Sayed R, Andon FT et al. Lactoperoxidase-mediated biodegradation of single-walled carbon nanotubes compromises the anti-bacterial function of lactoperoxidase. 2013 [manuscript in preparation].

27. Katwa P, Wang X, Urankar RN et al. A carbon nanotube toxicity paradigm driven by mast cells and the IL-33/ST2axis. *Small* 2012 Sep 24;8(18):2904–12.

28. Chen EY, Garnica M, Wang YC et al. A mixture of anatase and rutile TiO_2 nanoparticles induces histamine secretion in mast cells. *Part Fibre Toxicol.* 2012 Jan 19;9:2. doi: 10.1186/1743-8977-9-2.

29. Tahara K, Tadokoro S, Yamamoto H, Kawashima Y, Hirashima N. The suppression of IgE-mediated histamine release from mast cells following exocytic exclusion of biodegradable polymeric nanoparticles. *Biomaterials* 2012 Jan;33(1):343–51.

30. Maurer-Jones MA, Lin YS, Haynes CL. Functional assessment of metal oxide nanoparticle toxicity in immune cells. *ACS Nano*. 2010 Jun 22;4(6):3363–73.

31. Fadeel B. Clear and present danger? Engineered nanoparticles and the immune system. *Swiss Med Wkly*. 2012 Jun 26;142:w13609.

32. Deng ZJ, Liang M, Monteiro M, Toth I, Minchin RF. Nanoparticle-induced unfolding of fibrinogen promotes Mac-1 receptoractivation and inflammation. *Nat Nanotechnol*. 2011 Jan;6(1):39–44.

33. Walkey CD, Olsen JB, Guo H, Emili A, Chan WC. Nanoparticle size and surface chemistry determine serum protein adsorption and macrophage uptake. *J Am Chem Soc*. 2012 Feb 1;134(4):2139–47.

34. Kapralov AA, Feng WH, Amoscato AA et al. Adsorption of surfactant lipids by single-walled carbon nanotubes in mouse lung upon pharyngeal aspiration. *ACS Nano*. 2012 May 22;6(5):4147–56.

35. Andersen AJ, Hashemi SH, Andresen TL, Hunter AC, Moghimi SM. Complement: alive and kicking nanomedicines. *J Biomed Nanotechnol*. 2009 Aug;5(4):364–72.

36. Hamad I, Al-Hanbali O, Hunter AC et al. Distinct polymer architecture mediates switching of complement activation pathways at the nanosphere-serum interface: implications for stealth nanoparticle engineering. *ACS Nano*. 2010 Nov 23;4(11):6629–38.

37. Andersen AJ, Robinson JT, Dai H et al. Single-walled carbon nanotube surface control of complement recognition and activation. *ACS Nano*. 2013 Feb 26;7(2):1108–19.

38. Demento SL, Eisenbarth SC, Foellmer HG et al. Inflammasome-activating nanoparticles as modular systems for optimizing vaccine efficacy. *Vaccine* 2009;27:3013–21.

39. Tkach AV, Yanamala N, Stanlet S et al. Graphene oxide, but not fullerenes, targets immunoproteasomes and suppresses antigen presentation by dendritic cells. *Small* 2013 May;9(9–10):1686–90.

40. Boraschi D, Costantino L, Italiani P. Interaction of nano particles with immunocompetent cells:nano safety considerations. *Nanomedicine* 2012 Jan;7(1):121–31.

41. Mitchell LA, Lauer FT, Burchiel SW et al. Mechanisms for how inhaled multiwalled carbon nanotubes suppress systemic immune function in mice. *Nat Nanotechnol*. 2009;4:451–6.

42. Zhu M, Tian X, Song X, Li Y, Tian Y, Zhao Y, Nie G. Nanoparticle-induced exosomes target antigen-presenting cells to initiate Th1-type immune activation. *Small* 2012 Sep 24;8(18):2841–8.

43. Mohamed BM, Verma NK, Davies AM et al. Citrullination of proteins: a common post-translational modification pathway induced by nanoparticles in vitro and *in vivo*. *Nanomedicine* 2012;7:1181–95.

44. Proost P, Loos T, Mortier A et al. Citrullination of CXCL8 by peptidylarginine deiminase alters receptor usage, prevents proteolysis and dampens tissue inflammation. *J Exp Med*. 2008;205:2085–97.

45. Migliorini P, Pratesi F, Tommasi C et al. The immune response to citrullinated antigens in autoimmune diseases. *Autoimmun Rev*. 2005;4:561–564.

46. Vallhov H, Qin J, Johansson SM et al. The importance of an endotoxin-free environment during the production of nanoparticles used in medical applications. *Nano Lett*. 2006;6:1682–6.

47. Ishida T, Wang X, Shimizu T et al. PEGylated liposomes elicit an anti-PEG IgM response in a T cell-independent manner. *J Control Release* 2007;122:349–55.

48. Shimizu T, Ichihara M, Yoshioka Y et al. Intravenous administration of polyethylene glycol-coated (PEGylated) proteins and PEGylated adenovirus elicits an anti-PEG immunoglobulin M response. *Biol Pharm Bull*. 2012;35:1336–42.

49. Chen BX, Wilson SR, Das M et al. Antigenicity of fullerenes: antibodies specific for fullerenes and their characteristics. *Proc Natl Acad Sci USA*. 1998;95:10809–13.

50. Lee SC, Parthasarathy R, Duffin TD et al. Recognition properties of antibodies to PAMAM dendrimers and their use in immune detection of dendrimers. *Biomed Microdev.* 2001; 1:53–59.
51. Salvati A, Pitek AS, Monopoli MP et al. Transferrin-functionalized nanoparticles lose their targeting capabilities when a biomolecule corona adsorbs on the surface. *Nat Nanotechnol.* 2013 Feb;8(2):137–43.
52. Davis ME, Zuckerman JE, Choi CH et al. Evidence of RNAi in humans from systemically administered siRNA via targeted nanoparticles. *Nature* 2010 Apr 15;464(7291):1067–70.
53. Boraschi D, Aguado T, Dutel C et al. The gracefully aging immune system. *Sci Transl Med.* 2013 May 15;5(185):185–8.
54. Philbin VJ, Dowling DJ, Gallington LC et al. Imidazoquinoline Toll-like receptor 8 agonists activate human newborn monocytes and dendritic cells through adenosine-refractory and caspase-1-dependent pathways. *J Allergy Clin Immunol.* 2012;130:195–204.
55. Teig N, Moses D, Gieseler S et al. Age-related changes in human blood dendritic cell subpopulations. *Scand J Immunol.* 2002;55:453–7.
56. Semmler-Behnke M, Kreyling WG, Schulz H et al. Nanoparticle delivery in infant lungs. *Proc Natl Acad Sci USA.* 2012 Mar 27;109(13):5092–7.
57. Davis MM. A prescription for human immunology. *Immunity* 2008;29:835–38.
58. Huh D, Hamilton GA, Ingber DE. From 3D cell culture to organs-on-chips. *Trends Cell Biol.* 2011 Dec;21(12):745–54.
59. Costantino L, Boraschi D. Is there a clinical future for polymeric nanoparticles as brain-targeting drug delivery agents? *Drug Discov Today* 2011;17:367–78.
60. Cavadas M, González-Fernández A, Franco R. Pathogen-mimetic stealth nanocarriers for drug delivery: a future possibility. *Nanomedicine* 2011 Dec;7(6):730–43.
61. Konduru NV, Tyurina YY, Feng W et al. Phosphatidylserine targets single-walled carbon nanotubes to professional phagocytes in vitro and in vivo. *PLoS One* 2009;4(2):e4398.
62. Rodriguez PL, Harada T, Christian DA et al. Minimal "self" peptides that inhibit phagocytic clearance and enhance delivery of nanoparticles. *Science* 2013 Feb 22;339(6122):971–5.
63. Xiang SD, Scholzen A, Minigo G et al. Pathogen recognition and development of particulate vaccines: does size matter? *Methods* 2006;40:1–9.
64. Foged C. Subunit vaccines of the future: the need for safe, customized and optimized particulate delivery systems. *Ther Deliv.* 2011;2:1057–77.
65. Kool M, Fierens K, Lambrecht BN. Alum adjuvant: some of the tricks of the oldest adjuvant. *J Med Microbiol.* 2012;61:927–934.
66. Boraschi D, Rappuoli R, Del Giudice G. Optimising response to vaccination in the elderly: the case of influenza. In: Masoud A, Rezaei N, eds. *Immunology of Ageing.* Heildelberg, Germany: Springer-Verlag, 2013. [in press].
67. Champion CI, Kickhoefer VA, Liu G et al. A vault nanoparticle vaccine induces protective mucosal immunity. *PLoS One* 2009;4(4):e5409.
68. Hubbel JA, Thomas SN, Swartz MA. Materials engineering for immunomodulation. *Nature* 2009;462:449–460.
69. Pietroiusti A, Campagnolo L, Fadeel B. Interactions of engineered nanoparticles with organs protected by internal biological barriers. *Small* 2013;9:1557–1572.

17 Carbon Nanotubes and Cardiovascular Disease

Peter Møller, Cao Yi, Lise K. Vesterdal, Pernille H. Danielsen, Martin Roursgaard, Henrik Klingberg, Daniel V. Christophersen, and Steffen Loft

CONTENTS

17.1 INTRODUCTION

Exposure to carbon nanotubes (CNTs) causes concern for risk of cardiovascular diseases because inhalation of air pollution particles has been linked to increased risk of morbidity and mortality.[1] Cardiovascular diseases have been categorized by the International Classification of Disease (ICD), convened by the World Health Organization. Epidemiological studies use these classifications as disease outcome, whereas they are typically not used in particle toxicology. Table 17.1 lists the individual categories in the ICD-9 classification, which encompass diseases of the circulatory system in codes 390–459. Some of the classes are clearly not relevant for particle toxicology (e.g., codes 390–398), whereas diseases in the vasculature are relevant in regard to particle exposure. The development and use of CNTs is still relatively new and there are no epidemiological studies investigating the adverse health effects in exposed humans. Therefore, the hazards of CNTs are presently best assessed by animal exposure models, supported by observations from in vitro studies. Animal studies on associations between exposure to particles and cardiovascular effects have mainly focused on atherosclerosis in the peripheral blood vessels (code 440) and thrombosis (code 444).

TABLE 17.1

Classification of Cardiovascular Diseases and Comparison with Endpoints That Are Measured in Animal Experimental Models

Class (ICD-9 code)	Relevant Outcome in Particle Toxicology	Examples of Studies on Exposure to CNTs
Acute rheumatic fever (390–392)	Not directly relevant	Not applicable
Chronic rheumatic heart disease (393–398)	Not directly relevant	Not applicable
Hypertensive diseases (401–405)	Essential hypertension	Not studied
Ischemic heart disease (410–414)	Myocardial infarction, angina pectoris, coronary atherosclerosis	Increased ischemia/reperfusion injury mediated cardiac infarct size by MWCNTs[43]
Diseases of the pulmonary circulation (415–417)	Mainly relevant as secondary effect to cardiac effect	Not studied
Cerebrovascular disease (430–438)	Occlusion, hemorrhage	Not studied
Diseases of the arteries, arterioles, and capillaries (440–448)	Atherosclerosis, embolism, and thrombosis	Increased plaque progression by SWCNTs in *apoE*[-/-] mice[34] Increased thrombosis tendency in carotid artery by SWCNTs[57] Increased fibrinogen levels in plasma by SWCNTs[40]
Diseases of the veins and lymphatics, and other diseases of the circular system (451–459)	Deep vein thrombosis	Not studied

For the individual, there is a relatively low risk of cardiovascular disease associated with exposure to the limited levels of air pollution particles in ambient air in most areas. However, the whole population is potentially exposed to particles, resulting in a major public health concern. Combustion of fuel or wood is a major source of air pollution resulting in nanometer-sized particles of elemental carbon with surface-bound metals and organic compounds.[2] Elemental carbon in ambient air particles has been suggested to be the most relevant marker of risk.[3] Moreover, metals are also important contributors to the toxic effects of air pollution particles.[4] The mechanisms of action of cardiovascular disease caused or aggravated by exposure to combustion-derived particles are thought to involve oxidative stress and inflammation occurring in the lungs with secondary systemic effects, neuronal signaling from airway receptors, or after translocation of particles causing oxidative stress and inflammation in the circulation.[2,5] Particularly, diesel exhaust particles are considered to have adverse vascular effects shown in, for example, controlled human exposure studies.[5] CNTs also consist of elemental carbon and may have metal impurities and functional groups. In addition, CNTs have a high aspect ratio, which for some types of CNTs resemble the structure of other fibers such as asbestos, although the hazards may not be identical.

CNTs are grouped into single-walled carbon nanotubes (SWCNTs) and multi-walled carbon nanotubes (MWCNTs). They are typically developed to provide unique electrical, mechanical, and thermal properties to products. Human exposures have been considered to be workplace related.[6,7] However, it must be expected that the general population can also be exposed by handling or through the disposal of CNT-containing products.

17.2 RISK FACTORS AND DEVELOPMENT OF CARDIOVASCULAR DISEASES

The traditional risk factors for cardiovascular diseases include hypertension, hyperlipidemia, low daily consumption of fruits and vegetables, regular alcohol intake, diabetes, smoking, lack of exercise, and obesity.[8,9] The contribution of each of these risk factors to myocardial infarction, coronary insufficiency, and angina pectoris is likely to be substantially higher than the risk associated with exposure to CNTs. Individuals with several risk factors present at the same time have especially high risk and a reduced expected life span. Control of the traditional risk factors for cardiovascular diseases is a considerable public health concern as clearly evidenced by the fact that cardiovascular diseases are the leading cause of mortality and morbidity in the western countries. The mechanisms leading to cardiovascular health effects are thought to be mediated by oxidative stress and inflammation generating atheroma, vasomotor dysfunction, fibrinolytic imbalance, and reduced heart rate variability.[5] Pulmonary exposure to CNTs is associated with both oxidative stress and inflammation, which can be contributing factors to CNT-induced cardiovascular effects.[10] Figure 17.1 outlines the relationship between exposure to CNTs, intermediate endpoints (e.g., inflammation, oxidative stress, endothelial dysfunction, and prothrombotic tendency), and clinical outcomes.

The diagnosis of ischemic heart disease covers a condition of fatty acid accumulation in the vessel walls of coronary arteries, which causes a narrowing of the lumen and reduced blood flow to the myocardial cells. The narrowed lumen of the coronary arteries produces the symptoms of angina pectoris (chest pain) because of the oxygen deprivation to the myocardial cells. In the mild form of stable angina pectoris, patients typically experience a discomfort in the chest during physical activity and the symptoms alleviate at rest. However, unstable angina pectoris, which is classified as a type of acute coronary syndrome, is precipitated by the rupture of atherosclerotic plaque with partial thrombosis or vasospasms. Acute coronary syndrome is an umbrella term that includes the occlusion of a coronary artery, which results in a heart attack (i.e., myocardial infarction), in which the myocardium may be irreversibly damaged. Before such a grand heart attack, there may have typically been a number of silent ischemic episodes that went unnoticed unless they were detected by, for instance, routine medical surveillance. The exposure to CNTs may push the progression of atherosclerosis by increasing the atherogenesis and promoting the

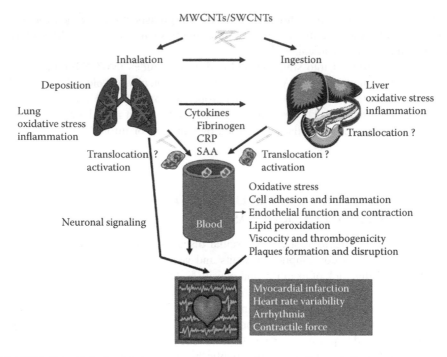

FIGURE 17.1 Relationship between exposure to carbon nanotubes and effect on intermediary endpoints (inflammation, oxidative stress, endothelial dysfunction, prothrombotic tendency) and clinical outcomes. CRP, C-reactive protein; SAA, serum amyloid A.

progression of plaque in the blood vessel, or rendering advanced plaque unstable. A destabilized atheroma may rupture and lead to myocardial ischemia and infarction.

17.3   ASBESTOS EXPOSURE AND ISCHEMIC HEART DISEASE

It is worth looking at some of the experiences learned from asbestos exposures. Asbestos was used, with concomitant risk of lung diseases for workers a long time ago, for building and insulation materials. The potential exposure has resurfaced as the buildings have been being renovated or demolished. Similarly, asbestos-containing brake pads emit fibers during wear. Inhaled asbestos-containing dusts can deposit in the airways, causing pulmonary inflammation, oxidative stress, and subsequently asbestosis, lung cancer, and mesothelioma (a cancer of the pleura that is rather specific for asbestos). Moreover, the pulmonary effects also can lead to cardiovascular disease. By analogy, exposure to CNTs is considered to be able to cause similar pulmonary health effects, although there are clear differences between asbestos fibers and CNTs as well as differences between SWCNTs and MWCNTs.[11]

It has been known for at least 20 years that exposure to asbestos is associated with increased risk of ischemic heart disease. Among the first studies was an investigation of employees from a Swedish shipyard where workers with impaired lung function (as a proxy measure of asbestos exposure) had a relative risk of 3.5 (95% CI: 2.2–5.7) of ischemic heart disease mortality.[12] This was supported by a cohort study of workers in a plant that produced asbestos in South Carolina, showing increased standardized mortality ratio (SMR = 1.20, 95% CI: 1.10–1.32) of ischemic heart disease.[13] However, this study had no direct measurement of smoking, leaving the possibility of confounding because smoking is strongly associated with cardiovascular diseases. Still, the association between asbestos exposure and increased risk of ischemic heart disease (SMR = 1.61, 95% CI: 1.38–1.87) was also observed in a cohort of British asbestos workers after adjustment for smoking.[14]

17.4 DEVELOPMENT OF ATHEROSCLEROSIS

Atherosclerosis is a progressive disease characterized by the accumulation of lipids and fibrous materials in the intima of mainly medium-sized and large arteries in a proinflammatory milieu where immune cells generate both reactive oxygen species (ROS) and cytokines.[15] The development of atherosclerosis typically starts with the expression of cell adhesion molecules on endothelial cells. The cell adhesion molecules bind to monocytes and assist their migration to the intima and differentiation to macrophages. These adhesion molecules include E-selectin and P-selectin as well as intercellular adhesion molecule 1 (ICAM-1) and vascular cell adhesion molecule 1 (VCAM-1). The early phase of atherosclerosis is characterized by the formation of fatty streaks consisting of lipid-laden macrophages (foam cells) and lymphocytes beneath the endothelium. Such fatty streaks are observed in young people, although they may not progress to an atheroma.[15] The development of atherosclerosis in adult life is a consequence of both environmental exposures, lifestyle and hereditary factors. The enlargement of the foam cells and the development of a necrotic core causes the protrusion of the endothelium into the lumen of the blood vessel, which is consequently narrowing and decreasing the blood flow.

The endothelium is the monolayer of cells that covers the inner surface of the blood vessels. These cells are involved in the regulation of blood flow and maintain the vascular wall in a quiescent state by inhibiting inflammation, cellular proliferation, and thrombosis.[15] Hampered ability of regulation of the lumen diameter is referred to as vasomotor dysfunction (or cardiac dysfunction when referring to dysfunction in the blood vessels in the heart). Prospective studies have shown that reduced endothelium-dependent vasodilation is associated with increased risk of hospitalization for cardiovascular events, myocardial infarction, and stroke in subjects with known risk factors for cardiovascular diseases and patients with coronary artery disease.[16] Reduced endothelium-dependent vasodilatation in subjects from the general population is also associated with increased risk of myocardial infarction, stroke, and vascular death after controlling for cardiovascular risk factors.[17]

Endothelial dysfunction is considered to be one of the earliest events in the formation of an atheroma and the magnitude of endothelial dysfunction correlates with the extent of atherosclerosis.[18] Endothelial dysfunction is also a predictor of adverse cardiovascular outcomes in humans.[19–22] The hallmark of endothelial dysfunction is an imbalance between vasodilating and vasoconstricting factors.[23] However, decreased release of plasminogen activator from the endothelium is also used as a marker of endothelial dysfunction.[19] Subjects with a smoking habits, dyslipidemia, diabetes, hypertension, coronary artery disease, or congestive heart failure have endothelial dysfunction.[23] In addition, the vascular function in humans and animals declines in a process described as vascular aging, which is associated with vessel intima thickening and stiffness as well as vasomotor dysfunction, resulting in reduced vaso-dilatation function.[24–27] This age-dependent decline in the endothelium-dependent vasomotor function is affected by atherosclerosis, which accelerates endothelial dysfunction and reduces the response in smooth muscle cells.

17.5   ATHEROSCLEROSIS IN ANIMALS EXPOSED TO FIBERS

In keeping with the observations from epidemiology, exposure to air pollution particles has consistently induced atherosclerosis progression in relevant animal models.[28] Similarly, it has been shown that inhalation of crysotile asbestos (5 mg/m^3, 6 hours/day) for 30 days was associated with a threefold increased level of atheroscle-rosis lesion size in the aorta sinus region.[29] Moreover, intratracheal (i.t.) instillation of Libby amphibole asbestos in rats increased the serum level of acute phase proteins and upregulated genes related to oxidative stress, thrombosis, and vasoconstriction biomarkers in aorta tissue, whereas the serum levels of tumor necrosis factor (TNF) and interleukin (IL) 6 were unaltered.[30,31] These fibers were collected from the Rainy Creek Complex near Libby, Montana, which is a mining area with increased inci-dence of cardiovascular diseases as well as traditional asbestos-related pulmonary disorders.[32] Collectively, the asbestos studies support the notion that fiber-mediated cardiovascular diseases are linked to systemic oxidative stress and/or inflammation. In addition, the mechanisms of CNT-mediated toxicity to cultured cells and multi-cellular organisms encompass oxidative stress reactions, although other nonoxidant effects of CNTs can also be relevant in relation to direct physical interaction between CNTs and biomolecules.[10]

Wild-type mice and rats do not develop atherosclerosis unless they are fed with a high-fat diet, subjected to pharmacological treatment or genetically altered to display a dyslipidemic phenotype. It has been shown that direct intravenous (i.v.) adminis-tration of MWCNTs (200 µg/kg body weight twice a week for 4 months) increased the plaque lesion area from 50% to 66% in the aorta, whereas lower doses (50 or 100 µg/kg) had no effect on plaque progression in rats that were rendered athero-sclerotic by injection with high doses of vitamin D and a 4-month cholesterol-rich diet before the MWCNT exposure.[33] The studies on particle-induced atherosclerosis in animals are usually carried out in susceptible models such as dyslipidemic apo-lipoprotein E (*apoE$^{-/-}$*) or low-density lipoprotein receptor (*LDLr$^{-/-}$*) knockout mice, or Watanabe hereditary hyperlipidemic rabbits.[28] It has been shown that exposure to SWCNT (20 µg/mouse) by i.t. instillation every other week for 8 weeks increased the atherosclerotic lesion area in *apoE$^{-/-}$* mice fed with a high-fat diet, whereas there was no difference between SWCNT-exposed mice and controls in the group of mice on a regular diet.[34] This may be an effect of oxidative stress, suggested by the increased expression of heme oxygenase 1 (HO-1) in the heart of HO-1 receptor transgenic mice and reduced glutathione concentration in aorta tissue after exposure to SWCNTs.[34] This is supported by a study showing signs of systemic inflammation and oxidative stress by increased gene expression of cytokines, HO-1 and E-selectin in aorta tissue at 4 hours after pharyngeal aspiration of SWCNTs or MWCNTs.[35] Exposure to these CNTs by pharyngeal aspiration also increased the serum and liver levels of acute phase proteins, including C-reactive protein (CRP), haptoglobin, and serum amyloid A (SAA).[36] Inhalation of 1 or 5 mg/m^3 of MWCNT for 14 days (6 hours/day) increased the expression of the antioxidant gene NQO1 in the spleen, whereas 0.3 mg/m^3 did not change the gene expression.[37] The same study also found increased gene expression level of IL-10 at 1 mg/m^3, which was confirmed by ELISA on protein level.[37] It is supported by a long-term inhalation study where rats had blood neutrophilia after exposure to 2.5 mg/m^3 of MWCNTs for 13 weeks (6 hours/day and 5 days/week), whereas animals exposed to doses of 0.1 or 0.5 mg/m^3 displayed no signs of systemic inflammation.[38] Another study using the same exposure scenario, although with other doses (0.1, 0.4, 1.5, 6 mg/m^3) and a different type of MWCNT, observed unaltered blood leukocyte counts after the 13-week exposure period.[39]

The exposure of spontaneous hypertensive rats to 0.6 mg/rat of SWCNT (corresponding to 2.4–2.7 mg/kg body weight) by i.t. instillation was associated with a thickening of arterial vessels, edema and leakage of erythrocytes, and myofiber degeneration in the heart, which occurred concomitantly with unaltered serum levels of CRP, TNF, and ICAM-1.[40] The same study also showed increased serum concentration of vasoconstricting factors endothelin 1 and angiotensin I converting enzyme.[40] However, a lower dose of SWCNT (0.5 mg/kg body weight) administered by i.t. instillation at 26 and 2 hours before sacrifice had no effect on endothelium-dependent vasodilation in aorta rings mounted in wire myograph, which might be due to the dose being below the threshold of effect as there was no sign of oxidative stress in the lungs.[41] This is to some extent similar to the observations in mice that were exposed to 10 or 40 µg/mouse of SWCNTs by oropharyngeal aspiration where the cardiac infarct size following ischemia was unaltered, whereas exposure to 40 µg/mouse of acid-functionalized SWCNT was associated with myocardial degeneration and increased infarct size.[42] Increased cardiac infarct size after ischemia/reperfusion injury has also been observed in rats after i.t. instillation of 100 µg/rat of MWCNTs, whereas lower doses (1 or 10 µg/rat) were not associated with this type of cardiac injury.[43] Another study has shown that i.t. instillation of SWCNTs (1 mg/kg) decreased the number of baroreflex sequences, indicating that the autonomic cardiovascular control regulation was altered immediately after the exposure.[44]

Collectively, the animal models indicate that pulmonary exposure to CNTs is associated with increased toxicity to cells in the vasculature, although it should be emphasized that there are several gaps in the current knowledge.

17.6 SYSTEMIC EFFECTS OF ORAL EXPOSURE TO CARBON NANOTUBES

Gastrointestinal exposure to CNTs could occur via mucociliary clearance of inhaled material or from food contaminated with CNTs from use in, for example, sensors or wrappings or from waste disposal via the food chain. In theory, CNTs could also be developed as carriers of food additives. So far, the vascular effects of oral exposure to CNTs have not been described. However, oral exposure to a single dose of 0.64 mg/kg SWCNT has been shown to cause oxidative stress with resulting oxidatively damaged DNA and altered gene regulation in the livers and lungs in rats.[45] Moreover, a similar oral dose of nanosized carbon black particles also induced oxidative stress, oxidatively damaged DNA and upregulation of both oxidative stress response and genes related to inflammation in the liver in rats.[46] With a similar and a 10-fold lower dose of nanosized carbon particles given orally once a week for 10 weeks, there was substantially reduced endothelial function in both lean and obese Zucker rats.[47] These results indicate that oral exposure to particles, and possibly also CNTs, can cause systemic effects including oxidative stress. It is likely that the systemic effects will also include inflammation and similar vascular effects as was seen with oral exposure to carbon nanoparticles. Long-term studies on atherosclerosis progression induced by oral exposure to CNTs are much warranted.

17.7 EFFECTS OF CARBON NANOTUBES ON VASCULAR ENDPOINTS IN CULTURED CELLS

It has been shown that human aortic and microvascular endothelial cells took up CNTs and increase release of proinflammatory cytokines (IL-8 and chemokine ligand 2), expression of cell adhesion molecules (ICAM-1 and VCAM-1), and increase the ROS production after the exposure to either SWCNT or MWCNT at concentrations in the range of 1.5–4.5 µg/mL.[48–50] In addition, the results showed that exposure to CNTs resulted in disruption of the cytoskeleton and cell–cell interactions, as well as decreased transendothelial electrical resistance.[48,50] This suggests that the direct exposure of epithelial and/or endothelial cell layers to CNTs is associated with increased permeability, possibly resulting in elevated translocation of particles across cell barriers. These in vitro studies have the huge asset of being able to investigate effects relatively quickly and cost effectively. In addition, it is possible to narrow the experiments to the study of specific exposure–effect relationships such as direct fiber-mediated expression of adhesion molecules without the effect of contributing factors (e.g., a general systemic inflammation response). However, several studies have shown that only a small fraction of nanoparticles translocate from the alveoli to the circulation[51–54] and so far no data are available for CNTs. This means that in vitro studies of vascular cells typically have high concentrations in the culture medium, which may not be relevant compared with the concentrations in the circulation of humans or animals after the inhalation exposure to CNTs. Still, with this dose–effect limitation in mind, the cell culture studies on specific cardiovascular endpoints support the animal models by showing increased atherogenic and proinflammatory responses after exposure to CNTs.

17.8 EFFECTS ON THROMBOSIS IN EXPERIMENTAL ANIMAL MODELS

The formation of a blood clot—a thrombus—is the final product of coagulation, which is a normal process in cases of injury to the blood vessels. The coagulation cascade can be initiated when blood coagulation factors come in contact with extracellular collagen or damaged cells express tissue factor, resulting in the formation of a thrombus consisting of platelets in a mesh of insoluble fibrin molecules. The thrombus may dislodge from the vessel wall and be transported with the blood as an embolus until it reaches a segment too narrow to pass and thus causes occlusion of the vessel. A prothrombolic or hypercoagulable state is a well-established cause of thrombosis in the venous system such as deep vein thrombosis.[55] However, the linkage between a prothrombotic state and arterial thrombosis has been considered somewhat elusive,[55] although recent evidence indicates a role of prothrombic factors in the development and progression of atherosclerosis.[56]

It has been shown that there was a systemic proinflammatory milieu and increased prothrombotic tendency in vivo during a 40-minute period in the carotid artery, which was photochemically injured at 24 hours after a single i.t. instillation of SWCNTs of either 200 or 400 μg/mouse.[57] This is supported by a study showing increased plasma levels of fibrinogen,[40] which is converted to fibrin by thrombin during the formation of blood clots. The same study showed that the plasma levels of von Willebrand factor were unaltered in spontaneous hypertensive rats after a single i.t. instillation of 0.6 mg/rat of SWCNT.[40] Fibrinogen is synthesized by the liver, whereas von Willebrand factor is produced constitutively in endothelial cells and megakaryocytes. This suggests that the prothrombotic effect of SWCNTs in this study was related to altered function in the liver or platelets rather than to direct damage to the endothelial cells. These observations are supported by a study of direct i.v. administration of 40, 200, or 1000 μg/mouse of SWCNT, showing a dose-dependent reduction in glutathione levels in the liver and lungs at day 90 after the exposure.[58] The effects were not related to systemic inflammation as evidenced by unaltered serum levels of TNF. The systemic oxidative stress is further supported by the observations that intraperitoneal injection of MWCNT (0.25–0.75 mg/kg body weight per day for 5 consecutive days) increased the ROS production in liver tissue homogenate and bone marrow cells.[59,60] Still, the results from a study on i.v. administration of MWCNTs (200 μg/kg body weight) indicated increased plasma concentrations of von Willebrand factor, possibly because of endothelial cell damage.[33] Another study has shown increased platelet aggregation in vitro after exposure to SWCNT or MWCNT.[61] The same type of CNTs also stimulated thrombosis in the carotid artery of rats after an i.v. injection of 25 μg/rat in the femoral vein.[61] In addition, a study on intra-arterial injection of SWCNT showed decreased thrombosis time in mice mesenteric arteries and cremasteric arterioles, which was in keeping with in vitro observations of increased expression of P-selectin on platelets and higher platelet aggregometry.[62] The higher platelet aggregation has been shown to be related to CNT-facilitated extracellular calcium influx,[63] although it has also been suggested that CNTs function as molecular bridges whereby they stimulate the aggregation of platelets.[61] Collectively, there is some evidence from animal models, supported by in vitro studies, that CNTs promote a prothrombotic state in the circulation.

17.9 CONCLUSION

Pulmonary exposure to CNTs is associated with accelerated atherosclerosis progression and prothrombotic state in the blood vessels. These effects occur in a milieu with proinflammatory stimuli and oxidative stress. Although there are still relatively few studies from animal models, there is supporting evidence from cultured cells on specific mechanistic endpoints such as expression of adhesion molecules on endothelial cells or increased platelet aggregation. The effects of inhalation exposure to CNTs in humans on the vascular system have not been investigated. In terms of chemical composition and size, CNTs are similar to combustion-derived air pollution particles for which epidemiology and toxicology consistently indicate important cardiovascular risks. In terms of CNTs with high aspect ratios, recent observations also indicate congruent associations between exposure to asbestos and increased risk of ischemic heart disease mortality and atherosclerosis in animal models. The observations of accelerated atherosclerosis in CNT-exposed mice, together with other mechanistic endpoints of endothelial cell activation, cause concern about the hazard of CNTs.

REFERENCES

1. Simeonova PP, Erdely A. Engineered nanoparticle respiratory exposure and potential risks for cardiovascular toxicity: predictive tests and biomarkers. *Inhal Toxicol* 2009;21 Suppl 1:68–73.
2. Brook RD, Rajagopalan S, Pope CA, III, Brook JR, Bhatnagar A, Diez-Roux AV, Holguin F et al. Particulate matter air pollution and cardiovascular disease: An update to the scientific statement from the American Heart Association. *Circulation* 2010;121:2331–78.
3. Janssen NAH, Hoek G, Simic-Lawson M, Fischer P, van Bree L, ten Brink H, Keuken M et al. Black carbon as an additional indicator of the adverse health effects of airborne particles compared with PM10 and PM2.5. *Environ Health Perspect* 2011;119:1691–9.
4. Møller P, Jacobsen NR, Folkmann JK, Danielsen PH, Mikkelsen L, Hemmingsen JG, Vesterdal LK, Forchhammer L, Wallin H, Loft S. Role of oxidative damage in toxicity of particulates. *Free Radic Res* 2010;44:1–46.
5. Mills NL, Donaldson K, Hadoke PW, Boon NA, MacNee W, Cassee FR, Sandstrom T, Blomberg A, Newby DE. Adverse cardiovascular effects of air pollution. *Nat Clin Pract Cardiovasc Med* 2009;6:36–44.
6. Donaldson K, Aitken R, Tran L, Stone V, Duffin R, Forrest G, Alexander A. Carbon nanotubes: A review of their properties in relation to pulmonary toxicology and workplace safety. *Toxicol Sci* 2006;92:5–22.
7. Lam CW, James JT, McCluskey R, Arepalli S, Hunter RL. A review of carbon nanotube toxicity and assessment of potential occupational and environmental health risks. *Crit Rev Toxicol* 2006;36:189–217.
8. Lloyd-Jones DM, Leip EP, Larson MG, D'Agostino RB, Beiser A, Wilson PW, Wolf PA, Levy D. Prediction of lifetime risk for cardiovascular disease by risk factor burden at 50 years of age. *Circulation* 2006;113:791–8.
9. Yusuf S, Hawken S, Ounpuu S, Dans T, Avezum A, Lanas F, McQueen M et al. Effect of potentially modifiable risk factors associated with myocardial infarction in 52 countries (the INTERHEART study): Case-control study. *Lancet* 2004;364:937–52.
10. Shvedova AA, Pietroiusti A, Fadeel B, Kagan VE. Mechanisms of carbon nanotube-induced toxicity: Focus on oxidative stress. *Toxicol Appl Pharmacol* 2012;261:121–33.
11. Mercer RR, Hubbs AF, Scabilloni JF, Wang L, Battelli LA, Schwegler-Berry D, Castranova V, Porter DW. Distribution and persistence of pleural penetrations by multi-walled carbon nanotubes. *Part Fibre Toxicol* 2010;7:28.
12. Sanden A, Jarvholm B, Larsson S. The importance of lung function, non-malignant diseases associated with asbestos, and symptoms as predictors of ischaemic heart disease in shipyard workers exposed to asbestos. *Br J Ind Med* 1993;50:785–90.

13. Hein MJ, Stayner LT, Lehman E, Dement JM. Follow-up study of chrysotile textile workers: cohort mortality and exposure-response. *Occup Environ Med* 2007;64:616–25.
14. Harding AH, Darnton A, Osman J. Cardiovascular disease mortality among British asbestos workers (1971-2005). *Occup Environ Med* 2012;69:417–21.
15. Hansson GK, Libby P. The immune response in atherosclerosis: A double-edged sword. *Nat Rev Immunol* 2006;6:508–19.
16. Widlansky ME, Gokce N, Keaney JF, Jr, Vita JA. The clinical implications of endothelial dysfunction. *J Am Coll Cardiol* 2003;42:1149–60.
17. Shimbo D, Grahame-Clarke C, Miyake Y, Rodriguez C, Sciacca R, Di TM, Boden-Albala B, Sacco R, Homma S. The association between endothelial dysfunction and cardiovascular outcomes in a population-based multi-ethnic cohort. *Atherosclerosis* 2007;192:197–203.
18. Ross R. Atherosclerosis—an inflammatory disease. *N Engl J Med* 1999;340:115–26.
19. Heitzer T, Schlinzig T, Krohn K, Meinertz T, Munzel T. Endothelial dysfunction, oxidative stress, and risk of cardiovascular events in patients with coronary artery disease. *Circulation* 2001;104:2673–8.
20. Lerman A, Zeiher AM. Endothelial function: cardiac events. *Circulation* 2005;111:363–8.
21. Rubinshtein R, Kuvin JT, Soffler M, Lennon RJ, Lavi S, Nelson RE, Pumper GM, Lerman LO, Lerman A. Assessment of endothelial function by non-invasive peripheral arterial tonometry predicts late cardiovascular adverse events. *Eur Heart J* 2010;31:1142–8.
22. Robinson SD, Ludlam CA, Boon NA, Newby DE. Endothelial fibrinolytic capacity predicts future adverse cardiovascular events in patients with coronary heart disease. *Arterioscler Thromb Vasc Biol* 2007;27:1651–6.
23. Deanfield J, Donald A, Ferri C, Giannattasio C, Halcox J, Halligan S, Lerman A et al. Endothelial function and dysfunction. Part I: methodological issues for assessment in the different vascular beds: A statement by the Working Group on Endothelin and Endothelial Factors of the European Society of Hypertension. *J Hypertens* 2005;23:7–17.
24. Lakatta EG, Levy D. Arterial and cardiac aging: major shareholders in cardiovascular disease enterprises: part I: aging arteries: A "set up" for vascular disease. *Circulation* 2003;107:139–46.
25. Yang YM, Huang A, Kaley G, Sun D. eNOS uncoupling and endothelial dysfunction in aged vessels. *Am J Physiol Heart Circ Physiol* 2009;297:H1829–36.
26. Blackwell KA, Sorenson JP, Richardson DM, Smith LA, Suda O, Nath K, Katusic ZS. Mechanisms of aging-induced impairment of endothelium-dependent relaxation: Role of tetrahydrobiopterin. *Am J Physiol Heart Circ Physiol* 2004;287:H2448–53.
27. Lesniewski LA, Connell ML, Durrant JR, Folian BJ, Anderson MC, Donato AJ, Seals DR. B6D2F1 Mice are a suitable model of oxidative stress-mediated impaired endothelium-dependent dilation with aging. *J Gerontol A Biol Sci Med Sci* 2009;64:9–20.
28. Møller P, Mikkelsen L, Vesterdal LK, Folkmann JK, Forchhammer L, Roursgaard M, Danielsen PH, Loft S. Hazard identification of particulate matter on vasomotor dysfunction and progression of atherosclerosis. *Crit Rev Toxicol* 2011;41:339–68.
29. Fukagawa NK, Li M, Sabo-Attwood T, Timblin CR, Butnor KJ, Gagne J, Steele C, Taatjes DJ, Huber S, Mossman BT. Inhaled asbestos exacerbates atherosclerosis in apolipoprotein E-deficient mice via CD4+ T cells. *Environ Health Perspect* 2008;1169:1218–25.
30. Shannahan JH, Schladweiler MC, Thomas RF, Ward WO, Ghio AJ, Gavett SH, Kodavanti UP. Vascular and thrombogenic effects of pulmonary exposure to Libby amphibole. *J Toxicol Environ Health A* 2012;75:213–31.
31. Shannahan JH, Alzate O, Winnik WM, Andrews D, Schladweiler MC, Ghio AJ, Gavett SH, Kodavanti UP. Acute phase response, inflammation and metabolic syndrome biomarkers of Libby asbestos exposure. *Toxicol Appl Pharmacol* 2012;260:105–14.
32. Larson TC, Antao VC, Bove FJ. Vermiculite worker mortality: estimated effects of occupational exposure to Libby amphibole. *J Occup Environ Med* 2010;52:555–60.

33. Xu YY, Yang J, Shen T, Zhou F, Xia Y, Fu Y, Meng J et al. Intravenous administration of multi-walled carbon nanotubes affects the formation of atherosclerosis in Sprague-Dawley rats. *J Occup Health* 2012; 54:361–9.

34. Li Z, Hulderman T, Salmen R, Chapman R, Leonard SS, Young SH, Shvedova A, Luster MI, Simeonova PP. Cardiovascular effects of pulmonary exposure to single-wall carbon nanotubes. *Environ Health Perspect* 2007;115:377–82.

35. Erdely A, Hulderman T, Salmen R, Liston A, Zeidler-Erdely PC, Schwegler-Berry D, Castranova V et al. Cross-talk between lung and systemic circulation during carbon nanotube respiratory exposure. Potential biomarkers. *Nano Lett* 2009;9:36–43.

36. Erdely A, Liston A, Salmen-Muniz R, Hulderman T, Young SH, Zeidler-Erdely PC, Castranova V, Simeonova PP. Identification of systemic markers from a pulmonary carbon nanotube exposure. *J Occup Environ Med* 2011;53:S80–6.

37. Mitchell LA, Gao J, Wal RV, Gigliotti A, Burchiel SW, McDonald JD. Pulmonary and systemic immune response to inhaled multiwalled carbon nanotubes. *Toxicol Sci* 2007;100:203–14.

38. Ma-Hock L, Treumann S, Strauss V, Brill S, Luizi F, Mertler M, Wiench K, Gamer AO, van Ravenzwaay B, Landsiedel R. Inhalation toxicity of multiwall carbon nanotubes in rats exposed for 3 months. *Toxicol Sci* 2009;112:468–81.

39. Pauluhn J. Subchronic 13-week inhalation exposure of rats to multiwalled carbon nanotubes: toxic effects are determined by density of agglomerate structures, not fibrillar structures. *Toxicol Sci* 2010;113:226–42.

40. Ge C, Meng L, Xu L, Bai R, Du J, Zhang L, Li Y, Chang Y, Zhao Y, Chen C. Acute pulmonary and moderate cardiovascular responses of spontaneously hypertensive rats after exposure to single-wall carbon nanotubes. *Nanotoxicology* 2012;6:526–42.

41. Vesterdal LK, Jantzen K, Sheykhzade M, Roursgaard M, Folkmann JK, Loft S, Møller P. Pulmonary exposure to particles from diesel exhaust, urban dust or single-walled carbon nanotubes and oxidatively damaged DNA and vascular function in *apoE$^{-/-}$* mice. *Nanotoxicology* 2012 (doi:10.3109/17435390.2012.750385).

42. Tong H, McGee JK, Saxena RK, Kodavanti UP, Devlin RB, Gilmour MI. Influence of acid functionalization on the cardiopulmonary toxicity of carbon nanotubes and carbon black particles in mice. *Toxicol Appl Pharmacol* 2009;239:224–32.

43. Thompson LC, Frasier CR, Sloan RC, Mann EE, Harrison BS, Brown JM, Brown DA, Wingard CJ. Pulmonary instillation of multi-walled carbon nanotubes promotes coronary vasoconstriction and exacerbates injury in isolated hearts. *Nanotoxicology* 2012 (doi:10.3109/17435390.2012.744858).

44. Legramante JM, Valentini F, Magrini A, Palleshi G, Sacco S, Iavicoli I, Pallante M et al. Cardiac autonomic regulation after lung exposure to carbon nanotubes. *Hum Exp Toxicol* 2009;28:369–75.

45. Folkmann JK, Risom L, Jacobsen NR, Wallin H, Loft S, Møller P. Oxidatively damaged DNA in rats exposed by oral gavage to C_{60} fullerenes and single-walled carbon nanotubes. *Environ Health Perspect* 2009;117:703–8.

46. Danielsen PH, Loft S, Jacobsen NR, Jensen KA, Autrup H, Ravanat JL, Wallin H, Møller P. Oxidative stress, inflammation and DNA damage in rats after intratracheal instillation or oral exposure to ambient air and wood smoke particulate matter. *Toxicol Sci* 2010;118:574–85.

47. Folkmann JK, Vesterdal LK, Sheykhzade M, Loft S, Møller P. Endothelial dysfunction in normal and prediabetic rats with metabolic syndrome exposed by oral gavage to carbon black nanoparticles. *Toxicol Sci* 2012;129:98–107.

48. Walker VG, Li Z, Hulderman T, Schwegler-Berry D, Kashon ML, Simeonova PP. Potential in vitro effects of carbon nanotubes on human aortic endothelial cells. *Toxicol Appl Pharmacol* 2009;236:319–28.

49. Vidanapathirana AK, Lai X, Hilderbrand SC, Pitzer JE, Podila R, Sumner SJ, Fennell TR, Wingard CJ, Witzmann FA, Brown JM. Multi-walled carbon nanotube directed gene and protein expression in cultured human aortic endothelial cells is influenced by suspension medium. *Toxicology* 2012;302:114–22.

50. Pacurari M, Qian Y, Fu W, Schwegler-Berry D, Ding M, Castranova V, Guo NL. Cell permeability, migration, and reactive oxygen species induced by multiwalled carbon nanotubes in human microvascular endothelial cells. *J Toxicol Environ Health A* 2012;75:112–28.

51. Kreyling WG, Semmler M, Erbe F, Mayer P, Takenaka S, Schulz H, Oberdörster G, Ziesenis A. Translocation of ultrafine insoluble iridium particles from lung epithelium to extrapulmonary organs is size dependent but very low. *J Toxicol Environ Health A* 2002;65:1513–30.

52. Mills NL, Amin N, Robinson SD, Anand A, Davies J, Patel D, de la Fuente JM et al. Do inhaled carbon nanoparticles translocate directly into the circulation in humans? *Am J Respir Crit Care Med* 2006;173:426–31.

53. Moller W, Felten K, Sommerer K, Scheuch G, Meyer G, Meyer P, Haussinger K, Kreyling WG. Deposition, retention, and translocation of ultrafine particles from the central airways and lung periphery. *Am J Respir Crit Care Med* 2008;177:426–32.

54. Wiebert P, Sanchez-Crespo A, Falk R, Philipson K, Lundin A, Larsson S, Moller W, Kreyling WG, Svartengren M. No significant translocation of inhaled 35-nm carbon particles to the circulation in humans. *Inhal Toxicol* 2006;18:741–7.

55. Chan MY, Andreotti F, Becker RC. Hypercoagulable states in cardiovascular disease. *Circulation* 2008;118:2286–97.

56. Borissoff JI, Spronk HMH, ten Cate H. The hemostatic system as a modulator of atherosclerosis. *N Engl J Med* 2011;364:1746–60.

57. Nemmar A, Hoet PH, Vandervoort P, Dinsdale D, Nemery B, Hoylaerts MF. Enhanced peripheral thrombogenicity after lung inflammation is mediated by platelet-leukocyte activation: role of P-selectin. *J Thromb Haemost* 2007;5:1217–26.

58. Yang ST, Wang X, Jia G, Gu Y, Wang T, Nie H, Ge C, Wang H, Liu Y. Long-term accumulation and low toxicity of single-walled carbon nanotubes in intravenously exposed mice. *Toxicol Lett* 2008;181:182–9.

59. Patlolla AK, Hussain SM, Schlager JJ, Patlolla S, Tchounwou PB. Comparative study of the clastogenicity of functionalized and nonfunctionalized multiwalled carbon nanotubes in bone marrow cells of Swiss-Webster mice. *Environ Toxicol* 2010;25:608–21.

60. Patlolla AK, Berry A, Tchounwou PB. Study of hepatotoxicity and oxidative stress in male Swiss-Webster mice exposed to functionalized multi-walled carbon nanotubes. *Mol Cell Biochem* 2011;358:189–99.

61. Radomski A, Jurasz P, Alonso-Escolano D, Drews M, Morandi M, Malinski T, Radomski MW. Nanoparticle-induced platelet aggregation and vascular thrombosis. *Br J Pharmacol* 2005;146:882–93.

62. Bihari P, Holzer M, Praetner M, Fent J, Lerchenberger M, Reichel CA, Rehberg M, Lakatos S, Krombach F. Single-walled carbon nanotubes activate platelets and accelerate thrombus formation in the microcirculation. *Toxicology* 2010;269:148–54.

63. Semberova J, De Paoli Lacerda SH, Simakova O, Holada K, Gelderman MP, Simak J. Carbon nanotubes activate blood platelets by inducing extracellular Ca^{2+} influx sensitive to calcium entry inhibitors. *Nano Lett* 2009;9:3312–7.

18 Single- and Multiwalled Carbon Nanotubes Toxicity and Potential Applications in Neuroregeneration

Fariborz Tavangarian, Guoqiang Li, and Yiyao Li

CONTENTS

18.1 INTRODUCTION

The use of nanomaterials in biomedical application and nanomedicine has increased dramatically in the past decade because of their unique properties and capabilities in the physiological environment. Some materials have shown an increase in their bioactivity and biodegradability as their crystallite size decreased from micron size to nanoscale.[1,2]

Toxicity can be defined as the degree to which a material can damage an exposed cell or a whole organ. It should be noted that toxicity is in direct relation to the amount of *damage*. All materials elicit some type of responses when they come into contact with living tissues but toxicity is a dose-dependent effect.[3] Therefore, when we evaluate the toxicity of a material we need to consider many parameters.

Materials may be either toxic or nontoxic and there is a need to screen these materials for toxicity before biomedical applications. If materials are toxic they may damage the surrounding tissue and should not be applicable as biomaterials.

Nontoxic materials may be divided into three subcategories: (1) biologically inert or inactive materials (or bioinert), in this condition a fibrous tissue capsule forms around the material; (2) biologically active materials (or bioactive), these materials bond with the surrounding tissue; (3) biologically dissolved materials (or biodegradable), in this case the material is replaced by the surrounding tissue.

In recent years, researchers have evaluated the potential of carbon nanotubes (CNTs) as a possible biomaterial. CNTs have excellent mechanical, electrical, and thermal properties, which make them a great candidate for biomedical applications. Particularly, using CNTs in neural scaffolds has become a hot subject. It has been shown that CNTs can boost cell viability and growth of neurons and neuronal cells.[4] Using these properties, we can produce a new generation of neural scaffolds that can be used for repairing large gaps in severed nerves. But before using CNTs in vivo, the toxicity should be evaluated. The toxicity of CNTs has become the subject of heated debate between different research groups. Many publications have reported the toxicity of CNTs, whereas others have shown that CNTs may be nontoxic both in in vivo or in vitro studies. In this chapter we discuss these inconsistencies and draw conclusions on the toxicity of CNTs.

18.2 NERVOUS SYSTEM

The nervous system consists of two main parts: central nervous system (CNS) and peripheral nervous systems (PNS) in which each part has its own repair procedures. In PNS, axons regenerate through the proliferation of Schwann cells that wrap around nerve fibers forming myelin sheaths and the phagocytosis of myelin by macrophages or monocytes, and the development of Büngner bands by the bundling of Schwann cells and recovering axons in the distal segment.[5] In CNS, recovery and regeneration of axons are much more difficult because axons are surrounded by inhibitory molecules that prevent neural growth.[6] After injury, a fluid fills the cavity which is surrounded by a dense glial scar. The presence of astrocytes, glycosaminoglycans, and other inhibitory molecules prevents neurons and other cells from infiltrating the injured site, which in turn may lead to the loss of axonal connections.[7] To treat a gap in a severed nerve, we can implant a donor nerve into the injured area. This method has several downsides, which include the loss of the donor nerve function, restricted availability of donor nerve, formation of potentially painful neuromas, size mismatch between the donor nerve and the injured nerve, and a lack of vascularization of the donor nerve.[8] Hence, there is a crucial need for a bridging scaffold that can serve as a nerve conduit with similar characteristics to that of a donor nerve. CNTs may be a promising candidate for neural tissue engineering applications because of their unique properties including biocompatibility, cell viability, cell growth, cell attachment and differentiation, as well as conductivity and excellent mechanical and electrical properties.[9] They can maintain cell viability and promote growth of cells and hence support their healings in severed nerves. Several studies have been performed on the use of CNTs as neural scaffolds in the nervous system. In Section 18.3, we discuss the most recent findings in this area.

18.3 CARBON NANOTUBES AS NEURAL SCAFFOLDS

Cell viability and cell growth are extremely important in nervous system injuries. Design and fabrication of new scaffolds, which affect these two parameters, can open a new horizon for neuroregeneration. Because neural cells can attach and grow on CNTs, they may serve as biomimetic scaffolds to guide axon regeneration and to improve neural activities.

There are several reports of interaction of CNTs and neural cells, which show the potential application of CNTs for neuroregeneration. Mazzatenta et al.[10] developed an integrated single-walled carbon nanotube (SWCNT)/neuron system and showed that electrical stimulation delivered via SWCNTs can induce neuronal signaling. Gheith et al.[11] showed that freestanding structures of modified SWCNTs can guide the outgrowth of neurites, support cell-to-cell communications and cell differentiation, and serve as a biocompatible platform for neuroprosthetic implants. Additional studies by Gheith et al. showed that they could externally stimulate NG108 neuroblastoma cells in culture using the electrical properties of SWCNT films.[12] They also found that SWCNTs can induce inner currents in NG108 neuroblastoma cells. Sorkin et al.[13] produced an SWCNT substrate and cultured them with rat cortical and hippocampal cells and found that neurites can grow on the substrates after 3 days. Ni et al.[14] found that proper treatment of neonatal rat hippocampal neurons grown on polyethyleneimine (PEI)-coated coverslips with water-soluble SWCNT graft copolymers can enhance the outgrowth of neurites. Liopo et al.[15] investigated the response of NG108 neuroblastoma cells to SWCNTs and found that NG108 cells can survive, attach, and grow in direct contact with SWCNT films.

Mattson et al.[16] produced PEI coated with multiwalled carbon nanotube (MWCNT) layers. They showed that rat hippocampal neurons can survive and grow on the prepared substrates. These results were also confirmed by Hu et al.[17] They studied the behavior of rat hippocampal neurons on MWCNTs and found that the positively charged MWCNTs had a higher number of growth cones and neurite branches. Also, they showed that the surface charges of MWCNTs can influence the length of neurites branching, and the number of growth cones. Lovat et al.[18] showed that nerve growth on CNTs is accompanied with a significant increase in network activity and that purified CNTs can improve neural signal transfer and cell adhesion.

All the above studies showed that CNTs have the potential to serve in the repair of nervous system injuries, although there are a few reports on the toxicity of CNTs. In Section 18.4, the toxicity issue of CNTs as well as its reason is discussed.

18.4 TOXICITY

Different researchers have evaluated the toxicity of CNTs. After detailed study of the literature it can be concluded that the toxicity of CNTs depends on many factors such as impurities (metal catalysts, graphite, amorphous carbon, etc.), dimension, distribution and concentration, crystal structure, degree of aggregation, influence of cell culture medium, many other secondary chemicals and pH values. Major limitation with all these studies is that there is no standardized procedure to define the toxicity of CNTs. However, it has been shown that there are conflicting reports on CNT toxicity. The same problems apply to exposure conditions such as various cell culture media, different amount of micronutrients, and different ratio between CNTs and medium.[3] To clarify this issue, we present a brief summary of the most important findings in Section 18.4.1.

18.4.1 TOXICITY OF SINGLE-WALLED CARBON NANOTUBES

Toxicity of SWCNTs was studied by Herzog et al.[19] on three cell models including the human alveolar carcinoma epithelial cell line (A549), normal human bronchial epithelial cell line (BEAS-2B), and in an immortalized human keratinocyte cell line (HaCaT) using the clonogenic assay. They found that SWCNTs elicit a strong cytotoxic response and that toxicity was different in the various cell lines. However, Davoren et al.[20] reported that exposure of A549 cells to a wide dose range of SWCNTs (1.56–800 μg/mL) for 24 hours revealed a low toxicity.

Shvedova et al.[21] also studied the cytotoxicity of SWCNTs in HaCaT cells and found that SWCNTs enhanced oxidative stress and cellular toxicity determined by the formation of free radicals, accumulation of peroxidative products, and a loss of cell viability. The prepared SWCNTs had a substantial amount of metal impurities that directly affected the oxidative stress and consequently resulted in a decrease in cell viability.

Giorgio et al.[22] evaluated the cytotoxicity of SWCNTs in mouse macrophage cell line RAW 264.7 and found SWCNTs altered cell membrane morphology after 24 hours and the toxic effects increased considerably after 72 hours.

Tong et al.[23] investigated the toxicity of SWCNTs in vivo in mice exposed by oropharyngeal aspiration to 10 or 40 μg of saline-suspended SWCNTs. After 24 hours, pulmonary inflammatory responses and cardiac effects were evaluated and compared with control animals. Studies showed that no inflammatory response was noted at the lower particle concentration and concluded that the toxicity of SWCNTs was directly related to their dose.

Yang et al.[24] investigated the long-term (3 month) accumulation and toxicity with intravenously dosed SWCNTs in the main organs of mice, such as liver, lung, and spleen, and observed low toxicity. Histological observations showed that slight inflammation and inflammatory cell infiltration occurred in the lung, but the serum immunological indicators (CH50l level and tumor necrosis factor-α [TNF-α] level) remained unchanged. The decreasing glutathione (GSH) level and increasing malondialdehyde (MDA) level suggested that the toxicity of SWCNTs might be due to oxidative stress.

Other studies by Casey et al.[25] investigated the cytotoxic effects of isolated and large-diameter bundled aggregates of SWCNTs and showed that the larger-diameter bundled aggregates had low toxicity compared with isolated SWCNTs because of internalization within the cell walls.

Alpatova et al.[26] studied the toxicity of SWCNTs noncovalently functionalized by a range of natural (gum arabic, amylose, Suwannee River natural organic matter) and synthetic (polyvinylpyrrolidone, Triton X-100) dispersants that bind to SWCNTs surface through different physiosorption mechanisms. They found no cytotoxic effects of these SWCNT suspensions on the prokaryotic, bacterial (*Escherichia coli*), or eukaryotic (WB-F344 rat liver epithelial) cell types. When the dispersant itself was toxic, losses of cell viability were observed. These results suggest a strong dependence of the toxicity of SWCNT suspensions on the toxicity of the solubilizing agent and point to the potential of noncovalent functionalization with nontoxic dispersants as a method for the preparation of aqueous suspensions of biocompatible CNTs.

Murray et al.[27] investigated the influences of SWCNTs both in vitro and in vivo. They found an increase in epidermal thickness and accumulation and activation of dermal fibroblasts in engineered skin in contact with SWCNTs that caused an increased collagen and release of proinflammatory cytokines. Also, exposure of JB6 P+ cells to SWCNTs (containing 30% iron) produced ESR-detectable hydroxyl radicals and caused a significant dose-dependent activation of AP-1. They found that topical exposure to SWCNTs induced free-radical generation, oxidative stress, and inflammation, thus causing dermal toxicity. On the other hand, they found that when partially purified SWCNTs (0.23% iron) were introduced to the cells no significant changes occurred in AP-1 activation.

Fiorito et al.[28] examined the toxicity of highly purified SWCNTs using murine and human macrophage cells in vitro. They evaluated the ability of SWCNTs to stimulate the release of nitric oxide (NO) by murine macrophages and to induce the phagocytic activity of human macrophages. They showed that highly purified SWCNTs do not induce the release of NO by murine macrophages and possess a low toxicity.

Ema et al.[29] reported the effects of SWCNTs on dermal and eye irritation using rabbits and skin sensitization using guinea pigs with the maximum allowable concentration for administration. They showed that SWCNTs were not irritating to the skin and eyes.

Other studies by Pulskamp et al.[30] investigated the cytotoxicity of SWCNTs on rat macrophages (NR8383) and human A549 lung cells. They found that SWCNTs were nontoxic in these cell lines; however, they observed a dose- and time-dependent increase of intracellular reactive oxygen species (ROS) and a decrease of the mitochondrial membrane potential with the commercial SWCNTs in both cell types, whereas incubation with the purified SWCNTs had no effect.

Incubation of SWCNTs with alveolar macrophages at low doses (0.38 mg/cm^2) decrease the process of phagocytosis.[31] Other studies reported that SWCNTs were nontoxic in human lung epithelial-like cells,[32] murine and human macrophages,[28] and rat cardiac muscle cells.[33] Studies in human epidermal keratinocytes (HEK) with functionalized CNT with 6-aminohexanoic acid ranging in concentration from 0.00000005 to 0.05mg/mL showed a decreased in viability and an increase in interleukin (IL)-8 at 24 and 48 hours. Transmission electron microscopy showed these functionalized CNT within the cytoplasmic vacuoles in the cells. Studies with Pluronic F127 surfactant caused a dispersion of these CNT in culture with less toxicity.[34]

18.4.2 TOXICITY OF MULTIWALLED CARBON NANOTUBES

Respiratory toxicity of MWCNTs was evaluated by Muller et al.[35] in which they administered MWCNTs via the trachea to rats and estimated the lung persistence, inflammation, and fibrosis by biochemical and histological techniques. They reported that MWCNTs were present in the lung after 60 days and induced inflammation and fibrosis. After 2 months, pulmonary lesions induced by MWCNTs were characterized by the formation of collagen-rich granulomas protruding in the bronchial lumen, in association with inflammation of the alveoli and in the surrounding tissues. These lesions were caused by the accumulation of large MWCNT agglomerates in the airways. In addition, MWCNTs stimulated the production of TNF-α in the lung of these rats.

Han et al.[36] investigated the potential toxicity of MWCNTs in C6 rat glioma cell lines. They found that exposure of C6 rat glioma cells to various sizes of MWCNTs concentrations between 25 and 400 mg/mL decreased the cell viability in a concentration- and time-dependent manner and resulted in a concentration-dependent cell apoptosis, G1 cell-cycle arrest, and an increase in oxidative stress levels. Their finding showed that small MWCNTs were more toxic than large MWCNTs.

Ellinger-Ziegelbauer and Pauluhn[37] evaluated pulmonary toxicity of MWCNTs on rats and found that pulmonary inflammation was concentration dependent with evidence of regression over time. The time course of pulmonary inflammation associated with retained MWCNTs was independent of the concentration of residual metal catalyst impurities, which supports the conclusion that the predominant response to inhaled MWCNTs could be related to the structure and not the catalyst impurities ($\leq 0.5\%$).

Simon-Deckers et al.[38] investigated the influence of MWCNTs on A549 human pneumocytes and found that MWCNTs were localized in the cytoplasmic vacuoles. These obtained results showed that MWCNTs were toxic and their toxicity was not dependent on their length or on the presence of impurities of the metal catalyst.

Deng et al.[39] injected mice with water-soluble MWCNTs with 60 or 100 mg/kg body weight on day 1 (and 4 hours after the first injection) and found no significant changes in the phagocytic activity of the reticuloendothelial system, activity of reduced GSH, superoxide dismutase, and MDA in splenic homogenate in 2 months. Histopathology showed no observable sign of damage in the spleen; however, the accumulated water-soluble MWCNTs gradually transferred from the red pulp to the white pulp over the exposure time, which could initiate the adaptive immune response.

Cheng et al.[40] evaluated the toxicity of unpurified MWCNTs on human macrophages. They observed that the cell viability decrease was due to necrosis. Also, they evaluated the potential toxicity of purified MWCNTs and found that the toxicity was not from the contaminant (Fe_2O_3). The authors also investigated the possible mechanism of interaction of MWCNTs with cell lines and realized that unpurified MWCNTs entered the cell actively and passively, and often penetrated through the plasma membrane into the cytoplasm and the nucleus, which suggests that MWCNTs may cause incomplete phagocytosis or mechanically pierce through the plasma membrane resulting in oxidative stress and cell death.

Young et al.[41] studied the influence of functionalized MWCNTs on *E. coli*. Chemiluminescence assays were used to measure the concentration of adenosine triphosphate (ATP) released from the damaged cells. These results showed that functionalized MWCNTs inhibited cell viability concluded from the increased ATP compared to the controls, indicating direct piercing of *E. coli* by functionalized

MWCNTs. The mechanism of MWCNTs on mouse macrophages was investigated by Giorgio et al.[22] Their findings showed that macrophages swallowed particles with their pseudopods and plasma membrane extensions and caused cell death by necrosis and MWCNTs caused an increase in ROS production simultaneously. Protracted oxidative stress may cause direct oxidative DNA damage, cytotoxicity, and necrosis. Moreover ROS contributed to cellular signaling, leading to apoptosis and cell-cycle arrest.

Prylutska et al.[42] evaluated the effect of MWCNTs on rat erythrocytes and thymocytes. They found that MWCNTs at concentrations up to 25 mg/mL were nontoxic, whereas at concentrations of 50 mg/mL caused an increase in erythrocyte hemolysis, a decrease in the number of viable thymocytes in suspension, and an inhibition of mitochondrial electron transport.

Magrez et al.[43] studied the toxic effect of MWCNTs in lung tumor cells as a function of their aspect ratio and surface chemistry and showed that toxicity was dependent on size.

Ding et al.[44] studied the human skin fibroblast cell populations in contact with MWCNTs. Their finding showed that certain concentrations of MWCNTs can influence the cellular growth and differentiation. Furthermore, they proposed that the regulation and expression of p38/ERK and epidermal growth factor receptor from fibroblasts may play a substantial role in cancer therapy, especially derived cancers.

Monteiro-Riviere et al.[45] studied the MWCNT interactions with HEK at several concentrations and up to 48 hours and showed that unfunctionalized MWCNTs were present in the cytoplasmic vacuoles of 59.1% of the cells at 24 hours and 84% of the cells at 48 hours. These MWCNTs also induced the release of the proinflammatory cytokine IL-8 in a time-dependent manner.

It can be concluded that the toxicity of CNTs is dependent on many factors, such as their structure (SWCNT vs. MWCNT), particle size, concentration, mobility in the environment, length and aspect ratio, surface area, degree of aggregation, chemical reactivity, extent of oxidation, surface topology, surface functionalization, bond functional group(s), and method of manufacturing (which can leave catalyst residues and impurities).[46–48]

18.5 PURIFICATION OF CARBON NANOTUBES

Prepared CNTs may contain impurities including graphite, amorphous carbon, metal catalyst, and smaller fullerenes.[49] Hence a purification step should be applied before using them in therapeutic devices. The purification methods can be divided into three main categories including chemical oxidation, physical-based purification, and multistep purification.[50–54] In chemical-based purification, which removes the amorphous carbon and metal impurities, a substantial amount of CNTs is lost and the structure of CNTs is destroyed. Physical-based purification maintains the structure of CNTs and can separate them according to their length or conductivity. However, the ability of this method in removing impurities is not as effective as chemical-based methods. Also, CNTs should be dispersed before purification and therefore a pretreatment operation is required. Multistep purification is a combination of several purification methods simultaneously in one purification process to achieve high-purity CNTs.[54–56] For instance, one of the most popular multistep purification methods is gas phase oxidation followed by acid treatment.[55–65] It should be noted that not all the purification methods are suitable for treatment of carbon nanostructures as biomedical devices. For example, some researchers reported that the cytotoxicity of MWCNTs was enhanced after acid treatment.[43]

18.6 CONCLUSION

On the basis of the literature cited in the text, we found that there is a lack of standardized methods to evaluate the toxicity of CNTs. CNTs may be reported as being toxic whereas in some literature they are not toxic. The lack of consistency in the experimental parameters and evaluating procedures prevents any accurate comparison between the results obtained by various researchers. All research performed according to a standard can be compared with each other and a rational conclusion could be drawn between the obtained results. Also, it seems that CNTs are inherently nontoxic and MWCNTs are intrinsically toxic and their toxicity comes from impurities or other compounds accompanied with them. Furthermore as many medicines, protein, and even some vitamins are toxic in overdose, CNTs can also be toxic in high dosages and may be used as a valuable medicine or therapeutic devices only in lower concentrations. However, it has been shown that the toxicity profiles can vary because of the type of viability assay used to assess the toxicity of nanomaterials such as CNTs. CNT interactions with dye-based assays can give varied results and may not be appropriate for assessing the toxicity of some types of nanomaterials. Therefore, there is a need for better high-throughput screening of CNTs so that they can be properly assessed for use in biomedical applications such as scaffolds for neuroregeneration.[66]

REFERENCES

1. Tavangarian F, Emadi R. Nanostructure effects on the bioactivity of forsterite bioceramic. *Mater Lett* 2011;65:740–3.
2. Tavangarian F, Emadi R. Improving degradation rate and apatite formation ability of nanostructure forsterite. *Ceram Int* 2011;37:2275–80.
3. Ren HX, Chen X, Liu JH, Gu N, Huang XJ. Toxicity of single-walled carbon nanotube: How we were wrong? *Mater Today* 2010;13:6–8.
4. Lobo AO, Corat MAF, Antunes EF, Palma MBS, Pacheco-Soares C, Garcia EE, Corat EJ. An evaluation of cell proliferation and adhesion on vertically-aligned multi-walled carbon nanotube films. *Carbon* 2010;48:245–54.
5. Zhang L, Webster TJ. Nanotechnology and nanomaterials: Promises for improved tissue regeneration. *Nano Today* 2009;4:66–80.
6. Nash HH, Borke RC, Anders JJ. Ensheathing cells and methylprednisolone promote axonal regeneration and functional recovery in the lesioned adult rat spinal cord. *J Neurosci* 2002;22:7111–20.
7. Kozlova EN. [Strategies to repair lost sensory connections to the spinal cord]. *Mol Biol* 2008;42:729–37.
8. Tavangarian F, Li Y. Carbon nanostructures as nerve scaffolds for repairing large gaps in severed nerves. *Ceram Int* 2012;38:6075–90.
9. Hummer G, Rasaiah JC, Noworyta JP. Water conduction through the hydrophobic channel of a carbon nanotube. *Nature* 2001;414:188–90.

10. Mazzatenta A, Giugliano M, Campidelli S, Gambazzi L, Businaro L, Markram H, Prato M, Ballerini L. Interfacing neurons with carbon nanotubes: Electrical signal transfer and synaptic stimulation in cultured brain circuits. *J Neurosci* 2007;27:6931–6.
11. Gheith MK, Sinani VA, Wicksted JP, Matts RL, Kotov NA. Single-walled carbon nanotube polyelectrolyte multilayers and freestanding films as a biocompatible platform for neuroprosthetic implants. *Adv Mater* 2005;17:2663–70.
12. Gheith MK, Pappas TC, Liopo AV, Sinani VA, Shim BS, Motamedi M, Wicksted JP, Kotov NA. Stimulation of neural cells by lateral currents in conductive layer-by-layer films of single-walled carbon nanotubes. *Adv Mater* 2006;18:2975–9.
13. Sorkin R, Gabay T, Blinder P, Baranes D, Ben-Jacob E, Hanein Y. Compact self-wiring in cultured neural networks. *J Neural Eng* 2006;3:95–101.
14. Ni Y, Hu H, Malarkey EB, Zhao B, Montana V, Haddon RC, Parpura V. Chemically functionalized water soluble single-walled carbon nanotubes modulate neurite outgrowth. *J Nanosci Nanotechnol* 2005;5:1707–12.
15. Liopo AV, Stewart MP, Hudson J, Tour JM, Pappas TC. Biocompatibility of native and functionalized single-walled carbon nanotubes for neuronal interface. *J Nanosci Nanotechnol* 2006;6:1365–74.
16. Mattson MP, Haddon RC, Rao AM. Molecular functionalization of carbon nanotubes and use as substrates for neuronal growth. *J Mol Neurosci* 2000;14:175–82.
17. Hu H, Ni Y, Montana V, Haddon RC, Parpura V. Chemically functionalized carbon nanotubes as substrates for neuronal growth. *Nano Lett* 2004;4:507–11.
18. Lovat V, Pantarotto D, Lagostena L, Cacciari B, Grandolfo M, Righi M, Spalluto G Prato M, Ballerini L. Carbon nanotube substrates boost neuronal electrical signaling. *Nano Lett* 2005;5:1107–10.
19. Herzog E, Casey A, Lyng FM, Chambers G, Byrne HJ, Davoren M. A new approach to the toxicity testing of carbon based nanomaterials—The clonogenic assay. *Toxicol Lett* 2007;174:49–60.
20. Davoren M, Herzog E, Casey A, Cottineau B, Chambers G, Byrne HJ, Lyng FM. In vitro toxicity evaluation of single walled carbon nanotubes on human A549 lung cells. *Toxicol In Vitro* 2007;21:438–48.
21. Shvedova AA, Castranova V, Kisin ER, Schwegler-Berry D, Murray AR, Gandelsman VZ, Maynard A, Baron P. Exposure to carbon nanotube material: Assessment of nanotube cytotoxicity using human keratinocyte cells. *J Toxicol Env Health A* 2003;66:1909–26.
22. Giorgio MLD, Bucchianico SD, Ragnelli AM, Aimola P, Santucci S, Poma A. Effects of single and multi walled carbon nanotubes on macrophages: Cyto and genotoxicity and electron microscopy. *Mutat Res* 2011;722:20–31.
23. Tong H, McGee JK, Saxena RK, Kodavanti UP, Devlin RB, Gilmour MI. Influence of acid functionalization on the cardiopulmonary toxicity of carbon nanotubes and carbon black particles in mice. *Toxicol Appl Pharm* 2009;239:224–32.
24. Yang S, Wang X, Jia G, Gu Y, Wang T, Nie H, Ge C, Wang H, Liu Y. Long-term accumulation and low toxicity of single-walled carbon nanotubes in intravenously exposed mice. *Toxicol Lett* 2008;181:182–9.
25. Casey A, Davoren M, Herzog E, Lyng FM, Byrne HJ, Chambers G. Probing the interaction of single walled carbon nanotubes within cell culture medium as a precursor to toxicity testing. *Carbon* 2007;45:34–40.
26. Alpatova AL, Shan W, Babica P, Upham BL, Rogensues AR, Masten SJ, Drown E, Mohanty AK, Alocilja EC, Tarabara VV. Single-walled carbon nanotubes dispersed in aqueous media via non-covalent functionalization: Effect of dispersant on the stability, cytotoxicity, and epigenetic toxicity of nanotube suspensions. *Water Res* 2010;44:505–20.
27. Murray AR, Kisin E, Leonard SS, Young SH, Kommineni C, Kagan VE, Catranova V, Shvedova AA. Oxidative stress and inflammatory response in dermal toxicity of single-walled carbon nanotubes. *Toxicology* 2009;257:161–71.

28. Fiorito S, Serafino A, Andreola F, Bernier P. Effects of fullerenes and single-wall carbon nanotubes on murine and human macrophages. *Carbon* 2006;44:1100–5.

29. Ema M, Matsuda A, Kobayashi N, Naya M, Nakanishi J. Evaluation of dermal and eye irritation and skin sensitization due to carbon nanotubes. *Regul Toxicol Pharmacol* 2011;61:276–81.

30. Pulskamp K, Diabate S, Krug HF. Carbon nanotubes show no sign of acute toxicity but induce intracellular reactive oxygen species in dependence on contaminants. *Toxicol Lett* 2007;168:58–74.

31. Jia G, Wang H, Yan L, Wang X, Pei R, Yan T, Zhao Y, Guo X. Cytotoxicity of carbon nanomaterials: Single-wall nanotube, multiwall nanotube, and fullerene. *Environ Sci Technol* 2005;39:1378–83.

32. Worle-Knirsch JM, Pulskamp K, Krug HF. Oops they did it, again! Carbon nanotubes hoax scientists in viability assays. *Nano Lett* 2006;6:1261–8.

33. Garibaldi S, Brunelli C, Bavastrello V, Ghigliotti G, Nicolini C. Carbon nanotube biocompatibility with cardiac muscle cells. *Nanotechnol* 2006;17:391–7.

34. Zhang LW, Zeng L, Barron AR, Monteiro-Riviere NA. Biological interactions of functionalized single-wall carbon nanotubes in human epidermal keratinocytes. *Int J Toxicol* 2007;26:103–13.

35. Muller J, Huaux F, Moreau N, Misson P, Heilier J, Delos M, Arras M, Fonseca A, Nagy JB, Lison D. Respiratory toxicity of multi-wall carbon nanotubes. *Toxicol Appl Pharmcol* 2005;207:221–31.

36. Han Y, Xu J, Li ZG, Ren GG, Yang Z. In vitro toxicity of multi-walled carbon nanotubes in C6 rat glioma cells. *NeuroToxicology* 2012;33:1128–34.

37. Ellinger-Ziegelbauer H, Pauluhn J. Pulmonary toxicity of multi-walled carbon nanotubes (Baytubes®) relative to α-quartz following a single 6 h inhalation exposure of rats and a 3 months post-exposure period. *Toxicology* 2009;266:16–29.

38. Simon-Deckers A, Gouget B, Mayne-L'Hermite M, Herlin-Boime N, Reynaud C, Carrière M. In vitro investigation of oxide nanoparticle and carbon nanotube toxicity and intracellular accumulation in A549 human pneumocytes. *Toxicology* 2008;253:137–46.

39. Deng X, Wu F, Liu Z, Luo M, Li L, Ni Q, Jiao Z, Wu M, Liu Y. The splenic toxicity of water soluble multi-walled carbon nanotubes in mice. *Carbon* 2009;47:1421–8.

40. Cheng C, Muller KH, Koziol KK, Skepper JN, Midgley PA, Welland ME, Porter AE. Toxicity and imaging of multi-walled carbon nanotubes in human macrophage cells. *Biomaterials* 2009;30:4152–60.

41. Young Y, Lee H, Shen Y, Tseng S, Lee C, Tai N, Chang H. Toxicity mechanism of carbon nanotubes on *Escherichia coli*. *Mater Chem Phys* 2012;134:279–86.

42. Prylutska SV, Grynyuk II, Matyshevska OP, Yashchuk VM, Prylutskyy YI, Ritter U, Scharff P. Estimation of multi-walled carbon nanotubes toxicity in vitro. *Physica E* 2008;40:2565–9.

43. Magrez A, Kasas S, Salicio V, Pasquier N, Seo JW, Celio M, Catsicas S, Schwaller B, Forro L. Cellular toxicity of carbon-based nanomaterials. *Nano Lett* 2006;6:1121–5.

44. Ding L, Stilwell J, Zhang T, Elboudwarej O, Jiang H, Selegue JP, Cooke PA, Gray JW, Chen FF. Molecular characterization of the cytotoxic mechanism of multiwall carbon nanotubes and nano-onions on human skin fibroblast. *Nano Lett* 2005;5:2448–64.

45. Monteiro-Riviere NA, Nemanich RJ, Inman AO, Wang YY, Riviere JE. Multi-walled carbon nanotube interactions with human epidermal keratinocytes. *Toxicol Lett* 2005;155:377–84.

46. Helland A, Wick P, Koehler A, Schmid K, Som C. Reviewing the environmental and human health knowledge base of carbon nanotubes. *Environ Health Perspect* 2007;115:1125–31.

47. Firme CP III, Bandaru PR. Toxicity issues in the application of carbon nanotubes to biological systems. *Nanomedicine* 2010;6:245–56.

48. Nimmagadda A, Thurston K, Nollert MU, McFetridge PS. Chemical modification of SWNT alters in vitro cell–SWNT interactions. *J Biomed Mater Res* A 2006;76:614–25.

49. Agboola AE, Pike RW, Hertwig TA, Lou HH. Conceptual design of carbon nanotube processes. *Clean Technol Environ Policy* 2007;9:289–311.

50. Strong KL, Anderson DP, Lafdi K, Kuhn JN. Purification process for single-wall carbon nanotubes. *Carbon* 2003;41:1477–88.

51. Chattopadhyay D, Galeska I, Papadimitrakopoulos F. Complete elimination of metal catalyst from single wall carbon nanotubes. *Carbon* 2002;40:985–8.

52. Andrews R, Jacques D, Qian D, Dickey EC. Purification and structural annealing of multiwalled carbon nanotubes at graphitization temperatures. *Carbon* 2001;39:1681–7.

53. Dillon AC, Gennett T, Jones KM, Alleman JL, Parilla PA, Heben MJ. A simple and complete purification of single-walled carbon nanotube materials. *Adv Mater* 1999;11:1354–8.

54. Hou PX, Liu C, Cheng HM. Purification of carbon nanotubes. *Carbon* 2008;46:2003–25.

55. Zhao N, He C, Li J, Jiang Z, Li Y. Study on purification and tip-opening of CNTs fabricated by CVD. *Mater Res Bull* 2006;41:2204–9.

56. Ismail AF, Goh PS, Tee JC, Mohd Sanip S, Abd Aziz M. A review of purification techniques for carbon nanotubes. *Nano* 2008;3:127–43.

57. Yan DW, Zhong J, Wang CR, Wu ZY. Near-edge X-ray absorption fine structure spectroscopy-assisted purification of single-walled carbon nanotubes. *Spectrochim Acta* B 2007;62:711–16.

58. Chen XH, Chen CS, Chen Q, Cheng FQ, Zhang G, Chen ZZ. Non-destructive purification of multi walled carbon nanotubes produced by catalyzed CVD. *Mater Lett* 2002;57:734–8.

59. Martinez MT, Callejas MA, Benito AM, Cochet M, Seeger T, Anson A, Schreiber J et al. Modifications of single-wall carbon nanotubes upon oxidative purification treatments. *Nanotechnology* 2003;14:691–5.

60. Zheng B, Li Y, Liu J. CVD synthesis and purification of single-walled carbon nanotubes on aerogel-supported catalyst. *Appl Phys* A 2002;74:345–8.

61. Delpeux S, Szostak K, Frackowiak E, Beguin F. An efficient two-step process for producing opened multi-walled carbon nanotubes of high purity. *Chem Phys Lett* 2005;404:374–8.

62. Moon JM, An KH, Lee YH. High-yield purification process of single-walled carbon nanotubes. *J Phys Chem* B 2001;105:5677–81.

63. Li HJ, Feng L, Guan LH, Shi Z, Gu Z. Synthesis and purification of single-walled carbon nanotubes in the cotton like soot. *Solid State Commun* 2004;132:219–24.

64. Mathur RB, Seth S, Lal C, Rao R, Singh BP, Dhami TL, Rao AM. Co-synthesis, purification and characterization of single- and multi-walled carbon nanotubes using the electric arc method. *Carbon* 2007;45:132–40.

65. Ramesh P, Okazaki T, Sugai T, Kimura J, Kishi N, Sato K, Ozeki Y, Shinohara H. Purification and characterization of double-wall carbon nanotubes synthesized by catalytic chemical vapor deposition on mesoporous silica. *Chem Phys Lett* 2006;418:408–12.

66. Monteiro-Riviere NA, Inman AO, Zhang LW. Limitations and relative utility of screening assays to assess engineered nanoparticle toxicity in a human cell line. *Toxicol Appl Pharmacol* 2009;234:222–35.

19 Current In Vitro Models for Nanomaterial Testing
The Reproductive System

Gary R. Hutchison and Bryony L. Ross

CONTENTS

19.1 INTRODUCTION

Over the last decade, engineered nanomaterials (ENMs) have emerged as a major new technology, whose novel physicochemical properties are being harnessed to provide huge benefits to society across a wide range of applications, from revolutionary medical solutions to advanced automotive parts, textiles, and building materials.[1]

However, understanding of their potential risk is not yet clear. As reproduction is the only certain constant within living animals, regulatory and societal acceptance of ENMs is only likely if risk assessment determines the integrity of this vital system is not compromised.

The reproductive system's interdependency on multiple organs and sensitivity to interference make it one of the most complex physiological processes to study. Reproductive toxicity is defined as a substance-induced adverse effect on male or female reproductive ability or capacity, including developmental effects on the offspring and effects on or via lactation.[2] Effects leading to impairment of reproductive function or capacity (fertility) may range from changes in libido and sexual behavior, alterations in oogenesis or spermatogenesis, to hormonal and physiological changes that interfere with fertilization or development of the fertilized ovum up to implantation. Developmental endpoints include nonheritable effects on the offspring, which interfere with normal development both peri- and postnatal; manifested, for example, via altered body weight, developmental defects, and impaired mental or physical development up to and including puberty.[3]

As a result, regulatory approaches to developmental and reproductive toxicity (DaRT) have been both complex and varied. Thus, there remains a large difference between the number of chemicals and substances in use by industry and those which have been evaluated fully for reproductive toxicity. This discrepancy is greater for ENMs as their novel properties mean that assays to evaluate toxicity must first be developed and validated to ensure they are fit for purpose. This is compounded by the sheer number of ENMs in production, a lack of full characterization data, and the fact that until very recently there existed no hard regulatory requirements for risk assessment of ENMs prior to marketing.

It is important to recognize the synergy that exists between the emerging field of nanomedicine and the development of ENMs for nonmedical applications. Nanomedicine can learn from and be shaped by the knowledge accumulated from wider toxicity studies of ENMs. Consolidation of such knowledge will accelerate the safe development of nanomedicine, balancing any potential risk to reproductive health against the potential therapeutic gain.

19.2 REPRODUCTIVE HEALTH: THE CURRENT POSITION OF THE DISCIPLINE

There is potential for exposure of humans to environmental contaminants, manmade or otherwise, which have the potential to elicit an adverse effect with varying degrees of severity as a result of the complexity of human reproductive physiology. Although problems relating to reproductive and developmental toxicity have been extant for centuries, they have received close attention and an increase in understanding of the scientific community within the last 100 years. Disasters involving environmental and medical toxicants, such as the Minimata incident in Japan, Wonder Wheat incident in Iraq, and the thalidomide tragedy, led to drug and chemical safety efforts being increased worldwide by both pharmaceutical and chemical regulators. These have developed over the years and are now represented within regulatory frameworks, typically using test guidelines from either the International

Committee on Harmonization (ICH) or the Organization for Economic Cooperation and Development (OECD). Work to date has reported chemicals that impair the function of the placenta during gestation, resulting in effects ranging from fetal growth retardation to fetal or neonatal death, and some effects that may not manifest until later in life.[4] Male reproductive health has also been shown to be adversely affected by a combination of environmental factors and lifestyle. Studies have demonstrated that reproductive health, in particular semen quality, is affected by air pollution and smoking and that reproductive development can be compromised by a number of environmental chemicals.[5–7] However, despite the presence of over 80,000 chemical substances on the market, only around 200 of these have been tested for DaRT to date.[8] Therefore, there may be opportunities for nanotoxicological insult through occupational or environmental exposure as well as through intentional administration which are yet to be identified.

19.2.1 Existing Approaches to DaRT In Vitro

The principles of testing for DaRT is a standard part of development and regulatory approval for any chemical substance, pharmaceutical or cosmetic product. The aim of DaRT is to identify possible adverse effects in animals resulting from exposure to exogenous substances (such as chemicals and pharmaceuticals) and to extrapolate the outcomes to human toxicity and risk assessment. Testing is either conducted as a series of tests to evaluate specific stages of the reproductive cycle (e.g., fertility, development) or as a single protocol generational test. Although the data based on human evidence are the preferred source of information on safety and efficacy of any product, there are major ethical implications of experimentation on human subjects and consequently such data are rarely available. Therefore, nonclinical information is gathered via in vivo animal or in vitro studies.

Development of DaRT testing within regulatory frameworks began in the late 1950s. Over the years further advances in understanding and the occurrence of additional large-scale incidents, which highlighted the potential for transplacental toxicity in utero and behavioral defects in children, have led to harmonization attempts by regulatory groups around the world. In 1988, the ICH process was formed to develop a single set of guidelines for reproductive and developmental toxicity testing for pharmaceuticals.[8] Its most recent guidelines propose an approach that splits the reproductive lifecycle into six stages from A to F. In relation to chemicals, the OECD has been publishing test guidelines on reproductive and developmental toxicity since 1983. Its guidelines was most recently updated in 2007, and these are approved for use within the European Union regulations for the registration, evaluation, authorization, and restriction of chemicals—REACH (ECHA, 2008).[3,9] The U.S. Food and Drug Administration first published a series of nested reproductive studies, known as segments I, II, and III.[10] This provided short-term evaluations of fertility (I), embryo–fetal development—"teratology" (II), and peri/postnatal evaluation (III). These approaches are regularly referred to within the literature and are outlined in brief according to reproductive stage within Figure 19.1.

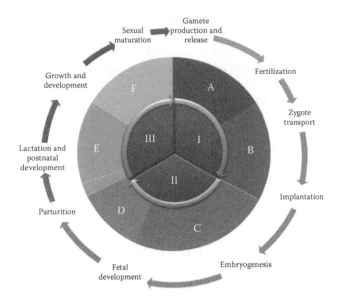

FIGURE 19.1 Stages of the reproductive and developmental cycle according to the two major regulatory approaches: U.S. Food and Drug Administration and the International Committee on Harmonization (ICH). Segments I (fertility), II (embryotoxicity/teratogenicity), and III (peri/postnatal toxicity) represent the FDA testing strategy. Stages A (premating to conception), B (conception to implantation), C (implantation to closure of the hard palate), D (closure of hard palate to end of pregnancy), E (birth to weaning), and F (weaning to sexual maturity) represent the ICH approach. (NB: ICH stage C spans both segment II and III, as indicated by the dotted white line). (Adapted from Spielmann, H., *Altern Lab Anim*, 37, 641–56, 2009.)

Traditionally, the tests used within each segment or stage have been heavily dependent on in vivo methods (Table 19.1). However, over recent years, for both chemical risk assessment and the safety testing of pharmaceutical and cosmetic ingredients, there has been increased consideration and effort put into developing nonanimal methods to assess toxicity, in line with the principle of the 3Rs—Reduction, Refinement, and Replacement of animals in scientific testing. This initiative has been strongly supported in major regulatory frameworks such as the REACH regulations and the Globally Harmonized System of Classification and Labelling of Chemicals.

19.2.2 Current State of the Art: DaRT In Vitro

The foregoing issues highlight the need to consider alternative testing approaches, particularly when investigating new and emerging technologies. Development of in vitro methods to assess reproductive toxicity should in theory allow for both a reduction in the number of animals used in studies and a much higher throughput in testing. However, given how the broad and complex the reproductive cycle is, it is very difficult to model the whole cycle in one in vitro system. In addition, the contributions of maternal factors or multiorgan influences on outcomes are not taken into account. However, those tests developed attempt to break the system down into selected biological components of relevance. Currently, early embryonic development is the

TABLE 19.1

Approaches to DaRT In Vivo

Approach			Name/Description	OECD Protocol Reference
In vivo	Development and reproductive toxicity		Prenatal Developmental Toxicity Study	OECD TG414
			Reproduction/Development Toxicity Screening Test	OECD TG421
			Combined Repeated Dose Toxicity Study with the Reproduction/Development Toxicity Screening Test	OECD TG422
			One-Generation Reproduction Toxicity Study	OECD TG415
			Two-Generation Reproduction Toxicity Study	OECD TG416
			Extended One-Generation Reproduction Toxicity Study	OECD TG443
			Development Neurotoxicity Study	OECD TG426
			Repeated Dose 28-Day Oral Toxicity Study in Rodents	OECD TG407
			Repeated Dose 90-Day Oral Toxicity Study in Rodents	OECD TG408
	Endrocrine disruption		Uterotrophic Bioassay in Rodents	OECD TG440
			Hershberger Bioassay in Rats	OECD TG441

only part of the reproductive cycle for which in vitro tests have been validated.[12] Development of tests to examine adult fertility and gamete production through to implantation is ongoing. These assays include intact rodent embryos in culture, primary cell cultures derived from embryos, mouse embryonic stem cell cultures, free-living embryos (vertebrate and invertebrate), and established cell cultures.

The European Centre for Validation of Alternative Methods (ECVAM) has endorsed three tests for reproductive toxicity. These are the whole rat (postimplantation) embryo culture (WEC), the micromass embryotoxicity assay (MM), and the embryonic stem cell test (EST). All these assays examine the early stages of development; there are currently no models covering the development from an early embryo to adulthood. It is recognized for all three methods that they are not yet suitable to replace animal studies in regulatory decision making, they can be of use in screening and contribute to weight-of-evidence assessments.[12] The in vitro and ex vivo tests currently considered to be of importance to the field are outlined below and summarized within Table 19.2.

19.2.3 Whole Embryo Tests

The WEC is an ex vivo test that covers the critical phase of organogenesis and is the only one of this type available for a mammalian embryo.[13] A nonvalidated test, the Frog Embryo Teratogenesis Assay: Xenopus (FETAX), identifies substances that may pose

TABLE 19.2

In Vivo Alternative Approaches to DaRT

Approach		Name/Description	Validation Stage and Protocol Reference (If Available)
Ex vivo	Whole embryo tests	Embryotoxicity testing in postimplantation WEC	ECVAM DB-ALM 123
		FETAX	Prevalidation
		DaRT	Prevalidation
		CHEST	Prevalidation
		MM test	ECVAM DB-ALM 122
	Pluripotent stem cell–based tests	EST	ECVAM DB-ALM 123
		ACDC assay	R&D
		Human ESTs	R&D
	Placental toxicity and transport	Placental perfusion assays	R&D
		Trophoblast cell assays	R&D
In vitro	Preimplantation toxicity using cell lines—male	Computer-assisted sperm analysis	R&D
		Leydig cell assays	R&D
		Sertoli cell assays	R&D
		ReProComet assay (repair proficient comet assay)	Prevalidation: DB-ALM 126
	Preimplantation toxicity using cell lines—female	Follicle culture bioassay (FBA)	R&D
		In vitro bovine oocyte maturation assay (bIVM)	Prevalidation: DB-ALM 129
		In vitro bovine fertilization test (bIVF)	Prevalidation: DB-ALM 128
		Mouse periimplantation assay (MEPA)	R&D
	In vitro tests for assessing effects on the endocrine system	H295R Steroidogenesis Assay	OECD TG456
		The stably transfected human estrogen receptor-alpha transcriptional activation assay for the detection of estrogenic agonist activity of chemicals	OECD TG455
		Ishikawa cell test	R&D

TABLE 19.2 (*Continued*)
In Vivo Alternative Approaches to DaRT

Approach		Name/Description	Validation Stage and Protocol Reference (If Available)
		Receptor-binding assays	R&D
		Transcriptional tests	R&D
		Tests assessing steroidogenesis	R&D
In silico	Computational modeling	SARs	R&D
		QSARs	R&D
		PBPK modeling	R&D

a developmental hazard to humans. It relies on the assumption that organogenesis is a highly conserved process phylogenetically; therefore, amphibians may be used as models for the process in mammals. Although not yet validated, the study is considered useful and has been used for the identification of hazards to human and environmental health at laboratory scale.[14] An additional in ovo system, the chicken embryotoxicity screening test (CHEST), which aims to cover later effects on embryo development, has been criticized for lack of consideration to maternal–fetal interactions and its inability to distinguish general toxicity from specific developmental effects.[13]

The zebrafish (*Danio rerio*) is also regularly used in general and developmental toxicity studies. It is small, cheap to maintain and breeds well throughout the year to yield embryos that develop outside of the mother and which are transparent. In addition, the seven developmental stages of zebrafish embryos are well documented and understood, as is their comparability to mammalian embryonic development, making them a potentially very useful model. The in vitro *Dario rerio* embryotoxicity test (DaRT) is the most common of these. Overall, both the FETAX and DaRT offer a simple animal model with a high number of endpoints, which may be analyzed and scored.

Unlike the WEC, the MM test is a validated approach that makes use of cell cultures of the limb bud of midorganogenesis rat embryos. However, although it does not require whole embryos like the WEC, the MM is a primary cell culture, and thus both tests still require intact animals as the source of embryos. The MM when tested for performance and predictability was found by ECVAM to be more accurate for strongly embryotoxic substances, in comparison to the WEC. However, the WEC encompasses a far more demanding procedure and may not be suitable as a routine first step in screening of a substance. It is proposed that the WEC has value as a second level/tier testing method where there are fewer compounds being evaluated for toxicity and to prioritize further in vivo testing. It could also be used alongside in vivo tests to assist in ranking the toxicity of chemicals within the same family.[15–17]

19.2.4 PLURIPOTENT STEM CELL–BASED IN VITRO TESTS

Stem cells offer the ability to shed light on the processes underlying organogenesis and thus are of great importance to the development of in vitro toxicity testing.[14] Use of assays which examine the interference of substances on stem cell differentiation have been developed and may be coupled with detailed genetic analysis to determine which genes are affected by each substance tested to research mechanisms of action exhaustively. The EST, which is based on cytotoxicity assessment as well as evaluating inhibition of differentiation into cardiomyocytes, is the furthest developed assay of this type, although a number of alternatives exist. For example, the adherent cell differentiation and cytotoxicity (ACDC) assay uses the single cell culture (rather than two as with the EST), thus offering a higher throughput. This test examines differentiation and proliferation of cells to cardiomyocytes using pluripotent mouse embryonic stem cells. Further investigation of this type of assay is required to provide appropriate and exact solutions for the determination of specific developmental fates. The recent establishment of human embryonic stem cell–based tests promises to bring a detailed understanding of mechanisms of toxicity. However, these tests are still very much in their infancy, and the EST is currently the only ECVAM-validated test. ECVAM confirms that the EST alone is not yet suitable for regulatory purposes, but may be used as part of an integrated strategy.

19.2.5 PLACENTAL TOXICITY AND TRANSPORT

Ex vivo placental perfusion models have been established to investigate the interface between maternal and fetal circulation during gestation. This assay is relatively preliminary in its development, and its application is limited due to placenta to placenta variations and the limited relevance of term placenta for the period of embryonic development.[14] The trophoblast assay, which uses the BeWo cell line (an immortalized trophoblastic line of human origin), presents another model of placental toxicity.

19.2.6 IN VITRO TOXICITY USING CELL LINES

The use of cell lines within reproductive toxicity testing has been growing. This method has the benefits of being easy to manipulate, offers economical savings in terms of cost per assay, and may be stored for long periods.[14] Thus, a number of tests may be conducted using cell lines representative of certain tissues and specific aspects of the reproductive cycle. Such tests can include assessment of endpoints of potential relevance such as reactive oxygen species production, oxidative stress, and cell signaling mediators.

In the case of male fertility, a number of assays and approaches exist. For example, cell lines such as the adult mouse Leydig (TM3, androgen-producing cells) and sertoli (TM4, cells which form the basis of the blood–testis barrier and nurture maturing sperm) cell lines or embryonic carcinoma (EC) cells are used to detect changes in morphology and proliferation. In the case of female fertility, a number of tests focusing on follicular cultures, oocyte maturation, the process of fertilization, and preimplantation of embryos are available. These tests are all in relatively early stages of development.

It must be noted that the use of established cell lines in developmental toxicity screening has not been successful on the whole, as assays tend only to be capable of demonstrating one relevant endpoint (e.g., proliferation), thus offering a particularly oversimplistic view of what is an immensely complex process.[12]

19.2.7 IN VITRO TESTS FOR ASSESSING THE ENDOCRINE SYSTEM

A number of other nonvalidated approaches are under development. These include tests which alter the expression of embryo implantation–associated target genes (Ishikawa Cell Test), cell proliferation–based assays, receptor-binding assays, transcriptional tests, and assays specifically assessing steroidogenesis. ECHA (2008) notes that testing approaches focusing on androgen and estrogen receptor binding and transcription are limited by the fact that they do not assess other mechanisms of action, hormone synthesis, and transport actions on other receptors or altered metabolism.[3]

Currently, there exist two OECD-validated in vitro methods to assess endocrine effects on the reproductive system. These include the stably transfected human estrogen receptor-alpha transcriptional activation assay for the detection of estrogenic agonist activity of chemicals and the H295R steroidogenesis assay.[18–19]

19.2.8 IN SILICO

Finally, the contribution of in silico methods to toxicity testing cannot be ignored. The complexity of reproductive processes makes the development of reliable methods difficult; however, there have been a number of efforts to bring together existing toxicological information electronically, such as the OECD eChemPortal and the U.S. EPA TOXREF database.[13]

Structure activity relationship approaches (SARs), such as DEREK, describe fragments of a molecule related to a particular effect and have been under development to include a reproductive toxicity endpoint. In addition, quantitative structure activity relationship (QSAR) approaches are under development and have received wide regulatory acknowledgment and implementation in safety assessment paradigms. Systems such as the commercial TopKat and MultiCASE systems and the freely available CESAR (Computer Assisted Evaluation of industrial chemical substances According to Regulations) developmental toxicity model group together compounds in terms of structural features to predict toxicity.[14] However, there still exist concerns over the ability of QSAR to predict complex endpoints such as developmental toxicity, and also that small differences between substances on a molecular level that can alter substance activity greatly (e.g., chemical bioisosteres) may not be flagged as a potential issue.[20] In addition, as evaluation of the hypothesis under investigation is reliant on a finite dataset, care must be taken to ensure that the QSAR is validated and that the risk of generating overfitted and thus useless interpretations on structural/molecular data is minimized.

Another area under development is that of physiologically based pharmacokinetic (PBPK) modeling. These represent one approach to examining Adsorption, Distribution, Metabolism, and Excretion (ADME) of substances, which show potential to provide a valuable contribution in the risk assessment of substances.[21] PBPK models are predictive, multicompartmental models, which represent mathematically physiological, anatomical, and chemical descriptors within the ADME process, incorporating information from multiple toxicity studies and taking into account parameters representing circulatory blood flow, ventilation, and organ perfusion rates to predict the ADME of substances within the body. PBPK modeling is still in its infancy, and techniques and models continue to be developed and improved (see Chapter 8).[22]

Although these systems currently hold limited use as predictors of reproductive toxicity due to their infancy and lack of robustness, development of further grouping strategies around endpoints identified from in vivo responses will eventually make them stronger tools for the testing of reproductive effects.

19.3   IMPACT OF ENMs ON REPRODUCTIVE AND DEVELOPMENTAL HEALTH

The literature indicates that ENMs may enter the body through inhalation, oral, injection, and dermal contact. It is known that ENMs pass through biological membranes, such as the blood–air barrier.[23–24] With respect to reproductive physiology, a number of studies have attempted to assess biodistribution of ENMs to the testes, using imaging techniques such as transmission electron microscopy, confocal microscopy, and mass spectrometry. To date, these studies have only been conducted in rodent models.[25–33] Biodistribution of ENMs to reproductive organs such as the testes has also been now been reported in fetal models. A number of studies have also demonstrated that blood-borne ENMs can reach the fetus of pregnant rats.[34] In addition, there is evidence of health effects in offspring exposed to diesel exhaust particles via the mother,[35–37] which includes effects on fetal testes sertoli cell development as well as more long-term effects such as impairment of spermatogenesis in adulthood.[38–40]

19.4   CURRENT STATE OF THE ART: REPRODUCTIVE NANOTOXICOLOGY IN VITRO

On account of increased data generated from biodistribution studies, a greater focus has been placed on examining mechanistic toxicity. This has resulted in ENM studies being conducted using in vitro reproductive models. Such models provide a useful platform to investigate the cellular interactions with ENMs and individual cell responses, something not easily assessed in vivo.

This research has predominantly relied on the use of cell lines. Key cell types are selected based on their importance to the maintenance of normal reproductive function; for example, Leydig, Sertoli, and germ cells are used to evaluate testicular function. Since the female reproductive tract contains many organs, several different models for each component have been developed. For example, there exist in vitro models of ovarian tissues, uterine endometrium, and trophoblast. A number of studies have used primary cells; however to date, these are more limited in number and have tended to focus on the impact of ENMs on human spermatozoa. To date, most reproductive studies in vitro have investigated the toxicological impact of ENMs on specific cells. These in vitro models may also prove useful if used to study the impact of ENMs on male and female reproductive function in relation to health or to explore targeted diagnosis and therapy.

19.4.1 Studies Relating to Male Physiology

The lining along the vagina is composed of a multilayered stratified epithelium, which provides the first line of defense against pathogens. As these cells form the surface that comes into contact with the external environment, these have become the focus of initial studies into ENM impacts on female reproductive physiology. In vitro, three-dimensional (3D) models of the vaginal lining using immortalized vaginal epithelial cell line V19I have been developed for high-throughput toxicity testing of microbicides,[41] and cervicovaginal mucous has been used to study ENM uptake kinetics for vaginal drug delivery.[42]

Models for ovarian surface epithelium (OSE) are also frequently used to study tumor development, as it is believed to be the source of most ovarian tumors.[5] Cell lines derived from tumors of the human OSE, like NIH:OVCAR-3 and SK-OV-3, have been used to study targeted molecular magnetic resonance imaging (MRI) using magnetic nanoparticles.[43]

In females, ovarian granulosa cells are involved in steroidogenesis, playing an important role in maintaining female fertility by regulating ovarian function and oocyte development. Primary granulosa cells extracted from the follicles are frequently used to study ovarian hormone secretion. Stelzer and Hutz[44] studied the effects of gold ENMs on primary rat granulosa cells and found that the particles entered granulosa cells and altered estradiol secretion. Similarly, Liu et al.[45] found that calcium phosphate nanoparticles increased apoptosis by interfering with the cell cycle of cultured primary human ovarian granulosa cells. The particles were able to enter the cells, but no change in the levels of either progesterone or estradiol was measured in the cell culture medium. Reproductive toxicity has also been shown for cadmium–telerium/zinc–telerium quantum dots, which interfered with the development of primary mouse follicles.[46]

Not all studies conducted to date have focused solely on hazard assessment; for example, the ovarian cancer cell line BG-1 has been used to study targeted diagnostics and therapy for ovarian cancer using ENMs. Core–shell hydrogel nanoparticles functionalized with peptides that specifically target ovarian carcinoma cell lines have proved to efficiently deliver siRNA to BG-1 cells and knockdown specific receptors without impairing cell viability.[47]

19.4.2 Studies Relating to Female Physiology

An important function of normal reproductive health in both male and female systems is the regulation of steroidogenic activity. Research into understanding the impact of ENMs on hormone regulation has thus far been overlooked. Only one in vitro study in males has undertaken an assessment of cells producing steroids. This used a murine Leydig cell line model (TM3) to assess the impact of ENMs on androgen synthesis and concluded that ENMs could affect steroidogenesis through the overexpression of genes such as the steroidogenic acute regulatory gene (StAR).[48]

19.4.3 IN VITRO MODELS OF FERTILIZATION AND IMPLANTATION

Most in vitro studies have focused their attention on assessing impact of nanomaterials on germ/sperm cells. An immortalized spermatogonial stem cell model (C18-4) has shown that regardless of the cell viability marker assessed (lactate dehydrogenase [LDH] leakage, mitochondrial function, apoptosis, or necrosis assay), the nanoparticulate form of an element (15 nm silver, 30 nm aluminum, and 30 nm molybdenum) displayed greater toxicity than the soluble form (silver carbonate, sodium molybdate, and aluminum chloride).[49] These results not only develop further our understanding of particle cell interactions but also have implications for the safe development of nanomedicines as each of the materials studied has relevance to potential nanomedicines. For example, silver powder is used in biocides and various consumer products to enhance antimicrobial properties, and nanoscaled aluminum powder is used in electronic circuits, as a scratch-resistant coating for plastic lenses, antimicrobial agents, and new tissue biopsy tools. Molybdenum nanoparticles have also been shown to have potential for use in medical imaging applications.

Another study by Asare et al.[50] used cells from a human testicular embryonic carcinoma Ntera2 (NT2) and isolated primary testicular cells from two strains of C57BL6 mice: one wild type and the other genetically modified to serve as a representative model for human male reproductive toxicity (mOgg1−/−). The cells were exposed to 21 nm titanium dioxide particles, 20 and 200 nm silver particles in a time and dose response study. Both sizes of silver ENMs were shown to inhibit normal cell function and cause greater cell death than the titanium dioxide. In this particular study, DNA damage in the human cells was increased in a concentration-dependent manner after exposure to the 200 nm silver particles. These cell-based studies thus allow nanotoxicologists to further investigate the mechanisms that underlie the physiological response.

This study built on the results of their earlier work using the spermatogonial stem cell line (C18-4) to examine the impact of nanosilver size and coating on toxicity.[7,49] It concluded that a significant decline in cell proliferation was dependent on both factors. The study evaluated traditional toxicity assays to assess the mechanism driving this effect; however, reactive oxygen species production (a known particle-driven effect) and/or apoptosis did not seem to play a major role. Using functional assays, the researchers were able to deduce that the silver nanoparticles were interacting with cellular pathways, specifically *Fyn kinase*, a downstream part of *Ret-* and *GDNF*-mediated signaling linked to cell proliferation.

A few in vitro studies have been conducted to assess the impact of nanomaterials on male fertility. These have focused on assessing human spermatozoa samples. One study assessed 9 nm gold nanoparticles at a concentration of 44 parts per million on healthy human sperm. Following exposure to gold ENMs, the control sperm motility rate of 95% was decreased to 75%. The study did not investigate any other indicators of toxicity; however, a morphological assessment of samples identified fragmentation of spermatozoa and the presence of gold nanoparticles in spermatozoa head and tail sections.[51] An earlier study using human spermatozoa found contradictory results, showing that nanoparticles had no effect on the mobility of spermatozoa nor their ability to initiate the acrosome reaction, two functions crucial for successful fertilization.[52] However, it must be noted that the 2009 study did have some biases and is thus difficult to interpret.

A study led by Makhluf et al.[53] demonstrated that iron oxide nanoparticles coated with amino-polyvinyl alcohol (PVA) are able to penetrate sperm cells without affecting motility or the spermatozoa's ability to undergo acrosome reaction. The magnetic properties of iron oxide nanoparticles offer the possibility of new diagnostic and biomedical applications such as cell-improved MRI, labeling and cell targeting, and drug delivery. Thus, this study demonstrates a potential opportunity to control movement and/or target using magnetic sperm cells within an animal body.

19.4.4 MODELS OF FERTILIZATION

Although this research shows great promise, little toxicological data are available to date on the impact of ENMs on fertilization and their presence in the fertilized oocyte. On fertilization, oocytes develop into a blastocyst before implantation in the uterine wall. The outer layer of the blastocyst consists of trophoblasts, which provide the early embryo with nutrients and continue to develop into a large part of the placenta. Therefore, trophoblast growth and migration are critical events during placental development and embryogenesis. A large number of trophoblast cell lines are available, derived from a variety of sources, including normal placenta, malignant tissue, embryonic carcinomas, and placental choriocarcinomas.[54] BeWo, JEG-3, and JAR cells derived from human placental choriocarcinoma are frequently used as models for the placental barrier, although the most established cell line for this purpose is BeWo.[4] BeWo cells have been used to assess placental transport of environmental pollutants,[55] as well as polystyrene nanoparticles.[56]

19.5 LIMITATIONS OF CURRENT IN VITRO SYSTEMS

As with in vivo testing systems, in vitro testing approaches also hold limitations. The critical question for in vitro testing strategies is how well the cell and culture reflects the cell within an organism, particularly with reference to differentiation and response patterns within an artificial environment. One of the biggest issues with in vitro systems is the lack of standard operating protocols to allow research groups to duplicate cell culture experimental techniques, or of quality assurance for the resulting outcomes. However, many other issues also exist. Additional studies provide a comprehensive review of the problems of current cell culture techniques, key aspects of which are outlined below.[57]

19.5.1 ORIGIN AND AUTHENTICITY OF CELLS

Cell lines usually originate from either cancer tissues or primary cells, which have undergone transformation in culture.[57] Tumor cells have been shown to have undergone thousands of mutations compared with their progenitors and thus possess drastically reduced expression of typical function.[58] Thus, the question arises whether it is possible to expect cells to resemble the normal tissue when the origin is already far from normal and when transformation is occurring as passage number increases.

19.5.2 Metabolic Competence of Cells

Often cell lines lack metabolic competence and although solutions are being developed to increase the maternal metabolic competence of cells within culture, they are not yet perfect. The contribution of cell metabolism to the toxicity of substances is not fully understood, although it is probably overestimated within the literature by a bias toward publication of positive results and focus on mechanisms of toxicity. In addition, species differences in metabolism of substances may lead to hazards relevant to humans being overlooked. Furthermore, it is common practice to standardize culture condition toward that of humans. However, when considering that many test methods use rodent cells, it is often forgotten that the body temperature of rodents is several degrees higher, the salt concentration of red blood cells is much higher than in human blood, and the pH of inflamed tissues may be as low as 4 (whereas within standard culture conditions a pH of 7 is often maintained). To ensure the most physiologically accurate test outcome, culture conditions should, therefore, be considered according to strain and circumstance.

19.5.3 Test System Additives

Often, test substances such as ENMs are not soluble in water and it is not possible to mimic uptake and transport of the substance within the organism. Solvents such as dimethyl sulfoxide may be used to assist, but these solvents themselves are often bordering on being toxic. Similarly, the use of antibiotics as a prophylactic within cell culture may also alter cell morphology and function.[59] Finally, the addition of serum, which is usually from a different species, can lead to binding of compounds to serum proteins and thus a decrease in the amount available to act on its cellular target.

19.5.4 Nanomaterial-Specific Considerations

Understanding how ENMs interfere with standard toxicological assays will be the key for the development of new in vitro reproductive systems and validation of current approaches for nanotoxicological investigations.

ENMs offer their own complications to in vitro studies such as induction of artifacts and assay interference. It is possible that many artifacts in cell culture are due to the enormous surface area of ENMs (which both allows them to absorb components and augments the chemical reactivity). In most cell cultures, serum albumin from fetal calf serum is abundant and is likely to coat many ENMs. Observations have shown that ENMs can absorb dyes, enzymes, or inflammation mediators and in some cases interact with substances used to test cells.[60–62] For example, it is known that 3-(4,5-dimethylthiazol-2-yl)-2,5-diphenyltetrazolium bromide (MTT), which is commonly used to test cell viability, interacts with ENMs producing assay interference.[60–61] Likewise, a recent evaluation of four common in vitro assays for oxidative stress, cell viability, cell death, and inflammatory cytokine production (dichlorofluorescein [DCF] assay, MTT, LDH, and IL-8 ELISA) using 24 different ENMs found that all tested were found to interfere with the optical measurement at concentrations of 50 $\mu g \cdot cm^{-2}$ and above.[63]

19.5.5 ACTUAL EXPOSURE OF CELLS TO ENMs

The actual exposure concentration of cells to ENMs needs to be ascertained after they are added to culture media. As cell culture monolayers provide a minimal cell to cell contact area, and cell density within culture is far lower than normal organ tissue, the resulting dosimetry is likely to be altered. It is easy to forget that test substances in vitro also exhibit kinetics; for example, they may be adsorbed, taken up by cells, precipitate or remain soluble, or be metabolized. Changing cell culture media may also alter these processes. In vitro biokinetics are not likely to be as complex as in vivo kinetics; however, the effective concentration of the dose applied must be understood to determine toxicity, and thus such kinetics must be taken into account. A major case in point is evident in soluble versus insoluble ENMs, as soluble ENMs introduce an additional kinetic component for cell exposure via dissolution.

19.6 CONCLUSION AND FUTURE THOUGHTS

Increasing availability of ENM bioaccumulation and biodistribution data, especially in women of childbearing potential, has led to an expectation that ENMs may persist in tissues and delayed exposure during pregnancy may be possible. The studies available to date provide an ever growing body of evidence; however, due to their low number and the lack of standardization in approach, it is somewhat difficult to draw any concrete conclusions, with some studies leading to contradictory findings.

Given that ENMs and submicron materials are progressively being used more in medicine, especially as a therapeutics or vaccine adjuvants,[64–66] there are too few studies at this time that examine their potential impact on reproductive health. Based on the models discussed, women are likely to be exposed to ENMs via relevant routes during therapeutic intervention with medically relevant ENMs (e.g., oral, intravenous, and/or vaginal application of ENM drugs) or for diagnostic purposes (e.g., injection of magnetic ENMs as contrast agents).

To achieve optimal targeted design and minimal toxic side effects, systematic characterization of ENMs used in in vitro and in vivo studies is essential so that observed effects can be directly related to particular ENM properties or combinations of properties. Important endpoint measurements for reproductive toxicity of ENMs are not only on effects on fertility (e.g., hormone secretion, oocyte maturation) and implantation but also on placental transport and toxicity. The testis provides researchers the opportunity to explore and study the blood–testis barrier, one of the few physiological barriers within humans, and to assess the impact of ENM exposure in relation to both the precautionary principles used within toxicology and hazard assessment, as well as the innovation-based questions of efficacy and distribution posed for development of nanomedicines.

To achieve this outcome, researchers in the field require robust and relevant in vitro models, particularly to study placental transport as there are very few animal models that mimic the structure of the human placenta. Further development of 3D cell culture and coculture platforms as opposed to standard two-dimensional models is very much recommended. Development of good practice and standard operating procedures for ENM testing would also be highly beneficial, so that results from different laboratories may be directly compared and collated. To progress beyond the state-of-the-art in DaRT testing, validation of alternative methods may be problematic, as in vitro or ex vivo methods cannot, for example, address the contribution of the multiple organs that influence reproduction in vivo, or maternal factors that may be at play. However, efforts to provide useful and relevant in vitro testing strategies for ENMs in reproductive toxicology are ongoing, and these will form a strong basis of support to more traditional in vivo approaches in line with the principle of the 3Rs.

REFERENCES

1. Aitken R., Ross B., Peters S., Geertsma R., Bleeker E., Wijnhoven S., Toufektsian M. C. & Nowack B. (2011). Nanotechnology EHS Landscape Document, Report on ObservatoryNANO FP7 project (Contract number 218528). ObservatoryNANO FP7: EC.
2. UNECE (2005). United Nations Economic Commission for Europe Globally Harmonized System for Classification and Labelling of Chemicals (GHS), Chapter 3.7. Reproductive Toxicity. New York/Geneva: United Nations.
3. ECHA (2008). Guidance on information requirements and chemical safety assessment, Chapter R. 7a: Endpoint specific advice. Helsinki, Finland: European Chemicals Agency.
4. Saunders M. (2009). Transplacental transport of nanomaterials. *Wiley Interdisciplinary Reviews: Nanomedicine and Nanobiotechnology.* 1(6):671–84.
5. Scully R. E. (1995). Early de novo ovarian cancer and cancer developing in benign ovarian lesions. *International Journal of Gynaecology and Obstetrics.* 49:S9–15.
6. Pincock S. (2004). BMA says smoking harms reproductive capability. *The Lancet.* 363(9409):628.
7. Sharpe R. M. & Irvine D. S. (2004). How strong is the evidence of a link between environmental chemicals and adverse effects. *British Medical Journal.* 328(7437):447–51.
8. Gupta R. C. (2011). Introduction. In:Ramesh C. G. (ed.) *Reproductive and Developmental Toxicology.* Academic Press, London.
9. OECD (2007). OECD No. 43 Draft guidance document on mammalian reproductive toxicity testing and assessment. *OECD Environment, Health and Safety Publications Series on Testing and Assessment.* Environment Directorate.
10. FDA (1966). *Guidelines for Reproduction and Studies for Safety Evaluation of Drugs for Human Use.* Rockville, MD: Bureau of Drugs.
11. Spielmann H. (2009). The way forward in reproductive/developmental toxicity testing. *Alternatives to Laboratory Animals.* 37:641–56.
12. Daston G. (2007). *Alternatives in Reproductive Toxicity: A Way Forward* [Online]. AltTox—Non-animal Methods for Toxicity Testing. Available: http://alttox.org/ttrc/toxicity-tests/repro-dev-tox/way-forward/daston/[Accessed 02/09/2012 2012].
13. Adler S., Broschard T., Bremer S., Cronin M., Daston G., Grignard E., Piersma P., Repetto G. & Schwarz M. (2010). Draft Report on Alternative (Non-Animal) Methods for Cosmetics Testing: Current Status and Future Prospects—2010: Chapter 5. Reproductive Toxicity. European Commission.
14. Pamies D., Martinez C., Sogorb M. & Vilanova E. (2011). Mechanism-based models in reproductive and developmental toxicology. In: Gupta R. C. (ed.) *Reproductive and Developmental Toxicology.* Academic Press, London.
15. ECVAM (2002a). Embryonic Stem Cell Test (EST) DB-ALM Protocol No. 113. ECVAM.
16. ECVAM (2002b). Embryotoxicity testing in post-implantation whole embryo culture (WEC), DB-ALM Protocol No. 123. ECVAM.

17. ECVAM (2002c). The Micromass Test, DB-ALM Protocol No. 122. ECVAM.
18. OECD (2009). Test No. 455: The Stably Transfected Human Estrogen Receptor-Alpha Transcriptional Activation Assay for Detection of Estrogenic Agonist-Activity of Chemicals, OECD Guidelines for the Testing of Chemicals, Section 4, OECD Publishing.
19. OECD (2011). Test No. 456: H295R Steroidogenesis Assay, OECD Guidelines for the Testing of Chemicals, Section 4, OECD Publishing.
20. Scialli A. R. (2008). The challenge of reproductive and developmental toxicology under REACH. *Regulatory Toxicology and Pharmacology*. 51:244–50.
21. Loizou G. (2008). Development of good modelling practice for physiologically based pharmacokinetic models for use in risk assessment: The first steps. *Regulatory Toxicology Pharmacology*. 50:400–11.
22. Maccalman L., Tran C. L. & Kuempel E. (2009). Development of a bio-mathematical model in rats to describe clearance, retention and translocation of inhaled nano particles throughout the body. *Journal of Physics: Conference Series*. 151:012028.
23. Nemmar A., Vanbilloen H., Hoylaerts M. F., Hoet P. H., Verbruggen A. & Nemery B. (2001). Passage of intratracheally instilled ultrafine particles from the lung into the systemic circulation in hamster. *American Journal of Respiratory Critical Care Medicine*. 164(9):1665–8.
24. Oberdörster G., Sharp Z., Atudorei V., Elder A., Gelein R., Lunts A., Kreyling W. & Cox C. (2002). Extrapulmonary translocation of ultrafine carbon particles following whole-body inhalation exposure of rats. *Journal of Toxicology and Environmental Health Part A*. 65(20):1531–43.
25. Araujo L., Kreuter J., Löbenburg R. & Sheppard M. (1999). Uptake of PMMA nanoparticles from the gastrointestinal tract after oral administration to rats: Modification of the body distribution after suspension in surfactant solutions and in oil vehicles. *International Journal of Pharmaceutics*. 176:209–24.
26. Bai Y., Zhang Y., Zhang J., Mu Q., Zhang W., Butch E. R., Snyder S. E. & Yan B. (2020). Repeated administrations of carbon nanotubes in male mice cause reversible testis damage without affecting fertility. *Nature Nanotechnology*. 5(9):683–9.
27. Balasubramanian S. K., Jittiwat J., Manikandan J., Ong C. N., Yu L. E. & Ong W. Y. (2010). Biodistribution of gold nanoparticles and gene expression changes in the liver and spleen after intravenous administration in rats. *Biomaterials*. 31(8):2034–42.
28. De Jong W. H., Hagens W. I., Krystek P., Burger M. C., Sips A. J. A. M. & Geertsma R. E. (2008). Particle size-dependent organ distribution of gold nanoparticles after intravenous administration. *Biomaterials*. 29(12):1912–19.
29. Kim J. S., Yoon T. J., Yu K. N., Kim B. G., Park S. J., Kim H. W., Lee K. H., Park S. B., Lee J. K., & Cho M. H. (2006). Toxicity and tissue distribution of magnetic nanoparticles in mice. *Toxicological Science*. 89:338–47.
30. Kwon J. T., Hwang S. K., Jin H., Kim D. S., Minai-Tehrani A., Yoon H. J., Choi M. et al. (2008). Body distribution of inhaled fluorescent magnetic nanoparticles in the mice. *Journal Occupational Health*. 50:1–6.
31. Lankveld D. P., Oomen A. G., Krystek P., Neigh A., Troost-de Jong A., Noorlander C. W., Van Eijkeren J. C., Geertsma R. E. & De Jong W. H. (2010). The kinetics of the tissue distribution of silver nanoparticles of different sizes. *Biomaterials*. 31(32): 8350–61.
32. Lee C. M., Jeong H. J., Yun K. N., Kim D. W., Sohn M. H., Lee J. K., Jeong J. & Lim S. T. (2012). Optical imaging to trace near infrared fluorescent zinc oxide nanoparticles following oral exposure. *International Journal of Nanomedicine*. 7:3203–9.
33. Park E. J., Bae E., Yi J., Kim Y., Choi K., Lee S. H., Yoon J., Lee B. C. & Park K. (2010). Repeated-dose toxicity and inflammatory responses in mice by oral administration of silver nanoparticles. *Environmental Toxicology and Pharmacology*. 30(2):162–8.

34. Semmler-Behnke M., Fertsch S., Schmid G., Wenk A. & Kreyling W. (2007). Uptake of 1.4 nm versus 18 nm gold particles by secondary target organs is size dependent in control and pregnant rats after intratracheal or intravenous application. In: *EuroNanoForum*, European Commission, Düsseldorf, 102.

35. Hougaard K. S., Jensen K. A., Nordly P., Taxvig C., Vogel U., Saber A. T. & Wallin H. (2008). Effects of prenatal exposure to diesel exhaust particles on postnatal development, behavior, genotoxicity and inflammation in mice. *Particle Fibre Toxicology.* 11(5):3.

36. Yoshida S., Ono N., Tsukue N., Oshio S., Umeda T., Takano H. & Takeda K. (2006). In utero exposure to diesel exhaust increased accessory reproductive gland weight and serum testosterone concentration in male mice. *Environmental Sciences.* 13:139–47.

37. Yoshida M., Yoshida S., Sugawara I. & Takeda K. (2002). Maternal exposure to diesel exhaust decreases expression of steroidogenic factor-1 and mullerian inhibiting substance in the murine fetus. *Journal of Health Science.* 48:317–24.

38. Ono N., Oshio S., Niwata Y., Yoshida S., Tsukue N., Sugawara I., Takano H. & Takeda K. (2008). Detrimental effects of prenatal exposure to filtered diesel exhaust on mouse spermatogenesis. *Archives Toxicology.* 82(11):851–9.

39. Ono N., Oshio S., Niwata Y., Yoshida S., Tsukue N., Sugawara I., Takano H. & Takeda K. (2007). Prenatal exposure to diesel exhaust impairs mouse spermatogenesis. *Inhalation Toxicology.* 19(3):275–81.

40. Izawa H., Kohara M., Watanabe G., Taya K. & Sagai M. (2007). Diesel exhaust particle toxicity on spermatogenesis in the mouse is aryl hydrocarbon receptor dependent. *Journal Reproductive Development.* 53(5):1069–78.

41. Hjelm B. E., Berta A. N., Nickerson C. A., Arntzen C. J. & Herbst-Kralovetz M. M. (2010). Development and characterization of a three-dimensional organotypic human vaginal epithelial cell model. *Biology of Reproduction.* 82(3):617–27.

42. Tang B. C., Dawson M., Lai S. K., Wang Y. Y., Suk J. S., Yang M., Zeitlin P., Boyle M. P., Fu J. & Hanes J. (2009). Biodegradable polymer nanoparticles that rapidly penetrate the human mucus barrier. *Proceedings of the National Academy of Sciences.* 106(46):19268–73.

43. Larsen B. A., Haag M. A., Serkova N. J., Shroyer K. R. & Stoldt C. R. (2008). Controlled aggregation of superparamagnetic iron oxide nanoparticles for the development of molecular magnetic resonance imaging probes. *Nanotechnology.* 19:265102.

44. Stelzer R. & Hutz R. J. (2009). Gold nanoparticles enter rat ovarian granulosa cells and subcellular organelles, and alter in-vitro estrogen accumulation. *Journal of Reproduction and Development.* 55(6):685–90.

45. Liu X., Qin D., Cui Y., Chen L., Li H., Chen Z., Gao L., Li Y. & Liu J. (2010). The effect of calcium phosphate nanoparticles on hormone production and apoptosis in human granulosa cells. *Reproductive Biology and Endocrinology.* 8:32.

46. Ding X., Guan H. & Li H. (2013). Characterization of a piRNA binding protein Miwi in mouse oocytes. *Theriogenology.* 79(4):610–5.e1.

47. Blackburn W. H., Dickerson E. B., Smith M. H., McDonald J. F. & Lyon A. L. (2009). Peptide-functionalized nanogels for targeted siRNA delivery. *Bioconjugate Chemistry.* 20(5):960–8.

48. Komatsu T., Tabata M., Kubo-Irie M., Shimizu T., Suzuki K., Nihei Y. & Takeda K. (2008). The effects of nanoparticles on mouse testis Leydig cells in vitro. *Toxicology In Vitro.* 22(8):1825–31.

49. Braydich-Stolle L., Hussain S., Schlager J. J. & Hofmann M. C. (2005). In vitro cytotoxicity of nanoparticles in mammalian germline stem cells. *Toxicological Sciences* 88(2):412–9.

50. Asare N., Instanes C., Sandberg W. J., Refsnes M., Schwarze P., Kruszewski M. & Brunborg G. (2012). Cytotoxic and genotoxic effects of silver nanoparticles in testicular cells. *Toxicology.* 291(1–3):65–72.

51. Wiwanitkit V., Sereemaspun A. & Rojanathanes R. (2009). Effect of gold nanoparticles on spermatozoa: The first world report. *Fertility and Sterility.* 91(1):e7–8.

52. Ben-David Makhluf S., Qasem R., Rubinstein S., Gedanken A. & Breitbart H. (2006). Loading magnetic nanoparticles into sperm cells does not affect their functionality. *Langmuir.* 22(23):9480–2.

53. Makhluf S. B., Abu-Mukh R., Rubinstein S., Breitbart H. & Gedanken A. (2008). Modified PVA-Fe3O4 nanoparticles as protein carriers into sperm cells. *Small.* 4(9):1453–8.

54. King A., Allan D. S., Bowen M., Powis S. J., Joseph S., Verma S., Hiby S. E., McMichael A. J., Loke Y. W. & Braud V. M. (2000). HLA-E is expressed on trophoblast and interacts with CD94/NKG2 receptors on decidual NK cells. *European Journal of Immunology.* 30(6):1623–31.

55. Correia Carreira S., Cartwright L., Mathiesen L., Knudsen L. E. & Saunders M. (2011). Studying placental transfer of highly purified non-dioxin-like PCBs in two models of the placental barrier. *Placenta.* 32(3):283–91.

56. Cartwright L., Poulsen M., Nielsen H. M., Pojana G., Knudsen L., Saunders M. & Rytting E. (2012). Invitro placental model optimization for nanoparticle transport studies. *International Journal of Nanomedicine.* 7(01):497.

57. Hartung T. & Sabbioni E. (2011). Alternative in vitro assays in nanomaterial toxicology. *Wiley Interdisciplinary Reviews Nanomedicine and Nanobiotechnology.* 3:545–73.

58. Ponten J. (2001). Cell biology of precancer. *European Journal of Cancer.* 37 (Suppl 8):S97–113.

59. Kuhlmann I. (1995). The prophylactic use of antibiotics in cell culture. *Cytotechnology.* 19:95–105.

60. Monteiro-Riviere N. A. & Inman A. O. (2006). Challenges for assessing carbon nanomaterial toxicity to the skin. *Carbon.* 44:1070–8.

61. Monteiro-Riviere N. A., Inman A. O. & Zhang L. W. (2009). Limitations and relative utility of screening assays to assess engineered nanoparticle toxicity in a human cell line. *Toxicology and Applied Pharmacology.* 234:222–35.

62. Zhang L. W., Zeng L., Barron A. R. & Monteiro-Riviere N. A. (2007). Biological interactions of functionalized single-wall carbon nanotubes in human epidermal keratinocytes. *International Journal of Toxicology.* 26:103–13.

63. Kroll A., Pillukat M. H., Hahn D. & Schnekenburger J. (2012). Interference of engineered nanoparticles with in vitro toxicity assays. *Archives of Toxicology.* 86:1123–36.

64. Oyewumi M. O., Kumar A. & Cui Z. (2010). Nano-microparticles as immune adjuvants: Correlating particle sizes and the resultant immune responses. *Expert Reviews of Vaccines.* 9(9):1095–107.

65. Singh M., Chakrapani A. & O'Hagan D. (2007). Nanoparticles and microparticles as vaccine-delivery systems. *Expert Reviews of Vaccines.* 6(5):797–808.

66. Peek L. J., Middaugh C. R. & Berkland C. (2008). Nanotechnology in vaccine delivery. *Advanced Drug Delivery Reviews.* 22, 60(8):915–28.

20 Current In Vitro Models for Nanomaterial Testing
Genotoxicity Issues

*Laetitia Gonzalez, Sara Corradi,
and Micheline Kirsch-Volders*

CONTENTS

20.1 INTRODUCTION

Nano-genotoxicology is a relatively new branch of toxicology that is on the rise. Both nanoparticle toxicologists and genotoxicologists are finding their way in this exciting new field, however, often overlooking some issues specific to each other's field. In this chapter, these nano-genotoxicity issues are addressed.

Up till now in vitro genotoxic effects of nanomaterials (NMs) have been investigated using a limited number of in vitro assays. The most commonly used assays are the bacterial Ames test, the mammalian micronucleus, alkaline comet, chromosomal aberration, and gene mutation assays.[1] All except the alkaline comet assay are Organisation for Economic Co-operation and Development (OECD) validated for new chemicals. For some of these assays adaptations for

NMs testing have been published. A short overview of these assays, measuring different endpoints, is given in Sections 20.1.1 through 20.1.5.

20.1.1 Bacterial Reverse Mutation Test or Ames Test

The bacterial reverse mutation test enables the detection of point mutations (substitution, addition, or deletion of one or few DNA base pairs) using mutant strains of *Salmonella typhimurium* or *Escherichia coli* that are deficient in the synthesis of an amino acid. Therefore, induced point mutations that revert the mutations in these amino acid requiring strains can be detected by their ability to grow in absence of the amino acid that is required by the parent strains. Many of the test strains have several features that make them more sensitive for the detection of mutations, including responsive DNA sequences at the reversion sites, increased cell permeability to large molecules, and elimination of DNA repair systems or enhancement of error-prone DNA repair processes. The specificity of the test strains can provide some useful information on the types of mutations that are induced by genotoxic agents.[2]

The validated OECD Test Guideline 471 has been adopted since 1997. The bacterial reverse mutation test is rapid, inexpensive, and relatively easy to perform; therefore, it is used as an initial screen for genotoxic activity. Because this assay uses prokaryotic cells, and not eukaryotic or mammalian cells, differing in uptake, metabolism, chromosome structure, and DNA repair mechanisms, it does not provide direct information about the mutagenic and carcinogenic potency in eukaryotic/mammalian systems. Therefore, this assay can be subject to false positives, in case factors increase sensitivity of the assay, and false negatives because of the specific endpoint measured, differences in metabolic activation, and differences in bioavailability. Furthermore, the assay might not be suited for specific compounds that are either highly bactericidal or interfere specifically with the mammalian replication system, such as topoisomerases.[3]

20.1.2 Alkaline Comet Assay

The comet or single-cell gel electrophoresis assay was developed during the late 1970s and 1980s. The main principle of the methodology is that, when single-strand (SS) and/or double-strand (DS) DNA breaks are induced, relaxation of the supercoiled DNA-forming DNA loops is increased. These relaxed negatively charged DNA loops migrate to a higher extent toward the positive pole compared to supercoiled DNA during electrophoresis, resulting in the characteristic "comet tails."[4]

There are several variations on the methodology. Most common methodology to date was described by Singh in 1988. This comet assay, also referred to as the alkaline comet assay, introduces electrophoresis at alkaline conditions (pH > 13). The alkaline comet assay enables the detection of SS and DS DNA breaks as well as alkali-labile sites. Other variations are the neutral comet assay with a lysis step at 50°C. Both variations are able to detect DS breaks. However, the lysis at high temperature disrupts the nuclear matrix, thereby eliminating interference of SS breaks.[5–8] The extent of DNA damage can be expressed in different ways, that is, tail length (TL), percentage of tail DNA (%TD) or tail moment. Tail moment is the TL multiplied by %TD. Several arguments are in favor of the use of %TD. De Boeck showed less inter-electrophoresis and inter-experimenter variability when using %TD compared to TL. Collins argues that TL can be useful at low DNA damage levels, but not at higher levels of DNA damage and that TL is more sensitive to background and threshold settings of the image analysis.[7,9] Furthermore, the %TD has a linear dose–response relationship with known DNA break-inducing agents.[7,8] Additional use of enzymes enables the detection of specific lesions. The most commonly used enzymes are endonuclease III (endoIII) for the detection of oxidized pyrimidines, formamidopyrimidine DNA glycosylase (FPG) and human 8-hydroxyguanine DNA glycosylase (hOGG1) for the detection of oxidized purines, T4 endonuclease V for the detection of UV-induced cyclobutane pyrimidine dimers, and Alk A for the detection of 3-methyladenines. Each of these enzymes introduces a strand break at the enzyme-sensitive site.[10] Smith et al.[10] found that hOGG1 detected oxidized purines with greater specificity and sensitivity compared with endoIII and FPG. Recently, the European Comet Assay Validation Group (ECVAG) performed a study for validation of the comet assay. The interlaboratory study retrieved dose–response relationships for oxidative DNA damage by assessment of FPG-sensitive sites in coded samples.[11–13] Furthermore, this assay allows the assessment of repair efficiencies by measuring the extent of DNA damage at different time points after treatment.[14] In contrast with the chromosomal aberrations, micronucleus and gene mutation assay, the alkaline comet assay does not detect fixed mutations.

20.1.3 Chromosomal Aberrations Test

Analyses of metaphases were applied in the beginning of the previous century for the detection of chromosomal structural changes in different plant and animal systems. However, its validation on human lymphocytes or cell lines for hazard assessment was translated into an OECD guideline only in 1997.

The purpose of the in vitro chromosomal aberration test is to identify agents that cause structural chromosome aberrations, either chromosome-type or chromatid-type aberrations, in cultured mammalian cells. Although this assay is not routinely used for detecting numerical aberrations (ploidy changes, aneuploidy), polyploidy can be used as a flag for induction of aneuploidy. The test principle[15] relies on the addition of colchicine, a drug blocking tubulin polymerization and therefore the transition from metaphase to anaphase, during the first mitotic wave following the treatment. Before fixation, cells undergo a hypotonic shock to allow swelling and chromosome spreading. Metaphases are stained and microscopically screened for chromosomal aberrations, such as deletions, fragments, intrachanges or interchanges, gaps and changes in ploidy. Chromosome painting or fluorescence in situ hybridization (FISH) with one or more probes can be applied to characterize structural interchromosomal aberrations and complex chromosome rearrangements as well as intrachromosomal exchanges, such as peri- and paracentric chromosome inversions.[16]

20.1.4 Micronucleus Assay

Micronuclei (MN) are small nuclear bodies arising from either chromosome/chromatid fragments or entire chromosomes/chromatids that did not segregate properly to the daughter nuclei during anaphase and are surrounded by a separate nuclear membrane during telophase. Detection of these MN is used as a measure for chromosome breakage and loss. MN can occur spontaneously or can be mutagen induced. These mutagens can be clastogens or aneugens. Exposure to a clastogen can lead to MN containing acentric chromosome or chromatid fragments through different mechanisms. Misrepair of double strand breaks, simultaneous base excision repair in close proximity and on opposite complementary DNA strands, and fragmentation of nucleoplasmic bridges (NPB) may lead to the formation of acentric chromosome/chromatid fragments.[17–19] Exposure to aneugens leads to MN containing entire chromosomes. Several mechanisms are responsible for aneuploidy. Hypomethylation of cytosine in centromeric and pericentromeric regions lead to chromosome malsegregation/loss probably because of defects in kinetochore assembly. Defects in spindle assembly, mitotic checkpoints, and centrosome amplification are also related to increased incidence of aneuploidy. Furthermore, dicentric chromosomes, when the centromeres are pulled to opposite poles, can detach from the spindle during anaphase and lead to chromosome-containing MN. Scoring of nuclear buds (NBUDs) and bridges may give interesting additional information.[17,18]

This assay can be performed either with or without the addition of cytochalasin-B. The use of cytochalasin-B, an actin inhibitor and an inhibitor of cytokinesis, allows the discrimination between cells that divide once, that is, binucleate cells or more (multinucleate cells), and cells that did not divide, that is, mononucleate cells during in vitro cell culture.[20] In addition, mononucleate cells with MN can be indicative of mitotic slippage.[21] In the absence of a functional spindle, cells can exit mitosis without chromatid segregation and immediately proceed to the next interphase, yielding tetraploid cells. This was shown by Elhajouji et al.[21] in lymphocytes after treatment with nocodazole.

The number of mono-, bi-, and multinucleated cells allows the calculation of the cytokinesis block proliferation index or CBPI, a measure for cell proliferation, which is a requirement for the expression of MN.

$$CBPI = \frac{No.\,of\,mononucleate\,cells \pm 2 \times No.\,of\,binucleate\,cells \pm 3 \times No.\,of\,multinucleate\,cells}{Total\,no.\,of\,cells}$$

In recent years, the in vitro MN assay developed into the cytokinesis block micronucleus cytome assay that detects additional biomarkers of DNA damage, NPBs and NBUDs.[17,18,22] To discriminate between clastogenic and aneugenic effects, FISH with pancentromeric probes can be performed. These probes hybridize with the pericentromeric region of the chromosomes and allow determination of the MN content (chromosome fragment or entire chromosome). To the best of our knowledge, the first study applying this methodology to investigate the effects of multiwall carbon nanotubes showed the induction of aneuploidy.[23]

At the regulatory level, this assay has been harmonized by the International Workshop on Genotoxicity Test Procedures (IWGTP) exercise[24,25] and validated by European Centre for the Validation of Alternative Methods (ECVAM) in 2006 based on a retrospective weight of evidence validation study.[26] After an interlaboratory effort to evaluate different measures of cytotoxicity/cytostasis when the in vitro micronucleus assay (MNvit) is performed without cytokinesis block,[27] the MNvit was finally accepted by the OECD in 2010. This assay has been validated by the OECD and its guideline is available since 2010 as OECD Test Guideline 487.[28]

20.1.5 Gene Mutation Assay

The in vitro mammalian cell gene mutation test[29] can be used to detect gene mutations induced by chemical substances. The most commonly used genetic endpoints measure mutation at thymidine kinase (TK) and hypoxanthine guanine phosphoribosyl transferase (HPRT), and a transgene of xanthine guanine phosphoribosyl transferase (XPRT). The TK, HPRT, and XPRT mutation tests detect different spectra of genetic events. The autosomal location of TK and XPRT may allow the detection of genetic events (e.g., large deletions) not detected at the HPRT locus on X-chromosomes. Mutations leading to deficiencies in TK, HPRT, or XPRT are detected by the cells ability to grow in presence of the pyrimidine analog trifluorothymidine, 6-thioguanine, or 8-azaguanine, respectively. Suitable cell lines include L5178Y mouse lymphoma cells, the CHO, AS52, and V79 lines of Chinese hamster cells, and TK6 human lymphoblastoid cells.

20.2 GENERAL CONSIDERATIONS FOR IN VITRO TESTING

For adequate assessment of the genotoxic hazard/risk of chemicals, but also of NMs, the following general considerations should be taken into account (Table 20.1).

In general, for hazard assessment the aim of every in vitro assay is to detect the lowest possible effect and to minimize false-negative and false-positive responses

---

TABLE 20.1
Issues in In Vitro Genotoxicity Testing for Hazard/Risk Assessment

General Genotoxicity Considerations	Nano-Specific Considerations
Human or human-derived cell types/lines	
Adequate levels of cytotoxicity for top dose selection (assay dependent)	
Avoid/reduce false negatives by correct assay/ battery choice covering the endpoints (gene, chromosome and genome mutations)	Unknown or new modes of action are possible that require development of new assays
Avoid/reduce false positives (conditions not reflecting intrinsic activity related to pH changes, osmolality, cytotoxicity)	
Negative controls with low background frequencies	
Positive controls appropriate for the endpoint assessed	If possible, nanoparticle positive control but only if appropriate for the endpoint assessed
Choice of cell type/line based on genetic background	Choice of cell type/line based on aim (most sensitive cell line for hazard, route of exposure) but always considering genetic background
Metabolic activation when tested compound needs metabolization for exerting its genotoxic effect	No metabolic activation needed (unless genotoxic effect is a consequence of impurities requiring metabolization)
	Dosimetry and topological distribution

when compared to relevant in vivo assays. Appropriate levels of cell viability should be assessed to avoid false positives/negatives; therefore adequate top doses and appropriate cell viability/proliferation assays should be taken into consideration.[27,30,31] False-positive results can also originate from changes in pH or osmolality. If the mode of action is unknown, a battery of tests should be applied according to validated protocols allowing the detection of all potential mutagenic endpoints. For risk assessment, where the mode of action is known, the most appropriate and sensitive test relevant for the concerned endpoint should be applied to avoid false-negative data.

Genotoxicity assays should always include negative and positive controls. As the search for an adequate nanoparticle positive control is still ongoing, chemical positive controls should be used. Furthermore if a good positive nanoparticle control would be found it is unlikely to be suitable for different assays looking at different endpoints/mechanisms. Therefore, the choice of an adequate positive control should rely on the genotoxic mechanism investigated by the assay of choice.

Knowledge about the genetic background of the used cellular system is a prerequisite for genotoxicity testing. p53 and caspase activity of the cells should be confirmed for allowing a normal cellular answer to a genotoxic insult.[31] Furthermore the antioxidant equipment of the cell could influence the outcome of an assay when performing assays that investigate oxidative lesions (e.g., the alkaline comet assay).

In any assay, primary cells and different cell lines can be used, but only by following the recommendation in the corresponding OECD guideline. Preference should be given to human or human-derived cells, to avoid interspecies extrapolation. The choice of cell type will influence the assay outcome; therefore, its selection is of major importance and should be justified. Selection can be based on the route of exposure, primary cells or cell lines, and so on, for genotoxicity testing. Knowledge about the genetic background of cells should always be taken into account irrespective of other choices. For hazard identification, the most sensitive cell system should be chosen. Besides monocultures, more complex in vitro cell systems combining complementary cell types are available or in development and aim at more accurately reflecting the in vivo situation. Coculture conditions have been shown to mimic the pulmonary response to particulate matter than monoculture conditions.[32] Also, genotoxicity assays are being developed on this type of complex in three-dimensional culture systems. Detailed protocols of the reconstructed skin micronucleus assay are available as part of a prevalidation project initiated by ECVAM.[33,34] In addition, cell media can be replaced or conditioned with fluids mimicking the in vivo situation such as surfactant for lung epithelial cells or artificial digestion fluids for the modelization of mouth, stomach, and gut environment when investigating the oral route of exposure.

The use of metabolic activation systems is recommended in OECD guidelines of most of the in vitro assays. The most commonly used system is the cofactor-supplemented postmitochondrial fraction (S9) prepared from livers of rodents treated with enzyme-inducing agents such as Aroclor 1254. When testing NMs, metabolic activation is not necessary, as NMs do not need metabolization to exert their genotoxic effects.

On the other hand, general NM-related issues should be considered such as the physicochemical characterization of the materials and the dosimetry. It is generally accepted within the nano-community that the physicochemical properties of NMs should be available for every study. Previously, minimal criteria for nanotoxicology testing were proposed, that is, information about method of synthesis and preparation, particle size and distribution, specific surface area, aggregation status in relevant media, composition and relevant coatings, crystallinity and purity of the sample.[35–38] With time and increasing knowledge, the properties that are relevant for toxicology become more clear and additional properties have been added such as charge and solubility.[39,40]

The importance of the dosimetry of NMs has become more evident with the appearance of publications trying to model the behavior of NMs in cell media.[41,42] The question arose whether nominal or administered dose versus cellular dose should be used. Lison et al. showed that because of convection, gravitation, and diffusion forces, the nominal dose can be used for the in vitro cytotoxicity assessments of monodisperse amorphous silica nanoparticles. This was shown by an experimental setup keeping the dose constant and conversely the particle number constant as shown in Figure 20.1 for human lung carcinoma A549 cells. Both cytotoxic effects and cellular dose were assessed.[43] Modeling of NM behavior in aqueous solutions (cell media) has improved based on this type of experiments.[42]

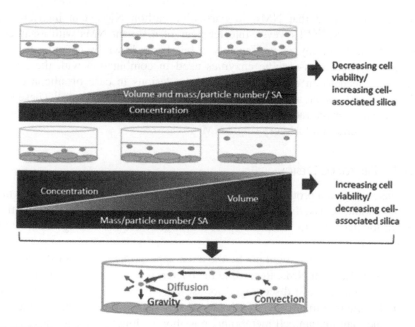

FIGURE 20.1 Schematic representation of experimental setup used to determine that the nominal or administered dose is adequate to describe cytotoxic effects induced by amorphous silica nanoparticles.[55] An experimental setup changing either the mass/particle number/surface area (SA) (upper panel) or concentration (middle panel) by altering the added cell media volume, demonstrates dependency of both cell viability and uptake. Therefore, besides diffusion and gravity, also convection forces lead to the adequacy of the nominal dose for in vitro cytotoxicity experiments. (From Lison, D. et al., *Toxicol. Sci.*, 104(1), 155–62, 2008. With permission.)

For genotoxic events the topological effect should also be considered. The dose delivered to each cell is dependent on the position effect (homogenous vs. heterogeneous distribution) and solubility of the NMs, as soluble or partially soluble NMs could affect more cells, but with a lower cellular dose compared to insoluble NMs. When considering cancer risk, this topological dose is of major importance as the risk is related to the emergence of a cell mutation rather than the effect on the whole cell population.[44]

20.3   TECHNICAL POINT OF VIEW

In general, in vitro testing of NMs entails the use of NM suspension, with their specific behavior and the possibility/probability of agglomerate/aggregate formation. These issues are not specific for genotoxicity testing but are valid for all branches of in vitro nanotoxicology.

The two main technical issues that are discussed in Sections 20.3.1 and 20.3.2 are the potential interference of NMs with assay compounds and the influence of the protocol choice on the assay outcome.

20.3.1   INTERFERENCE WITH ASSAY COMPOUNDS

It is well established that NMs, in particular carbon NMs, can interfere with cytotoxicity assays.[45–49] However, interferences of different NMs with assay compounds can occur in any assay. Recently, it has been shown by us and others that interference with lesion-specific enzymes used in combination with the alkaline comet assay lead to an underestimation of the lesions in case of silicon dioxide (SiO_2) nanoparticles but also silver and cerium oxide nanoparticles.[50,51] These types of interferences or interactions should be noted before a specific test is performed and should be checked for every single NM.

20.3.2   PROTOCOL DESIGN

To evaluate whether a protocol is suited for a certain assay, the following three main questions should be asked: Do these NMs reach their target? Are the exposure times adequate for the NMs to reach the specified target? Are the timings adequate for endpoint expression?

Bacterial cell systems might not be the most appropriate choice for the detection of mutagenic effects of NMs. Regarding the bacterial reverse mutation test or Ames test, several authors have discouraged its use because of the difference in barrier compared to mammalian cells. The bacterial cell wall is more impenetrable compared to the mammalian cell membrane, possibly leading to false-negative results. Recently it was shown that even though the Ames test seems not to be a good indicator for NM mutagenicity, some NMs did get taken up by the bacteria.[52] Taken together, mammalian gene mutation assays seem more appropriate for NM testing.[53]

Another example of how the choice of protocol can influence the assay outcome is the MNvit assay. Doak et al. showed that depending on the chosen protocol, that is, delayed co-treatment (cytochalasin-B addition after NM treatment), co-treatment (addition of cytochalasin-B at the same time of NM treatment), or no cytochalasin-B addition, the outcome of the assay differs. The genotoxic effects of superparamagnetic iron oxide nanoparticles and single-walled carbon nanotubes were shown to be more evident with the delayed co-treatment protocol.[54]

A literature review covering 21 publications using the MNvit assay as methodology enabled us to give some recommendations on how to perform this assay adapted for NM testing.[55] When testing NMs in the MNvit assay, one should consider the uptake mechanisms, in particular actin-dependent uptake of NMs when cytochalasin-B is used.[56] It is important to design a protocol including a treatment period in absence of cytochalasin-B. The length of this period is dependent on the uptake rate of the NMs.[55] Although the OECD draft guideline for the MNvit assay does not recommend scoring a greater number of cells in absence of cyotochalasin-B, expertise from different laboratories (and statistical evidence) indicate that scoring more cells when the assay is performed without cytochalasin-B should be taken into account.[28,55,57,58]

An increase of MN frequencies in mononucleates when performing the cytokinesis-blocked MN assay is an indication of mitotic slippage as an escape from a transitory metaphase block induced by aneugens. This phenomenon was shown by Elhajouji et al. and confirmed by an interlaboratory exercise on cell lines.[57,59] If, in the future, aneugenicity is confirmed to be a major mode of action of NMs, the use of cytochalasin-B (delayed co-treatment) should be advised to cover consequences of mitotic slippage as recommended by us earlier for chemical agents.[21,28,55,57,59] Nuclear uptake of NMs has been shown but has not been confirmed for all NMs and may not come into contact with the chromatin. It is critical that cells go through one round of mitosis during the treatment period allowing NM–chromatin contact as soon as nuclear membrane breakdown starts.[38,60] Exposure protocols should be adapted based on the average cell cycle of the cell lines used.[55]

20.3.3 MECHANISTIC POINT OF VIEW

Genotoxic agents can affect different cellular components (DNA and non-DNA targets) through different modes of action. When focusing on poorly soluble particles, their reactive oxygen species (ROS) producing capacity is central in the paradigm for particle (geno-)toxicology.[61–63] It is well documented that because of their small size and hence their high atom-to-surface ratio, NMs are characterized by a higher surface reactivity in comparison to their microsized counterparts and can produce comparatively more ROS than larger particles.[35,64] It might be suspected that NMs could carry a greater genotoxic hazard than their micron size counterparts. Additional mechanisms specific to NMs, such as mechanical interferences with cellular components, and other sources of genotoxic effects (e.g., metal release by NMs) need also to be considered.[38,65]

20.3.4 INTERACTIONS WITH PROTEINS

The interaction of NMs with proteins has become of greater interest.[66] It is conceivable that this interaction is one of the major modes of action leading to indirect genotoxic effects. This can occur extracellularly, at the cell membrane level, intracellularly, or within the nucleus (Figure 20.2).

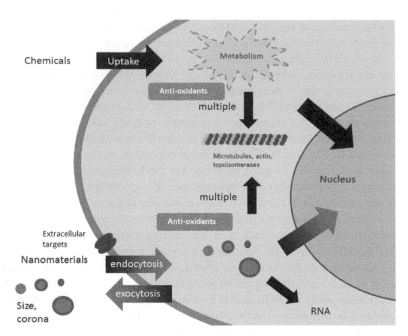

FIGURE 20.2 Schematic representation of the levels of interaction (extracellular, intracellular, and nuclear) with nanomaterials (NMs), potentially leading to indirect genotoxic effects, compared to chemical modes of action.

20.3.5 Serum or No Serum?

Another aspect to be considered, when NMs are involved, is the serum content included in the culture media during exposure. In fact the interaction between living cells and NMs depends dramatically on the behavior of the NMs in the biological fluid. When they interact with biological fluid, two effects should be considered: (1) formation of a "NM–protein corona" as consequence of serum protein interaction or (2) aggregation of NMs because of the kinetic processes that lead to the formation of NMs clusters of different sizes.

As the literature suggest, the majority of MNvit assays have been done in medium containing serum as proposed by the OECD guideline to ensure optimal growth condition to the cells, but several studies have shown that reducing or eliminating serum content can modulate the cytotoxicity and genotoxicity response. The presence of a NM–protein corona can heavily influence the uptake and the trafficking of the NMs into the cells showed that serum protein adsorption modulated cell toxicity of NMs in mouse RAW 264.6 macrophages.[67–69] Moreover, Petri-Fink et al. and later Clift et al. showed that the presence of 10% serum in superparamagnetic iron oxide and carboxylated polystyrene bead nanoparticles suspended media reduced the in vitro cytotoxicity.[70,71] Similar decreasing effect in cytotoxicity was found by us for titanium dioxide (TiO_2) and SiO_2 in the presence of serum in A549 human lung carcinoma cells.[44] Moreover, Corradi et al.[72] found that when lysine–silica were incubated with serum, the cytostasis of A549 human lung carcinoma cells was abolished.

The potential genotoxicity of NMs is affected by protein serum coating in which Doak et al.[54] showed treatment with single walled carbon nanotubes in BEAS-2B human lung epithelial cells, the MN frequency increased when the serum was reduced from 10% to 2%. Recently, Gonzalez et al.[44] showed that serum has an influence on different endpoints, respectively cell viability, cell cycle changes, and induction of MNs in A549 cells treated with monodisperse silica. Specifically, in absence of serum, cell viability decreased, cells arrest in G1 and S-phase, and MN frequency increased. In the study by Merhi et al.,[73] the effect of serum on cationic nanoparticles has been investigated and showed that cationic nanoparticles, interacting with serum protein, decrease its positive charge and reduces the toxic effect in human bronchial epithelial cell line 16HBE14o- by impairing their internalization. They had observed that in the presence of 10% serum, the endocytosis of nanoparticles was significantly reduced.

On the interaction of protein and NM surface, the corona is formed and this entity became the first one in entering in contact with the cells. This mechanism varies from the type of NMs. The presence of serum protein has been shown to diminish the aggregation state of NMs.[74,75] It has been shown that TiO_2 NMs in absence of serum tend to agglomerate and sediment rapidly. Corradi et al.[72] showed that the presence of 2% serum in the dispersion media, instead of serum-free medium, mitigated the agglomeration status of TiO_2 NMs. The protein corona can heavily modify the potential toxicity of nanoparticles and it alters the size and the interfacial composition of NMs, changing the biological identity that determines the response including signaling, kinetics, transport, accumulation, and toxicity.

Taking all these evidences together, the choice of serum/no serum became of major importance when hazard identification is the issue. The decision to take both options into consideration should be envisaged, as the assay outcome can be different, with higher sensitivity in absence of serum. This would imply for assays where cell division is a prerequisite to allow the expression of the genotoxic endpoint (e.g., MN assay), to choose a cell line that proliferates under serum-free condition.

20.3.6 DOSE–RESPONSES

The aforementioned considerations regarding dosimetry, uptake, and mechanisms lead to effects, reflected in dose–response relationships. Often when testing NMs, these curves do not present the classical shape expected from a single hit/single target, multiple hit/single target, or multiple hit/multiple target model.[24] Therefore, it is important to understand the modes of action to identify the complex dose–effect relationships that could lead to threshold.[44]

20.4 CONCLUSIONS

The aim of hazard identification is to assess whether a compound, here NMs, can induce irreversible genetic effects. Therefore, a combination of assays is required covering gene, chromosome, and genome mutations. The most sensitive situation needs to be adopted to identify the hazard, that is, the most sensitive cell type/line and the most sensitive experimental conditions (with or without serum), irrespective of the route of exposure that is the most probable. When performing risk characterization

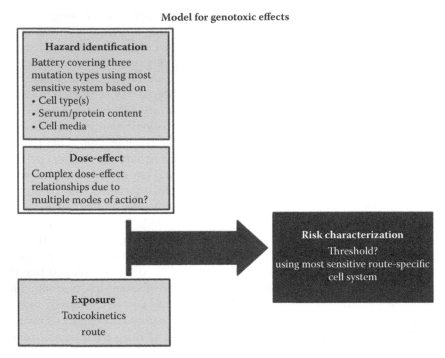

FIGURE 20.3   Strategy model for the hazard/risk assessment of genotoxic effects induced by NMs.

based on mutagenic effects, dose–effect responses, route of exposure and toxicokinetics should be taken into account. For this purpose, in an in vitro setting the most sensitive and route-specific cell line should be chosen to identify a certain threshold value (if possible depending on the mode of action) (Figure 20.3). It should be noted that a lack of primary genotoxicity, determined by in vitro testing, cannot exclude the induction of secondary genotoxicity. The latter occurs when a genotoxicant, in particular micro- or nanoparticles, induces a response in cells other than the target cells and can be elicited by inflammatory cells (e.g., macrophages and polymorphonuclear neutrophilic leukocytes). Negative in vitro tests with NMs may need in vivo confirmation to exclude secondary genotoxicity. Finally, in the future new in vitro models, such as complex culture systems or the use of tissue-specific media as well as new assays assessing NM-specific endpoints should be validated and implemented.

REFERENCES

1. Magdolenova Z, Collins AR, Kumar A, Dhawan A, Stone V, Dusinska M. Mechanisms of genotoxicity. Review of recent in vitro and in vivo studies with engineered nanoparticles. *Nanotoxicology* 2013; 1: 1–13.
2. Ames BN, Lee FD, Durston WE. An improved bacterial test system for the detection and classification of mutagens and carcinogens. *Proc Natl Acad Sci USA* 1973; 70(3): 782–6.
3. Organization for Economic Co-operation and Development (OECD). *OECD Guideline for Testing of Chemicals: (Test Guideline 471)*, 1997.

4. Collins AR, Oscoz AA, Brunborg G, Gaivao I, Giovannelli L, Kruszewski M, Smith CC, Stetina R. The comet assay: Topical issues. *Mutagenesis* 2008; 23(3): 143–51.
5. Singh NP, McCoy MT, Tice RR, Schneider EL. A simple technique for quantitation of low levels of DNA damage in individual cells. *Exp Cell Res* 1988; 175(1): 184–91.
6. Tice RR, Andrews PW, Hirai O, Singh NP. The single cell gel (SCG) assay: An electrophoretic technique for the detection of DNA damage in individual cells. *Adv Exp Med Biol* 1991; 283: 157–64.
7. Collins AR. The comet assay for DNA damage and repair: Principles, applications, and limitations. *Mol Biotechnol* 2004; 26(3): 249–61.
8. Moller P. The alkaline comet assay: Towards validation in biomonitoring of DNA damaging exposures. *Basic Clin Pharmacol Toxicol* 2006; 98(4): 336–45.
9. De Boeck M, Touil N, De Visscher G, Vande PA, Kirsch-Volders M. Validation and implementation of an internal standard in comet assay analysis. *Mutat Res* 2000; 469(2): 181–97.
10. Smith CC, O'Donovan MR, Martin EA. hOGG1 recognizes oxidative damage using the comet assay with greater specificity than FPG or ENDOIII. *Mutagenesis* 2006; 21(3): 185–90.
11. Johansson C, Moller P, Forchhammer L, Loft S, Godschalk RW, Langie SA, Lumeij S et al. An ECVAG trial on assessment of oxidative damage to DNA measured by the comet assay. *Mutagenesis* 2010; 25(2): 125–32.
12. Moller P, Moller L, Godschalk RW, Jones GD. Assessment and reduction of comet assay variation in relation to DNA damage: Studies from the European Comet Assay Validation Group. *Mutagenesis* 2010; 25(2): 109–11.
13. Ersson C, Moller P, Forchhammer L, Loft S, Azqueta A, Godschalk RW, van Schooten FJ et al. An ECVAG inter-laboratory validation study of the comet assay: Inter-laboratory and intra-laboratory variations of DNA strand breaks and FPG-sensitive sites in human mononuclear cells. *Mutagenesis* 2013; 28: 279–86.
14. Decordier I, Loock KV, Kirsch-Volders M. Phenotyping for DNA repair capacity. *Mutat Res* 2010; 705(2): 107–29.
15. Organisation for Economic Co-operation and Development (OECD). *OECD Guideline for Testing of Chemicals: (Test Guideline 473)*, 1997.
16. Mateuca RA, Decordier I, Kirsch-Volders M. Cytogenetic methods in human biomonitoring: Principles and uses. *Methods Mol Biol* 2012; 817: 305–34.
17. Fenech M, Kirsch-Volders M, Natarajan AT, Surralles J, Crott JW, Parry J, Norppa H, Eastmond DA, Tucker JD, Thomas P. Molecular mechanisms of micronucleus, nucleoplasmic bridge and nuclear bud formation in mammalian and human cells. *Mutagenesis* 2011; 26(1): 125–32.
18. Kirsch-Volders M, Decordier I, Elhajouji A, Plas G, Aardema M, Fenech M. In vitro genotoxicity testing using the micronucleus assay in cell lines, human lymphocytes and 3D human skin models. *Mutagenesis* 2011; 26(1): 177–84.
19. Kirsch-Volders M, Plas G, Elhajouji A, Lukamowica M, Gonzalez L, Vande Loock K, Decordier I. The in vitro MN assay in 2011: Origin and fate, biological significance, protocols, high throughput methodologies and toxicological relevance. *Arch Toxicol* 2011; 85(8): 873–99.
20. Fenech M, Morley AA. Measurement of micronuclei in lymphocytes. *Mutat Res* 1985; 147(1–2): 29–36.
21. Elhajouji A, Cunha M, Kirsch-Volders M. Spindle poisons can induce polyploidy by mitotic slippage and micronucleate mononucleates in the cytokinesis-block assay. *Mutagenesis* 1998; 13(2): 193–8.
22. Fenech M. Cytokinesis-block micronucleus cytome assay. *Nat Protoc* 2007; 2(5): 1084–104.

23. Muller J, Decordier I, Hoet PH, Lombaert N, Thomassen L, Huaux F, Lison D, Kirsch-Volders M. Clastogenic and aneugenic effects of multi-wall carbon nanotubes in epithelial cells. *Carcinogenesis* 2008; 29(2): 427–33.

24. Kirsch-Volders M, Aardema M, Elhajouji A. Concepts of threshold in mutagenesis and carcinogenesis. *Mutat Res* 2000; 464(1): 3–11.

25. Kirsch-Volders M, Sofuni T, Aardema M, Albertini S, Eastmond D, Fenech M, Ishidate M et al. Report from the in vitro micronucleus assay working group. *Mutat Res* 2003; 540(2): 153–63.

26. Corvi R, Albertini S, Hartung T, Hoffmann S, Maurici D, Pfuhler S, van Benthem J, Vanparys P. ECVAM retrospective validation of in vitro micronucleus test (MNT). *Mutagenesis* 2008; 23(4): 271–83.

27. Cytotoxicity measures in the in vitro micronucleus test. *Mutat Res* 2010; 702: 131–248 (special issue).

28. Organisation for Economic Co-operation and Development (OECD). *OECD Guideline for Testing of Chemicals: (Test Guideline 487)*, 2010.

29. Organisation for Economic Co-operation and Development (OECD). *OECD Guideline for Testing of Chemicals: (Test Guideline 476)*, 1997.

30. Galloway S, Lorge E, Aardema MJ, Eastmond D, Fellows M, Heflich R, Kirkland D et al. Workshop summary: Top concentration for in vitro mammalian cell genotoxicity assays; and report from working group on toxicity measures and top concentration for in vitro cytogenetics assays (chromosome aberrations and micronucleus). *Mutat Res* 2011; 723(2): 77–83.

31. Fowler P, Smith K, Young J, Jeffrey L, Kirkland D, Pfuhler S, Carmichael P. Reduction of misleading ("false") positive results in mammalian cell genotoxicity assays. I. Choice of cell type. *Mutat Res* 2012; 742(1–2): 11–25.

32. Alfaro-Moreno E, Nawrot TS, Vanaudenaerde BM, Hoylaerts MF, Vanoirbeek JA, Nemery B, Hoet PH. Co-cultures of multiple cell types mimic pulmonary cell communication in response to urban PM10. *Eur Respir J* 2008; 32(5): 1184–94.

33. Dahl EL, Curren R, Barnett BC, Khambatta Z, Reisinger K, Ouedraogo G, Faquet B et al. The reconstructed skin micronucleus assay (RSMN) in EpiDerm: Detailed protocol and harmonized scoring atlas. *Mutat Res* 2011; 720(1–2): 42–52.

34. Aardema MJ, Barnett BB, Mun GC, Dahl EL, Curren RD, Hewitt NJ, Pfuhler S. Evaluation of chemicals requiring metabolic activation in the EpiDerm™ 3D human reconstructed skin micronucleus (RSMN) assay. *Mutat Res* 2013; 750(1–2): 40–9.

35. Oberdorster G, Oberdorster E, Oberdorster J. Nanotoxicology: An emerging discipline evolving from studies of ultrafine particles. *Environ Health Perspect* 2005; 113(7): 823–39.

36. Powers KW, Brown SC, Krishna VB, Wasdo SC, Moudgil BM, Roberts SM. Research strategies for safety evaluation of nanomaterials. Part VI. Characterization of nanoscale particles for toxicological evaluation. *Toxicol Sci* 2006; 90(2): 296–303.

37. Warheit DB. How meaningful are the results of nanotoxicity studies in the absence of adequate material characterization? *Toxicol Sci* 2008; 101(2): 183–5.

38. Gonzalez L, Lison D, Kirsch-Volders M. Genotoxicity of engineered nanomaterials: A critical review. *Nanotoxicology* 2008; 2(4): 252–73.

39. Oberdorster G. Safety assessment for nanotechnology and nanomedicine: Concepts of nanotoxicology. *J Intern Med* 2010; 267: 89–105.

40. Fubini B, Ghiazza M, Fenoglio I. Physico-chemical features of engineered nanoparticles relevant to their toxicity. *Nanotoxicology* 2010; 4: 347–63.

41. Teeguarden JG, Hinderliter PM, Orr G, Thrall BD, Pounds JG. Particokinetics in vitro: Dosimetry considerations for in vitro nanoparticle toxicity assessments. *Toxicol Sci* 2007; 95(2): 300–12.

42. Hinderliter PM, Minard KR, Orr G, Chrisler WB, Thrall BD, Pounds JG, Teeguarden JG. ISDD: A computational model of particle sedimentation, diffusion and target cell dosimetry for in vitro toxicity studies. *Part Fibre Toxicol* 2010; 7(1): 36.

43. Lison D, Thomassen LC, Rabolli V, Gonzalez L, Napierska D, Seo JW, Kirsch-Volders M, Hoet P, Kirschhock CE, Martens JA. Nominal and effective dosimetry of silica nanoparticles in cytotoxicity assays. *Toxicol Sci* 2008; 104(1): 155–62.

44. Gonzalez L, Lukamowicz-Rajska M, Thomassen LCJ et al. Co-assessment of cell cycle and micronucleus frequencies demonstrates the influence of serum on the in vitro genotoxic response to amorphous monodisperse silica nanoparticles of varying sizes. Nanotoxicology. Doi:10.3109/17435390.2013.842266.

45. Monteiro-Riviere NA, Inman AO. Challenges for assessing carbon nanomaterial toxicity to the skin. *Carbon* 2006; 44: 1070–8.

46. Monteiro-Riviere NA, Inman AO, Zhang LW. Limitations and relative utility of screening assays to assess engineered nanoparticle toxicity in a human cell line. *Toxicol Appl Pharmacol* 2009; 234: 222–35.

47. Worle-Knirsch JM, Pulskamp K, Krug HF. Oops they did it again! Carbon nanotubes hoax scientists in viability assays. *Nano Lett* 2006; 6: 1261–8.

48. Casey A, Herzog E, Davoren M, Lyng FM, Byrne HJ, Chambers G. Spectroscopic analysis confirms the interactions between single walled carbon nanotubes and various dyes commonly used to assess cytotoxicity. *Carbon* 2007; 45: 1425–32.

49. Pulskamp K, Diabate S, Krug HF. Carbon nanotubes show no sign of acute toxicity but induce intracellular reactive oxygen species in dependence on contaminants. *Toxicol Lett* 2007; 168(1): 58–74.

50. Gonzalez L. *Exploring the Genotoxic Potential of Nanoparticles Using Amorphous Silica Nanoparticles*. 2011; VUB press, Brussels, Belgium, pp. 240.

51. Kain J, Karlsson HL, Moller L. DNA damage induced by micro- and nanoparticles—Interaction with FPG influences the detection of DNA oxidation in the comet assay. *Mutagenesis* 2012; 27(4): 491–500.

52. Clift MJD, Raemy DO, Endes C, Lehmann AD, Brandenberger C, Petri-Fink A, Wick P et al. Can the Ames test provide an insight into nano-object mutagenicity? Investigating the interaction between nano-objects and bacteria. *Nanotoxicology* 2012: 1–13.

53. Pfuhler S, Elespuru R, Aardema M, Doak SH, Maria Donner E, Honma M, Kirsch-Volders M et al. Genotoxicity of nanomaterials: Refining strategies and tests for hazard identification. *Environ Mol Mutagen* 2013; 54: 229–39.

54. Doak SH, Griffiths SM, Manshian B, Singh N, Williams PM, Brown AP, Jenkins GJ. Confounding experimental considerations in nanogenotoxicology. *Mutagenesis* 2009; 24(4): 285–93.

55. Gonzalez L, Sanderson BJ, Kirsch-Volders M. Adaptations of the in vitro MN assay for the genotoxicity assessment of nanomaterials. *Mutagenesis* 2011; 26(1): 185–91.

56. Gonzalez L, Corradi S, Thomassen LC, Martens JA, Cundari E, Lison D, Kirsch-Volders M. Methodological approaches influencing cellular uptake and cyto-(geno) toxic effects of nanoparticles. *J Biomed Nanotechnol* 2011; 7(1): 3–5.

57. Kirkland D. Evaluation of different cytotoxic and cytostatic measures for the in vitro micronucleus test (MNVit): Introduction to the collaborative trial. *Mutat Res* 2010; 702(2): 135–8.

58. Parry JM, Kirsch-Volders M. Special issue on in vitro MN trial. *Mutat Res* 2010; 702(2): 132–4.

59. Elhajouji A, Van Hummelen P, Kirsch-Volders M. Indications for a threshold of chemically-induced aneuploidy in vitro in human lymphocytes. *Environ Mol Mutagen* 1995; 26(4): 292–304.

60. Chen M, von Mikecz A. Formation of nucleoplasmic protein aggregates impairs nuclear function in response to SiO_2 nanoparticles. *Exp Cell Res* 2005; 305(1): 51–62, 304.
61. Schins RP. Mechanisms of genotoxicity of particles and fibers. *Inhal Toxicol* 2002; 14(1): 57–78.
62. Knaapen AM, Borm PJ, Albrecht C, Schins RP. Inhaled particles and lung cancer. Part A: Mechanisms. *Int J Cancer* 2004; 109(6): 799–809.
63. Schins RP, Knaapen AM. Genotoxicity of poorly soluble particles. *Inhal Toxicol* 2007; 19: 189–98.
64. Brown DM, Stone V, Findlay P, MacNee W, Donaldson K. Increased inflammation and intracellular calcium caused by ultrafine carbon black is independent of transition metals or other soluble components. *Occup Environ Med* 2000; 57(10): 685–91.
65. Gonzalez L, Decordier I, Kirsch-Volders M. Induction of chromosome malsegregation by nanomaterials. *Biochem Soc Trans* 2010; 38(6): 1691–7.
66. Nel AE, Madler L, Velegol D, Xia T, Hoek EM, Somasundaran P, Klaessig F, Castranova V, Thompson M. Understanding biophysicochemical interactions at the nano-bio interface. *Nat Mater* 2009; 8(7): 543–57.
67. Cedervall T, Lynch I, Lindman S, Berggard T, Thulin E, Nilsson H, Dawson KA, Linse S. Understanding the nanoparticle-protein corona using methods to quantify exchange rates and affinities of proteins for nanoparticles. *Proc Natl Acad Sci USA* 2007; 104(7): 2050–5.
68. Lundqvist M, Stigler J, Elia G, Lynch I, Cedervall T, Dawson KA. Nanoparticle size and surface properties determine the protein corona with possible implications for biological impacts. *Proc Natl Acad Sci USA* 2008; 105(38): 14265–70.
69. Dutta D, Sundaram SK, Teeguarden JG, Riley BJ, Fifield LS, Jacobs JM, Addleman SR, Kaysen GA, Moudgil BM, Weber TJ. Adsorbed proteins influence the biological activity and molecular targeting of NMs. *Toxicol Sci* 2007; 100(1): 303–15.
70. Petri-Fink A, Steitz B, Finka A, Salaklang J, Hofmann H. Effect of cell media on polymer coated superparamagnetic iron oxide nanoparticles (SPIONs): Colloidal stability, cytotoxicity, and cellular uptake studies. *Eur J Pharm Biopharm* 2008; 68(1): 129–37.
71. Clift MJ, Bhattacharjee S, Brown DM, Stone V. The effects of serum on the toxicity of manufactured nanoparticles. *Toxicol Lett* 2010; 198(3): 358–65.
72. Corradi S, Gonzalez L, Thomassen LC, Bilaničová D, Birkedal RK, Pojana G, Marcomini A, Jensen KA, Leyns L, Kirsch-Volders M. Influence of serum on in situ proliferation and genotoxicity in A549 human lung cells exposed to NMs. *Mutat Res* 2012; 745(1–2): 21–7.
73. Merhi M, Dombu CY, Brient A, Chang J, Platel A, Le Curieux F, Marzin D, Nesslany F, Betbeder D. Study of serum interaction with a cationic nanoparticle: Implications for in vitro endocytosis, cytotoxicity and genotoxicity. *Int J Pharm* 2012; 423(1): 37–44.
74. Murdock RC, Braydich-Stolle L, Schrand AM, Schlager JJ, Hussain SM. Characterization of nanomaterial dispersion in solution prior to in vitro exposure using dynamic light scattering technique. *Toxicol Sci* 2008; 101(2): 239–53.
75. Allouni ZE, Cimpan MR, Hol PJ, Skodvin T, Gjerdet NR. Agglomeration and sedimentation of TiO_2 nanoparticles in cell culture medium. *Colloids Surf B Biointerf* 2009; 68(1): 83–7.

21 Cell-Based Targeting of Anticancer Nanomaterials to Tumors

Matt Basel, Tej Shrestha, Hongwang Wang, and Stefan Bossmann

CONTENTS

21.1 INTRODUCTION

There are emerging and convincing data that certain cells have a natural propensity to migrate to tumors after systemic inoculation into preclinical models. This homing ability can be exploited for using cells as delivery vehicles for anticancer therapeutic or biological agents, which otherwise have major side effects or short half-life, precluding their systemic administration. Another potential use of tumor-seeking cells could be as stealth vehicles to ferry therapeutic nanomaterials to tumors to facilitate penetration, reduce potential toxicity, reduce potential unwanted immune consequences, and reduce removal by the reticuloendothelial system (RES).

21.2 CELLS CAN BE VECTORS FOR TARGETED CANCER THERAPY

It has been shown that primitive stem cells isolated from Wharton's jelly of umbilical cord can be used effectively as delivery vehicles for targeted cancer gene therapy.[1] These cells are easy to isolate noninvasively in large numbers, are not controversial, are relatively nonimmunogenic, and do not form teratomas.[2] When given intravenously to severe combined immunodeficiency (SCID) mice bearing human breast cancer metastases in the lung, human umbilical cord matrix stems cells were specifically localized adjacent to or within the metastatic tumors.[3–6] The tumor tissue tropism of human or rat Wharton's jelly mesenchymal stem cells (WJMSC) has also been shown for pancreatic tumor and lung cancer

393

rodent models.[7-9] WJMSC have been engineered to express *β-interferon* homed to and attenuated breast cancer, lung cancer, and bronchioalveolar carcinoma in mice.[6,8-10] An additive effect was observed after targeted delivery of a cytokine by WJMSC and low-dose delivery of 5-fluorouracil to mice with metastatic breast cancer lesions in their lungs.[10] Rat-derived WJMSC administered intravenously completely abolished orthotopic rat mammary carcinomas, and there was no evidence of recurrence or metastasis 100 days after tumor transplant.[5] Human WJMSC also have an intrinsic in vivo anticancer effect, but it is not as potent as that exhibited by the rat-derived cells.[4]

Other multipotent stem cells such as mesenchymal stem cells (MSCs) from bone marrow or fat have been used in this way as vectors for targeted therapy of preclinical models of breast cancer, pancreatic cancer, prostate cancer, and other cancers.[11-14]

Neural stem cells (NSCs) were actually the first cells to be described as potential delivery vehicles, when mouse NSCs were injected directly into the brain of mice with gliomas, and the ability of the transplanted cells to migrate from the contralateral hemisphere into the gliomas was sufficient to ignite great interest in using cells as delivery vehicles.[15] Since that report, NSCs, transplanted either directly into the brains of glioma-bearing mice or intravenously, have shown to be effective vectors for local gene therapy of glioma and other tumors.[16-21]

Cells other than stem cells can home to tumors and be used to deliver imaging or therapeutic agents. For example, Tie2-expressing monocytes have particular affinity for tumors, and monocytes engineered to express *interferon-α* under the Tie2 promoter were used to effectively deliver this cytokine to orthotopic human gliomas in SCID mice and spontaneous mouse mammary carcinomas with pronounced antitumor effects and reduction of metastasis.[22]

21.2.1 CELL DELIVERY AND ACTIVATION OF NANOMATERIAL-
PRODRUG COMBINATIONS

Irinotecan (CPT-11) is a potent anticancer prodrug that is converted by carboxylesterase (CE) to a topoisomerase I inhibitor, 7-ethyl-10-hydroxycamptothecin (SN-38).[23] CPT-11 has been clinically approved for treatment of colorectal cancer.[24] CPT-11 alone or in combination with other drugs showed promising results in clinical trials against metastatic colorectal cancer, metastatic small-cell lung cancer, and platinum-refractory epithelial ovarian cancer.[25-28] However, the use of this drug is limited because only 2%–5% of it is converted into active SN-38 and because of its undesirable side effects such as diarrhea, neutropenia, and interpatient variability of CE expression.[24,29-33] Poor solubility of SN-38 in water and physiological media further limits the use of this drug for cancer therapy.[24,34] To overcome these problems, different containment strategies have been used, such as binding in cationic peptides, encapsulation in vesicles like polymeric micelles, liposomes, thermosensitive liposomes, and multiarmed polyethylene glycol nanographene oxide.[34-39] Targeted delivery of prodrug to increase therapeutic efficacy and reduce the undesirable side effects, gene-directed enzyme/prodrug therapy

(GDEPT) and targeted cytotherapy are better alternatives to the use of potent SN-38.[40] Danks et al.[19] used a secretory form of rabbit CE, which is much more efficient than human CE for activation of CPT-11 to active SN-38. They showed that systemic administration of CPT-11 with NCS containing a secretory form of rabbit CE increased the survival of mice with neuroblastoma. Although rabbit CE is more efficient than human CE, use of rabbit CE to treat humans could induce an immune response causing deactivation of the enzyme or other undesirable side effects.[41] These unwanted side effects could be mitigated if both the prodrug and the prodrug-activating enzyme were delivered using a single targeted delivery system so that the enzyme is not activated until reaching the target. Tumor-tropic monocytes and macrophages are good candidates for this purpose. It has been shown that these cells can deliver their contents to various tumors, such as liposome-contained fluorescent markers to gastric tumors, adenovirus to prostate tumors, and gold nanoshells to gliomas.[42–44]

We used RAW264.7 monocyte/macrophage-like cells (Mo/Ma) for delivery of both prodrug and enzyme for GDEPT.[45] We engineered RAW264.7 Mo/Ma to be double stable (DS monocytes) using Tet-On® Advanced system for intracellular carboxylesterase (InCE) expression. This system has a double layer of protection. It sequesters both the prodrug, CPT-11 (tethered to dextran via an InCE cleavable bond), and prodrug-activating enzyme, InCE. Once the engineered cells reach the cancer site, systemic administration of doxycycline turns on InCE expression and activates CPT-11. Our laboratory demonstrated that CPT-11-loaded DS monocytes homed to lung melanoma in C57Bl/6 mice 1 day after administration. We have shown that this gene-regulated cytotherapy significantly reduced the tumor weight and number of tumors in lung melanoma in C57Bl/6 mice.

Although our GDEPT system proved effective, we have noticed that cells can easily remove (efflux) CPT-11 before activation. To overcome this problem and increase solubility, we designed an SN-38 dextran prodrug.[46] SN-38 was attached to dextran with an InCE cleavable bond. It has better cellular uptake and retention than the CPT-11 prodrug because dextran is a larger, water-soluble sugar-based molecule that incorporates multiple SN-38 moieties per molecule. This prodrug was tested in a murine disseminated peritoneal pancreatic cancer model using InCE engineered tumor homing mouse monocyte/macrophage cells as carriers for both prodrug and enzyme. This system significantly increased the survival time of mice bearing pancreatic tumors. Our laboratory has designed a magnetic nanoparticle-based SN-38 prodrug that is attached to the surface of magnetic nanoparticles (MNPs) with an InCE cleavable linker.[47] This modification has several advantages: the large surface area of the MNP can accommodate multiple SN-38 molecules, MNP are easily taken up by delivery cells, and this therapy can be combined with hyperthermia for a synergistic effect. This cell-based, enzyme-activating prodrug system provides multiple levels of specificity to preclude damage to normal tissues while localizing potent anticancer treatment modalities to the tumor. This system can be combined with other therapies for optimum effect. Because the syngeneic DS monocytes elicit no immune response, many potential side effects are eliminated, and the potential exists for applying this system to treat other diseases.

21.3 NANOPARTICLES AND CYTOTHERAPY

There are several reports describing the use of tumor-tropic cells as vectors for targeted delivery of imaging or therapeutic nanomaterials to tumors. This approach has the attractive feature of being an active (rather than passive) targeting strategy with potentially better penetration of the tumor. The nanoparticles (NPs) are also shielded from the RES system as well as the immune system.

Choi et al.[48] isolated peritoneal macrophages from mice and loaded them with liposomes containing doxorubicin. The preloaded cells were administered intravenously to mice with A549 lung or subcutaneous tumors. Tumor volume was reduced; however, this result was not statistically significant.[48]

MSCs loaded with NPs containing coumarin as an imaging agent were administered ipsilaterally or into the contralateral hemisphere of mice with intracranial gliomas. The tumors were successfully imaged after cell administration.[49]

When superparamagnetic NPs are subjected in a fast-switching magnetic field, their magnetization direction will switch quickly along the field directions. The physical rotation of an NP (Brownian relaxation), and the magnetization reversal within the NP (Neel relaxation), results in the conversion of magnetic energy to thermal energy.[50–52] The heating power of the NPs is determined by the frequency of the alternating magnetic field (AMF) and the ferromagnetic hysteresis area. The NP heating efficiency is measured by the specific absorption rate (W g^{-1}).[53,54]

Because of their chemical stability, superior magnetic properties, and low cytotoxicity, Fe_3O_4 MNPs are one of the most common magnetic nanomaterials produced for magnetic hyperthermia applications.[55] Fe_3O_4 NPs are produced through the thermal decomposition of Fe(acetyl acetone)$_3$ in the presence of reducing agent with oleic acid or oleylamine as ligands (surfactants). Fe_3O_4 NPs of 4–6 nm are obtained. In addition, larger Fe_3O_4 NPs can be synthesized through a seed-mediated growth, by further reductive decomposition of Fe(acetyl acetone)$_3$ with the small Fe_3O_4 NPs serving as seeds. The reaction results in MNP with controllable size between 5 and 20 nm range. The particles exhibit high crystallinity as well as uniform size distribution.[56–58]

Another method to prepare monodispersed Fe_3O_4 NPs is thermal decomposition of Fe(CO)$_5$ in the presence of surfactants followed by oxidation.[59] The nucleation step of the NPs is achieved by rapidly injecting precursor molecules Fe(CO)$_5$ into a hot solution heated above the decomposition temperature of the precursor. The solution contains coordinating ligands for stabilizing the particles, reducing the reactivity of the monomers thereby allowing slow incorporation of the monomers into the particles with reduced defect density and high crystallinity. It was found that the nucleation resulting from the thermal decomposition of Fe(CO)$_5$ takes place at relatively low temperature, whereas the growth derived from the decomposition of the iron oleate complex occurs at a higher temperature. A 1-nm-scale size-controlled synthesis of monodisperse magnetic iron oxide NPs was achieved by the addition of an iron oleate complex to previously made iron NPs under reflux condition, followed by oxidation of obtained iron NP with anhydrous trimethylamine N-oxide to produce highly crystalline iron oxide nanocrystals.[60]

Metallic Fe NPs are highly desired magnetic materials in magnetic hyperthermia application due to the very high magnetization of bulk iron (Ms = 218 A m^2 kg^{-1}).

One challenge in preparing stable metallic Fe NPs lies in how to stabilize these NPs against fast oxidation, because metallic Fe is chemically active, and the huge surface area of the Fe NPs makes it easily oxidized to various iron oxide NPs with much reduced magnetizations.[61] The Sun group demonstrated that monodisperse core/shell Fe/Fe_3O_4 NPs can be prepared by first thermal decomposition of $Fe(CO)_5$, then control oxidation of the metallic Fe NPs with $(CH_3)_3NO$ to form a crystalline iron oxide shell, thus protecting the amorphous Fe core from fast oxidation. These core/shell Fe/Fe_3O_4 NPs exhibit lower magnetization value (90.6 A m^2 kg^{-1} [Fe]) than that of bulk Fe.[62] The Sun group achieved a major advance in preparing metallic Fe NPs recently by adding hexadecylammonium chloride to the previously reported octadecene, oleylamine, $Fe(CO)_5$ system, in which body-centered cubic Fe can be made with drastically increased stability and magnetization (Ms = 164 A m^2 kg^{-1}[Fe]).[63]

We synthesized the stable core/shell Fe/Fe_3O_4 NPs by adapting the Sun group's method. By simple dopamine coating (ligand exchange with dopamine), this type of NP can be easily dissolved in aqueous media, and showed superior heating ability in an AMF. We further tested the cytotoxicity of cell-loading efficacy of the water-solubilized NPs and found that these NPs are not toxic to tumor-homing cells such as Mo/Ma cells, and can be loaded into Mo/Ma cells successfully. By targeting delivery of these highly magnetized, biocompatible core/shell Fe/Fe_3O_4 NPs via tumor homing cells, we expect to see better treatment of deep tumors through magnetic hyperthermia.

Another method for using cell delivery against cancer involves joining the forces of hyperthermia with cytotherapy. Because cancer cells have been shown to have an increase susceptibility to elevated temperatures compared to healthy tissue, hyperthermia has been used as a cancer therapy for decades (if not centuries).[64,65] Conventional hyperthermia, both extreme (>41.5°C) and fever-level (39°C–41°C), have been shown to be effective, but both have unwanted side effects.[66–72] Magnetic hyperthermia has been proposed as a method for targeting hypothermia directly to tumor tissue to alleviate side effects. In magnetic hyperthermia, magnetically active NPs, such as superparamagnetic Fe_3O_4 or core-shell Fe/Fe_3O_4 NPs, are used to absorb energy from an AMF and turn that energy into heat by either Brownian or Neel relaxation.[73–76] When the magnetically active NPs are concentrated specifically at the tumor site, localized hyperthermia can be created. It has been used successfully in many studies and has even progressed to clinical trials.[77–82] A similar method of therapy uses photoactive NPs, such as gold nanoshells, to absorb infrared radiation and to release it as heat. The primary limitation with magnetic hyperthermia or photoactive hyperthermia is that it requires injection of milligram amounts of NPs directly into the tumor, limiting the treatment to easily accessible tumors.

Cytotherapy can be used to overcome these limitations by specifically delivering NPs to the tumor site. Researchers at the University of Nevada have investigated the utility of using cell-delivered hyperthermia to treat gliomas. Macrophages were loaded with gold-coated silica nanospheres that were photoactive in the near infrared (NIR). Using neurospheres, we observed that the macrophages loaded with the gold nanoshells penetrated the gliomas in sufficient number to sensitize the gliomas to NIR light treatment. Neurospheres treated with nanoshell-loaded macrophages and NIR showed significant reduction in growth and even shrinkage depending on the dose of NIR given.[44] These nanoshell-loaded macrophages could be used to selectively deliver NIR sensitizers to gliomas, especially because they are known to be able to cross the blood–brain barrier.[83] Another group from Indiana with similar in vitro results showed that nanoshell-loaded macrophages can, in fact, infiltrate tumors in vitro when delivered intravenously.[84,85]

Another method for joining cytotherapy and hyperthermia has been shown by our group using NSCs to deliver MNPs to melanomas. Core/shell Fe/Fe_3O_4 NPs were loaded into NSCs. The heating ability of these NP-loaded NSCs was shown by heating them in an AMF; the cells were shown to increase the temperature 2.6°C in 10 minutes of exposure to AMF. This system was tested in vivo on a subcutaneous melanoma model in mice. Tumor-bearing mice were treated with intravenously injected, NP-loaded NSCs for 5 days after tumor insertion. The mice were treated with AMF for 10 minutes after 9, 10, and 11 days of tumor insertion. NP-loaded NSCs were able to traffic to the tumor and on treatment with AMF showed a 43% decrease in tumor size, demonstrating that cell-delivered NPs can be used in vivo to treat malignancies.[86]

Additional studies have shown hyperthermia/cytotherapy combinations to treat deep-seated tumors. Core/shell Fe/Fe_3O_4 NPs were loaded into RAW264.7 (Mo/Ma). To show the ability of these NP-loaded Mo/Ma to heat tumors to suitable levels, we injected them intratumorally into mice bearing subcutaneous pancreatic tumors. The mice were killed and the tumors were removed and exposed ex vivo to AMF for 15 minutes and a 3°C increase in temperature was obtained compared to controls. The NP-loaded Mo/Ma were then tested in vivo in a disseminated peritoneal pancreatic cancer model in mice. Tumor-bearing mice were treated with intraperitoneally injected, NP-loaded Mo/Ma 5, 9, and 13 days after tumor implantation. Mice were also treated with AMF for 20 minutes on 8, 12, and 15 days after tumor implantation. The NP-loaded Mo/Ma were able to traffic to the tumor and treated mice showed a 31% increase in life expectancy post-tumor insertion.[87] Showing the feasibility of cell-directed hyperthermia to treat nonaccessible tumors indicates promise for future development of the union of cytotherapy and hyperthermia.

Subsequent to this work, two other reports describing cell-based hyperthermia have been published. Ruan et al.[88] used MSCs to deliver iron oxide NPs to subcutaneous gastric tumors in mice. Intravenously injected MSCs trafficked well to the tumors delivering the NPs with high specificity. On treatment with AMF almost complete inhibition of tumor growth was achieved.[88] Brown et al. used tumor-associated phagocytes (TAP) to deliver iron oxide NPs to disseminated peritoneal ovarian cancer in mice. Iron oxide NPs were injected intraperitoneally into mice bearing tumors and the iron was taken up by soluble TAP. Soluble TAP were collected by peritoneal lavage the next day and injected intraperitoneally into other tumor-bearing mice. On treatment with AMF, cell death was induced specifically in tumors while healthy tissues and other organs were unaffected.[89]

21.4 CONCLUSIONS

There is a growing sense of urgency in developing effective targeted nanotheranostics for clinical trials. Heterogeneous distribution of therapeutics because of physiologic barriers attributable to abnormal tumor vasculature and interstitium remains a major issue.[90] Cell vectors have three distinct advantages compared to conventional chemotherapeutic strategies: (1) They are capable of penetrating tumors and metastases alike. (2) Depending on the selection of the cell vector, various regions of the tumors (e.g., microvasculature or tumor core) can be targeted. (3) As described here, numerous strategies for cell-based tumor targeting and treatment are being developed. We hope that cell-based methods will be adapted to the varying tumor biology of different cancers. In addition to using genetically engineered, actively tumor-homing cells capable of ameboid movement, the development of NPs that are contained within cells during transport, followed by drug release in response to external stimuli, will improve delivery and subsequent cancer treatment.

REFERENCES

1. Mitchell KE, Weiss ML, Mitchell BM, Martin P, Davis D, Morales L, Helwig B et al.: Matrix cells from Wharton's jelly form neurons and glia. *Stem Cells* 2003, 21(1): 50–60.
2. Weiss ML, Medicetty S, Bledsoe AR, Rachakatla RS, Choi M, Merchav S, Luo Y, Rao MS, Velagaleti G, Troyer D: Human umbilical cord matrix stem cells: Preliminary characterization and effect of transplantation in a rodent model of Parkinson's disease. *Stem Cells* 2006, 24(3): 781–792.
3. Ayuzawa R, Doi C, Rachakatla RS, Pyle MM, Maurya DK, Troyer D, Tamura M: Naive human umbilical cord matrix derived stem cells significantly attenuate growth of human breast cancer cells in vitro and in vivo. *Cancer Lett* 2009, 280(1): 31–37.
4. Chao KC, Yang HT, Chen MW: Human umbilical cord mesenchymal stem cells suppress breast cancer tumorigenesis through direct cell–cell contact and internalization. *J Cell Mol Med* 2012, 16(8): 1803–1815.
5. Ganta C, Chiyo D, Ayuzawa R, Rachakatla R, Pyle M, Andrews G, Weiss M, Tamura M, Troyer D: Rat umbilical cord stem cells completely abolish rat mammary carcinomas with no evidence of metastasis or recurrence 100 days post-tumor cell inoculation. *Cancer Res* 2009, 69(5): 1815–1820.
6. Rachakatla RS, Marini F, Weiss ML, Tamura M, Troyer D: Development of human umbilical cord matrix stem cell-based gene therapy for experimental lung tumors. *Cancer Gene Ther* 2007, 14(10): 828–835.
7. Doi C, Maurya DK, Pyle MM, Troyer D, Tamura M: Cytotherapy with naive rat umbilical cord matrix stem cells significantly attenuates growth of murine pancreatic cancer cells and increases survival in syngeneic mice. *Cytotherapy* 2010, 12(3): 408–417.
8. Matsuzuka T, Rachakatla RS, Doi C, Maurya DK, Ohta N, Kawabata A, Pyle MM et al.: Human umbilical cord matrix-derived stem cells expressing *interferon-beta* gene significantly attenuate bronchioloalveolar carcinoma xenografts in SCID mice. *Lung Cancer* 2010, 70(1): 28–36.
9. Maurya DK, Doi C, Kawabata A, Pyle MM, King C, Wu Z, Troyer D, Tamura M: Therapy with un-engineered naive rat umbilical cord matrix stem cells markedly inhibits growth of murine lung adenocarcinoma. *BMC Cancer* 2010, 10: 590.
10. Rachakatla RS, Pyle MM, Ayuzawa R, Edwards SM, Marini FC, Weiss ML, Tamura M, Troyer D: Combination treatment of human umbilical cord matrix stem cell-based *interferon-beta* gene therapy and 5-fluorouracil significantly reduces growth of metastatic human breast cancer in SCID mouse lungs. *Cancer Invest* 2008, 26(7): 662–670.
11. Loebinger MR, Eddaoudi A, Davies D, Janes SM: Mesenchymal stem cell delivery of TRAIL can eliminate metastatic cancer. *Cancer Res* 2009, 69(10): 4134–4142.
12. Cousin B, Ravet E, Poglio S, De Toni F, Bertuzzi M, Lulka H, Touil I et al.: Adult stromal cells derived from human adipose tissue provoke pancreatic cancer cell death both in vitro and in vivo. *PloS One* 2009, 4(7): e6278.

13. Ren C, Kumar S, Chanda D, Kallman L, Chen J, Mountz JD, Ponnazhagan S: Cancer gene therapy using mesenchymal stem cells expressing *interferon-beta* in a mouse prostate cancer lung metastasis model. *Gene Ther* 2008, 15(21): 1446–1453.

14. Kucerova L, Altanerova V, Matuskova M, Tyciakova S, Altaner C: Adipose tissue-derived human mesenchymal stem cells mediated prodrug cancer gene therapy. *Cancer Res* 2007, 67(13): 6304–6313.

15. Aboody KS, Brown A, Rainov NG, Bower KA, Liu S, Yang W, Small JE et al.: Neural stem cells display extensive tropism for pathology in adult brain: Evidence from intracranial gliomas. *Proc Natl Acad Sci USA* 2000, 97(23): 12846–12851.

16. Aboody KS, Najbauer J, Danks MK: Stem and progenitor cell-mediated tumor selective gene therapy. *Gene Ther* 2008, 15(10): 739–752.

17. Ahmed AU, Thaci B, Alexiades NG, Han Y, Qian S, Liu F, Balyasnikova IV, Ulasov IY, Aboody KS, Lesniak MS: Neural stem cell-based cell carriers enhance therapeutic efficacy of an oncolytic adenovirus in an orthotopic mouse model of human glioblastoma. *Mol Ther* 2011, 19(9): 1714–1726.

18. Brown CE, Vishwanath RP, Aguilar B, Starr R, Najbauer J, Aboody KS, Jensen MC: Tumor-derived chemokine MCP-1/CCL2 is sufficient for mediating tumor tropism of adoptively transferred T cells. *J Immunol* 2007, 179(5): 3332–3341.

19. Danks MK, Yoon KJ, Bush RA, Remack JS, Wierdl M, Tsurkan L, Kim SU et al.: Tumor-targeted enzyme/prodrug therapy mediates long-term disease-free survival of mice bearing disseminated neuroblastoma. *Cancer Res* 2007, 67(1): 22–25.

20. Gutova M, Najbauer J, Chen MY, Potter PM, Kim SU, Aboody KS: Therapeutic targeting of melanoma cells using neural stem cells expressing carboxylesterase, a CPT-11 activating enzyme. *Curr Stem Cell Res Ther* 2010, 5(3): 273–276.

21. Kim SK, Kim SU, Park IH, Bang JH, Aboody KS, Wang KC, Cho BK et al.: Human neural stem cells target experimental intracranial medulloblastoma and deliver a therapeutic gene leading to tumor regression. *Clin Cancer Res* 2006, 12(18): 5550–5556.

22. De Palma M, Mazzieri R, Politi LS, Pucci F, Zonari E, Sitia G, Mazzoleni S et al.: Tumor-targeted *interferon-alpha* delivery by Tie2-expressing monocytes inhibits tumor growth and metastasis. *Cancer Cell* 2008, 14(4): 299–311.

23. Chabot GG: Clinical pharmacokinetics of irinotecan. *Clin Pharmacokinet* 1997, 33(4): 245–259.

24. O'Reilly S, Rowinsky EK: The clinical status of irinotecan (CPT-11), a novel water soluble camptothecin analogue: 1996. *Crit Rev Oncol/Hematol* 1996, 24(1): 47–70.

25. Cunningham D, Pyrhonen S, James RD, Punt CJ, Hickish TF, Heikkila R, Johannesen TB et al.: Randomised trial of irinotecan plus supportive care versus supportive care alone after fluorouracil failure for patients with metastatic colorectal cancer. *Lancet* 1998, 352(9138): 1413–1418.

26. Aranda E, Valladares M, Martinez-Villacampa M, Benavides M, Gomez A, Massutti B, Marcuello E et al.: Randomized study of weekly irinotecan plus high-dose 5-fluorouracil (FUIRI) versus biweekly irinotecan plus 5-fluorouracil/leucovorin (FOLFIRI) as first-line chemotherapy for patients with metastatic colorectal cancer: A Spanish Cooperative Group for the Treatment of Digestive Tumors Study. *Ann Oncol* 2009, 20(2): 251–257.

27. Noda K, Nishiwaki Y, Kawahara M, Negoro S, Sugiura T, Yokoyama A, Fukuoka M et al.: Irinotecan plus cisplatin compared with etoposide plus cisplatin for extensive small-cell lung cancer. *N Engl J Med* 2002, 346(2): 85–91.

28. Bodurka DC, Levenback C, Wolf JK, Gano J, Wharton JT, Kavanagh JJ, Gershenson DM: Phase II trial of irinotecan in patients with metastatic epithelial ovarian cancer or peritoneal cancer. *J Clin Oncol* 2003, 21(2): 291–297.

29. Senter PD, Beam KS, Mixan B, Wahl AF: Identification and activities of human carboxylesterases for the activation of CPT-11, a clinically approved anticancer drug. *Bioconjug Chem* 2001, 12(6): 1074–1080.

30. Treinen-Moslen M, Kanz MF: Intestinal tract injury by drugs: Importance of metabolite delivery by yellow bile road. *Pharmacol Ther* 2006, 112(3): 649–667.
31. Guichard S, Terret C, Hennebelle I, Lochon I, Chevreau P, Fretigny E, Selves J, Chatelut E, Bugat R, Canal P: CPT-11 converting carboxylesterase and topoisomerase activities in tumour and normal colon and liver tissues. *Br J Cancer* 1999, 80(3–4): 364–370.
32. Charasson V, Haaz MC, Robert J: Determination of drug interactions occurring with the metabolic pathways of irinotecan. *Drug Metab Dispos* 2002, 30(6): 731–733.
33. Huang B, Desai A, Tang S, Thomas TP, Baker JR, Jr.: The synthesis of a c(RGDyK) targeted SN38 prodrug with an indolequinone structure for bioreductive drug release. *Org Lett* 2010, 12(7): 1384–1387.
34. Meyer-Losic F, Nicolazzi C, Quinonero J, Ribes F, Michel M, Dubois V, de Coupade C et al.: DTS-108, a novel peptidic prodrug of SN38: In vivo efficacy and toxicokinetic studies. *Clin Cancer Res* 2008, 14(7): 2145–2153.
35. Koizumi F, Kitagawa M, Negishi T, Onda T, Matsumoto S, Hamaguchi T, Matsumura Y: Novel SN-38-incorporating polymeric micelles, NK012, eradicate vascular endothelial growth factor-secreting bulky tumors. *Cancer Res* 2006, 66(20): 10048–10056.
36. Sadzuka Y, Takabe H, Sonobe T: Liposomalization of SN-38 as active metabolite of CPT-11. *J Control Release* 2005, 108(2–3): 453–459.
37. Lei S, Chien PY, Sheikh S, Zhang A, Ali S, Ahmad I: Enhanced therapeutic efficacy of a novel liposome-based formulation of SN-38 against human tumor models in SCID mice. *Anticancer Drugs* 2004, 15(8): 773–778.
38. Peng CL, Tsai HM, Yang SJ, Luo TY, Lin CF, Lin WJ, Shieh MJ: Development of thermosensitive poly(n-isopropylacrylamide-co-((2-dimethylamino)ethyl methacrylate))-based nanoparticles for controlled drug release. *Nanotechnology* 2011, 22(26): 265608.
39. Liu Z, Robinson JT, Sun X, Dai H: PEGylated nanographene oxide for delivery of water-insoluble cancer drugs. *J Am Chem Soc* 2008, 130(33): 10876–10877.
40. Rooseboom M, Commandeur JN, Vermeulen NP: Enzyme-catalyzed activation of anticancer prodrugs. *Pharmacol Rev* 2004, 56(1): 53–102.
41. Humerickhouse R, Lohrbach K, Li L, Bosron WF, Dolan ME: Characterization of CPT-11 hydrolysis by human liver carboxylesterase isoforms hCE-1 and hCE-2. *Cancer Res* 2000, 60(5): 1189–1192.
42. Matsui M, Shimizu Y, Kodera Y, Kondo E, Ikehara Y, Nakanishi H: Targeted delivery of oligomannose-coated liposome to the omental micrometastasis by peritoneal macrophages from patients with gastric cancer. *Cancer Sci* 2010, 101(7): 1670–1677.
43. Muthana M, Giannoudis A, Scott SD, Fang HY, Coffelt SB, Morrow FJ, Murdoch C et al.: Use of macrophages to target therapeutic adenovirus to human prostate tumors. *Cancer Res* 2011, 71(5): 1805–1815.
44. Baek S-K, Makkouk AR, Krasieva T, Sun C-H, Madsen SJ, Hirschberg H: Photothermal treatment of glioma; an in vitro study of macrophage-mediated delivery of gold nanoshells. *J Neurooncol* 2011, 104(2): 439–448.
45. Seo GM, Rachakatla RS, Balivada S, Pyle M, Shrestha TB, Basel MT, Myers C et al.: A self-contained enzyme activating prodrug cytotherapy for preclinical melanoma. *Mol Biol Rep* 2012, 39(1): 157–165.
46. Basel MT, Balivada S, Shrestha TB, Seo GM, Pyle MM, Tamura M, Bossmann SH, Troyer DL: A cell-delivered and cell-activated SN38-dextran prodrug increases survival in a murine disseminated pancreatic cancer model. *Small* 2012, 8(6): 913–920.
47. Wang H, Shrestha TB, Basel MT, Dani RK, Seo GM, Balivada S, Pyle MM et al.: Magnetic-Fe/Fe(3)O(4)-nanoparticle-bound SN38 as carboxylesterase-cleavable prodrug for the delivery to tumors within monocytes/macrophages. *Beilstein J Nanotechnol* 2012, 3: 444–455.

48. Choi J, Kim HY, Ju EJ, Jung J, Park J, Chung HK, Lee JS et al.: Use of macrophages to deliver therapeutic and imaging contrast agents to tumors. *Biomaterials* 2012, 33(16): 4195–4203.

49. Roger M, Clavreul A, Venier-Julienne MC, Passirani C, Sindji L, Schiller P, Montero-Meni C, Menei P: Mesenchymal stem cells as cellular vehicles for delivery of nanoparticles to brain tumors. *Biomaterials* 2010, 31(32): 8393–8401.

50. Lu JJ, Hunang HL, Klik I: Field orientations and sweep rate effects on magnetic switching of Stoner–Wohlfarth particles. *J Appl Phys* 1994, 76: 1726–1732.

51. Jordan A, Scholz R, Wust P, Fahling H, Felix R: Magnetic fluid hyperthermia (MFH): Cancer treatment with AC magnetic field induced excitation of biocompatible superparamagnetic nanoparticles. *J Magn Magn Mater* 1999, 201: 413–419.

52. Moroz P, Jones SK, Gray BN: Magnetically mediated hyperthermia: Current status and future directions. *Int J Hyperthermia* 2002, 18(4): 267–284.

53. Teran FJ, Casado C, Mikuszeit N, Salas G, Bollero A, Morales MP, Camarero J, Miranda R: Accurate determination of the specific absorption rate in superparamagnetic nanoparticles under non-adiabatic conditions. *Appl Phys Lett* 2012, 101(6): 062413–062413-4.

54. Huang S, Wang SY, Gupta A, Borca-Tascuic DA, Salon SJ: On the measurement technique for specific absorption rate of nanoparticles in an alternating electromagnetic field. *Meas Sci Technol* 2012, 23: 035701.

55. Lewinski N, Colvin V, Drezek R: Cytotoxicity of nanoparticles. *Small* 2008, 4(1): 26–49.

56. Sun SH, Zeng H: Size-controlled synthesis of magnetite nanoparticles. *J Am Chem Soc* 2002, 124(28): 8204–8205.

57. Sun SH, Zeng H, Robinson DB, Raoux S, Rice PM, Wang SX, Li GX: Monodisperse MFe2O4 (M = Fe,Co,Mn) nanoparticles. *J Am Chem Soc* 2004, 126(1): 273–279.

58. Niederberger M: Nonaqueous sol–gel routes to metal oxide nanoparticles. *Acc Chem Res* 2007, 40(9): 793–800.

59. Hyeon T, Lee SS, Park J, Chung Y, Na HB: Synthesis of highly crystalline and monodisperse maghemite nanocrystallites without a size-selection process. *J Am Chem Soc* 2001, 123(51): 12798–12801.

60. Park J, Lee E, Hwang NM, Kang MS, Kim SC, Hwang Y, Park JG et al.: One-nanometer-scale size-controlled synthesis of monodisperse magnetic iron oxide nanoparticles. *Angew Chem Int Ed Engl* 2005, 44(19): 2872–2877.

61. Dumestre F, Chaudret B, Amiens C, Renaud P, Fejes P: Superlattices of iron nanocubes synthesized from Fe[N(SiMe3)(2)](2). *Science* 2004, 303(5659): 821–823.

62. Peng S, Wang C, Xie J, Sun SH: Synthesis and stabilization of monodisperse Fe nanoparticles. *J Am Chem Soc* 2006, 128(33): 10676–10677.

63. Lacroix LM, Frey Huls N, Ho D, Sun X, Cheng K, Sun S: Stable single-crystalline body centered cubic Fe nanoparticles. *Nano Lett* 2011, 11(4): 1641–1645.

64. Hildebrandt B, Wust P, Ahlers O, Dieing A, Sreenivasa G, Kerner T, Felix R, Riess H: The cellular and molecular basis of hyperthermia. *Crit Rev Oncol Hematol* 2002, 43(1): 33–56.

65. Shecterle LM, St Cyr JA: Whole body hyperthermia as a potential therapeutic option. *Cancer Biother* 1995, 10(4): 253–256.

66. Hildebrandt B, Hegewisch-Becker S, Kerner T, Nierhaus A, Bakhshandeh-Bath A, Janni W, Zumschlinge R, Sommer H, Riess H, Wust P: Current status of radiant whole-body hyperthermia at temperatures >41.5 degrees C and practical guidelines for the treatment of adults. The German 'Interdisciplinary Working Group on Hyperthermia'. *Int J Hyperthermia* 2005, 21(2): 169–183.

67. Habash RW, Bansal R, Krewski D, Alhafid HT: Thermal therapy, part 2: Hyperthermia techniques. *Crit Rev Biomed Eng* 2006, 34(6): 491–542.

68. Sminia P, van der Zee J, Wondergem J, Haveman J: Effect of hyperthermia on the central nervous system: A review. *Int J Hyperthermia* 1994, 10(1): 1–30.
69. Vertree RA, Leeth A, Girouard M, Roach JD, Zwischenberger JB: Whole-body hyperthermia: A review of theory, design and application. *Perfusion* 2002, 17(4): 279–290.
70. Jia D, Liu J: Current devices for high-performance whole-body hyperthermia therapy. *Expert Rev Med Devices* 2010, 7(3): 407–423.
71. Kraybill WG, Olenki T, Evans SS, Ostberg JR, O'Leary KA, Gibbs JF, Repasky EA: A phase I study of fever-range whole body hyperthermia (FR-WBH) in patients with advanced solid tumours: Correlation with mouse models. *Int J Hyperthermia* 2002, 18(3): 253–266.
72. Jia D, Rao W, Wang C, Jin C, Wang S, Chen D, Zhang M, Guo J, Chang Z, Liu J: Inhibition of B16 murine melanoma metastasis and enhancement of immunity by fever-range whole body hyperthermia. *Int J Hyperthermia* 2011, 27(3): 275–285.
73. Koetitz R, Fannin PC, Trahms L: Time domain study of Brownian and Neel relaxation in ferrofluids. *J Magn Magn Mater* 1995, 149(1–2): 42–46.
74. Pakhomov AB, Bao Y, Krishnan KM: Effects of surfactant friction on Brownian magnetic relaxation in nanoparticle ferrofluids. *J Appl Phys* 2005, 97(10): 10Q305/301–10Q305/303.
75. Shapiro MG, Atanasijevic T, Faas H, Westmeyer GG, Jasanoff A: Dynamic imaging with MRI contrast agents: Quantitative considerations. *Magn Reson Imaging* 2006, 24(4): 449–462.
76. Jordan AW, Wust P, Fahling H, John W, Hinz A, Felix R: Inductive heating of ferrimagnetic particles and magnetic fluids—Physical evaluation of their potential for hyperthermia. *Int J Hyperthermia* 1993, 9(1): 51–68.
77. Shinkai M, Yanase M, Suzuki M, Honda H, Wakabayashi T, Yoshida J, Kobayashi T: Intracellular hyperthermia for cancer using magnetite cationic liposomes. *J Magn Magn Mater* 1999, 194(1–3): 176–184.
78. Le B, Shinkai M, Kitade T, Honda H, Yoshida J, Wakabayashi T, Kobayashi T: Preparation of tumor-specific magnetoliposomes and their application for hyperthermia. *J Chem Eng Jpn* 2001, 34(1): 66–72.
79. Ito A, Shinkai M, Honda H, Yoshikawa K, Saga S, Wakabayashi T, Yoshida J, Kobayashi T: Heat shock protein 70 expression induces antitumor immunity during intracellular hyperthermia using magnetite nanoparticles. *Cancer Immunol Immunother* 2003, 52(2): 80–88.
80. Jordan A, Scholz R, Wust P, Schirra H, Schiestel T, Schmidt H, Felix R: Endocytosis of dextran and silan-coated magnetite nanoparticles and the effect of intracellular hyperthermia on human mammary carcinoma cells in vitro. *J Magn Magn Mater* 1999, 194(1–3): 185–196.
81. Jordan A, Scholz R, Maier-Hauff K, van Landeghem FK, Waldoefner N, Teichgraeber U, Pinkernelle J et al.: The effect of thermotherapy using magnetic nanoparticles on rat malignant glioma. *J Neurooncol* 2006, 78(1): 7–14.
82. Jordan A, Scholz R, Wust P, Fahling H, Krause J, Wlodarczyk W, Sander B, Vogl T, Felix R: Effects of magnetic fluid hyperthermia (MFH) on C3H mammary carcinoma in vivo. *Int J Hyperthermia* 1997, 13(6): 587–605.
83. Madsen SJ, Baek S-K, Makkouk AR, Krasieva T, Hirschberg H: Macrophages as cell-based delivery systems for nanoshells in photothermal therapy. *Ann Biomed Eng* 2012, 40(2): 507–515.
84. Choi MR, Stanton-Maxey KJ, Stanley JK, Levin CS, Bardhan R, Akin D, Badve S et al.: A cellular trojan horse for delivery of therapeutic nanoparitcles into tumors. *Nano Lett* 2007, 7(12): 3759–3765.

85. Choi MR, Bardhan R, Stanton-Maxey KJ, Badve S, Nakshtri H, Stanz KM, Cao N, Halas NJ, Clare SE: Delivery of nanoparticles to brain metastases of breast cancer using a cellular Trojan horse. *Cancer Nanotechnol* 2012, 3: 47–54.

86. Rachakatla R, Balivada S, Seo G-M, Myers CB, Wang H, Samarkoon T, Dani R et al.: Attenuation of mouse melanoma by A/C magnetic field after delivery of bi-magnetic nanoparticles by neural progenitor cells. *ACS Nano* 2010, 4(12): 7093–7104.

87. Basel MT, Balivada S, Wang H, Shrestha TB, Seo GM, Pyle M, Abayaweera G et al.: Cell-delivered magnetic nanoparticles caused hyperthermia-mediated increased survival in a murine pancreatic cancer model. *Int J Nanomedicine* 2012, 7: 297–306.

88. Ruan J, Ji J, Song H, Qian Q, Wang K, Wang C, Cui D: Fluorescent magnetic nanoparticle-labeled mesenchymal stem cells for targeted imaging and hyperthermia of in vivo gastric cancer. *Nanoscale Res Lett* 2012, 7:309.

89. Toraya-Brown S, Sheen MR, Baird JR, Barry S, Demidenko E, Turk MJ, Hoopes J, Conejo-Garcia JR, Fiering S: Phagocytes mediate targeting of iron oxide nanoparticles to tumors for cancer therapy. *Integr Biol* 2013, 5(1): 159–171.

90. Jain RK, Stylianopoulos T: Delivering nanomedicine to solid tumors. *Nat Rev Clin Oncol* 2010, 7: 653–64.

22 Silver Nanoparticles in Biomedical Applications

Meghan E. Samberg and Nancy A. Monteiro-Riviere

CONTENTS

22.1 INTRODUCTION

There is perhaps no single field in scientific history that has developed so explosively and prolifically as that of nanotechnology. Intensive research over the past decade in the burgeoning field has resulted in nanomaterials that have unique properties with at least one dimension between 1 and 100 nm, of tightly controlled size and structure, composed of a variety of materials including metals, oxides, ceramics, organics and hybrids. Nanostructuring of materials results in an amplified ratio of reactive surface atoms to inert core atoms and subsequent increased surface reactivity.[1] The increased reactivity more often than not produces nanomaterials that exhibit properties and behaviors that differ drastically to corresponding bulk materials of the same chemical composition. Accordingly, current nanotechnology focuses on deliberately fabricating nanomaterials to take advantage of the enhanced magnetic, electrical, optical, mechanical, and biological properties.[2]

Owing to these heightened properties, many nanomaterials have commercial relevance in areas such as electronics, energy, and biotechnology, and the nanomaterials market is projected to grow at the annual rate of approximately 23% until 2016.[3–6] While carbon materials are perhaps the most researched nanomaterial, silver nanoparticles (AgNPs) are the single most manufacturer-identified material used in all nanotechnology products[7,8] (see Chapter 6). As is the case for most nanoparticles, Ag-nps exhibit unique physicochemical properties that differ from and provide distinct advantages over bulk Ag. Although Ag-nps exhibit unique optical and electrical properties at the nanoscale, the most remarkable and exploited nanoscale property of Ag-nps is biological: their enhanced antimicrobial activity. Accordingly, resurgence in the medical potential of Ag has been observed as a direct result of this augmented capacity. The professional and personal healthcare fields have taken advantage of this augmented property through the incorporation of Ag-nps into hundreds of products, specifically surgical and food handling, packaging and storage tools, water purifiers, textiles, cosmetics, contact lens cases, wound care products, implantable devices and catheters, and even children's toys.[8]

Despite the impressive breadth and seemingly endless number of applications for Ag-nps, they are used nearly exclusively for their heightened antimicrobial activity. Justifiable markets for the increased usage of antimicrobials exist in both public and private settings: the civic demand for increased antimicrobial usage may be due in parts to increased publicity, education, and a rising fear of transmitted infectious diseases; alternatively, private demands stem largely from hospitals due to the persistent incidence of nosocomial infections despite aggressive education for healthcare personnel.[9] This chapter will focus more on the latter setting, with sections including an introduction to Ag-np synthesis, incorporation into biomedical products for specialized applications, in vitro and in vivo toxicological information, and concluding with potential hazards associated with their misuse.

22.2 SILVER NANOPARTICLES

In its bulk form, Ag is a naturally occurring, soft, white, lustrous transition metal. Axiomatically, the everyday use of Ag has matured and continued over thousands of years, varying in function from jewelry and currency, to dental alloys, tableware and electrical wiring, and as an antimicrobial agent.[10] Ag has the highest electrical conductivity of any element and the highest thermal conductivity of any metal.[11] Along this line, the earliest publications to realize and take advantage of the augmented properties of Ag at the nanoscale were in the fields of spectroscopy and physics, owing to their remarkable surface plasmon.[12–14] Surface plasmon resonance occurs when an electromagnetic field drives the collective oscillations of a nanoparticle's free electrons into resonance. Interestingly, different Ag-np solutions of varying shapes yield multicolored plasmon resonances at visible wavelengths.[15] The plasmon resonant scattering from a single Ag-np is many orders of magnitude brighter than the signal from single fluorophores, fluorescent beads, or quantum dots in microscopic imaging applications.[16,17] In addition, the scattering signal does not suffer from photobleaching or blinking. These characteristics have made Ag-nps attractive as microscopic imaging labels, particularly for particle-tracking experiments and molecular contrast in tissues.[18]

While these enhanced optical and electrical properties of Ag at the nanoscale are of importance, the enhanced antibacterial activity of Ag has been the most valuable. Historically, the medicinal value of Ag is well documented: ancient Phoenicians placed Ag coins into water jugs as preservatives,[19] doctors administered Ag nitrate solutions to the eyes of newborns for the prevention of neonatal conjunctivitis,[20] and Ag sulfadiazine (SSD) creams are still considered the gold standard for the prevention of extensive bacterial colonization on burn patient's debrided skin. Although the majority of Ag-np predecessors such as Ag nitrate and SSD formulations were succeeded by the advent of penicillin and other contemporary antibiotics, advanced research into the medical potential of Ag continues, most recently at the nanoscale. In fact, as of the latest update to the consumer products inventory from The Project on Emerging Nanotechnology on March 10, 2011, Ag-nps are the single most manufacturer-identified material used in all of nanotechnology products.[8] Of the 1317 manufacturer-identified products that incorporate nanomaterials in some manner, the most frequent was Ag, followed by carbon nanotubes and fullerenes, titanium and titanium dioxide, silica, zinc and zinc oxide, and gold.

22.2.1 Synthesis and Functionalization of Silver Nanoparticles

With the increasing interest in Ag-nps, numerous publications have steadily emerged describing a variety of methods for Ag-nps synthesis. These methods of generation include spark discharging, electrochemical reduction, chemical reduction, solution irradiation, laser ablation, and cryochemical synthesis.[21,22] In the most recent publications, biosynthesis preparations have been described that use a range of ecofriendly materials such as fungi, leaf extracts, and sugars.[23–25] While the advantages of these methods include reduced occupational hazards, toxic byproducts, or generation of "dirty" Ag-np suspensions that require extensive washes, disadvantages include inferior uniformity and consistency. Furthermore, the method of synthesis can result in the production of Ag-nps of varied shapes such as spheres, rods, cubes, hollow tubes, multifacets, and films.[22] Transmission electron microscopy may be used to visualize the size, shape, and uniformity of Ag-nps, as seen in Figure 22.1 for 20 nm spherical Ag-nps.

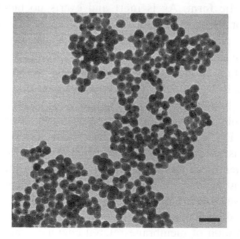

FIGURE 22.1   Transmission electron micrograph of 20 nm Ag-nps. Scale bar = 50 nm.

Functionalization of Ag-nps via grafting or internalization is an important step of synthesis to eliminate polydispersion, aggregation, and nonspecific binding issues, as well as to increase their specificity. For example, Ag-nps used for drug delivery and in vitro and in vivo imaging must be able to provide strong analytical signals, bind targets with high affinity and accuracy, and be adaptable.[26] Grafting of functional groups to the surfaces of synthesized nanoparticles has the advantage of altering the interfacial properties without affecting the bulk characteristics of materials. Methods of surface functionalization include plasma treatment, silanization, lipid self-assembly, and physical adsorption. For the majority of Ag-nps, synthesized in liquid phase, size and shape are often controlled by addition of organic stabilizers such as thiols, phosphates, amines, and/or surfactants, which are in some cases chemically bonded to the particle or, in other cases, simply adsorbed to its surface.[27] Comparatively, functionalization via internalization occurs during nanoparticle synthesis and is particularly useful for fluorescent dye and drug incorporation.

22.2.2 MICROCIDAL ACTIVITY OF SILVER NANOPARTICLES

Microscopic organisms are ubiquitous in nature, the majority of which are harmless to humans. However, a small percentage of microbes are pathogenic and result in infectious disease. A small list of common bacteria responsible for many nosocomial infections include *Escherichia, Pseudomonas, Salmonella, Vibrio, Bacillus, Clostridium, Enterococcus*, and vancomycin-resistant *Enterococcus, Listeria, Staphylococcus*, and methicillin-resistant *Staphylococcus aureus* (MRSA), and *Streptococcus*. Surfaces of everyday items are capable of harboring pathogens, particularly on high-contact surface areas such as computer keyboards, ATM buttons, indwelling catheters, handrails, and healthcare personnel uniforms. Their obvious presence coupled with the growing concern for antibiotic resistance has generated a renewed interest in developing products containing natural antimicrobial Ag. Although the microcidal use of Ag has been exploited for so long and tested so extensively, the exact mechanism of its activity has yet to be fully elucidated.

In its bulk metallic form, Ag is inert and exerts no biocidal action. In the presence of water or tissue fluids, surface atoms ionize to release antimicrobial Ag^+. Based on this phenomenon, the antimicrobial activity of Ag-nps may be explained by several mechanisms: (1) excessive binding of Ag ions and Ag-nps may prevent the uptake of essential nutrients to the bacterial cell; (2) excessive Ag^+ entry into the cell by competitive binding with essential metals such as Ca^{2+}, Mg^{2+}, and Mn^{2+}, without pumps for its removal; or (3) irreversible accumulation in the cell via complexation with ligands or substrates, to inhibit respiration or bind and condense DNA. Ultimately, the biocidal activity of Ag ions is likely caused by a synergistic effect between the binding of Ag ions to the cell wall, uptake and subsequent accumulation in the cell, and Ag interference with critical biomolecules within the cell. The Ag ions released from Ag-nps, as well as the Ag-nps themselves, bind to the functional groups present on the bacterial cell membranes such as disulfide, amino, carbonyl, and phosphate. This binding can result in the inactivation of membrane-related enzymes and denaturation of the bacterial cell envelope, impending its functional capacity to regulate the balance of solutes. The formation of reactive oxygen species has also been implicated in bacterial toxicity, and these are thought to further damage DNA and proteins as well as perturb cell membrane integrity.[28]

Ag-nps of various sizes, morphologies, surface conditions, and synthesis methods have proven to be effective against a broad spectrum of both Gram-positive and Gram-negative bacteria, and also against antibiotic-resistant bacterial strains and a variety of fungi.[29–31] As seen in the transmission electron micrograph (Figure 22.2), *S. aureus* exposed to 10 µg/mL of 20 nm Ag-nps displayed membrane integrity loss and ruptured cells with agglomerates of Ag-nps near the degenerated cells. The results have shown size-dependent antimicrobial activity of Ag-nps with diameters ranging from 1 to 450 nm against both Gram-negative and Gram-positive bacterial strains.[32] In addition to size and concentration, shape-dependent antimicrobial activity of Ag-nps show that between spherical, rod-shaped, and truncated triangular nanoplates, the truncated triangular Ag-nps displayed the strongest antibacterial activity.

Ag-nps offer a distinct advantage over traditional Ag nitrate and SSD therapies, by providing protracted release of microcidal Ag cations over the course of therapy. Furthermore, they provide a more efficient means for antimicrobial activity even at very low concentrations as a result of the increased surface area to volume ratio at decreased sizes. Because the majority of the data suggest that most pathogenic organisms are killed in vitro at Ag ion concentrations of 10–40 ppm, applications involving the Ag-np incorporation should determine the Ag-np concentration necessary to provide sufficient Ag cation release. This release is dependent on many factors such as Ag-np physicochemical properties and functionalization, extent of particle aggregation, degree of hydration, and solubility in the application environment.

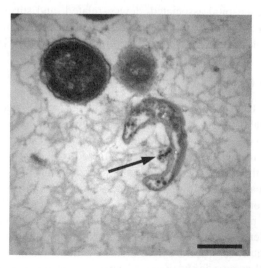

FIGURE 22.2 Transmission electron micrograph of *Staphylococcus aureus* treated with 10 µg/mL 20 nm Ag-nps. Bacterial cell walls appear compromised, and a whole cell has ruptured with agglomerates of Ag-nps attached to the remaining cell wall. Arrow depicts agglomerates of Ag-nps. Scale bar = 0.5 µm.

TABLE 22.1

Biomedical Applications of Ag-nps

Biomedical Application	Effect	Incorporation Method
Healthcare uniforms	Prevent bacterial spreading	Grafting
Surgical tools, indwelling devices, orthopedic implants	Prevent bacterial adhesion and biofilm formation	Coating, embedment
Biosensors	Increased resolution and contrast	Free nanoparticles

22.3 APPLICATIONS OF SILVER NANOPARTICLES

Depending on their application, nanomaterials may be delivered as free particles, immobilized on a surface or embedded within a matrix. Free nanoparticles are generally used as carriers because they are small enough to move within tissues and cells; furthermore, they may be functionalized to increase their specificity and function. For example, free magnetic nanoparticles and quantum dots have been frequently used in enzyme detection, contrast imaging agents, and excipients for drug delivery. Comparatively, nanoparticles used as coatings and embedded within matrices are more applicable for long-term applications such as antimicrobial technology.

Despite the vast number of publications that expound upon the limitless biomedical applications of differently synthesized Ag-nps, there are a shockingly small number of publications that have actually incorporated Ag-nps into testable biomedical products. The majority of application-specific publications featuring Ag investigate the antimicrobial efficacy of nanocrystalline Ag, and SSD, and these publications are currently driving the interest in Ag-np therapy. These therapies are most appropriate in instances where hospital-associated, device-related, and surgical-site infections are preventable. In Sections 22.3.1 through 22.3.5, the use of Ag-nps for a variety of biomedical applications will be explored and are summarized in Table 22.1.

22.3.1 SURFACE COATINGS

The colonization of hospital surfaces and medical devices by microorganisms, and the subsequent establishment of infection, is an increasing burden to the healthcare field that is owed to the increasing appearance of antibiotic-resistant and opportunistic organisms. Unfortunately, contact with contaminated surfaces can result in bacterial spreading 92% of the time.[33,34] Another example is the transfer of MRSA up to 65% of the time from healthcare uniforms when leaning over an infected patient.[35] Despite aggressive education to healthcare personnel, up to 70% of carefully cleaned hospital rooms of infected patients were found to still contain MRSA isolates.[36] It has been estimated by the Center for Disease Control (CDC) that approximately 1 in 20 hospital patients would acquire an infection from their stay alone, resulting in approximately 1.7 million hospital-acquired infections each year, and approximately 99,000 deaths.[37] Furthermore, depending on the percentage of preventable healthcare-associated infections (HAI), the CDC has reported that anywhere from $5.7 to $31.5 billion in potential annual benefits could be recouped through the prevention of HAI.

To alleviate this health and financial burden, Ag-nps may be coated onto high-contact surfaces such as bed and handrails, doorplates, and healthcare staff uniforms. Coating surfaces with a thin layer of Ag-nps is theoretically attractive because Ag exhibits broad antibacterial activity, reduces bacterial adhesion to devices in vitro, and blocks biofilm formation on medical device in vivo. Although clinical reports are currently lacking, surgical mesh coated with nanocrystalline Ag has demonstrated the capacity to inhibit bacterial growth in vitro, and could support future Ag-np coatings. Electrodeposition of a mixture of Ag-nps and lysozyme onto the surface of stainless steel surgical blades and syringe needles showed bacterial cell lysis within minutes of incision into agarose infused with *M. lysodeikticus* cells.[38] Promising for antimicrobial healthcare uniforms, cotton fabrics grafted with 10 nm Ag-nps have demonstrated in vitro bacterial burden reductions that persisted through 50 washing cycles.[39]

22.3.2 INDWELLING CATHETERS

In the United States, where central venous catheters (CVCs) in intensive care units have been estimated at one million catheter-days per year, approximately 80,000 cases of CVC-related bloodstream infections are reported annually.[40] Clinical studies have suggested that incorporation of catheters with antiseptics or antibiotics could decrease the rates of colonization and incidence of catheter-related bloodstream infection. Initial antimicrobial catheters included the coatings of SSD and Ag powders bonded with ceramic zeolite (ARROWg+ard Blue PLUS, TeleFlex, Research Triangle Park, NC; Multicath Expert, Vygon, Ecouen, France; Vantex, Edwards Lifesciences Corp., Irvine, CA). However, various reports regarding their respective abilities to prevent bacterial adhesion and bloodstream infection have not been confirmed.

Ag-nps have also been incorporated by way of coatings both on the inside and outside of the catheters. A recent study investigated whether Ag-nps embedded in common catheter plastics retained their antimicrobial activity. Similar to the results for poly(oxyethylene)-segmented imide, uniform distribution of 25–30 nm Ag-nps in polyurethane plastics at concentrations over 0.3% effectively inhibited the growth of pathogens commonly associated with the colonization of catheters such as *E. coli, S. epidermidis,* and *C. albicans*.[41] Hydrophilic coatings on catheters that were incorporated with 20 nm Ag-nps at either 8% or 15% dry mass showed a marked decrease in bacterial adhesion yet failed to completely inhibit its attachment.[42] Interestingly, the research indicated that the concentration of released Ag ions sufficient to inhibit bacterial adhesion also resulted in the activation of platelets, which could result in disrupted patient hemostasis. In another study, 4–6 layers of 50 nm Ag-nps were deposited onto the surfaces of catheters by way of chemical reduction.[43] Over a 10-day period, it was shown that the catheters released up to 14% of coated Ag, and was capable of inhibiting biofilm formation but incomplete inhibition of planktonic bacterial growth. A potential reason for disappointing clinical results of coated CVCs could be the obliteration or inactivation of the Ag coating by blood plasma, and the lack of durable activity inherent in coatings.

To combat this phenomenon, a more dynamic incorporation of Ag-nps could be used. Compared to Ag-np coatings, Ag-nps embedded within plastics have shown enhancements to their biostability, biocompatibility, and bioreactivity.[44] For example, the LogiCath AgTive® (MedeX Medical Inc., UK) is a CVC fabricated using poly-urethanes impregnated with Ag-nps. A number of studies have been performed to evaluate the efficacy of CVCs impregnated with Ag-nps, with mixed results. A multi-center study that investigated the efficacy of these catheters to reduce CVC coloniza-tion rates and catheter-associated infection rates showed no clear advantage for their use.[45] Comparatively, a single-center study showed significant reductions in coloni-zation (−37.7%) and catheter-associated infection (−71.3%) from Ag-np impregnated catheters.[46] The discrepancy between the two studies is likely due to differences in levels of infection risk and case types in the two study populations. In addition to the multicenter study, healthcare personnel were aware of the catheter type in each case, which may have resulted in biased decisions.

Ventilators are another indwelling medical device, and ventilator-associated pneumonia (VAP) causes substantial morbidity. A multicenter study that investi-gated the efficacy of an Ag-coated endotracheal tube (Agento I.C. endotracheal tube, C.R. Bard Inc., Covington, GA) found that patients intubated for 24 hour or longer were associated with a relative risk reduction of 36% and delayed occurrences of microbiological VAP.[47]

22.3.3 ORTHOPEDIC IMPLANTS

Infections originating from implanted devices such as orthopedic fixation hardware and artificial prosthetics are a persistent and serious health issue. Infection rates for traumatological procedures using stabilized devices such as nails and plates are approximately 2.5%, with increasing rates of up to 20% for patients with open frac-tures and external fixator pins. An important step toward reducing these infections is inhibiting bacterial colonization on subcutaneous device surfaces. Similar to the pre-vious sections, Ag-nps may be deposited onto implant surfaces or embedded within the orthopedic material to prevent infection.

In vitro tests of Ag-np coatings onto orthopedic implants have shown a signifi-cant reduction of Gram-positive and Gram-negative bacterial adhesion on treated surfaces.[48] Coating of orthopedic pins with Ag-nps or dispersing them throughout polymethyl-methacrylate bone cement could be used to prevent bacterial coloniza-tion.[49] However, care should be taken when introducing Ag-nps into bone cements. Nanosized abrasion particles have a notoriously bad reputation in orthopedics because they are known to be highly inflammatory, leading to osteolysis and are also able to migrate systemically with potentially adverse effects. Thus, introduc-ing Ag-nps in orthopedics would first require demonstrating convincingly that they would not behave as wear debris and compromise the integrity of the implant.

22.3.4 BANDAGES FOR TRAUMA AND DIABETIC WOUNDS

The continuing occurrence of chronic, nonhealing wounds across all health settings represents a major health and financial burden. The major etiologies of chronic wounds are venous, diabetic and pressure ulcers; the annual costs of treating leg ulcers alone are estimated at up to $1 billion, and do not include the cost of facility-acquired pressure ulcers. Over the past 10 years, the use of Ag in wound dressings has rapidly increased in response to marketing of Ag products, beginning with compresses soaked in 0.5% Ag nitrate applied to extensive burns.[50] During the past 40 years, SSD products have been the mainstay for superficial burn wound management, but the frequent required dressing changes due to its half-life of 10 hours is labor intensive to healthcare professionals and painful to patients. Comparatively, a one-time application using a product with Ag-nps could provide long-term antimicrobial therapy with decreased cost and pain.

Current Ag dressings exist in many forms, with each form exalting different advantages: powders, foams, hydrogels, hydrocolloids, polymeric films, and meshes. One such commercial dressing that is currently used in North America is Acticoat 7 (Smith and Nephew, Hull, UK). Acticoat is a three-ply dressing consisting of a fenestrated nonwoven rayon core embedded between layers of polyethylene mesh with nanosilver deposits that purportedly release Ag ions and Ag-nps. A randomized trial compared the effectiveness of two commonly used antimicrobials, nanosilver, and cadexomer iodine. The performance of each antimicrobial was comparable, yet Ag was associated with a quicker healing rate during the first 2 weeks of treatments and in wounds that were larger, older, and had more exudate.[51]

Similar to previous applications, the advancement of nanotechnology has recently allowed for the incorporation of Ag-nps into proposed wound dressings. In addition to traditional material incorporation, these dressings are fabricated from novel biofibers such as chitosan, chitin, silk, and starch but have not been tested in clinical settings so far. Chitosan–alginate nanoparticles have the potential for targeted drug delivery for acne therapy and showed good antimicrobial and anti-inflammatory properties for the treatment of cutaneous pathogens such as *P.acnes*.[52] Chitosan is frequently used because it is a natural polysaccharide biopolymer obtained from the exoskeleton of crustaceans. Recent studies suggest that due to chitosan's immunological properties of inhibiting proinflammatory cytokines, it potentially could be used as a wound healing accelerant. In addition to improved antimicrobial efficacy, some of the proposed composite fibers demonstrate good clottability and cell attachment, which is ideal for wound healing applications.

In general, the release of Ag ions from the Ag-nps in wound dressings is triggered by wound fluids and tissue exudates. Although it has been deemed that 10–40 ppm is required for appropriate antimicrobial action, the Ag content of traditional dressings currently available varies from <0.1 mg/cm^2 to more than 1 mg/cm^2, with the amount of Ag ions released ranging from <1 ppm to >70 ppm.[32]

22.3.5 Tissue Engineered Scaffolds

Every year, millions of people suffer tissue loss or end-stage organ failure. Tissue engineering uses bioresorbable scaffolds that are seeded with patient-derived cells to restore, maintain, or improve tissue function. Despite advances in tissue scaffold design and cell seeding, engineered scaffolds currently in clinical use often still suffer unnecessary failure due to contamination during the in vitro culture step and also surgical site infection during implantation.[53]

Tissue engineering scaffolds must be biocompatible to facilitate host cell adhesion and cell growth, be mechanically strong, capable of being fabricated into a desired shape, have a suitable degradation rate to meet the requirements of new tissue growth, produce nontoxic degradation products, and be easily handled and processed during manufacture.[54] More specific requirements are necessary on a per tissue basis; for example, skin replacements are required to be bacteriostatic, semipermeable to water, and cosmetically acceptable. Skin scaffolds should be elastic enough to withstand cyclic mechanical strains without any significant permanent deformation.[55–59]

The incorporation of antimicrobial Ag-nps into tissue engineering scaffolds would be pivotal in the reduction of implant associated, and surgical site infection incidences. Current nanofibrous scaffold systems that use Ag-nps have done so through the incorporation of Ag nitrate into the polymer solutions, subsequently subjected to an annealing process.[60,61]. In vitro studies conducted for 14 days on electrospun degradable poly (L-lactide-*co*-epsilon-caprolactone) copolymer scaffolds impregnated with Ag-nps showed no decrease in toxicity and good cell migration into the scaffolds. The studies suggested that Ag-np-incorporated scaffolds are biocompatible and suitable for skin tissue engineering grafts.[62] At present, only preliminary results have been published regarding Ag-np-coated thermoset biomaterials, and show good biocompatibility. Obviously, any clinically relevant benefit they may possess has yet to be evaluated.

22.4   SILVER NANOPARTICLE EXPOSURE

There currently exists little regulation regarding the manufacture of Ag-nps due to the extensive, historically safe use of Ag. Increases in the use of Ag-nps across multiple healthcare fields will undoubtedly increase exposure and the potential for deleterious interactions. The rate of Ag ion production, as well as the distribution of Ag in tissue, is dependent on the ability of Ag-nps to release ions into specific tissues, cell membranes, or cells. In addition, once a nanoparticle is located within a site, it has the ability for protracted release of Ag ions, thus increasing the potential toxic impact.

22.4.1   In Vitro Toxicity

Ideally, the concentration of Ag-nps required to inhibit and eradicate microbial growth should be less than the concentration required for human cell cytotoxicity. Because Ag acts indiscriminately against both prokaryotic and eukaryotic cells, the mechanism responsible for Ag-np toxicity to human cells is similar to the mechanism responsible for its antibacterial effects. It is likely to involve bound forms of Ag-nps on the surface of a target cell, the uptake and release of high concentrations of Ag ions into the cell and within cellular compartments, the disruption of integral cellular processes though catalytic or physical processes such as protein or enzyme binding and subsequent denaturation, and the generation of reactive oxygen species and induced oxidative stress. In vitro cell studies on free Ag-nps have demonstrated size-dependent cytotoxicity, decreased membrane integrity, and the generation of reactive oxygen species in a variety of human cell lines such as fibrosarcoma, skin carcinoma, lung fibroblast, glioblastoma, hepatoma, alveolar, and keratinocyte.[63,64] As seen in Figure 22.3a and b, 80 nm Ag-nps were internalized into human epidermal

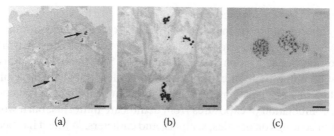

(a)   (b)   (c)

FIGURE 22.3 Transmission electron micrographs of Ag-nps in human epidermal keratinocytes in vitro (a), (b) and in vivo porcine skin (c). (a) Agglomerates of 80 nm Ag-nps localized within cytoplasmic vacuoles, scale bar = 2 μm; (b) higher magnification of the Ag-np agglomerates seen in (a), scale bar = 500 nm; (c) agglomerates of 50 nm Ag-nps in the uppermost layers of stratum corneum of in vivo porcine skin dosed daily for 14 days, scale bar = 300 nm. Arrows depict agglomerates of Ag-nps.

keratinocytes, and can be localized within cytoplasmic vacuoles. Furthermore, contaminants on the surface of Ag-nps or in their surrounding solutions can significantly influence their cytoticity.[63] The physicochemical and structural properties of Ag-nps play a major role in their interactions with cells, and variations in these properties among different Ag-nps may result in different toxicological effects.

However, the term Ag toxicity is used more in relation to clinical settings, whereas cytotoxicity is used more in relation to in vitro studies. Poon and Burd[65] demonstrated that Ag released from two products, a 0.5% Ag nitrate solution and Acticoat, was highly toxic to cultured monolayers of human epidermal keratinocytes and fibroblasts. Although the cells were susceptible to lethal damage when exposed to Ag, it was found that the toxic results were decreased as the cells were grown in more complex arrangements that closely simulate the clinical environment. This study concluded that Ag-based products cannot discriminate between healthy cells involved in wound healing and pathogenic bacteria, suggesting that Ag-based products should be used with caution where rapidly proliferating cells may be harmed. Another in vitro study concluded that out of seven Ag-containing dressings, Aquacel Ag (Convatec Inc., Skillman, NJ), Silvercel (Systagenix Gatwick, U.K.), and Polymem Silver (Ferris Mfg. Corp., Fort Worth, TX), had variable toxic effects on fibroblast contraction and viability.[66]

22.4.2 IN VIVO TOXICITY

The wide variety of uses of Ag-nps allows exposure through various routes of entry into the body. Because of their size, Ag-nps have the potential to impact the respiratory tract, gastrointestinal tract, and skin, and enter the systemic circulation or even the central nervous system. Correlating to the in vitro factors, several factors influence the ability of Ag-nps to produce toxic effect in vivo, such as their solubility, ability to bind to biological sites, and the degree to which complexes are formed and sequestered or metabolized and excreted. Theoretically, Ag can be deposited in any tissue of the human body, but the skin, brain, liver, kidneys, eyes, and bone marrow have received the greatest attention. Ag absorbed into human tissues from antiseptic respiratory sprays, implanted medical devices, wound dressings, or indwelling catheters can reach systemic circulation, primarily as a protein complex. The most common health effects associated with chronic exposure to Ag are a permanent grey or blue-grey discoloration of the skin known as argyria, found primarily in sun-exposed areas. Although argyria is most commonly reported clinically after excessive Ag ingestion, Ag deposition has been seen after SSD treatment of burned skin.[67]

The most common route of exposure to Ag-nps from biomedical applications is dermal; and for any given patient the contact may not be limited to just one instance, and rather a combination of exposures from healthcare uniforms, burn wound dressings, and from coatings on needles, scalpels, and catheters. While it has been deemed that 10–40 ppm is required for appropriate antimicrobial action, the Ag content of wound dressings currently available varies from <0.1 mg/cm^2 to more than 1 mg/cm^2 with the amount of Ag ions released ranging from <1 ppm to >70 ppm. Several studies have evaluated the toxicity of Ag from Ag-nps-containing wound dressings, and have shown elevated serum and liver enzyme levels as well as argyria. However, absorption through intact skin is low (<1 ppm) since much of the free ion is precipitated as Ag sulfide in the superficial layers of the stratum corneum. Daily repetitive dosing for 14 days with both freshly synthesized and thoroughly washed 20 and 50 nm Ag-nps topically applied onto the backs of pigs showed penetration of Ag-nps only in the superficial layers of the stratum corneum, as seen in Figure 22.3c.[63]

Very few large randomized prospective studies on the use of Ag dressings in human subjects have been published. Similar to the in vitro studies concerning Ag toxicity to dermal cell types, an in vivo study concluded that Acticoat significantly delayed reepithelialization on donor site wounds versus the faster healing rate with use of a nonsilver dressing.[68] The researchers speculated this was due to the high Ag concentration delivered to the wound surface. In this way, the characteristic quick release of Ag ions may not be suitable for acute wound models, and may be more appropriate for chronic wounds. This is supported by a randomized controlled study comparing the clinical efficacy and safety of Acticoat and SSD on 166 chronic burn wounds.[69] These authors concluded that Acticoat effectively and significantly improved the healing process of chronic burn wounds and restrained the bacterial growth more than in the SSD control group. The dermal absorption of Ag from burn patients wound dressings have been reported to be 0.1% or less.[70]

Over the centuries, Ag in any form has not shown to be toxic or carcinogenic to the immune, cardiovascular, nervous, or reproductive systems. The prevailing view is that the use of Ag compounds and Ag-nps in products is safe. Effective in 1999, the U.S. Food and Drug Agency (FDA) issued a *final rule* establishing that all over-the-counter (OTC) drug products containing Ag ingredients or Ag salts for internal or external use are not generally safe and effective. The reason for the final ruling is due to the rampant touting that OTC drug products containing colloidal Ag ingredients or Ag salts were being marketed for numerous serious disease conditions without substantial scientific evidence that supports their use for those disease conditions. The American government view is that any potential legitimate benefits of OTC drugs containing colloidal Ag or Ag salts have decreased dramatically over time and that what remains is a lack of established effectiveness and a potential for toxicity of these products. So far, the FDA has yet to issue a similar ruling for Ag-nps; however, the Environmental Protection Agency regulates all self-identified antimicrobials such as Ag-nps under the Federal Insecticide, Fungicide, and Rodenticide Act.

22.5 CONSEQUENCES OF SILVER NANOPARTICLE MISUSE

To provide antibacterial and antifungal action, Ag-nps are becoming widely used components in medical devices, healthcare products, and personal care items. Although experience has shown that metallic Ag and Ag compounds are generally safe to humans and effective in controlling pathogenic organisms, Ag-nps are capable of unprecedented entry into cellular compartments with protracted release of Ag ions. While this ability makes Ag-nps particularly valuable in eliminating harmful microbes, it also makes them potentially harmful. While some in vitro studies have shown toxicity related to Ag-nps, others have also shown an overall lack of toxicity. This may be due to the Ag-np synthesis method, concentration, or various physico-chemical properties or even the type of assay used to assess their toxicity. Therefore, it is important for future studies to determine the physicochemical properties of Ag-nps that make them toxic. Manufacturers of new products incorporating Ag-nps should envisage providing a balance between Ag ions released for antimicrobial purposes while maintaining a minimal human cell toxicity threshold. Furthermore, conclusive clinical data is difficult to obtain due to many factors including healthcare personnel bias, patient pathology, method of Ag-np incorporation, and data analysis.

Ag-resistant microorganisms have been found repeatedly in environments where Ag toxicity might be expected to select for resistance. An example is in the burn wards of hospitals where Ag salts are used as antiseptics to treat burns. The continued wide use of Ag products may result in more bacteria developing resistance, analogous to the worldwide emergence of antibiotic- and biocide-resistant bacteria.[71] Such resistant microbes would be detrimental to clinical uses that critically depend on the microcidal properties of Ag.

22.6 CONCLUSION

Ag-nps are potent and effective antimicrobial agents as a result of their protracted release of broad-spectrum bactericidal Ag cations. The varied uses of Ag-nps in the biomedical field include increased contrast and resolution in biosensing, coatings onto high contact surface areas, grafting onto healthcare professional clothing, coating onto indwelling medical devices, and incorporation into tissue-engineered scaffolds. The majority of the aforementioned biomedical applications are used in wound environments; therefore, the incorporation of Ag-nps into biomedical products should be approached with care due to their indiscriminate activity against both pathogenic organisms and proliferating human cells.

Although important topics have been touched upon in this chapter, many questions still need to be answered in future research: Are Ag-nps superior to other Ag forms in reducing microbial growth over time? Is one application with Ag-nps superior to repeated applications of traditional therapies when used clinically? Does the release of microcidal Ag cations into the wound milieu make a difference in healing outcomes? Does the benefit of using sustained, low-dose Ag-nps in multiple healthcare settings outweigh the risk of developing Ag-resistant organisms? Despite the overwhelming enthusiasm for Ag-np incorporation into biomedical applications, these questions demonstrate the need for large scaled, comprehensive studies to determine their validity as a superior antimicrobial.

REFERENCES

1. Auffan M, Rose J, Bottero J-Y, Lowry GV, Jolivet J-P, Wiesner MR. Towards a definition of inorganic nanoparticles from an environmental, health and safety perspective. *Nat Nanotechnol* 2009;4:634–41.
2. Nel A, Xia T, Mädler L, Li N. Toxic potential of materials at the nanoscale. *Science* 2006;311:622–27.
3. Korkin A, Gusev E, Labanowski JK, Luryi S, eds. *Nanotechnology for Electronic Materials and Devices.* New York, NY: Springer, 2007:1–368.
4. Nalwa HS. *Nanomaterials for Energy Storage Applications.* Valencia, CA: American Scientific Publishers, 2009:350.
5. Gouma P. *Nanomaterials for Chemical Sensors and Biotechnology.* Singapore: Pan Stanford Publishing, 2010:1–159.
6. Lucintel (Global Market Research Firm). Growth Opportunities in Global Nanomaterials Market 2011–2016. *Trend, Forecast, and Opportunity Analysis.* Dallas, TX: Research and Markets, 2011:1–153.
7. Woodrow Wilson Center (WWC). Project on Emerging Nanotechnologies. http://www.nanotechproject.org/inventories/consumer/analysis_draft/. Accessed April 29, 2013
8. The Project for Emerging Nanotechnology. Updated March 2011. www.nanotechproject.org Accessed April 4, 2013.
9. Coello R, Charlett A, Ward V, Wilson J, Pearson A, Sedgwick J, Borriello P. Device-related sources of bacteraemia in English hospitals—opportunities for the prevention of hospital-acquired bacteraemia. *J Hosp Infect* 2003;53(1):46–57.
10. Harmata M. *Silver in organic chemistry.* Hoboken, NJ: John Wiley & Sons. 2010:1–402.
11. Nordberg GF, Gerhardsson LS. In Seiler HG, Sigel H, eds. *Handbook on Toxicity of Inorganic Compounds*, New York, NY: Marcel Dekker, 1988:619–24.
12. Lehmann J, Merschdorf M, Pfeiffer W, Thon A, Voll S, Gerber G. Surface plasmon dynamics in silver nanoparticles studied by femtosecond time-resolved photoemission. *Phys Rev Lett* 2000;85(14):2921–24.
13. Haes AJ, Van Duyne RP. A nanoscale optical biosensor: Sensitivity and selectivity of an approach based on the localized surface plasmon resonance spectroscopy of triangular silver nanoparticles. *J Am Chem Soc* 2002;124(35):10596–604.
14. Bosbach J, Hendrich C, Stietz F, Vartanyan T, Träger F. Ultrafast dephasing of surface plasmon excitation in silver nanoparticles: Influence of particle size, shape, and chemical surrounding. *Phys Rev Lett* 2002;89(25):257404.
15. Liao H, Hehl CL, Hafner JH. Biomedical applications of plasmon resonant metal nanoparticles. *Nanomedicine* 2006;1(2):201–8.
16. Yguerabide J, Yguerabide EE. Light-scattering submicroscopic particles as highly fluorescent analogs and their use as tracer labels in clinical and biological applications. II. Experimental characterization. *Anal Biochem* 1998;262:157–76.
17. Schultz S, Smith DR, Mock JJ, Schultz DA. Single-target molecule detection with nonbleaching multicolor optical immunolabels. *Proc Natl Acad Sci USA* 2000;97:996–1001.
18. Schultz DA. Plasmon resonant particles for biological detection. *Curr Opin Biotechnol* 2003;14:13–22.
19. Angelotti N, Martini P. Treatment of skin ulcers and wounds through the centuries. *Minerva Med* 1997;88(1–2):49–55.
20. Crede CSF. Die verhutung der augenentzundung der neugeborenen [*The prevention of eye inflammation of the newborn*]. *Arch Gynakol* 1881;17:50–53.
21. Chen X, Schluesener HJ. Nanosilver: A nanoproduct in medical application. *Toxicol Lett* 2008;176:1–12.
22. Sun Y, Xia Y. Shape-controlled synthesis of gold and silver nanoparticles. *Science* 2002;298(5601):2176–79.

23. Prabakar K, Sivalingam P, Mohamed Rabeek SI, Muthuselvam M, Devarajan N, Arjunan A, Karthick R, Suresh MM, Wembonyama JP. Evaluation of antibacterial efficacy of phyto fabricated silver nanoparticles using Mukia scabrella (Musumusukkai) against drug resistance nosocomial gram negative bacterial pathogens. *Colloids Surf B Biointerfaces* 2013;104:282–288.

24. Raveendran P, Fu J, Wallen SL. Completely "green" synthesis and stabilization of metal nanoparticles. *J Am Chem Soc* 2003;125(46):13940–41.

25. Oluwafemi OS, Lucwaba Y, Gura A, Masabeya M, Ncapayi V, Olujimi OO, Songca SP. A facile completely 'green' size tunable synthesis of maltose-reduced silver nanoparticles without the use of any accelerator. *Colloids Surf B Biointerfaces* 2013;102:718–23.

26. Nakamura M. Approaches to the biofunctionalization of spherical silica nanomaterials. In Kumar CSSR, ed. *Nanotechnologies for the life sciences*. Verlag, Germany: Wiley-VCH, 2010.

27. Masala O, Seshadri R. Synthesis routes for large volumes of nanoparticles. *Annu Rev Mater Res* 2004;34:41–81.

28. Su HL, Chou CC, Hung DJ, Lin SH, Pao IC, Lin JH, Huang FL, Dong RX, Lin JJ. The disruption of bacterial membrane integrity through ROS generation induced by nanohybrids of silver and clay. *Biomaterials* 2009;30(30):5979–87.

29. Lok CN, Ho CM, Chen R, He QY, Yu WY, Sun H, Tam PK, Chiu JF, Chen CM. Proteomic analysis of the mode of antibacterial action of silver nanoparticles. *J Proteome Res* 2006;5:916–24.

30. Dror-Ehre A, Mamane H, Belenkova T, Markovich G, Adin A. Silver nanoparticle-E. coli colloidal interaction in water and effect on E. coli survival. *J Colloid Interface Sci* 2009;339:521–6.

31. Samberg ME, Orndorff PE, Monteiro-Riviere NA. Antibacterial efficacy of silver nanoparticles of different sizes, surface conditions and synthesis methods. *Nanotoxicology* 2011;5(2):244–53.

32. Wijnhoven SWP, Peijnenburg WJGM, Herberts CA, Hagens WI, Oomen AG, Heugens EHW, Roszek B et al. Nano-silver: A review of available data and knowledge gaps in human and environmental risk assessment. *Nanotoxicology* 2009;3(2):109–38.

33. Harvey MA. Critical-care-unit bedside design and furnishings: Impact on nosocomial infections. *Infect Control Hosp Epidemiol* 1998;19:597–601.

34. Noskin GA, Stosor V, Cooper I, Peterson LR. Recovery of vancomycin-resistant enterococci on fingertips and environmental surfaces. *Infect Control Hosp Epidemiol* 1995;16:577–81.

35. Boyce JM, Potter-Bynoe G, Chenevert C, King T. Environmental contamination due to methicillin-resistant *Staphylococcus aureus*: Possible infection control implications. *Infect Control Hosp Epidemiol* 1997;18:622–7.

36. Sexton T, Clarke P, O'Neill E, Dilland T, Humphreys H. Environmental reservoirs of methicillin-resistant *Staphylococcus aureus* in isolation rooms: Correlation with patient isolates and implications for hospital hygiene. *J Hosp Infect* 2006;62:187–94.

37. Klevens RM, Edwards JR, Richards Jr. CL, Horan TC, Gaynes RP, Pollock DA, Cardo MD. Estimating health care-associated infections and deaths in U.S. hospitals. *CDC Public Health Rep* 2002;122:160–6.

38. Eby DM, Luckarift HR, Johnson GR. Hybrid antimicrobial enzyme and silver nanoparticle coatings for medical instruments. *ACS Appl Mater Interfaces* 2009;1(7):1553–60.

39. Zhang D, Chen L, Zang C, Chen Y, Lin H. Antibacterial cotton fabric grafted with silver nanoparticles and its excellent laundering durability. *Carbohydr Polym* 2013;15;92(2):2088–94.

40. Mermel LA. New technologies to prevent intravascular catheter-related bloodstream infections. *Emerg Infect Dis* 2001;7(2):197–9.

41. Martínez-Gutiérrez F, Guajardo-Pacheco JM, Noriega-Trevino ME, Thi EP, Reiner N, Orrantia E, Av-Gay Y, Ruiz F, Bach H. Antimicrobial activity, cytotoxicity and inflammatory response of novel plastics embedded with silver nanoparticles. *Future Microbiol* 2013;8(3):403–11.

42. Stevens KN, Crespo-Biel O, van den Bosch EE, Dias AA, Knetsch ML, Aldenhoff YB, van der Veen FH, Maessen JG, Stobberingh EE, Koole LH. The relationship between the antimicrobial effect of catheter coatings containing silver nanoparticles and the coagulation of contacting blood. *Biomaterials* 2009;30(22):3682–90.

43. Roe D, Karandikar B, Bonn-Savage N, Gibbins B, Roullet JB. Antimicrobial surface functionalization of plastic catheters by silver nanoparticles. *J Antimicrob Chemother* 2008;61(4):869–76.

44. Chou CW, Hsu SH, Wang PH. Biostability and biocompatibility of poly(ether)urethane containing gold or silver nanoparticles in a porcine model. *J Biomed Mater Res A* 2008;84(3):785–94.

45. Antonelli M, De Pascale G, Ranieri VM, Pelaia P, Tufano R, Piazza O, Zangrillo A et al. Comparison of triple-lumen central venous catheters impregnated with silver nanoparticles (AgTive®) vs conventional catheters in intensive care unit patients. *J Hosp Infect* 2012;82:101–7.

46. Böswald M, Lugauer S, Regenfus A, Braun GG, Martus P, Geis C, Scharf J, Bechert T, Greil J, Guggenbichler JP. Reduced rates of catheter-associated infection by use of a new silver-impregnated central venous catheter. *Infection* 1999;27(Suppl 1):S56–S60.

47. Kollef MH, Afessa B, Anzueto A, Veremakis C, Kerr KM, Margolis BD, Craven DE et al. Silver-coated endotracheal tubes and incidence of ventilator-associated pneumonia: The NASCENT randomized trial. *JAMA* 2008;300(7):805–13.

48. Balasz DJ, Koerner E, Hossain MM, Hegemann D. Multi-functional nanocomposite plasma coatings-enabling new applications in biotechnology. *Proceedings of NanoEurope*, St. Gallen Switzerland, 2006.

49. Wagener M. Silver based nanocomposites for antimicrobial medical devices. *Proceedings of NanoEurope*, St. Gallen, Switzerland, 2006.

50. Moyer CA, Brentano L, Gravens DL, Margraf HW, Monafo WW Jr. Treatment of large human burns with 0.5% silver nitrate solution. *Arch Surg* 1965;90:812–67.

51. Miller CN, Newall N, Kapp SE, Lewin G, Karimi L, Carville K, Gliddon T, Santamaria NM. A randomized-controlled trial comparing cadexomer iodine and nanocrystalline silver on the healing of leg ulcers. *Wound Repair Regen* 2010;18(4):359–67.

52. Friedman AJ, Phan J, Schairer DO, Champer J, Qin M, Pirouz A, Blecher-Paz1 K et al. Antimicrobial and anti-inflammatory activity of chitosan-alginate nanoparticles: A targeted therapy for cutaneous pathogens. *J Invest. Dermatol* 2013;133:1231–9.

53. Langer R. Tissue engineering: Perspectives, challenges, and future directions. *Tissue Eng* 2007;13(1):1–2.

54. Tateishi T, Chen G, Ushida T. Biodegradable porous scaffolds for tissue engineering. *J Artif Organs* 2002;5:77–83.

55. Chung S, Ingle NP, Montero GA, Kim SH, King MW. Bioresorbable elastomeric vascular tissue engineering scaffolds via melt spinning and electrospinning. *Acta Biomaterialia* 2010; 6(6):1958–67.

56. Jeong SI, Kim SH, Kim YH, Jung Y, Kwon JH, Kim BS, Lee YM. Manufacture of elastic biodegradable PLCL scaffolds for mechano-active vascular tissue-engineering. *J Biomater Sci Polym Ed* 2004;15:645–60.

57. Jeong SI, Kwon JH, Lim JI, Cho SW, Jung Y, Sung WJ, Kim SH et al. Mechano-active tissue engineering of vascular smooth muscle using pulsatile perfusion bioreactors and elastic PLCL scaffolds. *Biomaterials* 2005;26:1405–11.

58. Kim B-S, Nikolovski J, Bonadio J, Mooney DJ. Cyclic mechanical strain regulates the development of engineered smooth muscle tissue. *Nat Biotechnol* 1999;17:979–83.

59. Kim B-S, Mooney DJ. Scaffold for engineering smooth muscle under cyclic mechanical strain conditions. *J Biomech Eng* 2000;122:210–15.
60. Jeon HJ, Kim JS, Kim TG, Kim JH, Yu W-R, Youk JH. Poly(ε-caprolactone)-based polyurethane nanofibers containing silver nanoparticles. *Appl Surf Sci* 2008;254:5886–90.
61. Liu D, Xiao P, Zhang Y, Garcia BB, Zhang Q, Guo Q, Champion R, Cao G. TiO_2 nanotube arrays annealed in N_2 for efficient lithium-ion intercalation. *J Phys Chem C* 2008;112(30):11175–80.
62. Samberg ME, Mente P, He T, King MW, Monteiro-Riviere NA. Incorporation of silver nanoparticles into a degradable poly(l-lactide-co-epsilon-caprolactone) copolymer scaffold for skin regeneration. *Toxicologist* 2013;132(1326):284.
63. Samberg ME, Oldenburg SJ, Monteiro-Riviere NA. Evaluation of silver nanoparticle toxicity in skin in vivo and keratinocytes in vitro. *Environ Health Perspect* 2010;118(3):407–13.
64. Christensen FM, Johnston HJ, Stone V, Aitken RJ, Hankin S, Peters S, Aschberger K. Nano-silver–feasibility and challenges for human health risk assessment based on open literature. *Nanotoxicology* 2010;4:1–12.
65. Poon VK, Burd A. In vitro cytotoxicity of silver: Implication for clinical wound care. *Burns* 2004;30(2):285–6.
66. Cochrane C, Walker M, Bowler P, Parsons D, Knottenbelt D. The effect of several silver-containing wound dressings on fibroblast function in vitro using the collagen lattice contraction model. *Wounds* 2006;18(2):29–34.
67. Temple RM, Farooqi AA. An elderly, slate-grey woman. *Practitioner* 1985;229:1053–54.
68. Innes ME, Umraw N, Fish JS, Gomez M, Cartotto RC. The use of silver coated dressings on donor site wounds: A prospective, controlled matched pair study. *Burns* 2001;27(6):621–27.
69. Huang D-M, Chung T-H, Hung Y, Lu F, Wu SH, Mou CY, Yao M, Chen YC. Internalization of mesoporous silica nanoparticles induces transient but not sufficient osteogenic signals in human mesenchymal stem cells. *Tox Appl Pharma* 2008;231:208–15.
70. Moiemen NS, Shale E, Drysdale KJ, Smith G, Wilson YT, Rapini R. Acticoat dressing and major burns: Systemic silver absorption. *Burns* 2011;37:27–35.
71. Gupta A, Silver S. Silver as a biocide: Will resistance become a problem? *Nat Biotechnol* 1998;16:888.

23 Cytotoxicity of Conjugated and Unconjugated Semiconductor and Metal Nanoparticles

Jay Nadeau

CONTENTS

23.1 INTRODUCTION

23.1.1 GENERAL PROPERTIES OF METAL AND SEMICONDUCTOR NANOPARTICLES

Over the past several decades, revolutions in wet-chemical synthesis have allowed for controlled production of specific sizes and shapes of metals, semiconductors, and insulators at the nanoscale. All materials show unique properties when they are reduced to nanometer sizes due to increased surface-to-volume ratios and reactivity. Other properties also manifest as the particles become small on the quantum mechanical scale, with resulting novel physics and chemistry. When these materials are introduced into biological systems, unexpected interactions with organisms, cells, and subcellular structures occur. Such interactions have only begun to be understood, but have important implications for human health and the environment.

A "quantum dot" (QD) is a particle of a semiconductor material that is comparable in size to the material's Bohr radius—several nanometers for most materials. Excitation by a photon slightly more energetic than the bandgap leads to the creation of an electron–hole pair, or exciton. Recombination of this exciton leads to emission of a photon (Figure 23.1). The confinement energy caused by restricting this exciton to a nanometer-sized particle results in a size-dependent broadening of the bandgap. Thus, when the semiconductor has a bandgap energy comparable to energies of photons of visible light, the particles exhibit size-dependent visible photoluminescence. Many materials, belonging to group II–VI, group IV–VI, or group III–V, can be used to produce QDs, although the most common materials for biological applications are cadmium (Cd) based: cadmium selenide (CdSe), cadmium telluride (CdTe), and cadmium sulfide. Emission from these particles spans the ultraviolet to visible range (Figure 23.2a). The precise bandgap energy and energetic positions of the band edges depend on not only QD material and size but also to some extent on solvent and pH.

Unfortunately, Cd is so toxic that the European Union has recently banned its use in all consumer electronics; the likelihood that it will ever be approved for in vivo use is nil. Consequently, alternative materials with bright, tunable emission have been sought. Indium phosphide (InP) is a highly fluorescent alternative to Cd-containing materials.[1] Although InP is toxic when inhaled at high doses, primarily causing lung irritation,[2] it is approved at low doses for medical imaging;

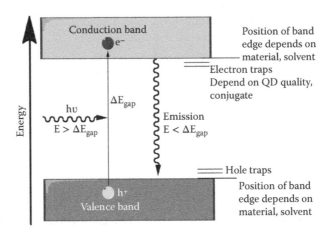

FIGURE 23.1 Photoluminescence by exciton recombination. A photon more energetic than the bandgap excites an electron–hole pair, which recombines to give a slightly less energetic photon. The energetic positions of the band edges are determined by material and solvent, and then shift according to quantum dot (QD) size as the particles become smaller (the conduction band edge shifts more than the valence band edge because an electron is lighter than a hole). Both electron and hole traps can exist and prevent QD emission, reducing quantum yield, and perhaps interacting with other molecules in solution. The synthesis procedure; exposure to light, water, or oxygen; the degree of shelling; conjugates; and other variables can influence trap states.

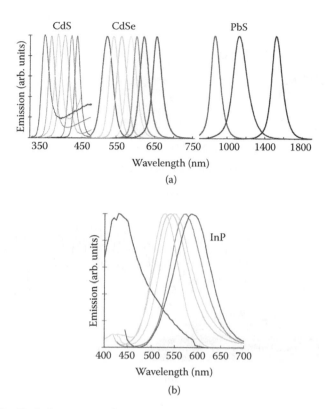

FIGURE 23.2 Emission spectra of common quantum dots (QDs). (a) Cadmium-containing QDs. (b) Cadmium-free indium phosphide (InP) QDs. Note the similarity of InP to cadmium selenide (CdSe), though the spectra tend to be somewhat broader.

trace amounts are found in some dietary supplements. InP has a bandgap similar to that of CdTe (937 nm) and thus also yields size-tunable emission in the near infrared and visible range (Figure 23.2b).

The core of a QD retains the lattice structure of the bulk semiconductor, but the chemistry is dominated by surface properties, especially in aqueous solution. The electron and/or hole from the exciton can transfer to the solvent, oxidizing or reducing the QD and leading to release of its component ions, and/or production of reactive oxygen species (ROS), primarily hydroxyl radicals and superoxide. To prevent this, QDs for biological applications are nearly always protected by both a shell and a cap. We will use "shell" to refer to an outer layer of another semiconductor material with a higher bandgap energy than the core. This layer may be a single monolayer thick or more, but usually contains defects that allow the solution to interact with the core directly. The "cap" is an organic layer that determines the QD solubility in water or organic solvent. It may be a simple molecule such as an alkanethiol, with the thiol group bound to the surface of the QD, or it may be a lipid, copolymer, or peptide (Figure 23.3). The size, charge, stability, and cell-permeating properties of this cap play a key role in determining QD cytotoxicity. Simple alkanethiols have been shown to be insufficiently stable for most biological applications; photooxidation

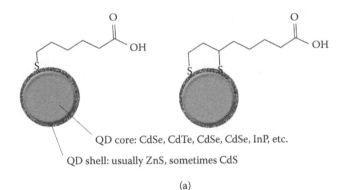

QD core: CdSe, CdTe, CdSe, CdSe, InP, etc.

QD shell: usually ZnS, sometimes CdS

(a)

(b)

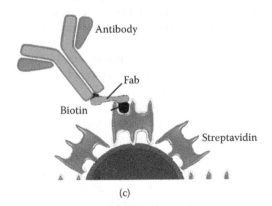

(c)

FIGURE 23.3 Methods of quantum dot (QD) functionalization for biological applications. Molecules are not to scale. (a) Alkanethiol (left) or dithiol (right) are the simplest approaches. (b) A peptide with three cysteine (Cys) residues and a short polyethylene glycol (PEG) chain has been reported to be highly stable and reduce nonspecific binding.[4] Also see Ref. 5 for a general method. (c) Most commercial QDs are coated with streptavidin, which then may be linked via biotin to antibodies or other molecules. This schematic shows an antibody linked via a Fab fragment as described in Ref. 6.

leads to detachment of the thiol and precipitation of a now-hydrophobic QD in a process called "cap decay."[3]

Nanoparticles of noble metals (silver [Ag] and gold [Au]) have been used as pigments for centuries; colloidal Au nanoparticles have been used in folk remedies for cancer and arthritis since antiquity. However, it is only recently that their physics and chemistry have been explored in a biological context. As with semiconductors, as a noble metal nanoparticle's physical dimension is reduced to the nanometer range, the optical and electronic properties become dramatically different from those of the bulk material. Rather than excitonic emission, noble metal nanoparticles have a peak in the absorption spectrum at a characteristic frequency known as the plasmon resonance frequency. This peak changes in position and width with particle size, so the apparent color of the particle is a function of diameter. The same principles of biofunctionalization that apply to QDs also apply to nanoparticles,[7,8] with a few differences. Au nanoparticles are often capped with thiols because of the strength of the Au–S bond; these particles are very stable in solution, without the issues of cap decay that occur with QDs. Nonetheless, it is often desirable to cap them with more complex ligands, such as peptides.[9] Both Ag and Au nanoparticles are frequently synthesized with a citrate cap, which is not highly stable, and many toxicity studies use these particles as is. It is important to take note of the particle cap when evaluating any toxicity studies. Ag nanoparticles, but not Au, will rapidly oxidize under air.

23.1.2 NANOPARTICLE VARIABILITY

One of the biggest issues in nanotoxicology is batch-to-batch variations in nanoparticle properties, including cytotoxicity. The particle surface quality, which varies among synthesis batches, determines the number of defect or trap states.[10] This in turn affects adsorption of water and oxygen and energy-transfer processes. When particles are shelled, the shell nearly always contains defects that are poorly reproducible. This directly affects interactions of the solution with the particle core. Several groups are now producing particles using robotic syntheses that are much more reproducible than past methods[11]; such standardization will help the nanotoxicology field. For now, it is sufficient to note that unexpected or contradictory results from the literature might arise from a single batch or synthesis method, and might not apply to nanoparticles produced in other laboratories or by different approaches.

23.1.3 PROBLEM OF REACTIVE OXYGEN SPECIES

Generation of ROS is believed to be a good predictor of nanoparticle cytotoxicity in cells, with the most damaging species being singlet oxygen. However, cell death from other species such as hydrogen peroxide (H_2O_2) and hydroxyl radicals ($\dot{O}H$) can also occur. The lifetimes and effective radii of action of each of these species differ. Thus, it is important to distinguish between different types of ROS when predicting effects on living systems, although making this distinction has been challenging with nanoparticles for several reasons. QDs generate ROS through different mechanisms. Excitation of the QD yields an exciton, of which both the electron and hole are reactive; the energy levels depend on the semiconductor material.[12] Cd can produce ROS in cells,[13,14] so oxidative toxicity from QDs is seen even in the absence of photoexcitation. With photoexcitation, both electron and hole transfer can occur to molecular oxygen, water, or biomolecules such as glutathione, catecholamines, or proteins. These mechanisms are well reviewed by Nel et al.[15] Thus, in any study reporting ROS from nanoparticles, it is important to consider

- Identity of the species produced.
- Degree of light exposure.
- Surrounding medium (buffer, cell contents, living cells).
- Possible false-positive or ambiguous results from using indirect measures of ROS generation. ROS-sensing dyes, such as dichlorofluorescein diacetate (DCFDA) and Amplex Red, can give false-positive results when incubated with nanoparticles since they can be directly oxidized by the nanoparticle itself.[16,17] Similarly, reagents used for electron paramagnetic resonance (EPR) spectroscopy may be directly oxidized or reduced by the nanoparticles rather than by intermediate oxygen species.[18]

Many cytotoxicity studies do not measure ROS at all, but rather infer its action by attempts to rescue cells with antioxidants or by examining expression of genes related to oxidative stress. All of these variables have led to a fuzzy picture of the role of ROS in QD cytotoxicity, one that has not been fully resolved to this day.

23.1.4 Problem of Complex Media

Nanotoxicology studies may be performed in water (rare), simple buffers such as phosphate-buffered saline, or cell culture medium with or without serum. Nanoparticles behave very differently in these media. Their colloidal stability is affected by electrolytes, even in simple buffers. All media, even serum-free media, contain redox-active molecules that can interact with particles. Serum-containing media are even more troublesome, as they rapidly interact with nanoparticles, changing many of their physical properties, especially size and charge. This has led to a good deal of confusion when comparing studies or comparing particles of nominal positive or negative charge. Adhesion of serum proteins occurs to both positively and negatively charged QDs very rapidly, giving both the same effective charge for cellular experiments.

Table 23.1 and Figure 23.4 summarize the particle-related elements that need to be considered in evaluating nanoparticle cytotoxicity studies. Section 23.1.5 will discuss cell-related variables and in vitro models.

23.1.5 In Vitro Models and Mechanisms of Cytotoxicity

There has been little standardization in the cells or cytotoxicity parameters among nanotoxicity studies. First of all, the choice of cells for assays varies widely from cancer cells to immortalized cell lines to primary cells (which may be terminally differentiated or not). Expression of antiapoptotic genes, sensitivity to metals, and degree of endocytosis of particles all vary among cell types. The degree of confluency of the cultures when the particles are applied is also important for cell survival and endocytosis.

TABLE 23.1

Nanoparticle-Related Variables to Be Taken into Account When Evaluating or Designing a Cellular Nanoparticle Toxicity Experiment

Variable	What It Affects
Particle core material	Reactivity, toxicity of released ions, position of band edges (if semiconductor)
Particle synthesis method	Particle homogeneity, crystal quality, stability, presence of chemical contaminants from the synthesis, particle reproducibility
Presence and composition of shell	Inorganic shell: ZnS, CdSe, other, or none. Shell thickness may also be a variable
Capping ligand	Solubility, stability in solution, interactions with cell membranes, release from endosomes. May be toxic in itself; if commercial QDs, may be proprietary
Solubilization method	Completeness of cap, stability of cap, presence of chemicals from solubilization reaction
Particle size	Reactivity, cellular uptake, access to organelles
Light exposure (yes or no, wavelength, power)	Generation of reactive electrons/holes (semiconductor); particle breakdown; cap decay
Solvent, electrolytes, pH	Particle stability, reactivity (band edge positions, traps), generation of ROS
Other molecules in solution (proteins, lipids, redox-active agents)	Electron transfer from electrons and holes, particle surface reactivity, particle stability, effective particle size (hydrodynamic radius), effective particle charge

Most studies report some type of cell proliferation assay, usually the 3-(4,5-dimethylthiazol-2-yl)-2,5-diphenyltetrazolium bromide (MTT) assay, but sometimes others such as the sulforhodamine B (SRB) assay. These are endpoint assays that give the final number of metabolically active cells (MTT) or cells in general (SRB) at a set time point after nanoparticle incubation. Apart from the choice of the time point, there are other variables that can cause disagreement among studies when such assays are used: how long nanoparticles are incubated with the cells before being washed off, and in what medium; the effects of the incubation medium itself on cell health, especially if it is serum free; whether incubation is done in the dark or under room light; variability in cell density and dividing times leading to application of particles during different stages of the cell cycle; and even differences in data analysis (particularly how normalization to 0% and 100% survival is performed). Some nanoparticles can also adhere to the reagents, leading to false-positive and seemingly absurd results (much greater survival with nanoparticles than with control). For example, we have observed strong adherence of SRB to aminocaproic acid–capped lanthanum fluoride nanoparticles (not shown). Because of all of these variables and limited test conditions, these endpoint assays are

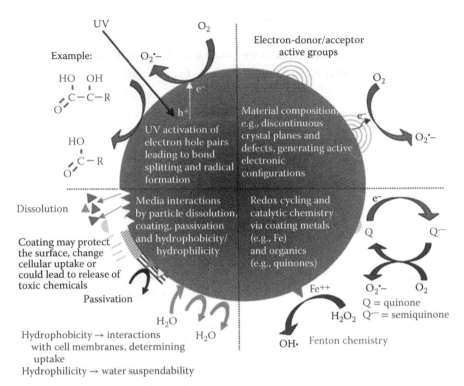

FIGURE 23.4 Possible mechanisms by which nanoparticles can interact with biological environments. (From Nel, A., Xia, T., Madler, L. and Li, N., *Science*, 311, 622–627 (2006). Reprinted with permission from AAAS.)

limited in their insight, and the need for alternatives is beginning to be appreciated in the nanotoxicology community.[19]

Other types of assays that are frequently performed are dye-based cell permeability assays (such as trypan blue or propidium iodide assays); fluorescent apoptosis assays using Annexin V or caspase; intracellular ROS assays; and gene expression assays for genes involved in metal processing, apoptosis, oxidative stress, and more. As discussed in Section 23.1.3, ROS dyes can interact with some nanoparticles to produce erroneous results. The same may be true for any fluorescent compound that comes in contact with a metal or semiconductor nanoparticle. Quenching, fluorescence resonance energy transfer (FRET), and other processes can alter the intensity and/or wavelength of the dye emission and obscure the results. We have observed FRET between CdTe and propidium iodide, leading to large wavelength shifts in the observed emission.[20]

In all of these experiments, the choice of time point for measurement and the choice of controls are all very important. Cell death unrelated to nanoparticles can yield signals of DNA damage, and controls that are sham exposed lack many of the steric or medium-related effects that are possible with nanoparticles. Controls should ideally include a reference nanoparticle with known properties.[21,22]

Table 23.2 summarizes the cell-related elements that need to be considered in evaluating nanoparticle cytotoxicity studies.

TABLE 23.2
Cell-Related Variables to Be Taken into Account When Evaluating or Designing a Cellular Nanoparticle Toxicity Experiment

Variable	What It Affects
Choice of cell line	Rate of cell division, degree of endocytosis of different particles, apoptosis, sensitivity to metals
Incubation time and concentration	Degree of particle uptake, cell death related to exposure to minimal media, cell death related to extreme particle concentrations
Incubation conditions (light, temperature, confluency, type of dish)	Degree of uptake, photoexcitation of particles, cell survival, erroneous results due to particles adhering to dye substrate
Wash steps	Degree of removal of particles at a given stage of the experiment; degree of cell loss (may vary in particle-exposed cells versus control cells)
Choice and parameters of survival assay	May measure mitochondrial activity, total protein, or other variable: usually done at a selected time point which may be critical: some dyes stick to some nanoparticles
Choice of membrane-permeability assay	Fluorescent dyes may react with nanoparticles
Choice of ROS assay	Fluorescent dyes may react with nanoparticles
Genotoxicity assays	Some are done in situ using DNA only; may not apply to whole cells
Choice of controls for normalization	Quantitative comparisons among conditions; behavior of some assays

23.2 CADMIUM-CONTAINING QUANTUM DOTS

The first cytotoxicity study on CdSe QDs appeared in 2004.[23] Not surprisingly, the study found that the major source of toxicity was release of Cd^{2+} ions from unshelled QDs; this was especially damaging to hepatocytes, which are sensitive to Cd. Many follow-up studies confirmed this mechanism by showing that a zinc sulfide (ZnS) shell reduces toxicity by reducing Cd leaching.[24] Although the exact amount of Cd^{2+} released from QDs varies from batch to batch and across materials, any Cd^{2+} release will render QDs severely cytotoxic.

Soluble Cd is only part of the picture, however, as most reports indicate that Cd-containing QDs are more toxic than their soluble Cd^{2+} concentration would imply, show different mechanisms of toxicity than Cd alone, and that toxicity occurs even in the absence of ion release. It is important to differentiate studies done with CdSe from those done using CdTe. CdTe has a much narrower bandgap than CdSe, and very different band edge energies. This means that it is less reactive to thiols than CdSe, and consequently more stable in cellular environments; it also generates different ROS than CdSe because of the different position of its band edges. The lack of shell also makes the particles' hydrodynamic radii very small. Thus, generalization from CdSe to CdTe or vice versa is not possible, and each should be considered separately when variables other than soluble Cd are discussed. Several reports of cytotoxicity have been published on CdTe, one of which showed nuclear entry of the smallest particles.[25,26]

Encapsulation of the QDs to prevent Cd release has been investigated using a wide variety of materials: polymers, lipids, and many more. The commonly used coatings on commercial QDs are degraded in endosomes,[27] so again ion release becomes a factor. When Cd release is prevented, the cytotoxic effects seen are strongly related to particle surface chemistry. Numerous studies have attempted to explain this phenomenon by examining mechanisms of particle uptake, cell death, and gene expression after QD exposure. In 2006, a thorough review of the findings was published stating that "QD size, charge, concentration, outer coating bioactivity (capping material and functional groups), and oxidative, photolytic, and mechanical stability have each been implicated as determining factors in QD toxicity."[28] An updated review was published in 2009, summarizing the results seen according to QD composition, size, concentration tested, and exposure time.[29] Quantitative results varied among all the studies, and there was little agreement on whether QDs were cytotoxic to cells under typical imaging conditions. Some more recent findings can help explain many of these discrepancies and begin to develop a more general picture.

Cytotoxicity depends on, and is proportional to, QD uptake by the cells. Although this may seem obvious, quantifying the amount of uptake in cultured cells is not straightforward. The cells need to be trypsinized and resuspended before most procedures, leading to possible errors due to particles on the dish surface. Degree of uptake is highly dependent on cell line, degree of confluence, concentration of QDs applied, incubation time, and medium of incubation. Thus, this is a simple factor that is easy to miscalculate or overlook. Recent studies have emphasized this by showing that particles whose shape prevents uptake do not show cytotoxicity.[30] Phagocytic cells also prefer QDs with different coatings than nonphagocytic cells, with degree of uptake correlated with toxicity in both cases.[31]

Another aspect that has been recently appreciated is that speaking of QD "surface charge" in complex media is often moot. Both negatively and positively charged particles associate with serum and develop a protein corona that makes them indistinguishable to the cells within 5–30 minutes.[32,33] However, if particles are incubated with cells in simple buffers, positively and negatively charged particles may indeed behave differently.

Autophagy has been increasingly recognized as a mechanism of QD-induced cell death. Even in the absence of metal-associated toxicity, nanomaterials can induce autophagic responses in porcine kidney cells; this has been suggested as a common mechanism of nanomaterial toxicity that is independent of ion release.[34,35] Interestingly, a comparison of autophagy induced by L-glutathione versus D-glutathione capped CdTe showed that the bioactive L-glutathione coating was significantly more toxic to liver cells than its inert D-glutathione counterpart.[36]

DNA damage has been widely reported from uncapped CdTe,[37] but this is likely due to release of Cd ions. CdSe/ZnS QDs can damage plasmid DNA when photoexcited, due to interactions of the photoexcited electron with oxygen.[38] However, these experiments were done in situ, so their role in toxicity to cells is not clear.

Mechanisms of Cd-based toxicity are reasonably consistent across many types of cells: mammalian cells, zebra fish cells, and simple eukaryotes.[39-41] Toxicity in bacteria must be considered separately from toxicity in eukaryotic cells for two key reasons:

- The electron transport chain of bacteria is on or near the cell surface, making it accessible to direct interactions with nanoparticles.[42]
- Bacteria can express a large variety of metal-processing genes, some of which are inducible, with some strains highly resistant to heavy metals also showing reduced toxicity with QDs.[43-45]

Nonetheless, bacterial bioassays for rapid estimation of QD toxicity have been developed[46] which may be useful for prescreening.

23.3   CADMIUM-FREE QUANTUM DOTS

InP/ZnS QDs are the most commonly used Cd-free QDs, but their relative novelty means that only a few toxicity studies have been published. Synthesis of biologically compatible InP QDs is more difficult than for the Cd-containing QDs, and has delayed development in this field. The core itself is oxidized very rapidly and cannot come in contact with biocompatible solutions without breaking down.[47] Thus, like CdSe, the particles must be capped with a higher bandgap material such as ZnS, but the differences in coordination strength between InP and ZnS makes this much more difficult than with CdSe. The growth of the first shell layer is a replacement reaction in which the InP QDs surface ligands and capping agents are exchanged for zinc and sulfur bonds. The strong coordinating strength of ligands to indium prevents these surface exchanges. With the different properties of InP and ZnS it is also more difficult to form an intermediate layer to fill the holes. A shell with holes will permit molecules in the solution to encounter and oxidize the QD, leading to breakdown and release of ions.

Using zinc carboxylates during the preparation of InP QDs may lead to more stable shells than other methods.[48] However, at least two atomic monolayers of ZnS are needed for a good quantum yield. In our recent work, we have performed a systematic study of InP toxicity in six different cell lines and found that InP exhibits some cytotoxicity that is proportional to the thickness of the ZnS shell and thus to the production of superoxide and perhaps hydroxyl radical (Figure 23.5). Cytotoxicity varies across cell lines as a function of degree of QD uptake. We have also found that alkanethiol-solubilized InP QDs associate strongly with cell membranes, including the nuclear membrane, much more so than CdSe/ZnS solubilized by the same methods (Figure 23.6). This may reflect differences in the shell structure, the number of thiols bound to each particle, and the strength of these bonds.

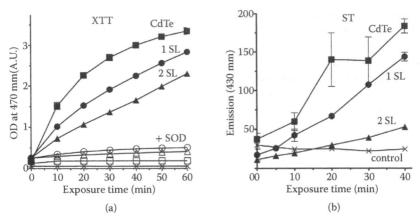

FIGURE 23.5 Generation of superoxide from InP/ZnS nanoparticles according to shell thickness, as a function of time of exposure to 440 nm light. (a) Superoxide anion detection using the sodium, 3'-(1-[phenylamino-carbonyl]-3,4-tetrazolium)-bis(4-methoxy-6-nitro) benzene-sulfonic acid hydrate (XTT) assay. Cadmium telluride is shown as a positive control, indium phosphide (InP) with a single monolayer ZnS shell (1SL), and InP/ZnS with a double monolayer shell (2 SL). To confirm that the species was superoxide, a control was performed containing 25 units/mL of superoxide dismutase (the open symbols with SOD correspond to the same quantum dots as the filled symbols). (b) The fluorescent hydroxyl radical sensor, sodium terephthalate (ST), showing a marked reduction in signal with 2SL InP versus 1SL InP. (Data previously shown in Chibli, H., Carlini, L., Park, S., Dimitrijevic, N. M. and Nadeau, J. L., *Nanoscale* 3, 2552–2559 [2011].)

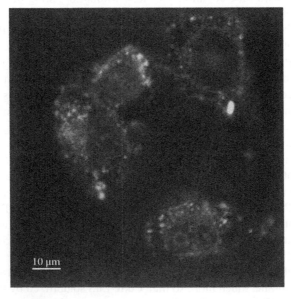

FIGURE 23.6 Distribution of InP/ZnS nanoparticles in B16 melanoma cells after 1 hour of incubation at 20 nM (Nadeau, J., unpublished data).

23.4   GOLD NANOPARTICLES

Au is an inert (noble) metal, and Au preparations have been ingested orally since antiquity for conditions such as arthritis and cancer.[50] The Food and Drugs Administration has approved Au preparations for some forms of arthritis. It might be expected that Au nanoparticles would not cause toxicity to cells in culture, but some experiments have shown varying levels of cytotoxicity in different cellular systems. Other studies see no toxicity. The studies are complicated by the wide variety of shapes, sizes, and surface coatings available in Au nanoparticle preparations (Figure 23.7). The earliest studies showing toxicity[51] were shown to result from cetyltrimethylammonium bromide remaining in the solution.[52,53] This drew attention to the need for ultrapure particles for biomedical applications. Several groups have developed methods for laser-assisted aqueous synthesis of Au nanoparticles with no chemical additives.[54]

Size of the nanoparticles also may play an important role in toxicity, especially at the smallest sizes. Several studies have shown that ultrasmall Au nanoparticles/ nanoclusters are able to penetrate into cell nuclei. Depending on their surface conjugate, they may then be able to interact with nuclear DNA and inhibit cell division. Au(55) nanoclusters, with a 1.4 nm radius, fit into the major groove of DNA.[55,56]

Even with larger Au particles that do not enter the nucleus, a number of recent studies have implicated exposure to Au with disturbances in cellular cytoskeleton. One such study showed that expression of actin and tubulin remains unchanged in exposed cells, but actin filaments are stretched and broken. Removal of the particles leads to complete recovery after 14 days.[57] A later study confirmed malfunction of actin and tubulin as well as focal adhesion kinase (FAK), with no change in FAK expression levels. The authors concluded that the effects of the particles

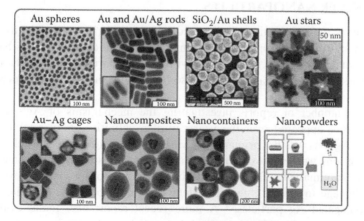

FIGURE 23.7 Examples showing some of the variety in shapes and sizes of gold (Au) and Au composite nanomaterials used in biological applications. (From Khlebtsov, N., Bogatyrev V., Dykman L., Khlebtsov B., Staroverov S., Shirokov A., Matora L., Khanadeev V., Pylaev T., Tsyganova N., Terentyuk G. et al., *Theranostics,* 3(3), 167–180, 2013.)

were indirect, probably resulting from steric hindrance of the cytoskeleton from particle-loaded endosomes and lysosomes.[58]

Oxidative cytotoxicity has also been reported with Au nanoparticles, but is less well described and less certain than with QDs. Although ROS has not been seen from the nanoparticles alone, signs of oxidative stress have been observed in lung fibroblasts exposed to 20 nm citrate-capped Au.[59] The previously cited cytoskeletal study also reported increased intracellular ROS as measured by CM-H$_2$DCFDA.[58] However, the possibility of contaminants, false positives from fluorescent reporters, or indirect effects have not been ruled out. A study from the National Institute of Standards and Technology reported standardized preparations of 10, 30, and 60-nm-sized Au nanoparticles capped with citrate. The particles were applied to cultured Hep G2 cells at levels expected to be physiologically relevant, as well as incubated with ct-DNA to investigate DNA damage. No significant cytotoxicity or DNA damage was seen. There was also no free radical generation from the particles as measured by EPR spectroscopy. These standardized materials were suggested as negative controls for Au nanoparticle toxicity studies, and several possible reasons for false-positive results were discussed: contamination of particles, for example with endotoxin; application of very large nanoparticle concentrations to cells in some studies; and mistaking DNA damage caused by cell death with nanoparticle-induced damage.[22]

A very recent study suggests lysosomal dysfunction as the underlying mechanism in autophagy and oxidative stress caused by nanoparticles[60] (Figure 23.8). Rupture of lysosomes can lead to permeabilization of mitochondrial outer membranes with ROS generation. This indirect mechanism seems to best explain the oxidative effects seen with Au, in the absence of any evidence that the nanoparticles themselves can generate free radicals.

23.5   SILVER NANOPARTICLES

Nanosilver is the most commercially used of all of all nanomaterials, usually for its known antibacterial properties. Clothing and medical devices such as bandages, catheters, and heart valves impregnated with Ag nanoparticles are widely available on the market.[61] Ag nanoparticles may also be used for their optical properties in a similar fashion to Au.[62,63]

A large number of reports investigating the mechanisms of Ag's antibacterial properties appeared over the years, until a definitive study established that release of Ag ions was the primary mechanism, and that Ag nanoparticles that did not release ions were nontoxic to bacteria.[62] There are studies of Ag nanoparticle toxicity in mammalian cells, especially those that control for ions and evidence of genetic damage in mouse embryonic stem cells and fibroblasts by water-soluble, nonbioconjugated Ag particles.[64] A more recent report also found evidence of genotoxicity of 46 nm unfunctionalized Ag particles in human mesenchymal stem cells after as little as 1 hour of incubation; however, this only occurred at very high particle concentrations.[65] A study that controlled for ions and used either tris(3-sulfonatophenyl)-phosphine or poly(N-vinylpyrrolidone) as particle capping

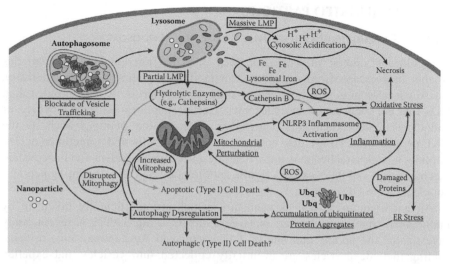

FIGURE 23.8 Mechanisms of autophagy and lysosomal dysfunction toxicity. The initiators of autophagy and lysosomal dysfunction toxicity, displayed in boxed text in the figure, include blockade of vesicle trafficking, lysosomal membrane permeabilization (LMP), and autophagy dysregulation. Nanoparticles could potentially cause autophagy dysfunction by overloading or directly damaging the lysosomal compartment, or altering the cell cytoskeleton, resulting in blockade of autophagosome-lysosome fusion. Nanoparticles could also directly affect lysosomal stability by inducing lysosomal oxidative stress, alkalization, osmotic swelling, or causing detergent-like disruption of the lysosomal membrane itself, resulting in LMP. Toxic effectors (lysosomal iron, cytosolic acidification, hydrolytic enzymes, reactive oxygen species, and the NLRP3 inflammasome) are displayed in ellipses. Conditions resulting from effector-mediated loss of homeostasis (oxidative stress, inflammation, ER stress, disrupted mitophagy, accumulation of ubiquitinated protein aggregates, and mitochondrial perturbation) are underlined. Finally, this loss of homeostasis can result in the cell death pathways necrosis, and Apoptotic (type I) and autophagic (type II) cell death (From S. T Stern, P. P. Adiseshaiah and R. M. Crist, *Particle and Fibre Toxicology* 2012, 9:20 [adapted under the terms of the Creative Commons Attribution License].)

agents reported toxicity associated with particle dissolution and ion release. Ag–Au alloy nanoparticles did not show the same toxicity as Ag alone.[66]

Biofunctionalization of Ag nanoparticles that prevents ion release also prevents toxicity. In antimicrobial applications, of course, this is not desired, but functionalized Ag nanoparticles have been developed for imaging, where reduction in toxicity to mammalian cells is needed. Biofunctionalization using simple adsorption of human serum albumin is sufficient to reduce toxicity in HeLa cells.[67] A review of biofunctionalization approaches to Ag may be found in a recent report,[68] and another review addressing applications of plasmonic particles discusses the imaging applications.[69] Although mammalian cells are more resistant to Ag ions than are bacteria and fungi, the safety of Ag nanoparticles in medicine has not been fully established.[70]

23.6 CONJUGATED PARTICLES

23.6.1 CONJUGATED PARTICLES THAT ESCAPE FROM ENDOSOMES

A significant amount of work has been done to functionalize nanoparticles so that they escape from endosomes, to facilitate imaging and drug delivery. These particles might be expected to show very different toxicity profiles than QDs that are sequestered in endosomes. Certain peptides assist endocytosis or endocytosis-independent cell penetration. Polyarginine peptides such as the TAT peptide from HIV-1 deliver nanoparticles into cells and sometimes nuclei. Peptides may be attached to nanoparticles by a variety of covalent and noncovalent mechanisms,[71] and there exists a significant literature on cell-penetrating peptides coupled to both QDs and Au.

Au nanostructures such as nanostars, which are inefficiently taken up by endocytosis, may be delivered in large numbers using TAT.[72] A recent study investigated the uptake and fate of TAT-conjugated Au nanoparticles in cells, finding that the particles are eventually collected into vesicles and expelled from the cells over a time course of 24–48 hours.[73] Detailed toxicity studies on Au-TAT have not been performed; however, all studies report no observed cytotoxicity within the parameters of the experiments. However, this deserves further investigation, since the peptide conjugation clearly affects the delivery and fate of nanoparticles.[74]

CdSe/ZnS and CdTe QDs have also been conjugated to TAT. The first study reporting these conjugates used CdSe/ZnS QDs capped with dihydrolipoic acid and conjugated to a spacer followed by the polyarginine sequence. The authors found that under conditions of acute exposure (1 hour before washing off the QDs), QD-TAT was identical in toxic effects to QDs alone using the MTT assay. However, with 24 h of incubation, QD-TAT was significantly more toxic. Since most QD-TAT is in endosomes, increased toxicity versus unconjugated QDs is likely due to increased uptake.[75] Another study reported that the peptide was released from CdTe QDs in endosomes, but did not study toxicity.[76] We have used TAT to deliver InP/ZnS QDs to B16 melanoma cell endosomes and nucleus. Although no toxicity was seen from the InP-TAT alone, nuclear delivery of the anti-DNA agent doxorubicin was not improved by the addition of TAT to the nanoparticles (Figure 23.9).

Some hyperbranched copolymers, such as polyethylenimine (PEI), are endosomolytic, and result in QD release into the cytoplasm after uptake. There have been only a few studies using such coatings, though PEI is used widely as a transfection reagent. The first study on QD-PEI showed that grafting of polyethylene glycol (PEG) onto the PEI was necessary to control intracellular release and prevent cytotoxicity (Figure 23.10). The number of grafted PEGs significantly affected fate and distribution of the QDs and toxicity measured in HeLa cells, illustrating that optimization is necessary for this type of approach. Observed cytotoxicity was attributed to the polymer and not to the properties of the QDs.[77] Another study used CdSe/ZnS on an immortalized adult human skin cell line (HaCaT), comparing QD-PEI with QDs solubilized with COOH or NH_2 groups. Only the

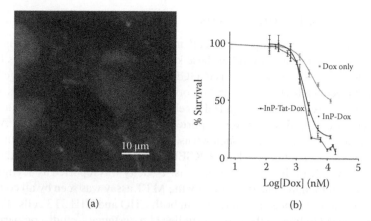

(a)   (b)

FIGURE 23.9   InP-TAT. (a) Nuclear delivery of InP/ZnS quantum dots(QDs) with TAT peptide. (b) Cytotoxicity of InP/ZnS-doxorubicin (Dox) versus InP/ZnS-Dox + TAT in B16 melanoma cells as a function of Dox concentration. Although Dox primarily acts on DNA, nuclear delivery of the QDs did not improve response (Nadeau, J., unpublished data).

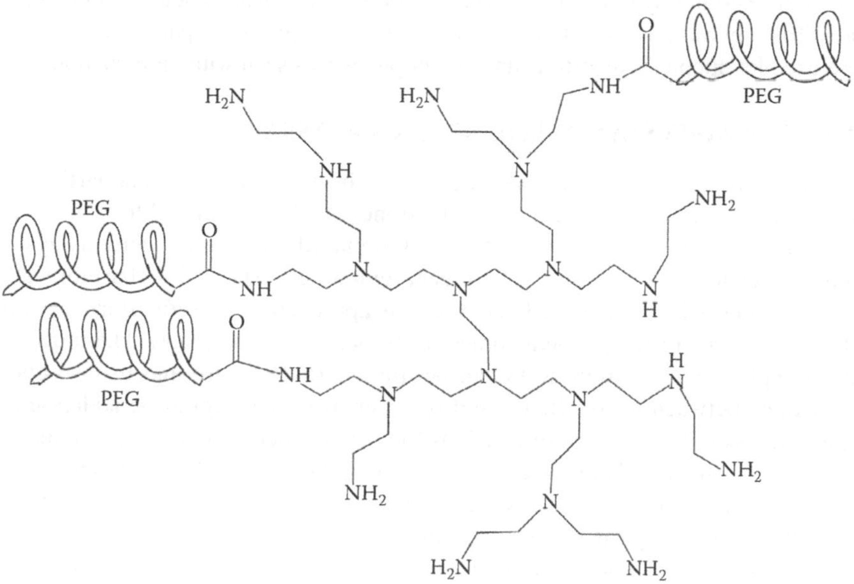

FIGURE 23.10   Polyethylene glycol (PEG)-grafted polyethylenimine (PEI) for endosomo-lytic, nontoxic quantum dot delivery.

PEI-coated QDs showed toxicity; those functionalized with COOH or NH_2 groups were nontoxic under the conditions of the study (MTT assay, 24 hours of incubation with particles; up to 400 nM concentration). Toxicity was not due to released Cd, but to mitochondrial depolarization and oxidative stress. PEI alone was not used as a control.[78]

23.6.2　NUCLEAR-TARGETED PARTICLES

Nanoparticles have been targeted to cell nuclei using nuclear localization signal (NLS) peptides. There are several of these known; ones used in QD studies include the NLS derived from the adenovirus (CGGFSTSLRARKA) and SV40 large T antigen (CGGGPKKKRKVGG). The first report used silanized CdSe/ZnS QDs and the SV40 NLS. No toxicity was seen in HeLa cells by MTT or colony counting assays at the concentrations used, which were reported as picomoles per 106 cells.[79] Another study compared four sequences: adenovirus, SV40, TAT, and the receptor-mediated endocytosis (RME peptide, CKKKKKKSEDEYPYVPN). All sequences delivered approximately 25% of the QDs to the nucleus. Toxicity was reported as negligible for all sequences, although some inhibition by the MTT assay was seen by all constructs at 8 nM, the highest concentration tested, in both CHO and NIH 3T3 cells. The QDs were not exposed to light in the toxicity studies.[80] A systematic study comparing QD phototoxicity and breakdown-related toxicity in nuclear versus endosomal QDs has not been done and remains an open question.

Ultrasmall Au clusters (10 or fewer atoms) conjugated to the SV40 NLS have been used for nuclear delivery. Because of the very small size of the particles, the conjugate retains the <40 nm hydrodynamic radius needed for efficient import through nuclear pores. Toxicity was not studied in the first report[81]; in a subsequent paper, the constructs were used for apoptosis reporting, and no apoptosis was seen without induction.[82]

23.7　TOWARD STANDARDIZATION OF ASSAYS

Enormous progress has been made since 2008 in understanding nanoparticle toxicity. However, a good deal remains to be done, much of it related to standardization of tests and results. The importance of standardizing nanotoxicity assays has been appreciated for some time.[15,83] A recent review suggested a flowchart for testing new nanomaterials.[21] (Figure 23.11). Such an approach is a useful start, but does not address some of the problems inherent in many of these assays. The most serious example is that of ROS. ROS can be misreported in vitro and in vivo due to interactions between nanoparticles and reporter dyes. This could be addressed by greater access to better measures, such as EPR spectroscopy and direct singlet oxygen detection by phosphorescence at 1271 nm. ROS can also be generated in cells by mechanisms such as autophagy that permeabilize mitochondria, independent of any ROS generated by the particles. Controls that account for this problem need to be performed in all studies. Another issue is that of endpoint assays. As technology develops, real-time cell growth assays are becoming available, and will eventually be a gold standard for cell growth. Finally, better choices of cell lines and appreciation of cell line differences are needed in these studies. The recent report on the HeLa cell genome[84] underscores the type of problem that can occur when using immortalized cancer cell lines. It is clear from many studies that different types of cells take up different amounts of differently coated nanoparticles; respond differently to metals; and have different thresholds for oxidative stress, to name only three possible differences. Standardization in cells and cell choice, including judicious use of primary cells, is needed to address these issues.

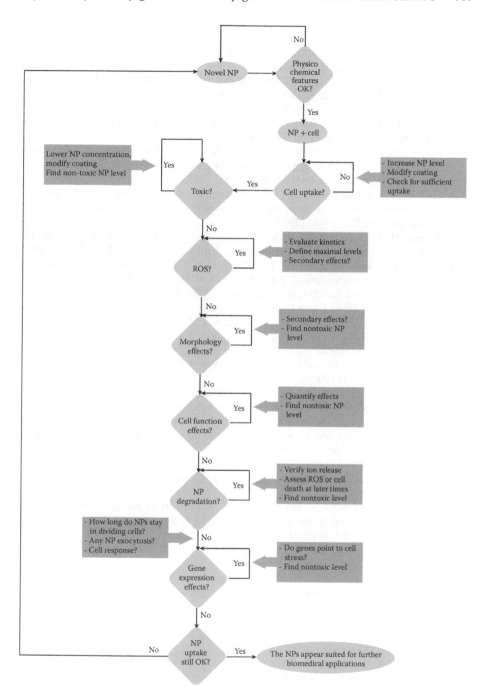

FIGURE 23.11 Flowchart for toxicity testing of nanomaterials. (Reprinted from *Nano Today*, 6(5), Soenen, S. J. et al., 446–465. Copyright 2011, with permission from Elsevier.)

REFERENCES

1. Carlini, L., Chibli, H., Ntumba, K. and Nadeau, J. "Cellular uptake of conjugated InP quantum dots," *Proc SPIE* 7575, 8 (2010).
2. National Toxicology Program. "Toxicology and carcinogenesis studies of indium phosphide (CAS No. 22398-90-7) in F344/N rats and B6C3F1 mice (inhalation studies)," *Natl Toxicol Program Tech Rep Ser*, 7–340 (2001).
3. Aldana, J., Wang, Y. A. and Peng, X. "Photochemical instability of CdSe nanocrystals coated by hydrophilic thiols," *J Am Chem Soc* 123, 8844–8850 (2001).
4. Dif, A., Boulmedais, F., Pinot, M., Roullier, V., Baudy-Floc'h, M. et al. "Small and stable peptidic PEGylated quantum dots to target polyhistidine-tagged proteins with controlled stoichiometry," *J Am Chem Soc* 131, 14738–14746 (2009).
5. Clarke, S., Tamang, S., Reiss, P. and Dahan, M. "A simple and general route for monofunctionalization of fluorescent and magnetic nanoparticles using peptides," *Nanotechnology* 22, 175103 (2011).
6. Bannai, H., Levi, S., Schweizer, C., Dahan, M. and Triller, A. "Imaging the lateral diffusion of membrane molecules with quantum dots," *Nat Protoc* 1, 2628–2634 (2006).
7. Jana, N. R., Earhart, C. and Ying, J. Y. "Synthesis of water-soluble and functionalized nanoparticles by silica coating," *Chem Mater* 19, 5074–5082 (2007).
8. Shi, X., Lee, I. and Baker, J. R. "Acetylation of dendrimer-entrapped gold and silver nanoparticles," *J Mater Chem* 18, 586–593 (2008).
9. Levy, R., Thanh, N. T., Doty, R. C., Hussain, I., Nichols, R. J. et al. "Rational and combinatorial design of peptide capping ligands for gold nanoparticles," *J Am Chem Soc* 126, 10076–10084 (2004).
10. Alivisatos, A. P., Gu, W. and Larabell, C. "Quantum dots as cellular probes," *Annu Rev Biomed Eng* 7, 55–76 (2005).
11. Einwächter, S. and Krüger, M. "A fully automated remote controllable microwave-based synthesis setup for colloidal nanoparticles with integrated absorption and photoluminescence online analytics," *MRS Proc* 1284, (2011).
12. Ipe, B. I., Lehnig, M. and Niemeyer, C. M. "On the generation of free radical species from quantum dots," *Small* 1, 706–709 (2005).
13. Wang, Y., Fang, J., Leonard, S. S. and Rao, K. M. "Cadmium inhibits the electron transfer chain and induces reactive oxygen species," *Free Radic Biol Med* 36, 1434–1443 (2004).
14. Hussain, T., Shukla, G. S. and Chandra, S. V. "Effects of cadmium on superoxide dismutase and lipid peroxidation in liver and kidney of growing rats: in vivo and in vitro studies," *Pharmacol Toxicol* 60, 355–358 (1987).
15. Nel, A., Xia, T., Madler, L. and Li, N. "Toxic potential of materials at the nanolevel," *Science* 311, 622–627 (2006).
16. Cooper, D. R., Dimitrijevic, N. M. and Nadeau, J. L. "Photosensitization of CdSe/ZnS QDs and reliability of assays for reactive oxygen species production," *Nanoscale* 2, 114–121 (2010).
17. Lyon, D. Y., Brunet, L., Hinkal, G. W., Wiesner, M. R. and Alvarez, P. J. "Antibacterial activity of fullerene water suspensions (nC60) is not due to ROS-mediated damage," *Nano Lett* 8, 1539–1543 (2008).
18. Chibli, H., Carlini, L., Park, S., Dimitrijevic, N. M. and Nadeau, J. L. "Cytotoxicity of InP/ZnS quantum dots related to reactive oxygen species generation," *Nanoscale* 3, 2552–2559 (2011).
19. Hoskins, C., Cuschieri, A. and Wang, L. J. "The cytotoxicity of polycationic iron oxide nanoparticles: Common endpoint assays and alternative approaches for improved understanding of cellular response mechanism, *J Nanobiotechnol* 10, 15 (2012).

20. Nadeau, J. "Quantum Dot Reactive Oxygen Species Generation and Toxicity in Bacteria: Mechanisms and Experimental Pitfalls," in *Quantum Dot Sensors: Technology and Commercial Applications*. J. Callan and F. Raymo, Eds., Pan Stanford Publishing, Boca Raton, FL (2013).

21. Soenen, S. J., Rivera-Gil, P., Montenegro, J. M., Parak, W. J., De Smedt, S. C. et al. "Cellular toxicity of inorganic nanoparticles: Common aspects and guidelines for improved nanotoxicity evaluation," *Nano Today* 6, 446–465 (2011).

22. Nelson, B. C., Petersen, E. J., Marquis, B. J., Atha, D. H., Elliott, J. T. et al. "NIST gold nanoparticle reference materials do not induce oxidative DNA damage," *Nanotoxicology* 7, 21–29 (2013).

23. Derfus, A. M., Chan, W. C. W. and Bhatia, S. N. "Probing the cytotoxicity of semiconductor quantum dots," *Nano Lett* 4, 11–18 (2004).

24. Kirchner, C., Liedl, T., Kudera, S., Pellegrino, T., Javier, A. M. et al. "Cytotoxicity of colloidal CdSe and CdSe/ZnS nanoparticles," *Nano Lett* 5, 331–338 (2005).

25. Cho, S. J., Maysinger, D., Jain, M., Roder, B., Hackbarth, S. et al. "Long-term exposure to CdTe quantum dots causes functional impairments in live cells," *Langmuir* 23, 1974–1980 (2007).

26. Lovric, J., Bazzi, H. S., Cuie, Y., Fortin, G. R., Winnik, F. M. et al. "Differences in subcellular distribution and toxicity of green and red emitting CdTe quantum dots," *J Mol Med (Berl)* 83, 377–385 (2005).

27. Soenen, S. J., Demeester, J., De Smedt, S. C. and Braeckmans, K. "The cytotoxic effects of polymer-coated quantum dots and restrictions for live cell applications," *Biomaterials* 33, 4882–4888 (2012).

28. Hardman, R. "A toxicologic review of quantum dots: Toxicity depends on physicochemical and environmental factors," *Environ Health Perspect* 114, 165–172 (2006).

29. Pelley, J. L., Daar, A. S. and Saner, M. A. "State of academic knowledge on toxicity and biological fate of quantum dots," *Toxicol Sci* 112, 276–296 (2009).

30. Zhang, Y., Tekobo, S., Tu, Y., Zhou, Q. F., Jin, X. L. et al. "Permission to enter cell by shape: Nanodisk vs nanosphere," *ACS Appl Mater Interfaces* 4, 4099–4105 (2012).

31. Frohlich, E. "The role of surface charge in cellular uptake and cytotoxicity of medical nanoparticles," *Int J Nanomedicine* 7, 5577–5591 (2012).

32. Casals, E., Pfaller, T., Duschl, A., Oostingh, G. J. and Puntes, V. "Time evolution of the nanoparticle protein corona," *ACS Nano* 4, 3623–3632 (2010).

33. Cedervall, T., Lynch, I., Lindman, S., Berggard, T., Thulin, E. et al. "Understanding the nanoparticle-protein corona using methods to quantify exchange rates and affinities of proteins for nanoparticles," *Proc Natl Acad Sci USA* 104, 2050–2055 (2007).

34. Stern, S. T., Zolnik, B. S., McLeland, C. B., Clogston, J., Zheng, J. W. et al. "Induction of autophagy in porcine kidney cells by quantum dots: A common cellular response to nanomaterials?" *Toxicol Sci* 106, 140–152 (2008).

35. Seleverstov, O., Phang, J. M. and Zabirnyk, O. "Semiconductor Nanocrystals in Autophagy Research: Methodology Improvement at Nanosized Scale," in *Methods in Enzymology: Autophagy in Mammalian Systems*, Vol 452, Pt B, Academic Press, San Diego, CA, pp. 277–296. D. J. Klionsky, Ed., (2009).

36. Li, Y. Y., Zhou, Y. L., Wang, H. Y., Perrett, S., Zhao, Y. L. et al. "Chirality of glutathione surface coating affects the cytotoxicity of quantum dots," *Angew Chem Int Ed Engl* 50, 5860–5864 (2011).

37. Wang, C., Gao, X. and Su, X. "Study the damage of DNA molecules induced by three kinds of aqueous nanoparticles," *Talanta* 80, 1228–1233 (2010).

38. Green, M. and Howman, E. "Semiconductor quantum dots and free radical induced DNA nicking," *Chem Commun (Camb)*, Issue 1, 121–123 (2005).

39. Domingos, R. F., Simon, D. F., Hauser, C. and Wilkinson, K. J. "Bioaccumulation and effects of CdTe/CdS quantum dots on *Chlamydomonas reinhardtii*: Nanoparticles or the free ions?," *Environ Sci Technol* 45, 7664–7669 (2011).

40. King-Heiden, T. C., Wiecinski, P. N., Mangham, A. N., Metz, K. M., Nesbit, D. et al. "Quantum dot nanotoxicity assessment using the zebrafish embryo," *Environ Sci Technol* 43, 1605–1611 (2009).

41. Xiao, Q., Qiu, T., Huang, S., Liu, Y. and He, Z. "Preparation and biological effect of nucleotide-capped CdSe/ZnS quantum dots on Tetrahymena thermophila," *Biol Trace Elem Res* 147, 346–353 (2012).

42. Dumas, E., Gao, C., Suffern, D., Bradforth, S. E., Dimitrijevic, N. M. et al. "Interfacial charge transfer between CdTe quantum dots and gram negative vs gram positive bacteria," *Environ Sci Technol* 44, 1464–1470 (2010).

43. Yang, Y., Mathieu, J. M., Chattopadhyay, S., Miller, J. T., Wu, T. P. et al. "Defense mechanisms of pseudomonas aeruginosa PAO1 against quantum dots and their released heavy metals," *ACS Nano* 6, 6091–6098 (2012).

44. Aruguete, D. M., Guest, J. S., Yu, W. W., Love, N. G. and Hochella, M. F. "Interaction of CdSe/CdS core-shell quantum dots and *Pseudomonas aeruginosa*," *Environ Chem* 7, 28–35 (2010).

45. Priester, J. H., Stoimenov, P. K., Mielke, R. E., Webb, S. M., Ehrhardt, C. et al. "Effects of soluble cadmium salts versus CdSe quantum dots on the growth of planktonic pseudomonas aeruginosa," *Environ Sci Technol* 43, 2589–2594 (2009).

46. Wang, L. L., Zheng, H. Z., Long, Y. J., Gao, M., Hao, J. Y. et al. "Rapid determination of the toxicity of quantum dots with luminous bacteria," *J Hazard Mater* 177, 1134–1137 (2010).

47. Jasinski, J., Leppert, V., Lam, S., Gibson, G., Nauka, K. et al. "Rapid oxidation of InP nanoparticles in air," *Solid State Commun* 141, 624–627 (2007).

48. Xu, S., Zieler, J. and Nann, T. "Rapid synthesis of highly luminescent InP and InP/ZnS nanocrystals," *J Mater Chem* 18, 2653–2656 (2008).

49. Chibli, H., Carlini, L., Park, S., Dimitrijevic, N. M. and Nadeau, J. L. "Cytotoxicity of InP/ZnS quantum dots related to reactive oxygen species generation," *Nanoscale* 3, 2552–2559 (2011).

50. Thakor, A. S., Jokerst, J., Zavaleta, C., Massoud, T. F. and Gambhir, S. S. "Gold nanoparticles: A revival in precious metal administration to patients," *Nano Lett* 11, 4029–4036 (2011).

51. Pernodet, N., Fang, X. H., Sun, Y., Bakhtina, A., Ramakrishnan, A. et al. "Adverse effects of citrate/gold nanoparticles on human dermal fibroblasts," *Small* 2, 766–773 (2006).

52. Alkilany, A. M., Nagaria, P. K., Hexel, C. R., Shaw, T. J., Murphy, C. J. et al. "Cellular uptake and cytotoxicity of gold nanorods: Molecular origin of cytotoxicity and surface effects," *Small* 5, 701–708 (2009).

53. Qiu, Y., Liu, Y., Wang, L. M., Xu, L. G., Bai, R. et al. "Surface chemistry and aspect ratio mediated cellular uptake of Au nanorods," *Biomaterials* 31, 7606–7619 (2010).

54. Besner, S., Kabashin, A. V. and Meunier, M. "Two-step femtosecond laser ablation-based method for the synthesis of stable and ultra-pure gold nanoparticles in water," *Appl Phys A* 88, 269–272 (2007).

55. Tsoli, M., Kuhn, H., Brandau, W., Esche, H. and Schmid, G. "Cellular uptake and toxicity of AU(55) clusters," *Small* 1, 841–844 (2005).

56. Shukla, M. K., Dubey, M., Zakar, E. and Leszczynski, J. "DFT investigation of the interaction of gold nanoclusters with nucleic acid base guanine and the Watson-Crick guanine-cytosine base pair," *J Phys Chem C* 113, 3960–3966 (2009).

57. Mironava, T., Hadjiargyrou, M., Simon, M., Jurukovski, V. and Rafailovich, M. H. "Gold nanoparticles cellular toxicity and recovery: Effect of size, concentration and exposure time," *Nanotoxicology* 4, 120–137 (2010).

58. Soenen, S. J., Manshian, B., Montenegro, J. M., Amin, F., Meermann, B. et al. "Cytotoxic effects of gold nanoparticles: A multiparametric study," *ACS Nano* 6, 5767–5783 (2012).

59. Li, J. J., Hartono, D., Ong, C. N., Bay, B. H. and Yung, L. Y. L. "Autophagy and oxidative stress associated with gold nanoparticles," *Biomaterials* 31, 5996–6003 (2010).

60. Stern, S. T., Adiseshaiah, P. P. and Crist, R. M. "Autophagy and lysosomal dysfunction as emerging mechanisms of nanomaterial toxicity," *Part Fibre Toxicol* 9, 20 (2012).

61. Henig, R. M. "Our silver-coated future," *OnEarth* 22–29 (2007, Fall).

62. Xiu, Z. M., Zhang, Q. B., Puppala, H. L., Colvin, V. L. and Alvarez, P. J. "Negligible particle-specific antibacterial activity of silver nanoparticles," *Nano Lett* 12, 4271–4275 (2012).

63. Nie, S. and Emory, S. R. "Probing single molecules and single nanoparticles by surface-enhanced Raman scattering," *Science* 275, 1102–1106 (1997).

64. Ahamed, M., Karns, M., Goodson, M., Rowe, J., Hussain, S. M. et al. "DNA damage response to different surface chemistry of silver nanoparticles in mammalian cells," *Toxicol Appl Pharmacol* 233, 404–410 (2008).

65. Hackenberg, S., Scherzed, A., Kessler, M., Hummel, S., Technau, A. et al., "Silver nanoparticles: Evaluation of DNA damage, toxicity and functional impairment in human mesenchymal stem cells," *Toxicol Lett* 201, 27–33 (2011).

66. Mahl, D., Diendorf, J., Ristig, S., Greulich, C., Li, Z. A. et al. "Silver, gold, and alloyed silver-gold nanoparticles: Characterization and comparative cell-biologic action," *J Nanopart Res* 14, 1154 (2012).

67. Shang, L., Dorlich, R. M., Trouillet, V., Bruns, M. and Nienhaus, G. U. "Ultrasmall fluorescent silver nanoclusters: Protein adsorption and its effects on cellular responses," *Nano Res* 5, 531–542 (2012).

68. Ravindran, A., Chandran, P. and Khan, S. S. "Biofunctionalized silver nanoparticles: Advances and prospects," *Colloids Surf B Biointerfaces* 105C, 342–352 (2013).

69. Sotiriou, G. A. "Biomedical applications of multifunctional plasmonic nanoparticles," *Wiley Interdiscip Rev Nanomed Nanobiotechnol* 5, 19–30 (2013).

70. Haboubi, H. N., Doak, S. H. and Jenkins, G. J. "Tomorrow's world: The jury is still out on the safety of silver nanoparticles," *BMJ* 346, f227 (2013).

71. Liu, B. R., Huang, Y. W., Chiang, H. J. and Lee, H. J. "Cell-penetrating peptide-functionalized quantum dots for intracellular delivery," *J Nanosci Nanotechnol* 10, 7897–7905 (2010).

72. Yuan, H., Fales, A. M. and Vo-Dinh, T. "TAT peptide-functionalized gold nanostars: Enhanced intracellular delivery and efficient NIR photothermal therapy using ultralow irradiance," *J Am Chem Soc* 134, 11358–11361 (2012).

73. Krpetic, Z., Saleemi, S., Prior, I. A., See, V., Qureshi, R. et al. "Negotiation of intracellular membrane barriers by tat-modified gold nanoparticles," *ACS Nano* 5, 5195–5201 (2011).

74. Oh, E., Delehanty, J. B., Sapsford, K. E., Susumu, K., Goswami, R. et al. "Cellular uptake and fate of PEGylated gold nanoparticles is dependent on both cell-penetration peptides and particle size," *ACS Nano* 5, 6434–6448 (2011).

75. Delehanty, J. B., Medintz, I. L., Pons, T., Brunel, F. M., Dawson, P. E. et al. "Self-assembled quantum dot-peptide bioconjugates for selective intracellular delivery," *Bioconjug Chem* 17, 920–927 (2006).

76. Xiong, R. L., Li, Z., Mi, L., Wang, P. N., Chen, J. Y. et al. "Study on the intracellular fate of tat peptide-conjugated quantum dots by spectroscopic investigation," *J Fluoresc* 20, 551–556 (2010).

77. Duan, H. W. and Nie, S. M. "Cell-penetrating quantum dots based on multivalent and endosome-disrupting surface coatings," *J Am Chem Soc* 129, 3333–3338 (2007).

78. Pathakoti, K., Hwang, H. M., Xu, H., Aguilar, Z. P. and Wang, A. "In vitro cytotoxicity of CdSe/ZnS quantum dots with different surface coatings to human keratinocytes HaCaT cells," *J Environ Sci(China)* 25, 163–171 (2013).

79. Chen, F. Q. and Gerion, D. "Fluorescent CdSe/ZnS nanocrystal-peptide conjugates for long-term, nontoxic imaging and nuclear targeting in living cells," *Nano Lett* 4, 1827–1832 (2004).
80. Kuo, C. W., Chueh, D. Y., Singh, N., Chien, F. C. and Chen, P. L. "Targeted nuclear delivery using peptide-coated quantum dots," *Bioconjug Chem* 22, 1073–1080 (2011).
81. Lin, S. Y., Chen, N. T., Sum, S. P., Lo, L. W. and Yang, C. S. "Ligand exchanged photoluminescent gold quantum dots functionalized with leading peptides for nuclear targeting and intracellular imaging," *Chem Commun (Camb)* 4762–4764 (2008).
82. Lin, S. Y., Chen, N. T., Sun, S. P., Chang, J. C., Wang, Y. C. et al. "The protease-mediated nucleus shuttles of subnanometer gold quantum dots for real-time monitoring of apoptotic cell death," *J Am Chem Soc* 132, 8309–8315 (2010).
83. Klaine, S. J., Alvarez, P. J., Batley, G. E., Fernandes, T. F., Handy, R. D. et al., "Nanomaterials in the environment: Behavior, fate, bioavailability, and effects," *Environ Toxicol Chem* 27, 1825–1851 (2008).
84. Callaway, E. "Most popular human cell in science gets sequenced," *Nature* doi:10.1038/nature.2013.12609 (2013).

Section VI

Risk Assessment

24 Issues Related to Risk Assessment of Nanomaterials

Maureen R. Gwinn

CONTENTS

24.1 INTRODUCTION

Nanomaterials are produced for a variety of uses, but the main parameters that make them so appealing may also make them more of a risk to human health and the environment.[1,2] Nanomaterials are lightweight yet durable, have an increased surface area, can be more porous, and can be more soluble. These factors, along with their small size, may give them the ability to interact with biological systems in ways not seen before. Exposure estimates or human health effects associated with exposure to nanomaterials during manufacturing or use as consumer products is not well investigated or known. There is a lack of information on the effects of nanomaterials as compared to many of the macroscale materials. There have been some occupational exposure studies performed, but these are small-scale studies limited to a particular production method or use of a specific nanomaterial.[3–5] This lack of exposure information in human populations limits the ability to adequately assess the risks associated with nanomaterials, making it necessary to extrapolate from studies based on ultrafine particles and fibers. This lack of understanding of the risks associated with nanomaterials further hinders the risk management of these materials. Many of the existing standards for regulation described here are based on materials in their macroscale forms but these may not be relevant to protect against the nanoscale counterparts.

24.2   RISK ASSESSMENT OF NANOMATERIALS

Various attempts to analyze nanomaterials have focused on combining the risk assessment paradigm with a life cycle framework. The use of a comprehensive environmental assessment (CEA) entails all life cycle stages in the context of hazard identification and risk assessment. CEA involves both qualitative and quantitative approaches to risk characterization of human or ecological risk.[6,7] Other efforts have focused on the use of multicriteria decision analysis to classify potential risks from nanomaterials.[8] This methodology allows for clustering of nanomaterials into various risk categories based on the toxicity and physicochemical characteristics of the materials. To be able to better understand the potential risks from nanomaterials, it is necessary to understand what aspects of nanomaterials may lead to adverse biological effects. In general, risk assessments are based on specific chemicals, regardless of formulation. In the case of nanomaterials, however, this may not be the case. Chemicals generally considered nontoxic in larger size ranges can become inherently toxic in the nanorange. The most studied example of this is titanium dioxide, which in the larger size range does not exhibit toxicity. Exposures to titanium dioxide have been associated with a variety of pulmonary effects in rats, including inflammation, pulmonary damage, fibrosis, and lung tumors.[9,10] Also, ultrafine titanium dioxide particles can impair macrophage function and increase pulmonary retention, enter the epithelium faster, and translocate to the subepithelial space more readily than fine particles.[11] These have also been shown to lead to mitotic disturbances, DNA damage, and apoptosis (please refer to other chapters for more detailed information). A key area of nanomaterial research is characterization of physicochemical properties. Although most researchers agree that characterization of nanomaterials is a necessary step in toxicological analysis, the level, type and timing of characterization is not as clear. The increasing usage of nanomaterials takes advantage of the differences between these materials and their larger scale counterparts. Research has focused on many of these characteristics and their potential role in toxicity, generally compared to their larger scale counterparts. Studies have examined the impact of their smaller size, increased surface area, and increased surface reactivity. By virtue of their small size, nanomaterials may have access to biological or ecological locations not generally exposed to macroscale-sized material of the same chemical components, and may remain suspended longer and travel farther in the groundwater. The increased surface area may lead to increased adsorption of materials for drug delivery and for remediation. On the other hand, it can also increase absorbance of biological molecules or chemicals that lead to potential adverse health effects in humans. Large surfaces also allow for increased interaction with contaminants and may lead to more rapid degradation. Increased surface area may also lead to increased surface reactivity and biological response in the target tissue. There is also lack of information related to exposure analysis of nanomaterials, as recently highlighted in a review of the current state of risk assessment of nanomaterials.[12] Recent workshops have addressed the issues of best practices for exposure assessment of nanomaterials, and for moving beyond occupational exposures. However, work is still needed on the methodology for exposure estimations in human populations, either occupational or environmental. Owing to many of these issues described earlier, there is a need to coordinate the research on nanomaterials, and in doing so encourage collaboration among researchers. For example, the Organization for Economic Cooperation and Development has established a Working Party of Manufactured Nanomaterials to collaborate internationally on the research of nanomaterials. This workgroup is a key example of the interdisciplinary nature of the work needed in nanomaterial research, including chemists, biologists, and regulators working on environmental health and safety, exposure, testing, and regulation of nanomaterials. Subcommittees of this working party are broken off based on the type of nanomaterial under study, with various countries serving as sponsors. The purpose of these subcommittees is to analyze the potential adverse human health effects of these nanomaterials with limited overlap. It is important to have a comprehensive approach to characterize materials given the effect of subtle differences on the potential toxic effects of nanomaterials.[13]

24.3 CHARACTERISTICS OF NANOMATERIALS

There are uncertainties in determining the role of specific nanomaterial characteristics in the toxicological response to exposure to nanomaterials. As described previously,[14] an important aspect of all nanomaterial research is a clear description of all preparation and characterization of the nanomaterials studied. There are research efforts underway to create reference nanomaterials for use in standardized procedures. The most studied aspects of nanomaterials in toxicological studies include size distribution, surface area and reactivity (chemistry and charge), shape, chemical composition, purity, aggregation state, crystal structure, and porosity. Analysis of these various characteristics is reviewed previously[15] and will not be reviewed in depth here. Complicating standardized method development, nanomaterials are known to alter in different environments; therefore the characteristics of the engineered nanomaterials at the point of production may be vastly different following transport, storage, and preparation for analysis. Generally, studies to date have examined the effect of different suspension media on aggregation of nanomaterials in relation to human health effects. Just as it is important to characterize the starting material in nanomaterial studies, it is also important to analyze the material in the form that will be analyzed. The lack of standardization of procedures has led to debates on the interpretation of toxicity data. A key example is the analysis of fullerene toxicity. Fullerenes are generally dissolved in different organic solvents, including tetrahydrofuran. The toxicity of fullerenes observed in earlier studies may in fact have been due to the contamination of these solvents. This is often used as a prime example for clear description and standardization of the preparation of nanomaterials in toxicity studies. Data analysis for risk assessment requires a knowledge and understanding of the effects of these materials in humans, animals, and the environment. Given the differences in how the current research is being performed (mass vs. surface area measurements, vehicles for exposure, etc.), future review of the present data will be unable to adequately describe the risks associated with nanomaterials exposure.

24.4   UNDERSTANDING POTENTIAL RISKS OF NANOMATERIALS

In the United States and Europe, there are coordinated research efforts designed to address the issues described earlier. A few of these are described as follows:

1. National Nanotechnology Initiative (NNI)[16]
   The NNI is a U.S. government initiative to coordinate federally funded nanomaterial research focusing on multiple topics, including metrology, environmental health and safety, and exposure to nanomaterials. Coordination of efforts is an important step in making sure the most useful research is done in an efficient manner.
2. Nanoscale Materials Stewardship Program (NMSP)[17]
   The NMSP of the U.S. Environmental Protection Agency (EPA) was launched in 2008 to voluntarily acquire information on the use and development of nanomaterials. Unfortunately, to date, limited information has been acquired through this program. Following the first year of the program, information on only approximately 100 nanomaterials has been submitted, mostly from smaller companies with limited health and safety data.
3. NanoSafety Cluster[18]
   To coordinate research in the area of nanosafety, the European Union (EU) research programs are organized as a "NanoSafety Cluster." This initiative includes over 30 funded research projects with the main goal of improving coherence of nanotoxicology studies and harmonizing methods for assessing nanotoxicology. These diverse groups interact regularly to address issues in nanosafety with broad discussion among a variety of disciplines, and to aid in reaching consensus on nanotoxicology in Europe.

24.5   POTENTIAL REGULATORY STATUTES FOR NANOMATERIALS

Currently, there are no specific regulations for nanomaterials. This is in part because of some of the issues of the risk analysis of these materials described earlier. Brief descriptions of potential regulatory statutes are described later, with more details in Table 24.1.[19]

24.5.1   TOXIC SUBSTANCES CONTROL ACT

The Toxic Substances Control Act (TSCA) is the main legislative authority under which EPA requires manufacturers of new chemicals to submit specific information on the effects of these materials.[20] Under TSCA, the EPA has taken steps to limit the use and exposure to nanomaterials, including carbon nanotubes. EPA has required the use of personal protective equipment and limitations on the use and environmental exposures of nanomaterials. The EPA's Office of Pollution Prevention and Toxics recently released a fact sheet describing the plans to study nanomaterials on a case-by-case basis as new chemicals, rather than existing chemicals.

TABLE 24.1

Legislative Authorities That May Encompass Nanomaterial Usage

Regulation	Description	Potential Use	Governing Body	Website
TSCA	TSCA provides EPA with the authority to require reporting, record-keeping and testing requirements, and restrictions relating to chemical substances and/or mixtures	Proposed case-by-case determination of nanomaterials as new materials	U.S. EPA	http://www.epa.gov/lawsregs/laws/tsca.html
FIFRA	FIFRA provides the basis for regulation, sale, distribution, and use of pesticides in the United States	Case-by-case determination of unreasonable adverse human health risks with the addition of nanomaterials to existing pesticides	U.S. EPA	http://www.epa.gov/opp00001/regulating/laws.htm
Safe Drinking Water Act	Federal law that ensures the quality of American's drinking water	No maximum contaminant level goals set for nanoparticles at time of publication	U.S. EPA	http://www.epa.gov/OGWDW/sdwa/
CERCLA/RCRA	Commonly known as Superfund, CERCLA was enacted in 1980 and created a tax on the chemical and petroleum industries and provides a broad Federal Authority to respond directly to releases or threatened releases of hazardous substances that may endanger public health or the environment.	Addresses the evaluation and control of nanomaterials in waste sites	U.S. EPA	http://www.epa.gov/superfund/policy/cercla.htm
CAAA/CWAA	The CAAA and CWAA both allow the EPA to set limits on certain air and water pollutants in the United States. Under these acts, the EPA also reviews and approves permit applications for industries or chemical processes	Release of nanomaterials into the air or water would fall under these acts	U.S. EPA	http://www.epa.gov/air/caa/peg/understand.html http://www.epa.gov/oecaagct/lcwa.html

(Continued)

TABLE 24.1 (*Continued*)

Legislative Authorities That May Encompass Nanomaterial Usage

Regulation	Description	Potential Use	Governing Body	Website
FDCA	The FDCA is a set of laws giving FDA the authority to oversee the safety or food, drugs and cosmetics in the United States	May cover regulation of the use of nanomaterials in food additives and cosmetics. Recent draft guidance released by FDA	U.S. FDA	http://www.fda.gov/ RegulatoryInformation/ Guidances/ucm257698.htm
Occupational Safety and Health Act of 1970 (29 US 654)	Section 5(a)(1) of the OSH Act requires employers to "furnish each of his employees employment and a place of employment which are free from recognized hazards that are causing or likely to cause death or serious physical harm to his employees"	Twenty-four states, Puerto Rico and the Virgin Islands have OSHA-approved state plans and have adopted their own standards and enforcement policies, which are in most cases similar to Federal OSHA standards	United States Occupational Safety and Health Administration (OSHA)	http://www.osha.gov/dsg/ nanotechnology/nanotech_ standards.html
Canada Environmental Protection Act	The New Substances Notification Regulations (Chemicals and Polymers) ensures that any new substance undergoes a risk assessment of its potential effects on the environment and human health	Proposed development of a regulatory framework for nanomaterials	Environment Canada/Health Canada	http://www.ec.gc.ca/ subsnouvelles-newsubs/default. asp?lang=En&n=FD117B60-1
REACH	REACH is European Community Regulation on chemicals and their safe use (EC 1907/2006) and went into force on June 1, 2007	Companies producing nanomaterials would need to report these under the same guidelines as other materials	EU	http://ec.europa.eu/environment/ chemicals/reach/pdf/ nanomaterials.pdf

Source: Adapted from Gwinn, M. R., Sokull-Kluttgen, B. Regulation and legislation, In: Fadeel, B. et al., (eds.), *Adverse Effects of Engineered Nanomaterials*, First Edition, Academic Press, 2012. With permission.

24.5.2 OTHER U.S. REGULATIONS

Nanomaterials added to pesticides would fall under the mandates of the Federal Insecticide, Fungicide, and Rodenticide Act (FIFRA) to determine whether their addition leads to adverse health risks (EPA 2008). Comprehensive Environmental Response, Compensation, and Liability Act and Resource Conservation and Recovery Act (CERCLA/RCRA) would be used to evaluate the risks of nanomaterials at waste sites. Clean Air Act Amendments (CAAA) (1990) and Clean Water Act Amendments (CWAA) (1996) would be used for regulation of nanomaterials in ambient air or ambient water and wastewater. For CAAA, the levels transported in air would be regulated based on their danger to public health. The CWAA would regulate the effluent limits of nanomaterials in the wastewater. The Food, Drug, and Cosmetic Act (FDCA) has been used by the Food and Drug Administration (FDA) to address the impact and use of nanomaterials. The FDCA requires premarket testing for food and color additives, but not for cosmetics. The FDA can require information on the identity and properties of materials, regardless of size of particles in the material. FDA also controls the regulations of cosmetics, including sunscreens. The FDA has recently released some draft guidance further describing potential review of nanomaterials.[21] The use of nanomaterials in consumer products in the United States is under the oversight of the Consumer Product Safety Commission (CPSC). The CPSC mission is protecting the public from unreasonable risks of serious injury or death from the products under their jurisdiction. Although there is no specific ruling for nanomaterials, if any of these products contain nanomaterials, they would fall under the regulation of CPSC.

24.5.3 ENVIRONMENT CANADA

Environment Canada issued a notice in early 2009 requiring companies producing nanomaterials to file federal reports on those materials describing their toxicity, volume produced, and other relevant and readily available data. This would be required for all materials produced in quantities greater than 1 kg. Canada also plans to launch a voluntary program much like the NMSP in the United States.

24.5.4 REGISTRATION, EVALUATION, AND AUTHORIZATION OF CHEMICALS

A group of nanomaterial experts selected by the EU debated, how best to handle nanomaterials under the Registration, Evaluation and Authorization of Chemicals (REACH) in early 2009. This group, known as the Competent Authorities Sub-Group on Nanomaterials, is a subgroup convened by the competent authorities responsible for overseeing the implementation of REACH in EU member states. Currently, there are no provisions in REACH specific to nanomaterials. Therefore, companies producing nanomaterials would be required to report these under REACH, the same as those producing their macroscale counterparts. On the basis of the REACH guidelines, this would mean that any company producing or importing more than 1 metric ton per year would be registered under REACH. One concern is that many of the nanoscale materials would be produced in much smaller volumes, and therefore not trigger the need for registration. This would lead to limited information on these materials for risk management. Discussions are ongoing as to whether this concern would lead to a requirement for registration of all nanoscale materials produced or imported, regardless of amount. This would give increased information on nanomaterials to be used in determining their human health risk.

24.6 CONCLUSION AND FUTURE RESEARCH

Nanotechnology is used in a broad range of applications, and plays a role in recent advances in medical diagnosis and treatment, electronics production, and environmental clean-up methods. As the use and potential benefits of nanomaterials increase rapidly, it is important that the ongoing research is designed to answer the risk assessment needs surrounding the manufacture, use, and disposal of nanomaterials in the world around us. Risk assessment of nanomaterials is complicated by limited exposure information, no clear dose metric, and the lack of characterization of nanomaterials in many early studies. Continuing efforts to address current data gaps and describe research needs for nanomaterial risk assessment must continue. There are clear data gaps in exposure assessment of nanomaterials, particularly moving beyond occupational exposures. Very limited information is available on the exposure to consumers through the transport, use, and disposal of multiple products. The lack of a dose metric for the analysis of toxic effects of nanomaterials is another large data gap. Continued harmonization of research in these and other areas is needed to inform future risk assessment of nanomaterials. Risk assessment of nanomaterials will aid in determination of appropriate risk management of these materials.

REFERENCES

1. Gwinn MR, Vallyathan V. Nanoparticles: Health effects-pros and cons. *Environ Health Perspect* 2006;114(12):1818–1825.
2. Warheit DB, Sayes CM, Reed KL, Swain KA. Health effects related to nanoparticles exposures: Environmental, health and safety considerations for assessing hazards and risks. *Pharmacol Ther* 2008;120:35–42.
3. Bergamaschi E. Occupational exposure to nanomaterials: Present knowledge and future development. *Nanotoxicology* 2009;3(3):194–201.
4. Murashov V. Occupational exposure to nanomedical applications. *Wiley Interdisc Rev Nanomed Nanobiotechnol* 2009;1(2):203–213.
5. Trout DB, Schulte PA. Medical surveillance, exposure registries, and epidemiologic research for workers exposed to nanomaterials. *Toxicology* 2010;269(2–3):128–135.
6. Davis JM. How to assess the risk of nanotechnology: Learning from past experience. *J Nanosci Nanotechnol* 2007;7(2):402–409.
7. Powers CM, Dana G, Gillespie P, Gwinn MR, Hendren CO, Long TC, Wang A, Davis JM. Comprehensive environmental assessment: A meta-assessment approach. *Environ Sci Technol* 2012;46(17):9202–9208.
8. Linkov I, Steevens J, Chappell M, Tervonen T, Figueira JR, Merad M. Classifying Nanomaterial Risks Using Multi-Criteria Decision Analysis. In Linkov I, Steevens J, eds. *Nanomaterials: Risks and Benefits*. Springer Science, The Netherlands, 2009;179–191.

9. Bermudez E, Mangum JB, Asgharian B, Wong BA, Reverdy E, Janszen D, Hext PM, Warheit DB, Everitt JI. Long-term pulmonary responses of three laboratory rodent species to subchronic inhalation of pigmentary titanium dioxide particles. *Toxicol Sci* 2002;70(1):86–97.

10. Bermudez E, Mangum JB, Wong BA, Asgharian B, Hext PM, Warheit DB, Everitt JI. Pulmonary responses of mice, rats, and hamsters to subchronic inhalation of ultrafine titanium dioxide particles. *Toxicol Sci* 2004; 77(2):347–357.

11. Churg A, Stevens B, Wright JL. Comparison of the uptake of fine and ultrafine TiO_2 in a tracheal explant system. *Am J Physiol* 1998;274(1 pt 1): L81–L86.

12. SCENIHR (Scientific Committee on Emerging and Newly Identified Health Risks). Risk Assessment of Products of Nanotechnologies. European Commission. 2009. http://ec.europa.eu/health/ph_risk/committees/04_scenihr/docs/scenihr_o_023.pdf. Accessed March 1, 2013.

13. OECD (Organization for Economic Co-operation and Development). Important Issues on Risk Assessment of Manufactured Nanomaterials. Safety of Manufactured Nanomaterials 2012;33. http://search.oecd.org/officialdocuments/displaydocumentpdf/?cote=env/jm/mono(2012)8&doclanguage=en. Accessed March 1, 2013.

14. Gwinn MR, Tran L. Risk management of nanomaterials. *Wiley Interdisc Rev Nanomed Nanobiotechnol* 2010;2(2):130–137.

15. Zuin S, Pojana G, Marcomini A. Effect-oriented physicochemical characterization of nanomaterials. In Monteiro-Riviere NA and Tran L, eds. *Nanotoxicology, Characterization, Dosing and Health Effects*. Informa Healthcare USA, Inc, New York, NY, 2007;19–58.

16. National Nanotechnology Initiative (NNI). http://www.nano.gov/ Accessed March 1, 2013.

17. U.S. EPA. Nanoscale Materials Stewardship Program. http://epa.gov/oppt/nano/stewardship.htm. Accessed March 1, 2013.

18. European Union. NanoSafety Cluster Compendium. http://www.nanosafetycluster.eu/home/european-nanosafety-cluster-compendium.html. Accessed March 1, 2013.

19. Gwinn MR, Sokull-Kluttgen B. Regulation and Legislation. In: Fadeel B, Pietroiusti A, Shvedova A, eds. *Adverse Effects of Engineered Nanomaterials*, First Edition, Academic Press, New York, NY, 2012;97–119.

20. U.S. EPA. Control of Nanoscale Materials under the Toxic Substances Control Act. http://epa.gov/oppt/nano/index.html. Accessed March 1, 2013.

21. FDA (Food and Drug Administration). Considering Whether an FDA-Regulated Product Involves the Application of Nanotechnology. http://www.fda.gov/RegulatoryInformation/Guidances/ucm257698.htm. Accessed March 1, 2013.

25 Risk Assessment of Engineered Nanomaterials

State of the Art and Roadmap for Future Research

Danail Hristozov, Laura MacCalman,
Keld A. Jensen, Vicki Stone,
Janeck Scott-Fordsmand, Bernd Nowack,
Teresa Fernandes, and Antonio Marcomini

CONTENTS

25.1 INTRODUCTION

Our understanding of the environmental and health risks associated with nanotechnology is still limited and may result in stagnation of growth and innovation. Some other technologies have revealed unexpected ecological and health effects only several years after their broader market introduction. In the worst cases this had tremendous cost implications for society and enterprise in the form of lock-in effects, overbalancing regulations, and loss of consumer trust.[1]

Several studies have shown that manufactured nanomaterials, understood as nano-objects, their aggregates and agglomerates (NOAA), are biologically more active than their bulk counterparts, and toxicity has been observed in animals for carbon nanotubes, fullerenes, nanoscale metals, and metal oxides.[2–21] This has raised awareness of the need to regulate the production and use of NOAA to ensure a high level of human and environmental protection.

Risk assessment (RA) is a central theme in the regulation of chemicals.[22–23] It is "a process intended to calculate or estimate the risk to a given target organism, system, or subpopulation, including the identification of attendant uncertainties, following exposure to a particular agent, taking into account the inherent characteristics of the agent of concern as well as the characteristics of the specific target system."[24] In the assessment of risk, it is important to differentiate between human health and ecological RA. In both cases, the process combines hazard and exposure assessment into risk characterization.

Hazard assessment consists of gathering and evaluating relevant environmental, health, and safety (EHS) information and performing ecotoxicity experiments. It also involves the characterization of the behavior of a chemical within the organism and its interactions with organs and cells, which includes the establishment of relationships between the observed biological responses and the physicochemical properties of a substance (European Chemicals Agency). It typically involves the estimation of human or environmental effect thresholds such as the derived no-effect level (DNEL) or the predicted no-effect concentration (PNEC). The principal aim of this step is to assess the intrinsic hazard of a chemical. It is the likelihood of impairment due to exposure that distinguishes risk from hazards. The exposure assessment is concerned with the estimation of the doses or concentrations, which humans or environmental species are or may be exposed to. It starts with the formulation of one or more exposure scenarios and includes the estimation of exposure either by direct measurements or by the application of models. This involves, for instance, the monitoring of indoor concentrations or the estimation of the amount of the substance coming into contact with the respiratory system, skin, or intestinal tract following this exposure. Similarly, in field conditions, assessment or derivation of exposure concentrations in environmental compartments is used in the assessment of risk to environmental species.

Risk characterization is the estimation of the incidence and severity of the adverse effects likely to occur in a human population or environmental compartment due to actual or predicted exposure to a substance, and may include risk estimation, that is, the quantification of that likelihood.[25] In this step, the estimated exposure concentrations are typically compared to the threshold effect levels to estimate risk. In many international regulatory frameworks, environmental risks are often expressed as ratios predicted environmental concentration (PEC)/PNEC, that is, as risk quotients (RQs). For human risks, a similar comparison between exposure and the DNEL is usually made. It should be noted that these ratios or comparisons provide no absolute measure of risks.

The RQ approach is generally used to assess risks from noncarcinogenic contaminants (i.e., threshold effects), whereas for carcinogens (i.e., nonthreshold effects) slope factors (SFs) are typically used. An SF is the 95% upper bound of the increase in cancer risk from a lifetime exposure via inhalation or ingestion. SFs are usually expressed in units of proportion (of a population) affected per milligram of substance per kilogram of body weight per day. They are generally derived from the low-dose region of the dose–response relationship, that is, for exposures corresponding to risks of less than 1 in 100.

It should be stressed that RAs are always associated with an uncertainty level. Our current knowledge does not allow us to adequately predict either the adverse effects on ecosystems (directly on species or on ecosystem services) or the sizes of the affected human populations, which results in generic and simplified risk analyses. In many cases, the available data are only sufficient for relative risk ranking, which enables us to compare chemicals in terms of "concern" rather than estimating actual risks.

This is especially valid for NOAA, which pose significant challenges to risk assessors since they dynamically transform during their life cycles (e.g., due to agglomeration or aggregation, corona formation or interaction with surrounding organic material, dissolution) and the life cycle of the products in which they are incorporated (from manufacture, use, and disposal), which leads to exposures to different forms of the same materials and consequently to different hazard profiles and risks. Since a multitude of different NOAA are expected to emerge on the market, full testing and risk analysis of all of them in all their forms and exposure scenarios will not be possible. Yet, safety with a reasonable degree of certainty should be demonstrated before introducing new nanomaterials in consumer or industrial settings. Hence, RA must be targeted to realistic exposure scenarios and should incorporate the grouping concept, as described below.

Considerable research efforts are currently focused on addressing the above challenges. However, NOAA risk analyses still face substantial remaining knowledge gaps with respect to mechanisms of biological uptake and toxicity modes of action as well as behavior in and between environmental compartments and potential bioaccumulation.[26] This work must also deal with the lack of nanospecific RA guidelines, which may either be considered a hindrance or something that is actually not needed.[27] The following sections discuss the state of the art in these areas and propose a roadmap for research by 2020 to address some critical uncertainties and allow quantitative longer-term RA of NOAA.

25.2 STATE OF THE ART

Several research projects have been concerned with evaluating the risks from the production and use of NOAA. Most of them have used pristine nanomaterials produced specifically for testing, which are not necessarily representative of real life cycle scenarios. Although European Union (EU) Seventh Framework Programme (FP7) projects such as NANOHOUSE (http://www-nanohouse.cea.fr) and SCAFFOLD (http://scaffold.eu-vri.eu/) focus on materials released from real products, they address only the construction sector. The SANOWORK project (http://www.sano-work.eu/) advanced a step further focusing on processing lines for nanomaterials in ceramics, textiles, and energy that encompass the production and handling only, but not on the use and end of life (EOL) stages of NOAA. In NANOSUSTAIN (http://www.nanosustain.eu/), the release of NOAA and toxicological tests have been analyzed for different nanomaterials and articles (paper, paint, glass, and polymer composite) at life cycle stages where highest exposure levels were to be expected, including EOL. Similarly, the project NANOMICEX (http://cordis.europa.eu /search/index.cfm?fuseaction = proj.document&PJ_RCN = 12796787) focuses on nanomaterials used in the pigment industry (inks and paints) and although it intends to address product life cycle aspects, it will not be able to address this area in full. These projects will, however, make inroads and provide data that will be essential for the progression with the development of the new RA framework.

25.2.1 ENVIRONMENTAL RELEASE, FATE, AND EXPOSURE

The behavior and fate of pristine nanomaterials have been significantly studied and are currently being addressed in the EU FP7 projects MARINA (http:// www .marina-fp7.eu/), NANOHOUSE, NANOSUSTAIN (http://www.nanosustain .eu), NANOPOLYTOX (http://www.nanopolytox.eu/), NANOMILE (http://www .nanomile.eu), NANOCHOP (http://nanochop.lgcgroup.com/), NEPHH (http://www .nephh-fp7.eu), and other projects.[28] As a result, much knowledge has been acquired. However, pristine NOAA undergo aging and transformation reactions during incorporation into products and when released into the environment.[29] Thus, at each life cycle stage of a nanoenabled product, there is a potential risk of exposure to NOAA with different physicochemical properties. Nearly, no data are available on the quantity and identity of NOAA released from actual products used by professional users and consumers. Although the U.S. International Life Sciences Institute NANORELEASE (Consumer Products) project (http://www.ilsi.org/ResearchFoundation/RSIA/Pages /NanoRelease1.aspx) identified it as the stage where significant release may occur, especially for product where the NOAA are bound in a matrix, the EOL stage (e.g., shredding, incineration, landfilling, recycling) of NOAA-containing products is far from well studied.[30,31] Some of the few results for textiles, paints, and nanocomposites suggest that the released particles undergo significant transformation and aging and exhibit different environmental behavior and effects compared to the pristine NOAA.[30,32,33] The EU FP7 project NANOHOUSE project has also shown that NOAA are released together with many other materials, whereas the Danish NANOKEM study revealed that during abrasion processes, addition of NOAA only made minor changes in the size distributions of sanding dust, resulting in limited release of free NOAA, and the toxicological effects of added NOAA were masked by the paint matrix.[34–36] Emerging results from toxicological tests carried out within the EU FP7 project NANOSUSTAIN further support these findings. Thus, comparison to a nano-free reference is crucial for assessing toxicity and ecotoxicity of any nano-containing materials released from products during their different life stages.

25.2.2 OCCUPATIONAL AND CONSUMER EXPOSURE

Although it is well known that nanomaterials are present in consumer products and new modeling studies have investigated consumer exposure to NOAA, no empirical data on consumer exposure measurements are available.[29,37–41]

In contrast, empirical data on workplace exposure are now emerging with increasing pace. Data are also slowly emerging on the emission characteristics and source strengths from different release scenarios in the production stage, such as synthesis, powder handling, simulated sanding, drilling, and cutting of nanocomposites.[42–46] However, experience from most previous studies has revealed that the data need to be improved; in particular, more contextual information is needed on process and use rates, duration of activity, room size, and ventilation. Improved strategies for exposure measurements are currently being discussed in both PEROSH (http://www.perosh.eu/) and the organization for Economic Cooperation and Development (OECD) Working Party on Manufactured Nanomaterials (WPMNM) (http://www.oecd.org/sti/nano/).

All current nanospecific models for consumer (though not based on empirical data) and occupational exposure assessment are qualitative.[47–51] Only the recent Stoffenmanager Nano and NanoSafer specifically consider personal exposure.[50,52] Only NanoSafer bases its assessment on time-resolved exposure estimates and allows the evaluation of both acute and chronic exposure. Schneider et al.[53] developed a conceptual model for the prediction of occupational exposure considering several important aerosol dynamic processes. Aerosol dynamic modeling is a major step forward in the NOAA exposure assessment, but is strongly constrained by the scarcity of data on source strength, workplace measurements, and contextual information.[53] In most models, the focus has been set on inhalation exposure. Nevertheless, dermal and oral exposure should not be neglected in a comprehensive approach corresponding to current RA strategies for traditional chemicals. New work is ongoing to improve the dermal exposure assessment for chemicals and consumer products in the DRESS project (http://www.cefic-lri.org/projects/27/21/LRI-B9-VITO-DRESS— -DeRmal-Exposure-aSsessment-Strategies/?cntnt01template = display_list_test). There is a pressing need to further develop models and to build up databases for inhalation, dermal, and oral exposure, as well as emission potentials related to different products, articles and process, and efficiency of engineered controls and personal protective equipment (PPE).

25.2.3 HUMAN AND ENVIRONMENTAL TOXICOLOGY AND EFFECTS ON ECOSYSTEM SERVICES

Toxicity data on NOAA are produced in many EU projects. For example, the central nervous response was investigated in NEURONANO (http://www.neuronano.eu), pulmonary, cardiovascular, hepatic, renal, and developmental effects were studied in NANOSH (http://www.ttl.fi/partner/nanosh/sivut/default.aspx), PARTICLE_RISK, and more recently in NANOTEST (http://www.nanotest-fp7.eu/) and ENPRA (http://www.enpra.eu). Immunotoxicity was investigated in NANOMMUNE (http://www.nanommune.eu), whereas the effects of the protein corona were studied in NANOINTERACT (http://nanointeract.nanosafetycluster.eu/). NANOSUSTAIN focused on the functional genomics activation of several response pathways. As a result, a coherent hazard profile of NOAA begins to emerge, demonstrating that reactive oxygen species, oxidative stress, and modified inflammatory responses play important roles in the animal and cellular toxicity of NOAA.[54–57] The results of the above projects also show relationships between key physicochemical characteristics, modes of action, and biological responses, indicating that factors such as surface area and reactivity, surface charge, solubility, shape, and composition are all keys.[58–61]

The main shortcomings in the state of the art include a lack of long-term inhalation studies essential for RA (in most projects, bolus instillation was used as a substitute for inhalation), and adequate comparison of in vitro and in vivo results to allow extrapolation and reduce animal testing as part of a 3Rs (Replacement, Reduction, Refinement) strategy. In addition, although the use of acute in vitro and in vivo data to predict long-term risks is not recommended as suitable nanospecific uncertainty factors for exposure duration extrapolation are presently unavailable, it is the case where the majority of data available are acute.[26] This is also true for the interspecies uncertainty factors and in result, current risk analyses use the less-conservative Margin of Exposure approach (avoiding extrapolation from test animal to human) rather than obtaining DNEL as recommended in the regulation, evaluation, authorization, and restriction of chemicals (REACH) guidelines.[62,63] The pulmonary exposure route has been investigated significantly more than ingestion, which is another important route of exposure, especially in consumer settings. In addition, most of the toxicity studies conducted so far involve pristine nanomaterials without inclusion in a complex matrix or aging. However, certain matrix components and transformations of NOAA due to weathering may significantly influence toxicity.

The state of the art in regard to the effects of NOAA in the environment, that is, ecosystems, is at an early stage. Published literature so far focuses almost entirely on standard short-term hazard testing of pristine NOAA. However, knowledge in this area will rapidly increase through major European projects, such as MARINA, NANOVALID, NANOSOLUTIONS, and NANOMILE, especially because they will provide a comprehensive set of standardized and comparable data. However, long-term effects following short-term or long-term low-concentration exposure are still missing. Next generation sequencing experiments with single species have provided evidences of long-term responses and changes being transferred along generations, causing epigenetic, mutational, or reproductive effects, but these techniques are not yet tailored to fully assess effects on ecosystems.[64] There are several sources of NOAA to the environment (Section 25.2.1). In regard to effluent from factories and households, studies have shown that NOAA can accumulate in the sludge of sewage treatment plants (STP), and consequently affect sediment and pelagic systems.[65–70] Since sewage sludge is used as fertilizer in many countries, it is an emerging source of NOAA entering soil affecting terrestrial organisms and potentially humans.[71] Runoff from surfaces to soils, sediment, and waters is also a relevant source of NOAA (Section 25.2.1), which will result in long-term repeated exposures of the environment.

Since it is likely that many nanomaterials are (designed to be) persistent, this may lead to long-term exposure and effects. However, long-term environmental effects are limited, especially for weathered NOAA released from real consumer products. Therefore, testing of long-term effects on multiple species should be developed, focusing on ecosystems subjected to multiple stressors (e.g., climate change, pollution) that provide services vital to mankind. This will result in valuable data to use with the tiered strategy currently developed in the MARINA project to achieve longer-term environmental RA of NOAA.

25.3 ROADMAP FOR FUTURE RESEARCH BY 2020

25.3.1 Environmental Release, Fate, and Exposure

We recommend to build on the results from the FP7 MARINA, NANOHOUSE, NANOSUSTAIN, NANOPOLYTOX, NANOMILE, NEPHH, NANOFATE, NANOCHOP, and other projects to assess the fate and behavior of NOAA released from actual products under real-life conditions over the whole life cycle, but mainly during the use and EOL stages. This should involve mapping hot spots for the release of nanomaterials during processes and uses in different stages of the life cycle to guide cost-effective strategies for release estimation. Released materials should be collected, characterized, and used in subsequent fate experiments and ecotoxicological tests. If the released nanomaterials are not sufficient in quantity for testing, they should be substituted for aged NOAA prepared in the laboratory specifically for this purpose. We should study in detail the physicochemical transformations of NOAA during release and in the environment and thus identify the properties of those NOAA that reach environmental organisms or humans. This can be achieved via an adaptation of the NANOCHOP measurement methods to a wider range of nanomaterials released from real consumer products. The comparison should always be made to similar pristine materials (i.e., same nanomaterials not incorporated in consumer products, as well as same pristine material at microscale).

Moreover, we recommend the development of (1) stochastic/probabilistic material flow techniques that have shown much potential to cope with high uncertainty and variability and link them to (2) new approaches for dynamic (time and geographically resolved) fate and exposure modeling.[72–74] This will reflect the NOAA behavior and transport mechanisms in natural compartments and will be linked to their mass balance dynamics. By using the aforesaid new experimental evidence, these material flow and fate/exposure models will be fed for the first time with empirical data, which will push the model output to a level beyond anything currently available. The fate models can be used to estimate PEC in different environmental compartments to combine them with PNEC for ecological RA.

25.3.2 Occupational and Consumer Exposure

We recommend building on the significant developments in the EU FP7 projects NANODEVICE, NANEX, ENPRA, NANOSUSTAIN, NANOVALID, MARINA, and the Swedish CEN/NEN Dermal Exposure Assessment to establish a versatile tiered qualitative to quantitative tool to assess inhalation, dermal and dermal-to-oral occupational and consumer exposure in specific life cycle stages.

The tiered model will build on existing nanospecific occupational exposure models, such as Stoffenmanager Nano and NanoSafer, the consumer risk categorization tool NanoRiskCat, and conventional tools such as Stoffenmanager and ConsExpo.[47,50,52,75] To enable high-tier near- and far-field occupational and consumer exposure assessments, these tools should be complemented by an advanced aerosol dynamic model as well as a dermal and dermal-to-oral exposure model.

For model development and refinement, we recommend developing an extensive set of high-quality, time-resolved occupational and consumer inhalation exposure data with detailed contextual information, as well as information on process- and product-specific emission source strengths and characteristics of relevant occupational (production, handling, and final treatment) practices and consumer uses.

Moreover, we recommend supplying the exposure assessment tool with a risk management module. To do this, we should collect and generate high-quality data on the efficiencies and protection factors for engineering controls and PPE, and establish dermal-to-oral NOAA exposure transfer factors considering accidental oral exposure. Moreover, dermal exposure efficiencies should be determined experimentally using tests for direct and indirect deposition from air to the skin and transfer from contaminated surface to skin, which are relevant for exposure to both powders, dust, and spray products. All data should be collected in a harmonized format to be easily shared through databases such as the nano exposure and contextual information repository (http://www.perosh.eu/p/0B560A3BFF7F1D30C12576190043FA2B) developed by PEROSH.

25.3.3 HUMAN AND ENVIRONMENTAL TOXICOLOGY AND EFFECTS ON ECOSYSTEM SERVICES

We recommend focusing the future research relevant for the RA of NOAA on the relationships among changes in their physicochemical characteristics, mechanisms of interactions with biological systems, and any subsequent longer-term effects.

Since inhalation is an important route of exposure, pulmonary studies are required. Instillation studies are not currently fully suitable for RA purposes due to the sudden delivery of a bolus dose, which may influence the uptake of NOAA in the lungs, their systemic distribution, and the resulting effects. Therefore, we recommend performing real inhalation studies with relevant models, as possible. Rodent models may not be sufficient because they do not mimic human exposure (see Chapter 14). A major constraint in investigating the long-term toxicity of NOAA is the high cost of chronic inhalation studies. To do this in a cost-efficient way, we recommend using the short-term inhalation studies (STIS) as devised in the European NANOSAFE2 (http://www.nanosafe.org/scripts/home/publigen/content /templates/show.asp?P = 56&L = EN) and the German NANOCARE (http://www .nanopartikel.info/cms/lang/en/Projekte/NanoCare) projects. STIS involve a 5-day inhalation exposure with examinations continuing for up to 28 days.[76] The 5-day protocol includes assessing progression or regression of effects, lung burden, and potential translocation to other tissues, which are aspects currently not covered in the OECD TG 412 and 413 testing guidelines.[76] Results of STIS are available for more than 20 pristine nanomaterials, including silica, titanium dioxide, multiwalled carbon nanotubes, and organic pigments. To investigate long-term ingestion risks, we recommend developing a novel reference short-term oral study (STOS) protocol, involving exposure via gavage for 5 days, followed by examinations continuing for up to 28 days. The STIS and STOS can be complemented by molecular, histological, and biochemical analysis of tissue responses, including the organ of exposure (lungs or intestine), liver (as the primary location of nanomaterial accumulation according to the results of ENPRA and INLIVETOX (http://www.inlivetox.eu/), Peyer's patches (very sensitive to both ingested and injected nanomaterials according to the results of INLIVETOX FP7 projects), lymph nodes, and spleen (to test the immune response). The measured molecular and biochemical markers can include indicators of oxidative stress, inflammation, and fibrosis.

To prioritize nanomaterials for the above in vivo studies and acquire an adequate dose range, we recommend using in vitro assays of acute cytotoxicity and proinflammatory activity (3Rs approach). Results from the FP7 ENPRA project suggest that most in vitro studies show no difference in cytotoxicity (i.e., cell death) among different cell types, except for macrophages, which are more sensitive to high aspect ratio NOAA.[77] Therefore, to reduce costs, we recommend using only macrophages and one nonimmune cell type (i.e., C3A hepatocytes) to screen NOAA and to allow benchmarking to previously studied nanomaterials. The C3A hepatocyte cell line represents the organ of highest exposure (i.e., the liver) and it reproduces well the response of primary human hepatocytes to NOAA.[78]

Screening tests should use standardized protocols such as the ones established in the FP7 projects ENPRA, NANOVALID (http://www.nanovalid.eu/), MARINA and QualityNano (www.qualitynano.eu/) projects to allow comparability of data with other studies.[79] Studied endpoints can include established measures of cytotoxicity as well as proinflammatory gene expression and genotoxicity.

In addition to conventional inductively coupled plasma mass spectroscopy determination of NOAA uptake in cells and tissues, when appropriate (depending on material type), in vitro 3-dimensional dermal models may be used to assess inflammation only supplemented by classic local toxicity approaches (histology and biochemical markers). However, they do not mimic normal human skin. These artificial skin models do not possess a "true lipid barrier" and therefore may not provide accurate data for absorption and skin penetration studies. Some of this data could be added to the growing datasets generated by projects such as ENPRA, MARINA, and INLIVETOX to allow better extrapolation from in vitro to in vivo and calculating long-term DNEL for human health RA. To do this, we recommend deriving nanospecific interspecies extrapolation factors using physiologically based pharmacokinetic/pharmacokinetic dynamic (PBPK/PD) modeling.

Computational PBPK models incorporate physicochemical and biochemical characteristics along with species-specific physiological properties to study the postexposure absorption, distribution, metabolism, and excretion (ADME) dynamics of chemicals or particles.[80,81] A specific, blood flow–limited PBPK model for quantum dots was introduced by Lee et al. (2009).[80] The first general PBPK model for NOAA was developed by the UK Institute of Occupational Medicine and the U.S. National Institute for Occupational Safety and Health in the context of the EU NANOMMUNE and ENPRA projects.[82] It is an adaptation and extension of an earlier PBPK model for larger particles, which was calibrated with data from the EU ENPRA, NANOMMUNE, and NANOTEST and will be further extended in the MARINA project. Such a model can be used to characterize not only the ADME profiles of the materials but also their biological interactions across a diverse range of species based on particle type and physicochemical properties.[80] Moreover, it can help to develop nanospecific assessment factors for interspecies differences (e.g., between rodents and humans), which may replace the default interspecies kinetic assessment factors suggested in the REACH guidelines for the calculation of DNELs.[63] Specifically, the PBPK models can be used to calculate peak rate of clearance/metabolization for both experimental animals (rodents) and humans. Then, the peak rate for the rodent can be divided by the peak rate for the human to derive a ratio representing the difference between a specified rodent strain and average human male or female, which can replace the default interspecies kinetic uncertainty factors suggested in the REACH guidelines.

Data on the toxicity of NOAA can be obtained faster, easier, and at lower cost from in vitro studies compared to in vivo studies. Hartung and Sabbioni[83] pointed out the need for alternative in vitro assays in nanotoxicology, given the limitations of the traditional animal-based approaches as well as the multitude of possible formulations of NOAA to be assessed for toxic properties. Hirsch et al.[84] discussed the reliability of current in vitro tests and proposed a series of recommendations (e.g., quality assurance systems, internal performance controls, interlaboratory comparisons, detailed material characterization, use of reference materials) to be implemented immediately to produce robust, reliable, and verified data from in vitro nanotoxicology tests. Dekkers et al.[85] concluded that even if in vitro data cannot always be directly used for RA of NOAA at the moment because of many limitations, they can be useful for giving an indication of the hazard or relative toxic potency by providing information to be considered in a weight of evidence approach and in dose range finding, improving fundamental knowledge on kinetics and modes of action.

Quantitative in vitro–in vivo extrapolations (IVIVEs) typically start with a dose–response modeling of raw continuous, quantal, or ordinal toxicity data.[86] Using an empirical approach, one can look for correlations between in vitro and in vivo dose–response relationships. Assuming that the experimental results are produced by standard protocols, the differences in the in vitro and in vivo data would be only due to variations in physicochemical properties in dispersion and/or in vivo cellular interactions that are not included in the in vitro model.[87] If careful analysis of the data excludes cellular interactions as the main reason behind differences in dose–response curves, a generic quantitative structure–activity relationship (QSAR)-like algorithm (i.e., a Quantitative Property [in vitro] Property [in vivo] Relationship [QPPR]) can be obtained. Using this QPPR, in vivo benchmark doses or effective concentrations can be derived from in vitro data.[87] The disadvantage of this method is that it requires standardized data for a large number of NOAA, which are generally difficult to acquire. In this context, an alternative mechanistic approach could be applied, which takes into account the complete toxicokinetic profiles of the NOAA characterized through PBPK modeling. Therefore, to facilitate the IVIVE of NOAA, there is a need to develop PBPK models, which can successfully predict the ADME behavior of NOAA in both experimental animals and humans.

For ecological RA, we recommend that, beside the important task of developing tools for estimating the short-term effects of pristine NOAA, there should be an emphasis on tools to estimate the long-term effects resulting from low concentrations and repeated dose exposures to NOAA released from products during their life cycle. This should additionally focus on possible effects on ecosystems exposed to multiple stressors (e.g., climate change, eutrophication, pollution), with shorter-term studies emphasizing on investigations on modes of action.

It is important to develop tools that can assess NOAA impairment of the biological function of STPs, since they provide an important ecosystem service.[88] Since ecosystems are also exposed (often repeatedly) to released NOAA in runoff from construction surfaces (Section 25.2.1) or through direct applications (e.g., zero-valent iron for environmental remediation), we recommend the development of tests and tools that can estimate the long-term effects on (1) terrestrial ecosystems, which are a primary target and are fundamental for vital ecosystem services (e.g., food production, waste recycling), (2) sediment organisms, where only short-term standard toxicity has been reported,[89] and (3) for pelagic organisms, where long-term in vivo fish bioaccumulation models should be developed.[14] In addition, we recommend developing single and multispecies models as useful tools for estimating long-term and multigeneration NOAA impacts.[90] Such studies should also include life history studies to enable full assessment at different stages and across generations and to estimate species interactions and secondary toxicity scenarios such as bioaccumulation and trophic transfer. Moreover, they can also identify whether particular species (or genus, etc.) or ecosystem functions are at elevated risks.

To enable a stronger focus on mode of action, these long-term studies should be combined with mechanistic short-term in vitro and in vivo studies. This will in the future allow the development of tools for a rapid prediction/estimation of longer-term ecological consequences, and support the development of mode of action studies useful also for human toxicity.[91–93] High-throughput (HTP) methods are available for predicting genetic and epigenetic consequences; hence, HTP gene expression–based tools should be used to investigate predicted genetic and epigenetic effects of NOAA.[94] Emphasis should also be set on exploring the suitability of the latter for predicting long-term effects based on shorter-term exposures.[95]

In addition to the more chronic lower concentration PEC level exposure studies, it will be crucial to perform repeated exposures over generations, studying any accompanying effects. These studies should partially aim at the population and community level, but also record potential genome-wide changes in DNA integrity and transcriptional activity.

Finally, in addition to long-term consequences including body burden, trophic transfer, and epigenetics tools, we recommend basing the estimation of PNECs on the new species sensitivity distribution (SSD) methodologies, which are well-suited to deal with high uncertainty and variability. Such SSD modeling procedures produce the biological responses of each single species of a target environmental system exposed to NOAA.[96] These single-species responses, also within multispecies exposures, can be combined to generate an SSD for a particular environmental compartment to enable cumulative probabilistic estimation of ecological risks.

25.4 EXPECTED IMPACT OF THE PROPOSED RESEARCH ROADMAP

Sustainable innovation in nanotechnology requires clear guidance on the safe manufacturing, use, and disposal of nanomaterials, informed by a rigorous understanding of their potential environmental and health impacts. Although significant progress in the nanosafety area has been achieved, the concept of safer product and process design, guiding risk prevention, and green manufacturing has not been sufficiently applied to nanotechnologies.

To apply this concept, it is essential to first understand the release of NOAA from composites/products. By doing this, the research proposed above will answer the pressing question of the public and regulators whether the processes studied in the laboratory have any relevance to real-world situations since it will help to determine if the materials released from real products are still in the nano range, and if so, what is the actual nanoscale fraction. This information is of utmost relevance for regulators to help them distinguish the nanomaterials from their bulk counterparts. Moreover, the generation of quality of release data is very important for industries to analyze the overall environmental impacts of their products to understand where they stand in terms of environmental performance and therefore refine their investment and marketing strategies. In addition, investigating the fate of NOAA released from real products following a life cycle approach will significantly support environmental risk analysis.

The research proposed above targets some of our most vulnerable ecosystems to provide data that can inform society about specific areas of safety. For example, it extends environmental RA to cover long-term realistic scenarios of ecosystems subjected to multiple stressors, including accumulation and contaminant transfer in the food chain up to humans. This will enable, for the first time, risk analysis of the insidious long-term low-dose accumulation of NOAA to assess lifelong hazard to wildlife and humans. It will also focus several important services that are essential to public health and society, including the effects of NOAA on sewage sludge treatments (e.g., aerobic and anaerobic treatment before agricultural use), which will provide data of great benefit for decision making by utility companies and regulators.

The above research roadmap will have a unique contribution to complex issues of repeated exposures. This is vital for industry, which is currently trapped by overregulated measures due to existing rules for industrial discharge consents, based on continuous exposure data. This would stifle production since industry cannot expand/innovate if it does not have the capacity to deal with its waste. Most industrial processes produce effluents intermittently that change in volume and composition with the weekly production cycle. We should enable site-specific assessment of the actual effluent flow so that companies can achieve maximum production while also complying with REACH. To do so, we should closely work together with the Commission services involved in the development of adaptations of REACH guidance documents concerning nanomaterials so that these new information on intermittent exposure are incorporated into guidance.

Most hazard studies with NOAA have focused on relatively traditional endpoints (e.g., inflammation, oxidative stress). No research projects have developed tools for long-term environmental risk analysis including genomic and epigenetic effects, although they have been identified as potential serious concerns with regard to nanomaterials. Our approach of using HTP techniques to identify long-term consequences has the potential to provide rapid and highly informative tools for the prediction and management of ecological risks.

Most recent nanosafety projects have been focused on investigating hazard rather than exposure. Consumer and environmental exposures have only received limited attention, even though their investigation is a major activity in NANOREG. The proposed research roadmap significantly progresses beyond the state of the art as it focuses on both worker and consumer inhalation, dermal, and dermal-to-oral (accidental ingestion) exposure and on environmental release in different manufacturing, use, and disposal scenarios. By harmonization of the existing strategies for control banding and higher-tier exposure assessment (e.g., Stoffenmanager Nano and NanoSafer), the proposed tiered exposure assessment approach will have a significant impact on the health risk analysis and management.

The above tools can help to identify the critical attributes of NOAA that stimulate their release or cause ecotoxicity, which may have a significant impact on developing standard measurement and testing methods to quantitatively assess these critical attributes. The current lack of standards for NOAA significantly constrains effective regulation and therefore the impact of potential advancements in this area will be huge.

To maximize their impact, it is essential to tune all proposed activities with the NanoSafety Cluster agenda to facilitate research integration and cohesion. The obtained scientific results should be shared also with other research networks such as PEROSH and the European Technology Platform(s), NANOfutures (http://www.nanofutures.info/), infrastructure and cluster activities of FP7 projects, the ERAnet, and national research programs such as NanoCare2 and NanoNature as well as with the OECD WPMNM.

25.5 SUMMARY AND CONCLUSIONS

Considerable research efforts have focused on evaluating the health and ecological risks of NOAA; however, risk analyses still face remaining knowledge gaps concerned with longer-term effects and exposures, mechanisms of biological uptake, bioaccumulation and toxicity action modes as well as behavior and fate in and between environmental compartments. In this chapter, we propose a research roadmap by 2020, intended to advance significantly in the current state of the art. It involves the development of methods and tools to predict NOAA longer-term exposure and effects on the human health and/or ecosystem integrity/services, especially in environments subjected to multiple stressors, diffuse pollution, and global changes.

The proposed approach will analyze how NOAA transform as they change from one form to another in different life cycle stages to identify their properties that reach environmental organisms or humans. To do so, it will first focus on the nature of their primary release characteristics and then on their environmental fate (including transformation and degradation) to answer the question: Are NOAA released directly from the products or later as a consequence of degradation of nanocomposite debris? This will be achieved through the development of standardized, intercomparable, and reliable methods for release estimation and characterization of NOAA in end products and in biological/environmental matrices. Released materials will be collected, characterized, and used in subsequent ecotoxicological tests. If they are not sufficient in quantity for testing, they will be substituted for NOAA artificially aged under laboratory conditions.

Moreover, we suggest developing stochastic/probabilistic material flow techniques and link them to new approaches for mass balance dynamic fate and exposure modeling. The fate models will be used to estimate PECs in different environmental compartments to combine them with PNECs based on SSD for probabilistic ecological RA.

To derive realistic PNECs, we suggest developing tests to estimate the long-term effects of NOAA released from products and composites, focusing on the possible effects on ecosystems potentially exposed to multiple stressors such as climate change and pollution.

These tests may include estimation of species interactions and secondary toxicity scenarios such as bioaccumulation and trophic transfer. Specifically, we suggest emphasizing on the development of tests for studying the long-term effects on sediment and pelagic organisms. These tools can be combined with short-term in vitro (worm and fish) studies to provide fast and reliable HTP methods that can be also used to investigate genetic and epigenetic effects of NOAA and to explore long-term mechanisms of response based on short-term exposures.

To derive realistic DNELs for health RA, we suggest deriving nanospecific interspecies extrapolation factors by means of PBPK/PD modeling. It is too uncertain to estimate long-term DNEL from the currently available acute inhalation data, and chronic studies are very expensive; hence, we recommend applying STIS in the near term to generate subacute data. Moreover, to enable long-term ingestion RA, we suggest developing STOS: the oral route alternative of the STIS protocol. To prioritize NOAA for STIS and STOS and identify adequate dose ranges, we suggest using in vitro assays of acute cytotoxicity and proinflammatory activity. The resulting in vitro data may then also be used for RA if successfully extrapolated to in vivo.

The proposed roadmap involves bottom–up generation of environmental, health, and safety data and methods for RA. We recommend that this approach is combined with a top–down decision support system for both manufacturers and regulators.

It will guide nanomanufacturers into the design and application of the most appropriate risk analysis and prevention/mitigation strategy for a specific product. The correct implementation of this system will ensure that the risks associated with the production and downstream use of a nanotechnology have been appropriately evaluated and mitigated to an acceptable level, according to the most recent knowledge at the time of implementation. This approach will enable more sustainable nanomanufacturing processes, will result in more solid risk prevention and mitigation strategies, and will be easily applicable to different materials and industrial settings.

REFERENCES

1. Koehler AR, Som C. Environmental and health implications of nanotechnology: Have innovators learned the lessons from past experiences? *Human and Ecological Risk Assessment.* 2008;14:512–31.
2. Lam CW, James JT, McCluskey R, Hunter RL. Pulmonary toxicity of single-wall carbon nanotubes in mice 7 and 90 days after intratracheal instillation. *Toxicological Sciences.* 2004;77(1):126–34.

3. Shvedova AA, Kisin ER, Mercer R, Murray AR, Johnson VJ, Potapovich AI, Tyurina YY et al. Unusual inflammatory and fibrogenic pulmonary responses to single-walled carbon nanotubes in mice. *American Journal of Physiology—Lung Cellular and Molecular Physiology.* 2005;289(5):L698–708.

4. Poland CA, Duffin R, Kinloch I, Maynard A, Wallace WAH, Seaton A, Stone V, Brown S, MacNee W, Donaldson K. Carbon nanotubes introduced into the abdominal cavity of mice show asbestos-like pathogenicity in a pilot study. *Nature Nanotechnology.* 2008;3(7):423–8.

5. Takagi A, Hirose A, Nishimura T, Fukumori N, Ogata A, Ohashi N, Kitajima S, Kanno J. Induction of mesothelioma in p53+/− mouse by intraperitoneal application of multi-wall carbon nanotube. *The Journal of Toxicological Sciences.* 2008;33(1):105–16.

6. Warheit DB, Laurence BR, Reed KL, Roach DH, Reynolds GAM, Webb TR. Comparative pulmonary toxicity assessment of single-wall carbon nanotubes in rats. *Toxicological Sciences.* 2004;77(1):117–25.

7. Donaldson K, Murphy F, Duffin R, Poland C. Asbestos, carbon nanotubes and the pleural mesothelium: A review of the hypothesis regarding the role of long fibre retention in the parietal pleura, inflammation and mesothelioma. *Particle and Fibre Toxicology.* 2010;7(1):5.

8. Mercer R, Hubbs A, Scabilloni J, Wang L, Battelli L, Schwegler-Berry D, Castranova V, Porter D. Distribution and persistence of pleural penetrations by multi-walled carbon nanotubes. *Particle and Fibre Toxicology.* 2010;7(1):28.

9. Aschberger K, Johnston HJ, Stone V, Aitken RJ, Hankin SM, Peters SAK, Tran CL, Christensen FM. Review of carbon nanotubes toxicity and exposure—Appraisal of human health risk assessment based on open literature. *Critical Reviews in Toxicology.* 2010;40(9):759–90.

10. Chen HHC, Yu C, Ueng TH, Chen S, Chen BJ, Huang KJ, Chiang LY. Acute and subacute toxicity study of water-soluble polyalkylsulfonated C60 in rats. *Toxicologic Pathology.* 1998;26(1):143–51.

11. Oberdörster E, Zhu S, Blickley TM, McClellan-Green P, Haasch ML. Ecotoxicology of carbon-based engineered nanoparticles: Effects of fullerene (C60) on aquatic organisms. *Carbon.* 2006;44(6):1112–20.

12. Saitoh Y, Xiao L, Mizuno H, Kato S, Aoshima H, Taira H, Kokubo K, Miwa N. Novel polyhydroxylated fullerene suppresses intracellular oxidative stress together with repression of intracellular lipid accumulation during the differentiation of OP9 preadipocytes into adipocytes. *Free Radical Research.* 20101;44(9):1072–81.

13. Ogami A, Yamamoto K, Morimoto Y, Fujita K, Hirohashi M, Oyabu T, Myojo T et al. Pathological features of rat lung following inhalation and intratracheal instillation of C60 fullerene. *Inhalation Toxicology.* 2011;23(7):407–16.

14. Fraser TWK, Reinardy HC, Shaw BJ, Henry TB, Handy RD. Dietary toxicity of single-walled carbon nanotubes and fullerenes (C60) in rainbow trout (Oncorhynchus mykiss). *Nanotoxicology.* 2011;5(1):98–108.

15. Lansdown ABG. Critical observations on the neurotoxicity of silver. *Critical Reviews in Toxicology.* 2007;37(3):237–50.

16. Wijnhoven SWP, Peijnenburg WJGM, Herberts CA, Hagens WI, Oomen AG, Heugens EHW, Roszek B et al. Nano-silver—A review of available data and knowledge gaps in human and environmental risk assessment. *Nanotoxicology.* 2009;3(2):109–38.

17. Christensen FM, Johnston HJ, Stone V, Aitken RJ, Hankin S, Peters S, Aschberger K. Nano-silver—Feasibility and challenges for human health risk assessment based on open literature. *Nanotoxicology.* 2010;4(3):284–95.

18. Simon M, Barberet P, Delville MH, Moretto P, Seznec H. Titanium dioxide nanoparticles induced intracellular calcium homeostasis modification in primary human keratinocytes. Towards an in vitro explanation of titanium dioxide nanoparticles toxicity. *Nanotoxicology.* 2011;5(2):125–39.

19. Warheit DB, Reed KL, Sayes CM. A role for surface reactivity in TiO_2 and quartz-related nanoparticle pulmonary toxicity. *Nanotoxicology.* 2009;3(3):181–7.

20. Christensen FM, Johnston HJ, Stone V, Aitken RJ, Hankin S, Peters S, Aschberger K. Nano-TiO_2—Feasibility and challenges for human health risk assessment based on open literature. *Nanotoxicology.* 2011;5(2):110–24.

21. Landsiedel R, Ma-Hock L, Van Ravenzwaay B, Schulz M, Wiench K, Champ S, Schulte S, Wohlleben W, Oesch F. Gene toxicity studies on titanium dioxide and zinc oxide nanomaterials used for UV-protection in cosmetic formulations. *Nanotoxicology.* 2010;4(4):364–81.

22. European Parliment and the Council. Regulation (EC) No. 1907/2006 on the Registration, Evaluation, Authorisation and Restriction of Chemicals (REACH). Brussels, Belgium. *Official Journal of the European Union.* 2006

23. U.S. Congress. *Toxic Substances Control Act.* Washington: Congressional Research Service, Library of Congress, 1976.

24. OECD. *Description of Selected Key Generic Terms Used in Chemical Hazard/Risk Assessment. Joint Project with the International Programme on Chemical Safety (IPCS) on the Harmonization of Hazard/Risk Assessment Terminology.* Paris, France: OECD Publishing, 2003.

25. Van Leeuwen C, Vermeire T. *Risk Assessment of Chemicals: An Introduction.* Dordrecht: Springer, 2007.

26. DG ENV. *Joint Meeting of the Chemicals Committee and the Working Party on Chemicals, Pesticides and Biotechnology.* Brussels, Belgium: Directorate General Environment of the European Commission, 2012.

27. Hansen SF, Baun, A. European regulation affecting nanomaterials—review of limitations and future recommendations. *Dose Response* 2012;10(3):364–83.

28. European Nanosafety Cluster. *Compendium of Projects in the European NanoSafety Cluster.* 2013.

29. Owack B, Ranville JF, Diamond S, Gallego-Urrea JA, Metcalfe C, Rose J, Horne N, Koelmans AA, Klaine SJ. Potential scenarios for nanomaterial release and subsequent alteration in the environment. *Environmental Toxicology and Chemistry.* 2012;31(1):50–9.

30. Gottschalk F, Nowack B. The release of engineered nanomaterials to the environment. *Journal of Environmental Monitoring.* 2011;13(5):1145–55.

31. Asmatulu E, Twomey J, Overcash M. Life cycle and nano-products: End-of-life assessment. *Journal of Nanoparticle Research.* 2012;14(3):1–8.

32. Labille J, Feng J, Botta C, Borschneck D, Sammut M, Cabie M, Auffan M, Rose J, Bottero JY. Aging of TiO_2 nanocomposites used in sunscreen. Dispersion and fate of the degradation products in aqueous environment. *Environmental Pollution.* 2010;158(12):3482–9.

33. Auffan M, Pedeutour M, Rose J, Masion A, Ziarelli F, Borschneck D, Chaneac C et al. Structural degradation at the surface of a TiO_2-based nanomaterial used in cosmetics. *Environmental Science and Technology.* 2010;44(7):2689–94.

34. Koponen I, Jensen, KA, Schneider, T. Comparison of dust released from sanding conventional and nanoparticle-doped wall and wood coatings. *Journal of Exposure Science and Environmental Epidemiology.* 2011;21:408–18.

35. Saber AT, Koponen IK, Jensen KA, Jacobsen NR, Mikkelsen L, Møller P, Loft S, Vogel U, Wallin H. Inflammatory and genotoxic effects of sanding dust generated from nanoparticle-containing paints and lacquers. *Nanotoxicology.* 2012;6(7):776–88.

36. Saber A, Jacobsen NR, Mortensen A, Szarek J, Jackson P, Madsen AM, Jensen KA et al. Nanotitanium dioxide toxicity in mouse lung is reduced in sanding dust from paint. *Particle and Fibre Toxicology.* 2012;9:4.

37. Weir A, Westerhoff P, Fabricius L, Hristovski K, von Goetz N. Titanium dioxide nanoparticles in food and personal care products. *Environmental Science and Technology.* 2012;46(4):2242–50.
38. Windler L, Lorenz C, von Goetz N, Hungerbühler K, Amberg M, Heuberger M, Nowack B. Release of titanium dioxide from textiles during washing. *Environmental Science and Technology.* 2012;46(15):8181–8.
39. Hansen S, Michelson E, Kamper A, Borling P, Stuer-Lauridsen F, Baun A. Categorization framework to aid exposure assessment of nanomaterials in consumer products. *Ecotoxicology.* 2008;17(5):438–47.
40. Pasricha A, Jangra SL, Singh N, Dilbaghi N, Sood KN, Arora K, Pasricha R. Comparative study of leaching of silver nanoparticles from fabric and effective effluent treatment. *Journal of Environmental Sciences.* 2012;24(5):852–9.
41. Nazarenko Y, Zhen H, Han T, Lioy PJ, Mainelis G. Potential for inhalation exposure to engineered nanoparticles from nanotechnology-based cosmetic powders. *Environmental Health Perspectives.* 2012;120(6):885–92.
42. Schneider T, Jensen K. Relevance of aerosol dynamics and dustiness for personal exposure to manufactured nanoparticles. *Journal of Nanoparticle Research.* 2009;11(7):1637–50.
43. Koponen IK, Jensen KA, Schneider T. Comparison of dust released from sanding conventional and nanoparticle-doped wall and wood coatings. *Journal of Exposure Science and Environmental Epidemiology.* 2011;21(4):408–18.
44. Vorbau M, Hillemann L, Stintz M. Method for the characterization of the abrasion induced nanoparticle release into air from surface coatings. *Journal of Aerosol Science.* 2009;40(3):209–17.
45. Bello D, Wardle B, Yamamoto N, deVilloria R, Garcia E, Hart A, Ahn K, Ellenbecker M, Hallock M. Exposure to nanoscale particles and fibres during machining of hybrid advanced composites containing carbon nanotubes. *Journal of Nanoparticle Research.* 2009;5(1):231–49.
46. Bello D, Wardle B, Zhang J, Yamamoto N, Santeufemio C, Hallock M, Virji M. Characterization of exposures to nanoscale particles and fibers during solid core drilling of hybrid CNT advanced composites. *International Journal of Occupational Environmental Health.* 2010;16:434–50.
47. Hansen SF, Baun A, Alstrup-Jensen K. *NanoRiskCat—A Conceptual Decision Support Tool for Nanomaterials. Danish Ministry of Environmental Protection Agency, Copenhagen (Denmark)* 2011.
48. Genaidy A, Sequeira R, Rinder M, A-Rehim A. Risk analysis and protection measures in a carbon nanofiber manufacturing enterprise: An exploratory investigation. *Science of the Total Environment.* 2009;407(22):5825–38.
49. Paik SY, Zalk DM, Swuste P. Application of a pilot control banding tool for risk level assessment and control of nanoparticle exposures. *Annals of Occupational Hygiene.* 2008;52(6):419–28.
50. Duuren-Stuurman B, Vink S, Brouwer D, Kroese D, Heussen H, Verbist K, Telemans E, Niftrik MV, Fransman W. *Stoffenmanager Nano: Description of the Conceptual Control Banding Model.* Zeist, Netherlands: Netherlands Organisation for Applied Scientific Research (TNO), 2011.
51. Giacobbe F, Monica L, Geraci D. Risk assessment model of occupational exposure to nanomaterials. *Human and Experimental Toxicology.* 2009;28(6–7):401–6.
52. Jensen K, Saber A, Kristensen H, Koponen I, Wallin H. NanoSafer: Web-based precautionary risk evaluation of dust exposure to manufactured nanomaterials using first order modeling (in preparation).

53. Schneider T, Brouwer D, Koponen I, Jensen K, Fransmann W, van Duuren-Stuurman B, van Tongeren M, Tielemans E. Conceptual model for assessment of inhalation exposure to manufactured nanoparticles. *Journal of Exposure Science and Environmental Epidemiology.* 2011;21(5): 450–63.

54. Wilson MR, Lightbody JH, Donaldson K, Sales J, Stone V. Interactions between ultrafine particles and transition metals in vivo and in vitro. *Toxicology and Applied Pharmacology.* 2002;184(3):172–9.

55. Stone V, Shaw J, Brown DM, MacNee W, Faux SP, Donaldson K. The role of oxidative stress in the prolonged inhibitory effect of ultrafine carbon black on epithelial cell function. *Toxicology in Vitro.* 1998;12(6):649–59.

56. Li XY, Brown D, Smith S, MacNee W, Donaldson K. Short-term inflamatory responses following intratracheal instillation of fine and ultrafine carbon black in rats. *Inhalation Toxicology.* 1999;11(8):709–31.

57. Brown D, Wilson M, MacNee W, Stone V, Donaldson K. Size-dependent proinflammatory effects of ultrafine polystyrene particles: A role for surface area and oxidative stress in the enhanced activity of ultrafines. *Toxicology and Applied Pharmacology.* 2001;175:191–9.

58. Duffin R, Tran L, Brown D, Stone V, Donaldson K. Proinflammogenic effects of low-toxicity and metal nanoparticles in vivo and in vitro: Highlighting the role of particle surface area and surface reactivity. *Inhalation Toxicology.* 2007;19(10):849–56.

59. Lockman PR, Koziara JM, Mumper RJ, Allen DD. Nanoparticle surface charges alter blood–brain barrier integrity and permeability. *Journal of Drug Targeting.* 2004;12(9–10):635–41.

60. Luoma S. *Silver Nanotechnologies and the Environment: Old Problems or New Challenges.* Washington, DC: The Pew Charitable Trusts and Woodrow Wilson International Center for Scholars. Project on Emerging Nanotechnologies, 2008.

61. Brown DM, Kinloch IA, Bangert U, Windle AH, Walter DM, Walker GS, Scotchford CA, Donaldson K, Stone V. An in vitro study of the potential of carbon nanotubes and nanofibres to induce inflammatory mediators and frustrated phagocytosis. *Carbon.* 2007;45(9):1743–56.

62. US EPA. *Decision Document: Conditional Registration of HeiQ AGS-20 as a Materials Preservative in Textiles.* Washington, DC: Environmental Protection Agency, Office of Pesticide Programs, Antimicrobials Division, 2011.

63. European Chemicals Agency. *Guidance on Information Requirements and Chemical Safety Assessment. Chapter R.8: Characterisation of Dose [concentration]-Response for Human Health.* Helsinki, Finland: European Chemicals Agency (ECHA), 2008.

64. Pluskota A, Horzowski E, Bossinger O, Mikecz Av. In *Caenorhabditis elegans* nanoparticle-bio-interactions become transparent: Silica-nanoparticles induce reproductive senescence. *PLoS ONE.* 2009;4(8):6622.

65. Tiede K, Boxall ABA, Wang X, Gore D, Tiede D, Baxter M, David H, Tear SP, Lewis J. Application of hydrodynamic chromatography-ICP-MS to investigate the fate of silver nanoparticles in activated sludge. *Journal of Analytical Atomic Spectrometry.* 2010;25(7):1149–54.

66. Kaegi R, Voegelin A, Sinnet B, Zuleeg S, Hagendorfer H, Burkhardt M, Siegrist H. Behavior of metallic silver nanoparticles in a pilot wastewater treatment plant. *Environmental Science and Technology.* 2011;45(9):3902–8.

67. Shafer MM, Overdier JT, Armstong DE. Removal, partitioning, and fate of silver and other metals in wastewater treatment plants and effluent-receiving streams. *Environmental Toxicology and Chemistry.* 1998;17(4):630–41.

68. Hou L, Li K, Ding Y, Li Y, Chen J, Wu X, Li X. Removal of silver nanoparticles in simulated wastewater treatment processes and its impact on COD and NH4 reduction. *Chemosphere.* 2012;87(3):248–52.

69. Yang Y, Chen Q, Wall JD, Hu Z. Potential nanosilver impact on anaerobic digestion at moderate silver concentrations. *Water Research.* 2012;46(4):1176–84.
70. Wang Y, Westerhoff P, Hristovski K. Fate and biological effects of silver, titanium dioxide, and C60 (fullerene) nanomaterials during simulated wastewater treatment processes. *Journal of Hazardous Materials.* 2012;201–202:16–22.
71. Wiechmann B, Dienemann C, Kabbe C, Brandt S, Vogel I, Roskoch A. Klärschlammentsorgung in Der Bundesrepublik Deutschland [*Sludge Disposal in the Federal Republic of Germany*]. Bonn: Umweltbundesamt, 2012.
72. Gottschalk F, Scholz RW, Nowack B. Probabilistic material flow modeling for assessing the environmental exposure to compounds: Methodology and an application to engineered nano-TiO2 particles. *Environmental Modelling & Software.* 2010;25(3):320–32.
73. Gottschalk F, Ort C, Scholz RW, Nowack B. Engineered nanomaterials in rivers— Exposure scenarios for Switzerland at high spatial and temporal resolution. *Environmental Pollution.* 2011;159(12):3439–45.
74. Praetorius A, Scheringer M, Hungerbühler K. Development of environmental fate models for engineered nanoparticles—A case study of TiO_2 nanoparticles in the Rhine River. *Environmental Science and Technology.* 2012;46(12):6705–13.
75. Schneider T, Brouwer DH, Koponen IK, Jensen KA, Fransman W, Van Duuren-Stuurman B, Van Tongeren M, Tielemans E. Conceptual model for assessment of inhalation exposure to manufactured nanoparticles. *Journal of Exposure Science and Environmental Epidemiology.* 2011;21:450–63.
76. Klein C, Wiench K, Wiemann M, Ma-Hock L, Ravenzwaay B, Landsiedel R. Hazard identification of inhaled nanomaterials: Making use of short-term inhalation studies. *Archives of Toxicology.* 2012;86(7):1137–51.
77. HWU. *ENPRA Deliverable 4.1.* Edinburgh: Heriot-Watt University, 2012.
78. Kermanizadeh A, Gaiser B, Hutchison G, Stone V. An in vitro liver model—Assessing oxidative stress and genotoxicity following exposure of hepatocytes to a panel of engineered nanomaterials. *Particle and Fibre Toxicology.* 2012;9(1):28.
79. NASA. The dialogue continues. *Nature Nanotechnology* 2013;8(2):69.
80. Lee HA, Leavens TL, Mason SE, Monteiro-Riviere NA, Riviere JE. Comparison of quantum dot biodistribution with a blood-flow-limited physiologically based pharmacokinetic model. *Nano Letters.* 2009;9(2):794–9.
81. Riviere JE. Pharmacokinetics of nanomaterials: An overview of carbon nanotubes, fullerenes and quantum dots. *Wiley Interdisciplinary Reviews: Nanomedicine and Nanobiotechnology.* 2009;1(1):26–34.
82. Tran L, editor. Risk assessment of ENM: The results from FP7 ENPRA. Symposium on safety issues of nanomaterials along their lifecycle, May 2011, Barcelona, Spain.
83. Hartung T, Sabbioni E. Alternative in vitro assays in nanomaterial toxicology. *Wiley Interdisciplinary Reviews: Nanomedicine Nanobiotechnology.* 2011;3:545–73.
84. Hirsch C, Roesslein M, Krug HF, Wick P. Nanomaterial cell interactions: Are current in vitro tests reliable? *Nanomedicine* (London). 2011;6(5):837–47.
85. Dekkers S, Casee F, Heugens E, Baun A, Pilou M, Asbach C, Dusinska M, Nickel C, Riediker M, Heer Cd. Consensus Report Risk Assessment of Nanomaterials: In vitro-in vivo extrapolation. NanoImpactNet Deliverable 3.5., 2010.
86. Slob W. Dose-response modeling of continuous endpoints. *Toxicological Sciences.* 2002;66(2):298–312.
87. Bessems J, McCalman L, Gosens I. ENPRA dose-response modeling. In: Hristozov D, editor. *Bilthoven.* 2011. Personal Communication.
88. Kaegi R, Voegelin A, Sinnet B et al. Behavior of metallic silver nanoparticles in a pilot wastewater treatment plant. *Environmental Science and Technology.* 2011;45:3902–8.

89. Bradford AH, Handy RD, Readman JW, Atfield A, Mühling M. Impact of silver nanoparticle contamination on the genetic diversity of natural bacterial assemblages in estuarine sediments. *Environmental Science and Technology.* 2009;43(12):4530–6.

90. Scott-Fordsmand JJ, Maraldo K, van den Brink PJ. The toxicity of copper contaminated soil using a gnotobiotic Soil Multi-species Test System (SMS). *Environment International.* 2008;34(4):524–30.

91. Fernández-Cruz ML, Lammel T, Connolly M, Conde E, Barrado AI, Derick S, Perez Y, Fernandez M, Furger C, Navas JM. Comparative cytotoxicity induced by bulk and nanoparticulated ZnO in the fish and human hepatoma cell lines PLHC-1 and Hep G2. *Nanotoxicology.* 2013;7:935–52.

92. Gaiser B, Fernandes T, Jepson M, Lead J, Tyler C, Baalousha M, Biswas A et al. Interspecies comparisons on the uptake and toxicity of silver and cerium dioxide nanoparticles. *Environmental Toxicology and Chemistry.* 2012;31(1):144–54.

93. Hayashi Y, Engelmann P, Foldbjerg R, Szabó M, Somogyi I, Pollák E, Molnár L et al. Earthworms and humans in vitro: Characterizing evolutionarily conserved stress and immune responses to silver nanoparticles. *Environmental Science and Technology.* 2012;46(7):4166–73.

94. Muñoz A, Costa M. Elucidating the mechanisms of nickel compound uptake: A review of particulate and nano-nickel endocytosis and toxicity. *Toxicology and Applied Pharmacology.* 2012;260(1):1–16.

95. Gomes SIL, Novais SC, Scott-Fordsmand JJ, De Coen W, Soares AMVM, Amorim MJB. Effect of Cu-nanoparticles versus Cu-salt in Enchytraeus albidus (Oligochaeta): Differential gene expression through microarray analysis. *Comparative Biochemistry and Physiology Part C: Toxicology & Pharmacology.* 2012;155(2):219–27.

96. Gottschalk F, Nowack B. A probabilistic method for species sensitivity distributions taking into account the inherent uncertainty and variability of effects to estimate environmental risk. *Integrated Environmental Assessment and Management.* 2013;9(1):79–86.

Index